ADVANCES IN ENZYMOLOGY

AND RELATED AREAS OF MOLECULAR BIOLOGY

Volume 78

ADVANCES IN ENZYMOLOGY

AND RELATED AREAS OF MOLECULAR BIOLOGY

Founded by F. F. NORD

Edited by Eric J. Toone

DUKE UNIVERSITY, DURHAM, NORTH CAROLINA

VOLUME 78

A JOHN WILEY & SONS, INC., PUBLICATION

Published by John Wiley & Sons, Inc., Hoboken, New Jersey.
Published simultaneously in Canada.

Library of Congress Cataloging-in-Publication Data is available.

ISBN 978-1-118-01428-8

Printed in the United States of America

oBook ISBN: 978-1-118-10577-1
ePDF ISBN: 978-1-118-10580-1
ePub ISBN: 978-1-118-10578-8

10 9 8 7 6 5 4 3 2 1

CONTENTS

CONTRIBUTORS

WILLIAM ANDRÉ, Unité Propre de Recherche 2228 du Centre National de la Recherche Scientifique, Régulation de la Transcription et Maladies Génétiques, Université Paris Descartes, Paris, France

MARC ANTONYAK, Department of Molecular Medicine, College of Veterinary Medicine, Cornell University, Ithaca, NY, USA

CARLO M. BERGAMINI, Department of Biochemistry and Molecular Biology, University of Ferrara, Italy

RICHARD A. CERIONE, Department of Molecular Medicine, College of Veterinary Medicine, Cornell University, Ithaca, NY, USA

NICOLAS CHABOT, Department of Chemistry, University of Montréal, Montréal, QC, Canada

ATHAR H. CHISHTI, Department of Molecular Physiology and Pharmacology, Tufts University School of Medicine, Boston, MA, USA

RUSSELL J. COLLIGHAN, School of Life and Health Sciences, Aston University, Birmingham, UK

MANUELA D'ELETTO, Department of Biology, University of Rome "Tor Vergata", Rome, Italy

PHILIPPE DJIAN, Unité Propre de Recherche 2228 du Centre National de la Recherche Scientifique, Régulation de la Transcription et Maladies Génétiques, Université Paris Descartes, Paris, France

LAURA FALASCA, Department of Biology, University of Rome "Tor Vergata", Rome, Italy

MARIA GRAZIA FARRACE, Department of Biology, University of Rome "Tor Vergata", Rome, Italy

LÁSZLÓ FÉSÜS, Department of Biochemistry and Molecular Biology, Faculty of Medicine, University of Debrecen, Hungary; Apoptosis and Genomics Research Group of the Hungarian Academy of Sciences, Debrecen, Hungary

VITTORIO GENTILE, Department of Biochemistry and Biophysics, Medical School, Second University of Naples, Naples, Italy

MARTIN GRIFFIN, School of Life and Health Sciences, Aston University, Birmingham, UK

GUYLAINE HOFFNER, Unité Propre de Recherche 2228 du Centre National de la Recherche Scientifique, Régulation de la Transcription et Maladies Génétiques, Université Paris Descartes, Paris, France

JEFFREY W. KEILLOR, Department of Chemistry, University of Montréal, Montréal, QC, Canada

ANWAR A. KHAN, Department of Pharmacology, University of Illinois College of Medicine, Chicago, IL, USA

SOO-YOUL KIM, Division of Cancer Biology, Research Institute, National Cancer Center, Kyonggi-do, Republic of Korea

RÓBERT KIRÁLY, Department of Biochemistry and Molecular Biology, Faculty of Medicine, University of Debrecen, Hungary; Apoptosis and Genomics Research Group of the Hungarian Academy of Sciences, Debrecen, Hungary

ILMA KORPONAY-SZABÓ, Department of Pediatrics, Faculty of Medicine, University of Debrecen, Hungary

OLIVIER LEOGANE, Department of Chemistry, University of Montréal, Montréal, QC, Canada

BO LI, Department of Molecular Medicine, College of Veterinary Medicine, Cornell University, Ithaca, NY, USA

OKSANA LOCKRIDGE, Eppley Institute, University of Nebraska Medical Center, Omaha, NE, USA

LASZLO LORAND, Department of Cell and Molecular Biology, Feinberg Medical School, Northwestern University, Chicago, IL, USA

GEORGE D. MARKHAM, Institute for Cancer Research, Fox Chase Cancer Center, Philadelphia, PA, USA

AMINA MULANI, Department of Chemistry, University of Montréal, Montréal, QC, Canada

ZOLTÁN NEMES, Department of Psychiatry and Signaling and Apoptosis Research Group of Hungarian Academy of Sciences Research Center for Molecular Medicine, University of Debrecen, Debrecen, Hungary

MARÍA A. PAJARES, Instituto de Investigaciones Biomédicas “Alberto Sols” (CSIC-UAM), Madrid, Spain

S. N. PRASANNA MURTHY, Department of Cell and Molecular Biology, Feinberg Medical School, Northwestern University, Chicago, IL, USA

CHRISTOPHE PARDIN, Department of Chemistry, University of Montréal, Montréal, QC, Canada

MAURO PIACENTINI, Department of Biology, University of Rome “Tor Vergata”, Rome, Italy

CARLO RODOLFO, Department of Biology, University of Rome “Tor Vergata”, Rome, Italy

ISABELLE ROY, Department of Chemistry, University of Montréal, Montréal, QC, Canada

ZSUZSA SZONDY, Department of Biochemistry and Molecular Biology, Faculty of Medicine, University of Debrecen, Debrecen, Hungary

AMANDINE VANHOUTTEGHEM, Unité Propre de Recherche 2228 du Centre National de la Recherche Scientifique, Régulation de la Transcription et Maladies Génétiques, Université Paris Descartes, Paris, France

ZHUO WANG, School of Life and Health Sciences, Aston University, Birmingham, UK

WEIHUA XUE, Eppley Institute, University of Nebraska Medical Center, Omaha, NE, USA

PREFACE

Transglutaminases, first described in 1957, are a large, widely distributed family of enzymes canonically responsible for the amidation/transamidation of protein side chains. The extraordinary diversity of names associated with various enzymatic activities now recognized and aggregated as transglutaminase bears witness to the remarkable diversity of biological roles associated with the activity, including myriad human diseases. Eight transglutaminases have been identified—transglutaminases 1 through 7, plus the coagulation factor XIII—as well as a structurally related but catalytically inactive relative, designated protein band 4.2. Transglutaminases are ubiquitous in higher organisms, but have also been identified in various lower forms of life, including single-celled organisms, plants, and worms. Transglutaminases catalyze a variety of amidations to both protein and nonprotein amines, creating chemically robust cross-links and durable (but reversible) protein modifications. They are expressed in some instances as zymogens, showing differential activity based on proteolytic activation. Varying requirements for Ca^{2+} provide a molecular mechanism by which to effect Ca^{2+} signaling. At least some transglutaminases bind nucleoside triphosphates, and so act as G-proteins and kinases. Protein disulfide isomerase activity has also been attributed to transglutaminases.

Despite the remarkable ubiquity of the group, clear physiological roles for most of transglutaminases are lacking. The combination of biological ubiquity, clear implication in various human disease states, and poorly identified native physiological roles make transglutaminases ripe for comprehensive review, and this volume of *Advances in Enzymology* considers transglutaminases from the perspective of biochemistry (structure and activity), human disease, and inhibition for therapeutic intervention. The first chapter, by Bergamini et al., serves as an overview and introduction to the topic, covering protein biochemistry, including structure and regulation, and clearly identified physiological roles for transglutaminases. Chapters 2 and 3 consider the biochemical basis for the involvement of transglutaminases in human disease. Together these chapters, by Gentile, and Hoffner et al., respectively, consider the amidation activity of transglutaminases as well as the various additional activities ascribed to this class of proteins, including protein disulfide isomerase, G-protein, and kinase activity, and the biochemical basis for transglutaminase involvement in human disease.

The second broad portion of this volume, encompassing Chapters 4 to 9, considers in detail the role(s) of transglutaminases in human disease. The transcription factor NF-κB is responsible, in part, for the cellular stress response to a variety of external stimuli, and its function is intimately involved in a wide range of human diseases. In Chapter 4, Soo-Youl Kim considers the role of transglutaminase 2 in the regulation of NF-κB, providing a unifying mechanism for the activity of this biological effector. The regulation of programed cell death is intimately involved in myriad diseases, including inflammatory diseases, cancer, and protein accumulation-related disorders. In Chapter 5, Piacentini et al. describe the role of transglutaminase 2 as both a pro- and an antiapoptotic factor and the means by which disregulation of these activities lead to disease progression. The disregulation of transglutaminase-based cell survival, propagation, and migration in the context of cancer is considered by Cerione and Antonyak in Chapter 6. Autoimmune diseases are among the most important associated with transglutaminase 2, and in Chapter 7 Szondy, Korponay-Szabó, Király, and Fésüs consider the role of transglutaminase in celiac disease. Remarkably, both gain and loss of function by this protein triggered disease, from deamidation of gluten peptides in the former case, or by defective apoptotic clearance of cells in the latter. Protein aggregation diseases—amyloid-β, τ, α-synuclein, and Huntingtin—are, in many instances, induced by the formation of resistant protein cross-links that lead to both persistence and insolubility of unnatural protein aggregates. In Chapter 8, Zoltán Nemes reviews the role of transglutaminase-mediated cross-link formation in the progression of neurodegenerative disease. Finally, in Chapter 9, Lorand et al. describe the role of transglutaminase in remodeling of the erythrocyte membrane and its relevance to the induction of nonharitable genetic erythrocyte diseases such as Koln disease.

With the remarkable diversity of biological function, continuing interest in additional activities, and clear relevance to human disease, it comes as no surprise that the development of highly selective reversible inhibitors of transglutaminase would play an important role in transglutaminase biochemistry, both for the preparation of biological probes and the development of human therapeutics. In Chapter 10 of this volume, Jeffrey Keillor et al. describe various approaches to both reversible and irreversible inhibitors of transglutaminase activity.

The final chapter of this volume, by Pajares and Markham, describes the structure and mechanism of *S*-adenosylmethionine synthetase. This ubiquitous and remarkably important enzyme is also involved in a variety of human diseases, including cognitive and neurodegenerative disorders, demyelination diseases, and cancer, and the role of SAM-synthetase in these and other disorders is described.

Eric J. Toone

STRUCTURE AND REGULATION OF TYPE 2 TRANSGLUTAMINASE IN RELATION TO ITS PHYSIOLOGICAL FUNCTIONS AND PATHOLOGICAL ROLES

CARLO M. BERGAMINI
RUSSELL J. COLLIGHAN
ZHUO WANG
MARTIN GRIFFIN

CONTENTS

I. INTRODUCTION

Transglutaminases (TGs) are enzymes that catalyze the posttranslational modification of proteins at the amide moiety of the side chain of glutamine

Advances in Enzymology and Related Areas of Molecular Biology, Volume 78.
Edited by Eric J. Toone.

residues. In the case of mammalian TGs, this reaction is absolutely dependent on the availability of calcium ions, which behave as essential activators of the enzyme. Designated as "protein–glutamine γ-glutamyltransferase" by the Enzyme Commission (since they catalyze the posttranslational modification of proteins at glutamine residues through acyl-transfer reactions), they are represented by E.C. number 2.3.2.13. In this reaction, the primary amide group of the peptidyl-glutamine substrate is converted into a secondary amide through an isopeptide bond, involving either a low molecular weight amine (most frequently a polyamine) or the ε-amino group of a lysine residue belonging to the same or, more likely, to another protein acting as acyl-acceptors to establish an intra- or an intermolecular cross-link in the substrate protein(s). Therefore, products are either "polyamidated" or cross-linked aggregated proteins. For the interested reader, a history of TGase research is available from Beninati et al. [1].

As we and others have detailed elsewhere [2–4], TGs are widely distributed in nature from bacteria to plants and animals. Mammalian TGs are characterized by their absolute requirement for calcium ions for activity. In this chapter, we will limit the discussion to the properties of mammalian TGs, focusing on type 2 TGase, commonly referred to as tissue TGase (tTG or TG2), which is the most widely distributed isoform in animal tissues, but is also the enzyme that is least understood. We hope that it will become evident from our discussion that the peculiar interest in this protein stems from its nature as a multifunctional protein, which is involved in the control of several functions in resting and in stressed cells. In particular, along with its calcium-dependent cross-linking activity, TG2 can also act as a protein disulphide isomerase (PDI) and as a serine/threonine protein kinase. TG2 also binds guanine nucleotides that control (i.e., inhibit) the transamidating reaction and convert the enzyme into a G-like protein for transduction of extracellular signals. In addition, a fraction of the cellular TG2 is bound to the plasma membranes or deposited into the extracellular matrix (ECM) where it serves to stabilize cell adhesion, vascular permeability, and cellular interactions with the ECM. In this perspective, it is presumed to be capable of transmitting extracellular signals to the intracellular milieu.

A. DEFINITION AND REACTIONS CATALYSED. THE FAMILY OF TGs

Kinetic properties of TGs when acting as calcium-dependent protein cross-linking enzymes were investigated extensively in the late 1960s in the laboratory of Jack Folk at NIH by means of classic two-substrates enzyme kinetics whereby the enzyme reaction proceeds by a ping–pong mechanism. In this mechanism, enzyme reacts via an essential cysteine residue with the first substrate, a peptide containing a glutamine residue, which can be specifically recognized at the enzyme active site. During this first part of the catalytic cycle, ammonia is released with formation of an acyl enzyme intermediate between the enzyme active site thiol and the reactive glutamine residue. The intermediate then undergoes nucleophilic attack by a second substrate (usually a primary amine), which acts as the final acyl-acceptor, releasing an isopeptide-modified final product. The careful kinetic analysis performed at that time demonstrated that the general scheme is actually that of a "modified" ping–pong mechanism, since the covalent acylenzyme intermediate can also undergo hydrolysis, employing water as the acceptor nucleophile. This alternative hydrolytic reaction is favored in the absence of the amine stronger acceptor, releasing a peptide in which the recognized glutamine has undergone hydrolysis to glutamate (Figure 1). This latter reaction has proved highly important in the pathology of celiac disease, where TG2 is the major autoantigen [6].

Formation of the acyl-intermediate is a rapid step in the catalytic cycle, and the aminolysis is the preferred process of regeneration of the active enzyme with rates that greatly exceed those of hydrolysis. The prevalent cleavage of the thioester by aminolysis rather than by hydrolysis is probably also favored by a general base catalytic mechanism whereby the incoming amine is "activated" by deprotonation by a histidine residue, which belongs to the enzyme catalytic triad as it will be detailed. Access to the active site is by means of channels lined with hydrophobic residues, which disfavors the availability of water to the thioester intermediate.

Early studies demonstrated that the properties of TGs differed significantly when isolated from different tissues, in terms of catalytic activity, proteolytic processing, and molecular size [7], suggesting the existence of different isoenzymes. This question was solved in the 1980s when the first enzyme TG2 was cloned and its gene structure elucidated [8], ultimately leading to the identification of the existence of several isoenzymes representing a family of homologous proteins. Now it is clear that at least 9 isoforms exist, which are blood clotting factor XIII present in plasma (and platelets and monocytes), 7 cellular forms (TGs type 1 to type 7), and an additional form, protein 4.2, present in the erythrocyte membrane, which is devoid of catalytic activity because it lacks the cysteine residue

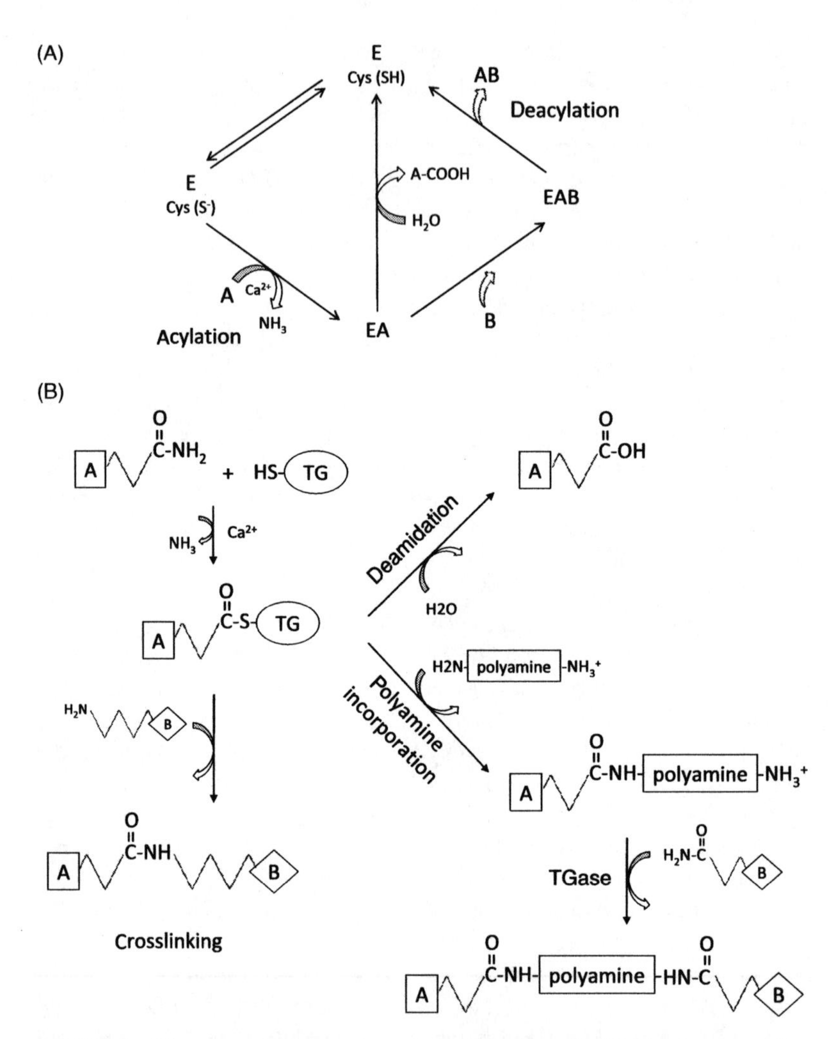

FIGURE 1. The transamidating reaction of transglutaminase 2. (A) Kinetic scheme. The transamidating reaction takes place through a modified ping–pong mechanism. In the first step, the thiolate anion of Cys 277 of the free enzyme (E) interacts, in the presence of calcium ions, with the amide moiety of the peptidyl glutamine substrate (A) to form the covalent thioester acyl intermediate (EA) and to release ammonia. This intermediate can regenerate the free enzyme through either (1) hydrolysis of the thioester linkage, releasing the deamidated glutamyl derivative of the first substrate, or (2) interaction with the second amine B substrate, releasing the enzyme at the end of the catalytic cycle. The prevalence of the aminotransfer over the hydrolytic pathway depends on the availability of the stronger amine nucleophile

TABLE 1
Mammalian Transglutaminases and Their Functions

Protein Name	Synonyms	Gene	Prevalent Function
Factor XIII a-subunit	Factor XIII a; Plasma TGase	*F13A1*	Stabilization of fibrin clots and wound healing
TGK	Keratinocyte TGase; TGase type 1	*TGM1*	Cell envelope formation during terminal differentiation of keratinocytes
TGC	Tissue TGase; TGase type 2; Gah	*TGM2*	Cell death/survival, cell differentiation, matrix stabilization, adhesion protein
TGE	Epidermal/hair follicle TGase; TGase type 3	*TGM3*	Cell envelope formation during terminal differentiation of keratinocytes
TGP	Prostate TGase; TGase type 4; Dorsal prostate protein 1	*TGM4*	Reproductive function involving semen coagulation
TGX	TGase type 5	*TGM5*	Epidermal differentiation
TGY	TGase type 6	*TGM6*	Not characterized
TGZ	TGase type 7	*TGM7*	Not characterized
Band 4.2 protein	Erythrocyte protein band 4.2	*EPB42*	Hematopoiesis; Not active TGase

essential for activity. This information is summarized schematically in Table 1.

On the basis of firmly established published data, the main goal of our discussion is to analyze the structure and regulation of the multifunctional enzyme TG2 and to explain its role in a physiologic and pathologic perspective. Another review focusing on initial data collected in vivo, which

FIGURE 1. (*Continued*) rather than water. (Redrawn with slight modifications from [5].) (B) Reaction mechanism. In a mechanistic way, the enzyme reacts with the peptidyl glutamine substrate to form the thioester intermediate, which can undergo hydrolysis (right, upper arrow) or nucleophilic displacement either by a second protein bearing a free ε-amino lysine group (left, downward arrow) to form protein polymers stabilized by an isopeptide bond, or by a free low-molecular weight amine, usually a polyamine in the reaction of "polyamidation" of the A protein (right, lower trace). This initial product can be acted upon in an additional catalytic cycle, serving through the distal amine group as the nucleophile of an additional transglutaminase thioester intermediate. In this final instance, cross-linked proteins are formed, stabilized through a polyamine bridge. (Redrawn with slight modifications from [2].)

can aid the reader to explore these issues, is published recently by Iismaa et al. [9].

B. TG2 TISSUE AND CELL DISTRIBUTION

The concept of ubiquity for TG2 in mammalian tissues is probably not completely correct since the widespread distribution of the protein in tissues is probably not, or not only, derived from enzyme present in the parenchymal cells but rather from enzyme present in vascular endothelia and in smooth muscle cells, which express it at high levels [10].

At the cellular level, the vast majority of the enzyme is present in the cellular cytosolic compartment, but appreciable fractions are also recovered within purified subcellular fractions, enriched in subcellular organelles. For instance, TG2 was detected in the plasma membrane, ECM, nuclei, and also mitochondria. Although it has been suggested that the mitochondrial location of TGase activity is not related to TG2 but to another isoenzyme [11], ample evidence for the presence of considerable amounts of TG2 in mitochondria, where it may also act as a PDI, is discussed by Piacentini in another chapter in this volume.

Even more convincing evidence is available for the association of TG2 with the plasma membrane. The translocation of TG2 to the plasma membrane apparently takes place, leading to binding of TG2 to the internal membrane surface. As it will become apparent later on, TG2 also has sites for binding GTP but with an affinity that is lower than that typical of G-proteins. In this inner membrane location when bound to GTP, the classic transamidating activity of TG2 is inhibited [12] and the enzyme displays GTPase activity acting as a G-protein in the activation of phospholipase C, leading to hydrolysis of phosphatidyl-inositol in response to hormonal signaling, e.g., by the α1-adrenergic receptors [13]. Interestingly, the membrane-bound form of TGase is not only adsorbed on the internal surface, but some of it is also translocated to the external surface, by a process that is not fully understood since the enzyme lacks any of the characteristic features classically associated with protein secretion [14]. Although some of the exposed TG2 can also be released into the extracellular fluids, it is likely that most of the external enzyme is retained on the cell surface or deposited into the extracellular matrix (ECM), a process that may involve several mechanisms (see later), which may include its interaction with fibronectin (FN), heparan sulfates, integrins, collagen, and possibly additional proteins of the ECM [15, 16].

II. STRUCTURE AND REGULATION

A. STRUCTURE AND REGULATION OF ACTIVITY AT THE PROTEIN LEVEL

Regulation of the transamidating and GTPase activities of the TG2 in the tissues is achieved via allosteric conformational changes in enzyme structure following the binding of Ca^{2+} and GTP (see Section II.B). The first TGase structure to be resolved was the intact zymogen of the catalytic subunit (subunit A) of blood factor XIII [17]. The crystal structures of subunit A of factor XIII demonstrate that each peptide chain contains four domains (designated β-sandwich, core domain, β-barrel 1, and β-barrel 2) arranged in pairs in a hairpin conformation, so that the active centre, present deeply buried in the core domain, is protected from the interaction with the substrate by the β-barrel domains that overlay the core domain itself.

All other forms of animal TGase, whose structures have been investigated at high resolution, share this overall organization of the peptide chain in four domains, although they are monomers of a peptide chain with high homology to the subunit of factor XIIIA. They include TGase 1 from keratinocytes (structural studies were performed by homology building in the perspective of understanding effects of spontaneous mutations associated with lamellar ichthyosis), TG2 from sea bream [18] and humans (crystallographic patterns have been obtained for the inactive enzyme in the presence of GDP and in a transition state, with a peptide substrate covalently tethered at the active site) [19, 20], and TG3 [21], which is the form present in hair follicles. In this last instance, investigations included the zymogen, the proteolyzed enzyme alone, and in the presence of calcium ions. Additional members of the TGase family include type 4 TGase (present in the prostate), type 5 TGase in the skin, and type 6 and type 7 TGase, which have been identified as gene products but not fully characterized at the protein level (see [2]).

TG2 consists of 687 amino acids organized, as mentioned, in four domains, 1 to 4, spanning amino acids 1–139, 140–454, 479–585, and 586–687 (see Figure 2). The first domain is a β-sandwich; the second contains virtually all helical structure elements of the protein; the third and the fourth are organized as β-barrels. The active site is present in the second ("core") domain and involves a catalytic triad reminiscent of that of pronase and of other cysteine proteinases, including cysteine 277, histidine 335, and aspartate 358, arranged in a charge-relay system. The third and fourth β-barrel domains have an inhibitory action in relation to the transamidating (i.e., cross-linking) activity, which is kept latent probably because of steric hindrance and blockage of the essential cysteine 277, either through formation

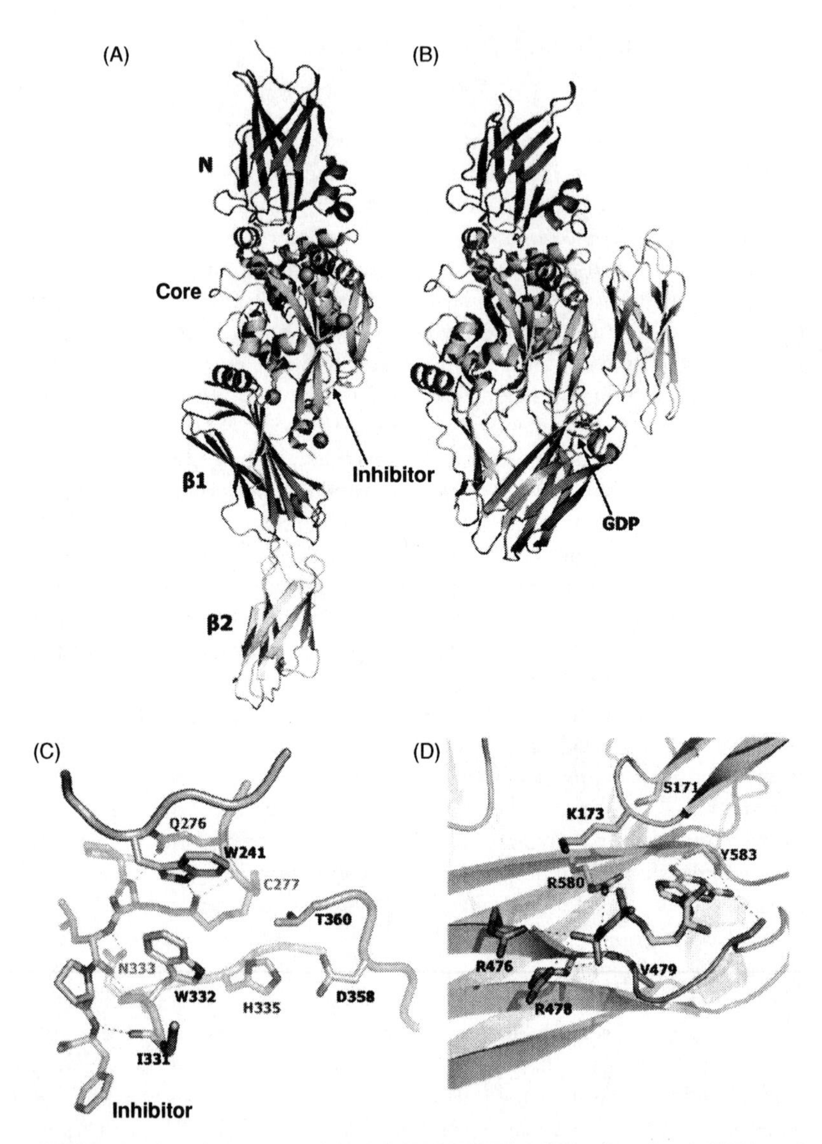

FIGURE 2. Crystal structures of Ac-P(DON)LPF-NH2 inhibitor-bound and GDP-bound TG2 represented as ribbons. The *N*-terminal sandwich domain (N) is shown in red, the catalytic core domain (core) in green, β-barrel 1 (β1) in magenta, and β-barrel 2 (β2) in cyan. (A) Open conformation of inhibitor-bound TG2 [20] showing the proposed positions of Ca^{2+} binding

of a disulfide with a neighbor cysteine (Cys 336) or hydrogen bonding to tyrosine 516. Upon activation of the enzyme by calcium ions, cysteine 277 becomes extremely reactive and mediates the catalytic step towards the first peptidyl-glutamine substrate, forming a thioester intermediate, in a transition state that is stabilized by interaction with an additional essential amino acid, tryptophan 241 [23]. From the mechanistic point of view, the formation of the acyl-enzyme intermediate occurs through a channel pointing to the active site from the enzyme surface, allowing access of the glutamine substrate and appropriate formation of the acyl-intermediate. The existence of this channel, which has been identified definitively only in TG3 [21], is controlled in a coordinated way by guanine nucleotides and by calcium ions, also allowing access of the amine substrate from either the upper or the lower face, for formation of the isopeptide product.

In relation to the enzyme specificity, most studies indicate preferential labeling of proteins related to cytoskeleton and cell mobility, to stabilization of ECM proteins, along with a few crucial enzymes involved in energy metabolism [2, 24]. Recognition patterns are apparently strictly selective towards glutamine residues, since only a tiny fraction of surface-exposed residues are recognized in native proteins, while many more are available for reaction in denatured proteins. In contrast, selection of lysine residues (or of low molecular weight amines) is much more relaxed. An important emerging point is represented by the distinct patterns of labeling of glutamine residues in intact proteins and in model peptides. Although this suggests that flanking sequences might be important keys to understanding minimum requirements for recognition of different glutamine residues by distinct TGs, rules that govern real mechanisms of selection of substrate glutamines are still partially understood and several studies devoted to this topic could provide only a

FIGURE 2. (*Continued*) (orange spheres) according to Kiraly et al. [22]. (B) Closed conformation of GDP-bound TG2 [19] showing the degree of movement of β1 and β2 after nucleotide binding. (C) Structure of the active site of inhibitor-bound TG2 showing putative contacts between Ac-P(DON)LPF-NH2 active site residues. The hydrophobic tunnel for primary amine substrate access to C277 is formed by W241, W332, and T360, which disfavors entry of water. (D) The GTP binding site of TG2 showing bound GDP according to the crystal structure [19]. Putative contacts between GTP and TG2 are represented by dashed lines. The side chains of V479 and Y583 have been omitted for clarity. Residue S171 is involved in stabilizing H-bonds between the core domain and β-barrel 1 and also between neighboring residues in the loop that it resides in. The unfavorable proximity of K173 to R580 is thought to be important in control of TG2 conformation, being stabilized in the presence of guanine nucleoside phosphates. (See insert for color representation.)

limited accepted consensus sequence [25, 26] around the peptide *p*Q*x*(P)*l*, where *p*, *x*, and *l* represent polar, any and hydrophobic amino acid, and Q and P glutamine and proline, respectively. This sequence is consistent with that recognized in gliadin peptides by intestinal TG2, a process that is relevant in the pathogenesis of celiac disease (CD) [6]. What, however, is apparent is that although this sequence is a required criterion, it is not a sufficient one, since analogous sequences are not labeled in other contexts.

Among additional factors that are likely to be involved in substrate recognition and pattern of labeling of glutamine residues is the influence of the local organization in tertiary structure around the reactive glutamines. This has been thoroughly investigated by Fontana and associates [27], employing a bacterial calcium-independent TGase and model proteins (IL-2, human growth hormone, and apomyoglobin) as glutamine donors. The authors observed that only glutamine present in regions of high crystallographic temperature factor (factor B) could act as TGase substrate. These criteria serve to identify unstructured regions largely represented by mobile exposed loops, which are also the preferential site of cleavage by proteinases because the flexible regions allow appropriate fitting of the protein substrate into the active site. Therefore, these findings point out the importance of the fitting of flexible substrates in the active site in determining the posttranslational modification of glutamine residues by TGs.

Regulation of the transamidating activity is achieved through a conformational change that is triggered in an allosteric way by calcium binding at multiple sites in the protein [20, 22, 28, 29], leading to exposure of the active site region for interaction with the glutamine substrate (Figure 2). The drastic conformational change results in movement of β-barrel 1 and β-barrel 2 away from the core domain, increasing the radius of hydration of TG2 from 3.7 to 4.1 nm [20, 30], sufficient to be resolved by nondenaturing polyacrylamide gel electrophoresis.

The Ca^{2+} activation step is prevented by the guanine nucleotide GTP that binds to the protein at the interface between the core domain (close to the active site) and β-barrel 1, tethering them together. The GTP-binding site of TG2 is unique amongst GTP-binding proteins, sharing no homology with other G-proteins, which delayed its identification until mutagenesis studies showing that GTP binding was dependent on the core domain and required residues Ser171 and Lys173 [31, 32]. With the GDP-bound crystal structure of TG2 solved [19], the GTP-binding site was confirmed (Figure 2D). It has been proposed that GTP-binding masks Arg579 in β-barrel 1, perhaps preventing its clash with neighboring Lys173 in the core domain, thereby allowing TG2 to adopt its compact conformation. In doing so, the compact conformation is also stabilized by a hydrogen bond between Tyr516 on an extended loop of β-barrel 1 and the active site Cys277 residue [33]. When Ca^{2+} binds, a conformational change is transmitted between the Ca^{2+}-binding site and the GTP-binding site, although the exact nature of this is unknown. It has been postulated that this step requires a process of domain movement across a hinge represented by a mobile unstructured superficial loop spanning amino acids 455–478. The mobile 455–478 loop is crucial in controlling the catalytic cycle of TGs as demonstrated by the linear correlation between residual activity and residual mass of intact protein during in vitro proteolysis studies [29], but its role may be different in several isoforms, since TG2 is inactivated by cleavage of the loop, while cleavage of the homolog in the TG3 zymogen is required to produce the active enzyme [34]. A more recently proposed mechanism is that calcium binding disrupts the interaction between peptide Ile416-Ser419 and peptide Leu577-Glu579 in β-barrel 1, causing a destabilizing of the GTP-binding site [19]. A combination of these effects, at least in part, is responsible for the reciprocal regulatory effects of Ca^{2+} and GTP.

Nucleotide is probably released as the enzyme interacts with protein substrates during catalysis, since no nucleotide is present in the enzyme stably modified by irreversible gluten peptide mimetic inhibitor Ac–P(DON)LPF-NH_2 and crystallized in a fully open conformation [20]. In this fully open conformation of TG2, the GTP-binding site is destroyed by the movement of β-barrel 1 away from the core domain (see Figure 2). Interestingly, this structure also possesses a vicinal disulphide bond between residues Cys370 and Cys371, which may be partly responsible, along with a putative Cys277–Cys336 disulphide bridge in the nucleotide bound form [18], for the requirement of reducing agents to maintain TG2 activity in purified preparations. The large conformational change observed between the structures of nucleotide-bound and substrate-bound TG2 to expose the active site for substrate requires a large amount of energy in the order of 11 kcal/mol. The energy provided by Ca^{2+}-binding alone is insufficient since no alteration in structure is observed in the Ca^{2+}-bound crystal structure of factor XIII [35]. The presence of two energetically unfavorable nonproline *cis*-peptide bonds in TG2 between Lys273 and Tyr274 and between Lys387 and Tyr388 have been postulated to isomerize to the more stable trans forms providing some of the required energy, in a mechanism conserved amongst all TGs [36]. However, the X-ray crystal structure of TG2 in the open conformation showed that these peptide bonds were still in the cis conformation, although this does not preclude that they did not undergo a *cis–trans–cis* isomerization during the conformational change with the energy for the transition potentially provided by formation of the Cys370–C371 vicinal disulphide bond [20]. The energy requirements for restoration to the closed form could be provided by reducing agents. If this were true, then under normal physiological conditions, TG2 released from cells would only be catalytically active for one cycle and would then be present in the ECM in an inactive oxidized form, which is consistent with recent in vivo observations [37].

B. REGULATION OF TRANSAMIDATING ACTIVITY IN THE INTRACELLULAR COMPARTMENT

In vitro, the cross-linking activity of TG2 is normally measured by the incorporation of labeled primary amines into substrate proteins. In this assay, the activity is absolutely dependent on availability of high concentrations of calcium ions. The plots of activity against the concentration of calcium using this can be sigmoidal with half saturation values for Ca^{2+} calculated in the region of 20–200 μM, which clearly exceed those that are encountered physiologically in the intracellular compartment. However, most of these assays are undertaken with *N,N′*-dimethylcasein as the substrate protein that sequesters free Ca^{2+} via its phosphate groups, making the actual amount of free Ca^{2+} much lower than that calculated. When undertaken with dephosphorylated casein using clamped Ca^{2+}-EGTA buffers, the half saturation for Ca^{2+} reduces to around 2–3 μM [38], making TG2 a potential Ca^{2+} target. However, the other significant modulator of transamidating activity is GTP, which is a potent inhibitor at low but not at high concentrations of Ca^{2+}. The combination of these two regulators, i.e., the inhibition by GTP, which is particularly effective at suboptimal (and presumably physiologic) concentrations of Ca^{2+}, is believed to effectively limit intracellular activity, as far as we can extrapolate data in vitro [28, 39] to in vivo conditions. We must underline that albeit studies on intact or permeabilized cells, manipulated to alter their relative content of calcium and/or GTP [40–42], have suggested that TG2 is inactive at normal physiological levels of Ca^{2+} and GTP (see also [43]), no real evidence has been collected that the enzyme is truly active in the transamidating reaction under basic cellular conditions. As already mentioned, GTP is hydrolyzed after binding so that the nucleotide that is firmly bound to the enzyme at the time of purification is the diphosphate form GDP, which has lower inhibitory efficiency as demonstrated by the enhanced inhibition brought about by supplementing the assay mixture with external GTP. These properties are believed to be physiologically relevant since they are probably related to the shift between the transamidating and the signaling activity, and to the translocation of the enzyme to the membrane. Furthermore, as discussed later in Section IV, there are reports about the existence of GTP insensitive forms of TGase, which are synthesized by a mechanism of alternative splicing at exons 12 and 13 and are characterized by an altered sequence due to expression of noncoding DNA and truncation involving amino acids at the *C*-terminal region in domain 4 [44, 45]. These forms are characterized by a reduced sensitivity to GTP inhibition and tend to display higher calcium-stimulated activity.

At the present time, we cannot yet be sure that all factors relevant to regulation in vivo, particularly in the intracellular environment, have been identified and correctly evaluated since TG2 might interact with other cellular proteins and is known to be sensitive to regulatory effects of phospholipids like sphingosylphosphocholine or to nitric oxide, which respectively activate and inhibit the enzyme under conditions close to the physiologic ones. This pattern of intracellular regulation is certainly complex and has resulted into contrasting results even if there are compelling indications of the latency of the intracellular TG2. For instance, in the presence of sphingosylphosphocholine, the enzyme displays definitely high affinity for Ca^{2+} [46] during the in vitro assays, resulting into a stimulation of activity under conditions that are ineffective to activate the enzyme in the absence of this rare phospholipid, whose turnover is not completely clear [47] also in relation to signal transduction pathways.

With regard to the effects of NO on TG2 function, it is known that NO, which is a powerful oxidant, acts both by controlling activity of guanylate cyclase (which is crucial in several critical processes as, e.g., vascular responsiveness and immunologic responses) and in a more direct way by promoting protein posttranslational modification by nitrosylation prevalently at cysteine [48] but also at tyrosine residues [49]. The initial report indicated that NO regulates the susceptibility of cells to undergo apoptosis, a process probably related to the activity of TG2 [50], as well as the activity of factor XIII [51]. Subsequent studies by Lai et al. [52] demonstrated that TG2 is affected mainly at Cys residues that are modified by S-nitrosylation, employing NO–Cys as an NO donor. Seven out of 18 thiols are modified in the absence of calcium ions, without effects on the transamidating activity, while additional two residues are modified in the presence of calcium with total loss of catalytic activity, either by formation of a disulfide or by direct nitration of Cys277, since no evidence was obtained for peculiar conformational changes that might account for the loss in activity. Notably, the GTPase and ATPase are not deranged in the nitrosylated TG2, but the modified enzyme is more sensitive to inhibition by GTP than the native protein. Therefore, this appears to be an efficient mechanism contributing to maintaining TG2 in the inactive state. It is also noteworthy that the nitro groups can be released from nitrosylated TGase upon addition of calcium ions, so that NO–TG2 can also act as a store of NO particularly in relation to vascular function, e.g., to inhibit platelet aggregation. These phenomena are also probably relevant in the control of TG2 activity in the extracellular compartment where nitrosylation of TG2 not only inhibited its extracellular cross-linking activity but also prevented its deposition into the matrix [53], an observation that parallels the decrease of NO synthesis during fibrosis where TG2 is reputed to be important [54].

Further contributions to the complexity of regulation of TG2 are brought about by changes in the properties of the enzyme in relation to its subcellular distribution. This emerged clearly as authors tried to solve an apparent contradiction in the action of TG2 on the cell cycle since the enzyme can promote both cell survival or cell death [55], depending chiefly on its localization in the cytosol or in the nucleus [56]. Employing cells transfected with wild type and mutant TG2 bearing a nuclear localization sequence, Johnson et al. [55] showed that in the cytoplasm transamidating activity is found only under extreme stress conditions when the cell is already committed to cell death. In contrast, translocation of TG2 to the nucleus is likely involved in the transcription of genes promoting cell survival in a nontransamidating manner. The overall conclusion of these studies was that in normal cells, TG2 is mainly protective and only in certain types of extreme stress conditions when death is unavoidable is TG2 transamidating activity increased. As previously mentioned, an additional alternative intracellular location of TG2 arises through its interaction with the internal surface of the membrane, where it can serve as a monomeric G-protein, denoted G_h. In the membrane, signaling activity is raised, while the transamidating activity is inhibited (although probably still partially active), through the reciprocal shift triggered by the guanine nucleotide exchange between solution GTP and bound GDP. The nucleotide exchange is also likely to play a role in its membrane translocation, as proven by changes in the surface hydrophobicity of the protein and by hydrophobic chromatography (Bergamini, personal observations). Among the multiple actions played in its location at the membrane internal surface, mention must be made of its role as a G_h activity, triggered by interactions with extracellular activated receptors such as α_1-adrenoreceptors, thromboxane A_2, and oxytocin receptors to activate downstream transducers such as PLCδ [57, 58], and the ability to modify, usually dimerize, membrane-bound receptors, as it happens in the case of the extracellular angiotensin II receptor AT2 [59].

C. REGULATION OF TG2 EXPRESSION

Expression of TG2 is under control of the TG2 promoter. Briefly, the TG2 gene consists of 13 exons, with two possible alternative splicing sites within exons 6 and 10 [60]. Upstream of the ORF, classic TATAA and CAAT boxes are found along with sequences to bind transcription factors NF-κB and Sp-1, and retinoid responsive elements to bind retinoids through the RAR and RXR receptors. A 3.8 kb sequence 5′ upstream of the initiation site is involved in directing the tissue-specific expression of the gene. Within this same region, there are also the elements that convey the retinoid specificity to the induction of the transcription of the TG2 gene [61]. In addition, it has been reported that the TG2 promoter can be highly methylated at GC islands [62], and it is proposed that the hypermethylation of these sequences is related to the silencing of the gene in relation to the tissue differentiation, resulting in the rather selective expression of the enzyme in endothelial and smooth muscle cells. As already noted, virtually all animal tissues contain some level of TG2 in the intra- and in the extracellular compartments. The most likely reason for this finding is represented by the elevated expression of the protein in cells (fibroblasts, endothelial, and smooth muscle cells) that are widely distributed in all organs [10]. In addition, erythrocytes also contain remarkably high levels of TG2. Several folds higher levels of expression of TG2 can be obtained in stromal and parenchymal cells in many organs as the result of induction by treatment with a large number of transcriptional inducers that have been identified in the recent years. Among them are the inflammatory cytokines, such as TGFβ and TNFα [63], which can induce expression either directly through a responsive element at the TG2 promoter or through activation of the NF-κB system, interleukin 1 and 6 (IL-1 and IL-6), and interferons; several steroid hormones (chiefly glucocorticoids and progesterone); vitamin D (particularly in macrophages); and notably retinoids, which are the most effective inducers known to date [61, 64]. These effects have been referred to the onset of inflammation or of apoptosis, as both conditions are accompanied by a massive increase in enzyme tissue levels. Also, certain drugs have been reported to augment the expression of TG2, as it is the case of statins [65].

Much attention has also been paid to the induction of TG2 by retinoids because of the medical applications in cancer prevention and treatment, as in the case of acute promyelocytic leukemia [66], for which a limited number of therapeutic cycles is curative, specifically triggering apoptosis in the neoplastic cells. Differentiative antineoplastic therapies based on the induction of TG2 as a crucial element of apoptosis have also been proposed for other kinds of neoplasia, including breast cancer and hepatomas [67, 68].

Apparently, as outlined above, the mechanism of induction is related to a direct interaction of the retinoids in complex with their receptors with a retinoic acid responsive element [61]. Under these conditions, possibly also through the recruitment of several transcription factors (e.g., NF-κB, AP-1, and Sp-1), synthesis of TG2 is triggered to increase the tissue levels of the native protein.

D. SECRETION AND REGULATION OF TG2 IN THE EXTRACELLULAR COMPARTMENT

The mechanism of secretion itself may be expected to form part of the regulatory function of an extracellular protein. However, the mechanism by which TG2 is secreted still remains a mystery, although several mechanisms have been proposed and some clues may be gained from the literature. It is known that TG2 is released from cells after wounding and that although leakage from cells whose membrane integrity has been compromised undoubtedly occurs, the process of secretion may not be solely due to this, because TG2 is found on the surface and in the matrix of intact healthy cells where its expression is increased [69, 70]. In addition, TG2 mutants have been made that do not affect its intracellular distribution, but prevent its externalization [14, 71].

Release of TG2 from mechanically damaged lung cancer cells actually promoted resealing of the membrane [72], in a process that required, at least in part, transamidation activity since cystamine was inhibitory. In response to cell wounding/tissue injury, TG2 appears in the surrounding matrix and this has been observed in a number of in vitro studies [70, 73, 74], where it plays a crucial role in maintaining ECM integrity and cell survival. The regulated release of TG2 by normal cells is thought crucial to normal ECM homeostasis; however, this process is as yet still unknown. TG2 and FXIIIA are known to be secreted via an uncharacterized nonclassical secretory pathway [70, 75]. They both lack an ER-targeting signal sequence and there is no evidence of their glycosylation [76, 77]. In terms of the structural requirement for TG2 secretion, membrane translocation and deposition into the ECM requires an intact *N*-terminal β-sandwich domain and an intact transamidation active site [14, 70]. In addition, several mutations that restrict the conformation of TG2 also affect secretion. Mutation K173L, which abolishes GTP binding, prevents secretion, and C277S, which blocks TGase activity, prevents deposition into the ECM [14, 71]. However, both mutations may be expected to hold TG2 into its open conformation. Mutation Y274A, which is thought to prevent the transition between open and closed conformations, also prevents secretion [14, 71]. This latter mutation does not result in loss of TGase activity [71], suggesting that conformation rather than TGase activity may be critical for secretion. Other data suggests that TG2 secretion is in some way associated with integrin β1 [78]. Recently, it has been suggested that binding to heparan sulphate proteoglycans is a requirement for secretion of TG2 onto the cell surface and subsequent transamidating activity [79], although loss of the stabilizing effects of heparan sulphate binding on TG2 towards membrane associated MMPs [80] could also account for this observation in the syndecan 4 knockout model used. Irrespective of the mode of release into the ECM, TG2 is thought to be retained there by virtue of its interaction with FN [70], although the addition of exogenous collagen reduces its matrix deposition [74]. Since the extracellular conditions of high calcium and absence of GTP are conducive to an active TG2, it would be expected that release from cells would liberate a transamidation active enzyme that would participate in random cross-linking of ECM components. Indeed, TG2 activity at the cell surface is easily measurable [69] and its presence in the matrix can be detected using labeled amines in cells induced to overexpress TG2. However, these observations are normally undertaken where forced overexpression of the enzyme has taken place, or in the case of measurement of cell surface TG2, where the cell is first removed from its substratum using either trypsin or EDTA and placed onto FN, which may be construed as a stressful event. Moreover, it has also been observed that TG2 is deposited into the matrix in a controlled fashion and that transamidation activity is not always apparent [37]. These authors suggested that TG2 is present in the ECM in a latent form, which is then activated after disruption of the ECM, rather than released from damaged cells. This activity was observed to last for a relatively short time period 12 hours after wounding and did not correlate with a loss of TG2 antigen. This effect of latency of TG2 activity under conditions (high concentrations of calcium and low concentrations of nucleotide triphosphates) prevalent in the extracellular space has been ascribed to inhibition brought about by anchorage of the enzyme to integrins and to ECM-associated FN, although it cannot be excluded that additional factors (e.g., enzyme nitrosylation or oxidation) can participate in these effects [53]. The transient activation would thus require either the release of free soluble enzyme or the neutralization and reversal of the oxidized/nitrosylated form.

It is tempting to speculate about the structural basis of the prospect of latent TG2 in the matrix, because as indicated above, TG2 is known to undergo oxidation and nitrosylation, leading to its reversible inactivation; moreover, many of its functions as a cell adhesion protein do not require transamidating activity (see later). Nitrosylation of TG2 does not affect its role as a cell adhesion protein and aids in its adhesive function by confining the enzyme to the cell surface by limiting its deposition into the matrix [53]. Perhaps, a transient release of reducing agents, such as glutathione from damaged cells, may satisfy its reduction requirements and/or relieve its inhibition from nitrosylation of essential cysteines [81].

III. PHYSIOLOGIC FUNCTIONS OF TG2

A. TG2 IN THE INTRACELLULAR COMPARTMENT

Evidence so far suggests that the distribution of TG2 in distinct cell compartments is likely linked to different functional activities. In the intracellular environment, the main processes so far published in which TG2 is involved are represented by protein transamidation and by signaling activity, which predominate over each other when the enzyme is present in the cytosol or is associated with the membrane internal surface, provided that the conditions prevailing in the local environment are capable to shift the protein from the resting to the active state [12]. In addition to these roles, evidence is increasing to suggest that TG2 may also influence transcriptional or signaling events by acting as a scaffolding protein through its association with a number of intracellular proteins involved in physiologically important processes. For example, acting in its antiapoptotic role, it has been suggested that TG2 can protect against neuronal cell death in ischemia and stroke by interacting with the hypoxia-inducible factor 1*B* (HIF 1*B*), leading to attenuation of the HIF1 hypoxic response signaling pathway. This association is not dependent on transamidating activity [56]. Other evidence indicating TG2 acts as a scaffolding protein is in EGF-stimulated cancer cell migration, where by virtue of its active site, it binds to actin, where it is reputed to act as a scaffold for the recruitment of proteins that influence actin polymerization at the leading edge of cells. Again transamidating activity is not required in this function [82]. Following on from this report is the recent observation that TG2 binds to and regulates the GTPase activating activity of Bcr, which in turn regulates Rac GTPase, resulting in increased membrane ruffling. Interestingly, transamidating activity is also not required for this function, but the GTP-binding mutants R580A and S171E that are locked into the closed form were not effective [83].

In relation to transamidating activity, TG2 is clearly a protein-modifying enzyme and this is suggestive that its diverse functional effects might relate to the induction of new properties in the proteins that are modified during the reaction. This possibility is more likely in the case where it involves local selective changes at glutamine residues (coupling with polyamines or hydrolysis of the amide) and not massive protein aggregation. However, it is a general rule that when protein modifying enzymes exert their physiologic functions by triggering new properties in targeted proteins, these processes are functionally relevant when they are reversible events, e.g., in protein phosphorylation in signal transduction. This is not the case for TGs since both hydrolysis of the substrate glutamine and formation of isopeptide bonds in substrate proteins are essentially irreversible. In the case of cleavage of the isopeptide bond, it cannot be cleaved either by proteases which do not recognize it [77, 84] or by other enzymes—the isopeptidases—which are of very limited distribution and are more likely involved in de-ubiquitinating proteins to prevent undue degradation by proteosomes. Apparently, there are only two ways to remove from tissues proteins that have been modified by TGs, their complete degradation either by proteolysis or by the specific cleavage of the isopeptide through the reversal of the reaction of the TGs themselves. This back-reaction has been demonstrated in the case of both factor XIII and TG2 [85] but to the limited extent of using proteins modified by labeling with low molecular weight fluorescent probes, not of cross-linked proteins.

Therefore, the reactions catalyzed by TGs are expected to be virtually irreversible and to alter deeply the surface properties of the substrates, since either their solubility is heavily affected in the case of protein cross-linking or their surface charge is modified by conversion of a neutral glutamine residue into a positively charged secondary amine (following polyamidation by spermidine or spermine) or into a negatively charged glutamate residue (see Figure 1). In this perspective, it is likely that TGs in general are kept as latent enzymes to be activated only for special local purposes, e.g., following extreme insult to a cell where Ca^{2+} homeostasis is perturbed, in fibrin stabilization and wound healing and in the terminal differentiation pathway of keratinocytes in skin or in extreme pathologic conditions, e.g., during fibrogenesis. However, we cannot ignore reports particularly in a number of drug-resistant metastatic cells where intracellular TG2 transamidating activity is reputed to be active in a protective survival function, e.g., in the activation of NF-κB by the modification of IkBα into polymers that reduces the binding of this inhibitory protein to the NF-κB [86, 87]. In addition in cystic fibrosis (CF), TG2 transamidating activity is reputed to be involved in modulating inflammation via PPAR modulation [88] and in pancreatic cancer in the modulation of PTEN expression [89]. Transamidating activity is also reputed to be required in ovarian cancer cells for the expression and function of MMP-2, which is associated with an increased migratory phenotype [90]. Moreover, in normal cells where intracellular Ca^{2+} levels are perturbed, e.g., in neuronal cells following excitotoxic shock, activation of NF-κB via transamidating activity may also occur as a means of preventing cell death [91].

However, the overall role of TG2 in cell death is still not clear, since it has been reported to either facilitate or prevent apoptosis depending on the nature of the cell type and the trigger for cell death. The precise nature of TG2 involvement is also poorly understood, although several mechanisms have been proposed. This uncertainty is compounded by the fact that $TG2^{-/-}$ mice do not display any phenotype attributable to altered apoptosis [92, 93], although the functions of TG2 may be compensated by another TGase isoform. Induction of apoptosis in human promonocytic cell line U937 by treatment with all-*trans*-retinoic acid and calcium ionophore A23187 leads to a reduction in Bcl-2 and an increase in TG2 expression that results in the terminal cross-linking of intracellular proteins. Inhibition of TG2 in this model by transfection with an antisense gene produced a drastic reduction in apoptosis stimulated with either A23187, cycloheximide, or calphostin C [94]. One of the cellular targets of TG2 transamidation has been suggested as dual leucine zipper-bearing kinase (DLK) in calphostin C treated NIH-3T3 cells [95]. DLK is a proapoptotic kinase that is integral to the JNK signaling pathway. TG2-dependent oligomerization of DLK was found to increase its kinase activity and subsequent activation of the JNK pathway.

The presence of a truncated isoform of TG2 as a consequence of alternate mRNA splicing has been observed in cytokine-induced cells [44, 96–98]. This short isoform, designated TGase-S, lacks a *C*-terminal region related in GTP binding. Consequently, TGase-S has a greatly reduced affinity for

GTP and would be activated intracellularly at much lower calcium concentrations than the full length TG2. However, it has been shown recently that TGase-S exerts its proapoptotic effect not through its transamidation activity, but more likely due to its propensity to form intracellular aggregates that are themselves cytotoxic [99]. As already mentioned, the main mechanisms that have been identified as involved in the regulation of transamidation in the intracellular space are related to the complex interplay between activation by calcium and inhibition by GTP, although additional factors that might also contribute to this phenomenon are represented by variations in pH (data obtained in vitro with the membrane-associated TGase indicate that optimum pH for the transamidating and the GTPase activities are pH 9.0 and pH 7.0, respectively) [100] and by protein nitrosylation [52].

As previously mentioned, GTP (but not ATP) plays a major regulatory role in controlling activation of transamidating reaction by calcium, preventing activity at low concentrations of the cation. However, the subsequent discovery that TG2 has intrinsic protein kinase activity, employing ATP as phosphoryl donor, is suggestive that the ATP-binding site on TG-2 site is that involved in the kinase activity, which targets serine and threonine residues in basic proteins as histones, insulin-like growth factor-binding protein-3 (IGFBP-3) and oncoprotein p53 [101]. The protein kinase activity is inhibited by calcium, suggesting that under normal physiological conditions where [Ca^{2+}] is low, it should be active as demonstrated by the above reports, but its physiologic and regulatory features are still elusive.

Additional activities that have been discovered for TG2 and that are not fully understood are represented by a PDI activity. The PDI activity has been detected through the classic RNAase refolding assay and has been confirmed in relation to the derangement of mitochondrial function that can occur in relation to autophagy and apoptosis [102], as discussed extensively in other chapters in this volume. The mechanism that underlies the disulfide exchange is related to the high number of free sulfhydryl groups of the enzyme, but as yet cannot be ascribed to a specific reaction centre.

B. TG2-MEDIATED SIGNALING AT THE PLASMA MEMBRANE

As mentioned earlier, recruitment of the TG2 protein to the cell membrane brings about a shift from protein cross-linking to GTPase-dependent signaling, to transduce extracellular signals originated either by hormones or by extracellular proteins involved in cell–matrix interactions. Once bound at the membrane, TG2 can remain associated with the internal surface or eventually be exported to the external surface by a mechanism involving its active conformation since extrusion does not occur with inactive mutants lacking the critical active site cysteine 277. However, details are lacking. The prevalent activity of TG2 in the membrane-associated form is represented by the GTPase activity, since the cross-linking activity is inhibited under these conditions [12]. In these conditions, the enzyme acts as a G-protein, interacting with external G-protein coupled receptors (GPCR) to activate a final intracellular transducing system, frequently phospholipase Cδ1. The observation that the GTPase activity was linked to a signal transduction activity was performed by Nakaoka et al. [103] a few years after the inhibition of the cross-linking activity by GTP had been discovered. In the original experiments, Nakaoka et al. noted that α1 adrenergic receptors from rat liver membranes copurified with a large molecular weight G-protein (originally designated Gh_{α}), whose sequencing yielded peptides identical to the sequence of TG2. Transfection experiments with cloned Gh_{α} further proved its identity with intact TG2, and its interaction with an additional 50-kD protein, Gh_{β}, which is now assumed to be calreticulin [104]. Gh-mediated activation of the adrenergic receptor was coupled with hydrolysis of phosphatidyl-inositols, leading to increased cell calcium. The signaling activity of Gh to transduct external signals, which is independent of the transamidating activity, has been reported also for receptors for other hormones, notably oxytocin, thromboxane A2, and FSH in Sertoli cells [105] as well as for other complex integrated processes like control of calcium-activated large conductance K^+ channels [106] or the activity of the type 1 angiotensin II receptor in macrophages from hypertensive individuals, through intracellular cross-linkage with contributions by factor XIII [107].

The signaling events mediated by TG2 are not limited to its activity as intracellular Gh protein in response to endocrine-like signals, but are also involved in organization of the actin cytoskeleton (see above); in modification (e.g., dimerization) of extracellular ligands [107] and receptors; in control of growth and metastatic spreading of tumoral cells, as it happens in melanoma cells upon triggering by the "orphan" GPCR for the ligand GRP56 [108]; in terminal differentiation of specialized tissues (as in cartilage and bone) [71, 109, 110]; and in modulation of local inflammatory responses [111]. In mineralized tissues, these processes are usually linked to the metabolism of pyrophosphate and rely on TGase-dependent and -independent steps [112].

C. TG2 IN CELL ADHESION AND MATRIX ASSEMBLY

In keeping with some of the intracellular roles of TG2 as a scaffolding protein not requiring transamidating activity, TG2 can also function in a similar mode in the extracellular environment either at the cell surface where it may act as integrin coreceptor for the binding of FN via the 42-kDa gelatin fragment or following its deposition into the ECM where it acts as a heterocomplex with FN in the binding of cell surface heparan sulphate proteoglycans. When present at the cell surface, the enzyme can associate with integrins beta 1, beta 3, and beta 5 to facilitate cell adhesion via binding to FN [70, 78]. Acting as a scaffolding protein, TG2 is also reputed to be able to induce integrin clustering and promote activation of the GTPase RhoA via suppression of the Src-p190RhoGAP signaling pathway (Figure 3B, [15]). As a consequence, the enzyme has the ability to fine-tune integrin-mediated signaling via focal adhesion kinase (FAK) and extracellular signal-regulated kinase (ERK) pathways following a cells interaction with its substratum. A more recent report has also indicated that cell surface TG2 may promote functional collaboration between growth factors, i.e., between platelet-derived growth factor (PDGF) and integrins by bridging the two receptors at the cell surface, thus promoting PDGF binding and signaling (Figure 3A), which may be important in many of the pathological processes that TG2 is reputed to be involved in, such as tissue fibrosis, inflammation, and tumor metastasis [113]. Following deposition into the ECM, once bound to its high affinity binding partner FN, TG2 can also facilitate cell adhesion and spreading via its direct binding to the heparan sulphate proteoglycan syndecan-4. Unlike its cell surface role outlined above, this mechanism is an RGD-independent mechanism. Hence even if the integrin is occupied by RGD-containing peptides, as commonly found during matrix turnover following proteolytic degradation, TG2 via binding to cell surface syndecan-4 can mediate cell adhesion and cell survival via a signaling pathway involving activation of PKC alpha (Figure 3C) and integrin beta 1 through an inside–out signaling mechanism. Interestingly, cell adhesion studies looking at TG2 null fibroblasts transfected with TG2 mutants also found that inhibition of PKC alpha led to the abrogation of the adhesion and migration effects mediated by TG2 [115]. In this study, it was suggested that TG2 was acting in an intracellular manner in cell adhesion but in an extracellular manner in ECM remodeling. ECM remodeling involving FN deposition can also take place by either involving a transamidating or nontransamidating mechanism. In the latter mechanism induced by TGF beta, cell surface TG2 was demonstrated to

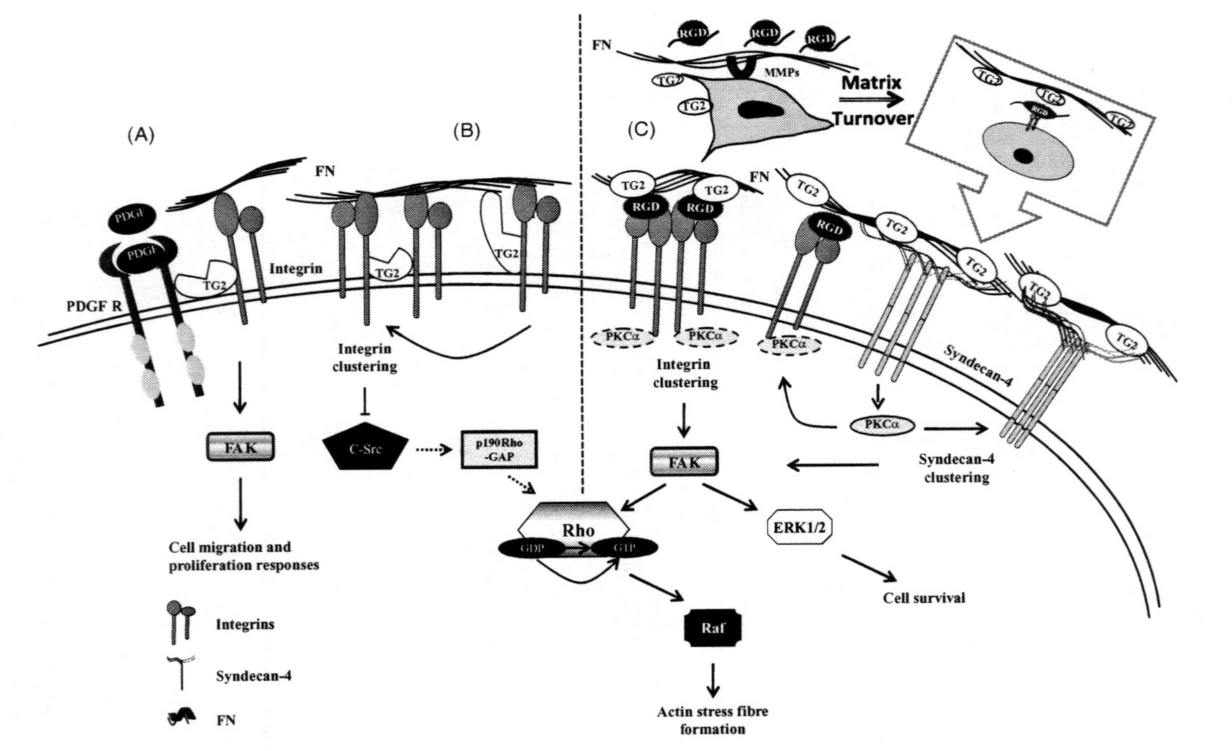

FIGURE 3. The role of TG2 as a scaffolding protein in extracellular environment. (A) Cell surface TG2 promotes and stabilizes the interaction between integrins and PDGFR by acting as a bridging protein, further enhancing the PDGFR-related signaling transduction in regulating cell migration and proliferation responses [113]. (B) Cell surface TG2 can also mediate the integrin clustering process, as a fibronectin (FN) coreceptor for integrins. TG2 is thought to enhance cell adhesion via inhibiting Src-p190RhoGAP leading to activation of RhoA signaling [15]. (C) Once deposited into the ECM and bound to FN, forming a TGase–FN heterocomplex, this complex can also enhance cell adhesion and also rescue cells from the anoikis induced by the RGD peptides released from the MMPs-digested matrix proteins during matrix turnover. FN bound TG2 via its direct interaction with the cell surface syndecan-4 receptor can activate the syndecan-4 and β1 integrin cosignaling pathway through their intracellular downstream molecules PKCα, FAK, Raf, and ERK1/2 to promote cell survival and actin cytoskeleton formation [114].

cooperate with but unable to substitute for α5β1 integrin in FN fibril assembly [75]. The role of TG2 in cell adhesion and migration is also dependent on the cell substratum. For example, the cross-linking of collagen by TG2 promotes both cell adhesion and proliferation of both fibroblasts and smooth muscle cells [116], while proteolytic degradation of TG2 by MTI–MMP in a cancer cell line stimulated migration on collagen but inhibited migration on FN [117].

The importance of TG2 transamidating activity in matrix deposition is now well documented particularly during matrix turnover following tissue injury, during wound healing [16, 115, 118, 119], and during matrix assembly in bone maturation [120, 121], where TG2 expression is upregulated and evidence of its extracellular presence is evident. Essentially, TG2 may increase deposition of ECM proteins in at least two ways. Firstly, by activation of matrix-bound TGFβ1 through the cross-linking of the large latent TGFβ1-binding protein [122, 123], leading to increased levels of active TGFβ, which by either autocrine or paracrine stimulation can increase expression of both matrix proteins and TG2 and decrease the expression levels of matrix-degrading metalloproteinases (MMPs), thus tipping the balance towards matrix deposition. Activation of NF-κB by TG2 in this cycle of events can also lead to increased expression of both TGFβ1 and TG2, thus prolonging the sustained response that facilitates matrix deposition [53, 124, 125], again favoring matrix accumulation, which under pathological conditions, i.e., fibrosis, becomes a progressive process eventually leading to organ failure. A recent report has suggested that loss of extracellular regulation of the enzyme via a reduction in NO levels commonly found in damaged tissues like kidney lead to deregulation of activity, again tipping the balance towards matrix accumulation. TG2 cross-linking of collagen can also facilitate increased cell adhesion and proliferation of cells such as fibroblasts, osteoblasts [116], and smooth muscle cells [126], but in contrast can inhibit vascular tube formation in endothelial cells [127, 128] (see Figure 4).

IV. DISRUPTION OF TG2 FUNCTIONS IN PATHOLOGIC CONDITIONS

In a recent and comprehensive review, Iismaa et al. [9] discussed the involvement of TGs in human pathologies, giving evidence of the involvement of TG2 in (1) acute and chronic inflammatory processes (including fibrosis, tissue remodeling, apoptosis, and autoimmune reactions), (2) neurodegenerative diseases, and (3) neoplasia. Reconnecting with our previous discussion

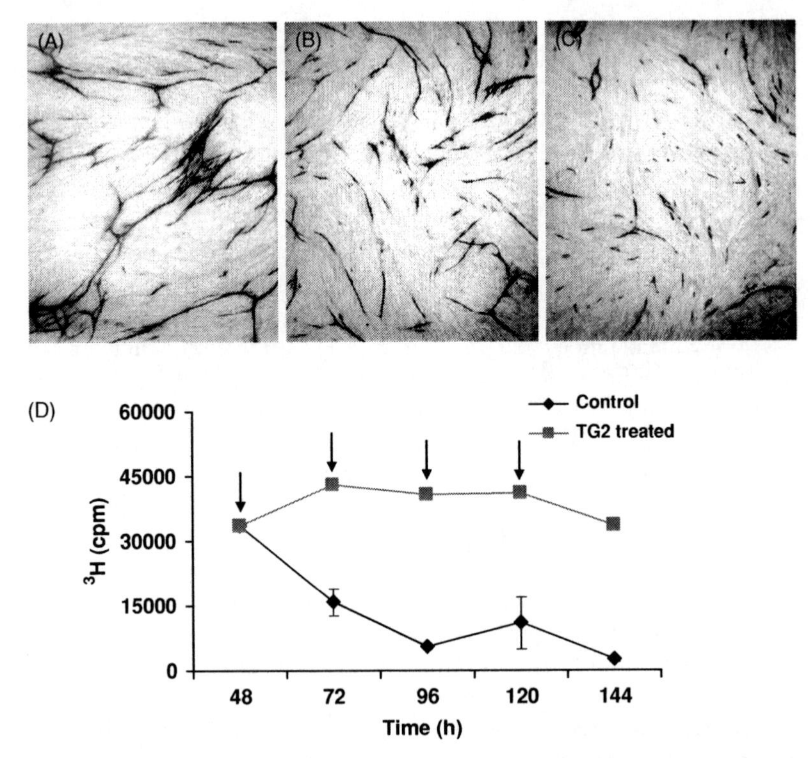

FIGURE 4. Inhibition of angiogenesis in a HUVEC cell coculture model by the addition of exogenous TG2. TG2 additions were made on days 2, 5, 7, and 10 of culture. Vessel formation was observed by staining for von Willebrand factor (vWF). (A) untreated, (b) TG2 treatment after day 5, (C) TG2 treatment after day 10. (D) Total ECM turnover following ^{3}H amino acid mixture pulse labeling. Arrows indicate addition of exogenous TG2 [adapted from ref. 127]. (See insert for color representation.)

on the regulation of TG2 tissue levels, we will now discuss separately some of the conditions related to the disruption of the functional activities of the enzyme in the intra- and/or extracellular compartments.

As we have already mentioned, tissue levels of TG2 are increased in vitro as well as in vivo by exposition to hormones, cytokines, and retinoids through transcriptional stimulation of protein expression. In general, conditions of "altered" expression of TG2 are characterized frequently by an increased activity either due to augmented expression of the normal protein, whose activation is triggered by increase in calcium and decrease in GTP levels in the intracellular space or by the expression of mutant forms of TG2, insensitive to the normal regulatory effects and thus constitutively active. Increased expression (and activity) of TG2 in the intracellular compartment are frequent features of inflammatory conditions and are possibly due to stimulation of protein expression by cytokines or other exogenous factors. For instance, this is the case that is observed during exposure of cultured cells to physiological stress or trauma, e.g., TNF-α,TGF-β, or excitotoxic or hyperosmotic stress [63, 124, 129, 130], a process that can lead to both intra- and extracellular cross-linking and intracellular aggregate formation as found in neurodegenerative diseases or increased accumulation of ECM as found in tissue fibrosis of lung, liver, and kidney (see Section III.C) [125, 131–133]. Both the intra- and the extracellular TG2 pools are often involved in this phenomenon because of the presumed quantitative and functional correlates between them.

Increased TG2 activity involvement in chronic neurodegenerative diseases can also be due to the expression of mutated forms of TG2, which have a decreased sensitivity to the inhibitor GTP, or eventually to overexpression of substrate proteins. The phenomenon of the expression of GTP-insensitive forms of TG2 was originally explored by Monsonego et al. [44], who proved the expression of a short form of the enzyme lacking part of the *C*-terminal domain 4 in astrocytes treated in vitro with the cytokines IL-1β and TNF-α. Analogous results were also reported by Festoff et al. [97], who demonstrated the occurrence of short forms of TG2 in vivo in an experimental model of spinal cord injury and in samples of neocortex and of hippocampus from brains obtained postmortem from patients affected by Alzheimer's disease [134]. Although the pathogenesis of AD is not yet completely clear, results have been interpreted as an indication of the expression of short GTP-insensitive forms of TG2 following in vivo exposition of neuronal cells to stimulation by endogenous cytokines, as a consequence of chronic neuronal inflammation occurring in these patients. Similar mechanisms would be operative in severe traumatic lesions and in other diseases related to chronic inflammation in the CNS as in Parkinson's disease and possibly in multiple sclerosis. A modified mechanism is probably involved in other neurodegenerative diseases (notably in Hungtinton's disease) that are characterized by the expression of altered substrates, with expanded poly-glutamine tails that affect protein solubility and tendency to polymerization by TG2. In these instances, the increased in vivo polymerization of proteins triggered by deregulated activity of the mutant short forms of TG2, which are insensitive to inhibition by GTP, or by the intrinsic tendency of mutant substrate proteins to aggregate leads to accumulation of intracellular protein inclusion bodies and of polymerized proteins in the extracellular or the intraneuronal space. The former consist prevalently of inclusion bodies in AD, which represent a landmark of the disease, largely consisting of polymerized TAU protein [135].

The importance of increased expression of TG2 in the chemoresistance of cancers, including ovarian, breast, pancreatic, and melanomas, has already been mentioned above, where the prime role of the enzyme appears to be in the cell survival mechanisms commonly associated with drug resistance commonly found in highly aggressive tumors (reviewed in [87, 136]). However, the overall role and function of TG2 in cancer progression is not always that clear (reviewed in [137]).

Given its well-established role in matrix stabilization and deposition, which in turn can lead to inhibition of angiogenesis [127] (Figure 4), it is not surprising to see that in a number of reports that during cancer progression prior to metastasis, TG2 levels actually decrease. Such a process would favor destabilization of the ECM, thus facilitating dysplasia and invasion of the surrounding environment, leading to the malignant phenotype. In support of this hypothesis is the finding in two recent separate reports that tumor progression and metastasis is significantly increased following subcutaneous implantation of the mouse B16 melanoma [127, 138].

Increased TG2 is also reported to be present in cystic fibrosis (CF), but the regulatory mechanism that maintains this increase is different to that reported in other pathologies involving the enzyme. CF is a disease caused by mutations in the CF transmembrane conductance regulator gene that is characterized by chronic inflammation and sustained bacterial lung infections. In a recent report, it was shown that SUMOylation of the enzyme leads to inhibition of its ubiquitination and as a consequence its subsequent proteosome degradation. This in turn leads to reduced turnover of the enzyme and hence its sustained activation. Increased levels of the enzyme were shown to lead to increased activation of NF-κB and to an uncontrolled inflammatory response. It was suggested by the authors that this mechanism for prolonging the sustained but unwanted activation of TG2 might occur in other diseases where TG2 is involved, such as in metastatic cancer and neurodegeneration [139].

Probably the best reported example of TG2 involvement in autoimmune disease is in coeliac disease (CD), which is a lifelong autoimmune condition of the gastrointestinal tract, affecting the small intestine of genetically susceptible individuals [6, 9, 140, 141]. Glutens and gliadins, which are the storage protein of wheat, are the inducers of the disease, with similar structurally related molecules also found in barley and rye. In CD, TG2 is the major autoantigen, and the subepithelial compartment becomes highly decorated with the enzyme [142] following induction of its expression by released inflammatory cytokines like TNFα and TGFβ. A 33 amino acid peptide rich in glutamine and proline is released after the proteolytic digestion of the glutens and gliadins, corresponding to amino acids 57–89 (that is resistant to luminal digestion by gastric, pancreatic, and intestinal brush border proteases), and is reactive to deamidation by TG2. This peptide is thought to play a pivotal role in disease induction since it is able to bind with high avidity to the key positions on gliadin peptide T-cell epitopes, to HLA-DQ2/8 molecules. Presentation of the antigen to CD4 T lymphocytes, in the lamina propria results in Th1 cell-type activation and subsequent release of inflammatory cytokines. The major role of TG2 in CD is in deamidation but reports now suggest that the activity of the enzyme can be further modified by the autoantibodies directed against it, which may also play an important role in celiac progression as well as in other autoimmune diseases [128, 143].

V. PERSPECTIVES FOR PHARMACOLOGIC INTERVENTIONS

The awareness that TGs are involved in several processes relevant to human pathology stimulated interest in modulating its function in vivo, in terms of both gain/loss of activity.

Interest in increasing TG2 expression is related to oncologic therapies upon observations that sensitivity of solid cancers to chemotherapy is related in a complex way to altered expression of TG2, since the enzyme can either promote or suppress cellular proliferation depending on experimental conditions [55, 144]. As already mentioned, best results in cancer therapy were obtained through induction of TG2 and stimulation of apoptosis in hematologic malignancies such as acute PML [66]. In all instances, favorable effects of expanded enzyme expression were reconnected to the in situ activation of the overexpressed protein and to switch-on of the apoptotic machinery. It cannot be excluded that additional signaling mechanisms (e.g., control of activity of transcriptional factors) might contribute to the final effects [87], and this apparently is the case for the inhibition of tumor cells survival reported for treatment in vitro with glucosamine, which has the opposite effect to that of inhibition of TG2; this would in turn block the in situ activity of NF-κB or other cell survival mechanisms requiring TG2 (see Section II.B; [145, 146]). Consequently, blockage of TG2 activity would limit the availability of active NF-κB and make the tumor more susceptible to normal chemotherapy. Initial studies in animal models for orthotopically growing pancreatic ductal adenocarcinomas using TG2 siRNA delivered in lipid vesicles has so far shown promising results [136].

Although early studies used competitive substrates of TGs such as dansylcadaverine and cysteamine, which were developed as tools to investigate kinetic properties of TGs, the potential effects of these nonspecific compounds was a problem. Cysteamine has also been utilized in vivo, improving significantly survival and motor symptoms of murine experimental Huntington's disease [147]. In this, as in other chronic neurologic diseases, excessive TGase activity might contribute to the pathologic pattern through accumulation of abnormal cross-linked proteins because of either expression of a deregulated enzyme or overexpression of substrate proteins [148]. The "curative" effect of cysteamine was originally ascribed to inhibition of the in vivo activity of TG2 [149], but this hypothesis has later been questioned because cysteamine can also affect other enzymes, notably cysteine proteinases, which share with TGs a similar organization of the active site [150]. The search for specific inhibitors to TG2, in particular active site directed irreversible inhibitors, has been the subject of recent research by several groups (see [151] for a recent review).

Recently identified competitive substrates have been based on the peptide sequence around reactive residues in natural protein substrates. The first examples were peptides derived from proelafin, which were employed for experimental control of keratinocyte proliferation and in modulation of phospholipase A_2 activity for therapy of skin scars and allergic conjunctivitis [111, 152]. Interest is now extending to explore the properties and applications of additional peptidomimetics derived from the sequence of other TGase substrates (including gliadin, the protein responsible for CD). These peptides can be employed either nonmodified or after introduction of chemical reactive groups (e.g., aldehyde, maleimide, or diazo groups), which convert them into suicide substrates reactive to the active site C277.

Among the first "specific" mechanism-based inhibitors that were developed to control TGase activity, a major interest rose around halogenated dihydroisoxazoles [153], which are effective inhibitors of TG1 and TG2, despite their modest solubility. Additional compounds we have also employed while studying thermodynamics of ligand binding to TG2 and which have proved highly useful in cell culture studies are the derivatives of 2-[(2-oxopropyl)-thio]imidazolium [154]. However, it must be underlined that the thioimidazolium derivatives of the type that has been reported are not suitable for therapeutic use in vivo because their powerful inhibition of plasma factor XIIIa would bring about unacceptable risks and potential side effects. The mechanism of inactivation involves acetonylation of Cys 277 with release of the aromatic portion of the reagent.

For the successful therapeutic application of TGase inhibitors, it is essential that they are specific only to the TG2 isoform to prevent unwanted side effects. For this reason, the peptide-based inhibitors based upon, or related to either *N*-benzoyloxycarbonyl-protected phenylalanine or *N*-benzyloxycarbonyl–protected *L*-glutaminylglycine [155], a commonly used dipeptidyl acyl donor [155, 156] or gliadin-derived peptides [157] have been used. The warheads utilized with these compounds, which are normally separated by a spacer group and which target the active site thiol of TGs, include reactive groups such as epoxides, α,β unsaturated amides, aldehydes, 1,2,4-thiadiazoles, cinnamoyl derivatives, and dihydroisoxazoles. In some instance, care was taken to achieve high water solubility so that the 6-diazo-5-oxo-*L*-norleucine-based peptidyl derivatives terminating in the dimethylsulphonium methylketone-based warhead could not get access to the intracellular compartment [158] and thus serve as selective inactivator of the extracellularly exposed pool of TG2.

There is now much interest in the evaluation of all these new potential drugs in the experimental therapies of TGs-related diseases [9], but little real progress to a clinical application has been obtained. In the case of hypertrophic scarring, where excessive TG2-mediated matrix deposition of collagen leads to excessive and unsightly scarring, the application of topical TGase inhibitor (putrescine) led to an improvement in clinical outcome [159].

Some efforts have also now been dedicated to CD [160], in which for instance deamidation of gliadin by pretreatment with bacterial TGs has been reported to be effective in controlling reactivity of gluten peptides to T-lymphocytes. TG2 inhibitors (R281) directed towards the ECM as outlined above [158] have been tested in animal models of renal scarring and diabetic nephropathy, where up to a 90% reduction in scarring and a parallel reduction in protein cross-link occurred (Figure 5). A significant protection of kidney function was also observed and there is no obvious signs of toxicity [161, 162]. In addition, animal models with subcutaneous transplanted murine DBT glioblastomas treated with dihydroisoxazole-based TGase inhibitors KCA075 and KCC009 combined with the chemotherapeutic agent BCNU have led to increased chemosensitivity and tumor cell death [163]. In addition to the use of site-directed inhibitors in cancer studies, TG2 silencing using siRNA delivered in liposomes has also led to increased

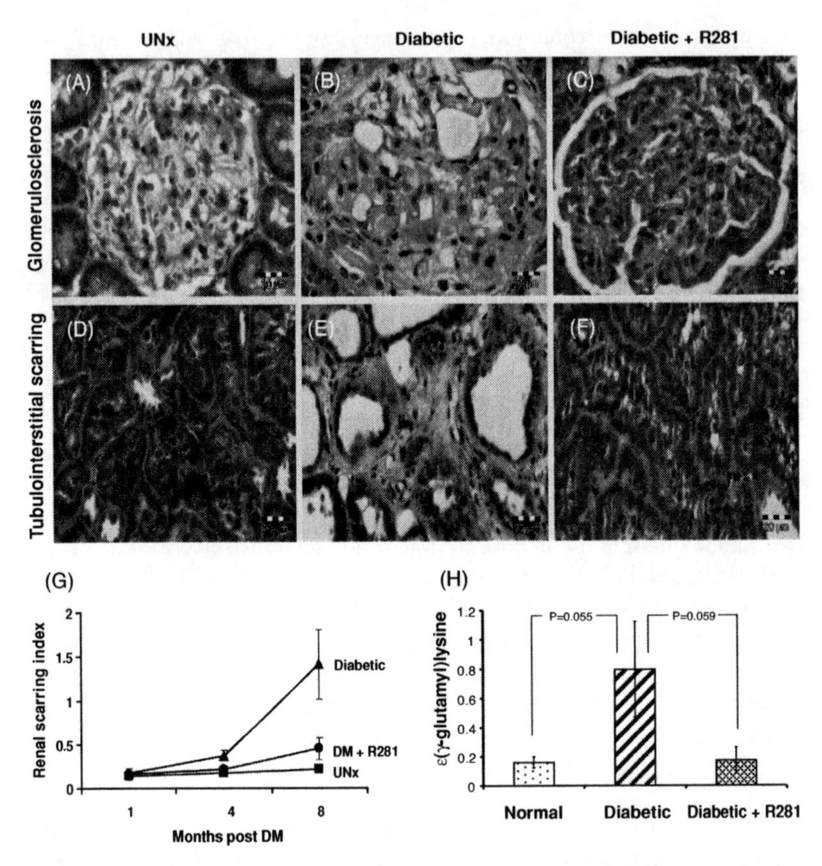

FIGURE 5. Inhibition of collagen deposition and ε(γ-glutamyl)-lysine cross-links into the ECM by TGase inhibitors in a rat streptozotocin-induced model of diabetic kidney disease. Rats were subjected to unilateral nephrectomy followed by streptozotocin injection after 7 days, with blood glucose controlled between 10 and 25 mM using insulin implants (diabetic or DM). Control animals were subjected to unilateral nephrectomy, but did not receive streptozotocin or blood glucose control (UNx). TG2-inhibitor-treated rats (diabetic + R281 or DM + R281) had R281 [157] infused directly into the kidney for the duration of the experiment. (A–F) Masson's trichrome stained kidney sections showing degree of glomerulosclerosis (A–C) and tubulointerstitial scarring (D–F). Collagenous material is stained blue, with nuclei and cytoplasm red/pink. (G) Renal scarring index determined by analysis of Masson's trichrome stained sections. (H) ε(γ-glutamyl)-lysine cross-link levels in kidney sections. (Adapted from [161].) (See insert for color representation.)

chemosensitivity and tumor regression in mice transplanted with pancreatic ductal adenocarcinoma [133], while intratumor injection of TG2 has been used to induce tumor regression by induction of fibrosis [125].

A recent interesting method of modulating the action of TG2 in the extracellular compartment is by the application of antibodies against TG2. The commonly used monoclonal antibody CUB7402 is inhibitory to the activity of TG2 and has been applied in studies to reduce the activity of extracellular TG2 [69, 142]. It is also well known that CD autoantibodies against TG2 can be inhibitory to the transamidating activity of TG2 and that these could be used as potentially therapeutic inhibitory agents [142]. In this study, the effect of CUB7402, CD patient serum, and monoclonal antibodies derived from a CD patient were investigated in vitro and also in human umbilical cord, demonstrating their ability to reduce extracellular TGase activity and the incorporation of biotinylated dansyl cadaverine into the matrix, respectively. A human monoclonal antibody against TG2 has been demonstrated to reduce the expression and deposition of collagen in rat kidney fibrosis following unilateral ureteral obstruction, showing the in vivo efficacy of this therapeutic strategy [164].

Perhaps more interestingly, a small subset of autoantibodies against TG2 derived from CD patient serum have been observed to be inhibitory to angiogenesis in an in vitro model, reducing endothelial sprouting and the migration of both endothelial and vascular mesenchymal cells, suggesting that an enhancement of TG2 activity may be responsible [165]. Phage antibody libraries were created from CD patient lymphocyte DNA and single-chain antibody fragments (scFv) to tTG were isolated and recombinantly expressed as scFv-Fc fusions [166]. These "miniantibodies" to TG2 were able to inhibit angiogenesis in vitro and that chemical inhibition of TG2 using irreversible inhibitors (from [158]) restored angiogenesis to normal levels [143]. This suggested that the CD autoantibodies were capable of activating the transamidating activity of TG2 by direct interaction. The differential effects of CD autoantibodies on TG2 activity pose an interesting question about the nature of their interaction and the relevance to CD pathology (see Section IV). They also offer an approach to the modulation of TG2 activity in a positive or negative manner dependent on the desired effects.

VI. CONCLUDING COMMENTS

Our laboratories first collaborated on TG2 in the early 80s, a time in which investigations on TGs were carried out largely undertaken from an enzymological point of view focusing on kinetics and specificity in recognition of substrates, as recollected in a prominent review by Folk in 1980 [167]. Although there was a few hints towards the relevance of TGase in a few pathologic conditions, e.g., lung fibrosis and cataracts [132, 168], at that particular time, the major biomedical interest was dealing with factor XIII [169]. In that particular era, the concept of TGase isoenzymes was at its dawn and only emerged as important issue in TGase research when the effects of retinoids began to be investigated. The real explosion in research on the medical importance of TG2 was brought about by a report that described TG2 as the main autoantigen in celiac disease [170]. This concept was further extended to other diseases with an autoimmune pathogenesis. The explosion of interest in TG2 and biomedicine is far from complete and the enzyme is likely to offer us additional surprises, as it has already with the demonstration that it is a multifunctional protein and that its transamidase catalytic activity is not the main function of the protein. We hope to have clarified this concept in the composition of this review, which has tried to detail the functions of TG2 in the intra- [2] and extracellular [37] compartments. In our opinion, the transamidating activity of TG2 under normal physiological conditions appears mainly as a terminal epiphenomenon, while the functionally relevant properties of the enzyme should be represented by its signaling and scaffolding activities, both of which need to be investigated much more extensively, along with the enzyme protein kinase and the PDI activities.

ACKNOWLEDGEMENTS

The senior authors CMB and MG express their gratitude to the several junior coworkers who contributed during the years to the progress of research on TG2.

REFERENCES

1. Beninati, S., Bergamini, C. M., and Piacentini, M. (2009) An overview of the first 50 years of transglutaminase research, *Amino Acids 36*, 591–598.
2. Griffin, M., Casadio, R., and Bergamini, C. M. (2002) Transglutaminases: nature's biological glues, *Biochem. J. 368*, 377–396.
3. Fesus, L. and Piacentini, M. (2002) Transglutaminase 2: an enigmatic enzyme with diverse functions, *Trends Biochem. Sci. 27*, 534–539.
4. Lorand, L. and Graham, R. M. (2003) Transglutaminases: crosslinking enzymes with pleiotropic functions, *Nat. Rev. Mol. Cell Biol. 4*, 140–156.

5. Folk, J. E. (1969) Mechanism of action of guinea pig liver transglutaminase. VI. Order of substrate addition, *J. Biol. Chem. 244*, 3707–3713.
6. Sollid, L. M. (2000) Molecular basis of celiac disease, *Ann. Rev. Immunol. 18*, 53–81.
7. Chung, S. I. (1975) Multiple molecular forms of transglutaminases in human and guinea pig. In: *Isoenzymes: Molecular Structure*, ed Markert, C. L. Academic Press, San Diego; Vol I, pp. 259–273.
8. Ikura, K., Nasu, T., Yokota, H., Tsuchiya, Y., Sasaki, R., and Chiba, H. (1988) Amino acid sequence of guinea pig liver transglutaminase from its cDNA sequence, *Biochemistry 27*, 2898–2905.
9. Iismaa, S. E., Mearns, B. M., Lorand, L., and Graham, R. M. (2009) Transglutaminases and disease: lessons from genetically engineered mouse models and inherited disorders, *Physiol. Rev. 89*, 991–1023.
10. Thomazy, V. and Fesus, L. (1989) Differential expression of tissue transglutaminase in human cells. An immunohistochemical study, *Cell Tissue Res. 255*, 215–224.
11. Krasnikov, B. F., Kim, S. Y., McConoughey, S. J., Ryu, H., Xu, H., Stavrovskaya, I., Iismaa, S. E., Mearns, B. M., Ratan, R. R., Blass, J. P., Gibson, G. E., and Cooper, A. J. L. (2005) Transglutaminase activity is present in highly purified nonsynaptosomal mouse brain and liver mitochondria, *Biochemistry 44*, 7830–7843.
12. Park, H., Park, E. S., Lee, H. S., Yun, H. Y., Kwon, N. S., and Baek, K. J. (2001) Distinct characteristic of G alpha(h) (transglutaminase II) by compartment: GTPase and transglutaminase activities, *Biochem. Biophys. Res. Commun. 284*, 496–500.
13. Baek, K. J., Kang, S., Damron, D., and Im, M. (2001) Phospholipase Cdelta1 is a guanine nucleotide exchanging factor for transglutaminase II (Galpha h) and promotes alpha 1B-adrenoreceptor-mediated GTP binding and intracellular calcium release, *J. Biol. Chem. 276*, 5591–5597.
14. Balklava, Z., Verderio, E., Collighan, R., Gross, S., Adams, J., and Griffin, M. (2002) Analysis of tissue transglutaminase function in the migration of swiss 3T3 fibroblasts—the active-state conformation of the enzyme does not affect cell motility but is important for its secretion, *J. Biol. Chem. 277*, 16567–16575.
15. Janiak, A., Zemskov, E. A., and Belkin, A. M. (2006) Cell surface transglutaminase promotes RhoA activation via integrin clustering and suppression of the Src-p190RhoGAP signaling pathway, *Mol. Biol. Cell 17*, 1606–1619.
16. Collighan, R. J. and Griffin, M. (2009) Transglutaminase 2 cross-linking of matrix proteins: biological significance and medical applications, *Amino Acids 36*, 659–670.
17. Yee, V. C., Pedersen, L. C., Le Trong, I., Bishop, P. D., Stenkamp, R. E., and Teller, D. C. (1994) Three-dimensional structure of a transglutaminase: human blood coagulation factor XIII, *Proc. Natl. Acad. Sci. U.S.A. 91*, 7296–7300.
18. Noguchi, K., Ishikawa, K., Yokoyama, K., Ohtsuka, T., Nio, N., and Suzuki, E. (2001) Crystal structure of red sea bream transglutaminase, *J. Biol. Chem. 276*, 12055–12059.
19. Liu, S. P., Cerione, R. A., and Clardy, J. (2002) Structural basis for the guanine nucleotide-binding activity of tissue transglutaminase and its regulation of transamidation activity, *Proc. Natl. Acad. Sci. U.S.A. 99*, 2743–2747.

20. Pinkas, D. M., Strop, P., Brunger, A. T., and Khosla, C. (2007) Transglutaminase 2 undergoes a large conformational change upon activation, *PLoS Biol. 5*, 2788–2796.
21. Ahvazi, B., Boeshans, K. M., and Rastinejad, F. (2004) The emerging structural understanding of transglutaminase 3, *J. Struct. Biol. 147*, 200–207.
22. Király, R., Csosz, E., Kurtán, T., Antus, S., Szigeti, K., Simon-Vecsei, Z., Korponay-Szabó, I. R., Keresztessy, Z., and Fésüs, L. (2009) Functional significance of five noncanonical Ca^{2+}-binding sites of human transglutaminase 2 characterized by site-directed mutagenesis, *FEBS J. 276*, 7083–7096.
23. Iismaa, S. E., Holman, S., Wouters, M. A., Lorand, L., Graham, R. M., and Husain, A. (2003) Evolutionary specialization of a tryptophan indole group for transition-state stabilization by eukaryotic transglutaminases, *Proc. Natl. Acad. Sci. U.S.A. 100*, 12636–12641.
24. Esposito, C. and Caputo, I. (2005) Mammalian transglutaminases. Identification of substrates as a key to physiological function and physiopathological relevance, *FEBS J. 272*, 615–631.
25. Sugimura, Y., Hosono, M., Wada, F., Yoshimura, T., Maki, M., and Hitomi, K. (2006) Screening for the preferred substrate sequence of transglutaminase using a phage-displayed peptide library—identification of peptide substrates for TGase 2 and factor XIIIA, *J. Biol. Chem. 281*, 17699–17706.
26. Keresztessy, Z., Csosz, E., Harsfalvi, J., Csomos, K., Gray, J., Lightowlers, R. N., Lakey, J. H., Balajthy, Z., and Fesus, L. (2006) Phage display selection of efficient glutamine-donor substrate peptides for transglutaminase 2, *Protein Sci. 15*, 2466–2480.
27. Fontana, A., Spolaore, B., Mero, A., and Veronese, F. M. (2008) Site-specific modification and PEGylation of pharmaceutical proteins mediated by transglutaminase, *Adv. Drug Deliv. Rev. 60*, 13–28.
28. Bergamini, C. M. (1988) GTP modulates calcium binding and cation-induced conformational changes in erythrocyte transglutaminase, *FEBS Lett. 239*, 255–258.
29. Casadio, R., Polverini, E., Mariani, P., Spinozzi, F., Carsughi, F., Fontana, A., Polverino de Laureto, P., Matteucci, G., and Bergamini, C. M. (1999) The structural basis for the regulation of tissue transglutaminase by calcium ions, *Eur. J. Biochem. 262*, 672–679.
30. Mariani, P., Carsughi, F., Spinozzi, F., Romanzetti, S., Meier, G., Casadio, R., and Bergamini, C. M. (2000) Ligand-induced conformational changes in tissue transglutaminase: Monte Carlo analysis of small-angle scattering data, *Biophys. J. 78*, 3240–3251.
31. Iismaa, S. E., Chung, L. P., Wu, M. J., Teller, D. C., Yee, V. C., and Graham, R. M. (1997) The core domain of the tissue transglutaminase gh hydrolyzes GTP and ATP, *Biochemistry 36*, 11655–11664.
32. Iismaa, S. E., Wu, M. J., Nanda, N., Church, W. B., and Graham, R. M. (2000) GTP binding and signaling by G(h)/transglutaminase II involves distinct residues in a unique GTP-binding pocket, *J. Biol. Chem. 275*, 18259–18265.
33. Begg, G. E., Carrungton, L., Stokes, P. H., Matthews, J. M., Wouters, M. A., Husain, A., Lorand, L., Iismaa, S. E., and Graham, R. M. (2006) Mechanism of allosteric activation of transglutaminase 2 by GTP. *Proc. Natl. Acad. Sci. USA. 103*, 19683–19688.

34. Kim, H. C., Lewis, M. S., Gorman, J. J., Park, S. C., Girard, J. E., Folk, J. E., and Chung, S. I. (1990) Protransglutaminase E from guinea pig skin. Isolation and partial characterization, *J. Biol. Chem. 265*, 21971–21978.

35. Yee, V. C., Le Trong, I., Bishop, P. D., Pedersen, L. C., Stenkamp, R. E., and Teller, D. C. (1996) Structure and function studies of factor XIIIa by x-ray crystallography, *Semin. Thromb. Hemost. 22*, 377–384.

36. Weiss, M. S., Metzner, H. J., and Hilgenfeld, R. (1998) Two non-proline cis peptide bonds may be important for factor XIII function, *FEBS Lett. 423*, 291–296.

37. Siegel, M., Strnad, P., Watts, R. E., Choi, K., Jabri, B., Omary, M. B., and Khosla, C. (2008) Extracellular transglutaminase 2 is catalytically inactive, but is transiently activated upon tissue injury, *PLoS One 26*(*3*), e1861.

38. Hand, D., Bungay, P. J., Elliott, B. M., and Griffin, M. (1985) Activation of transglutaminase at calcium levels consistent with a role for this enzyme as a calcium receptor protein, *Biosci. Rep. 5*, 1079–1086.

39. Achyuthan, K. E. and Greenberg, C. S. (1987) Identification of a guanosine triphosphate-binding site on guinea pig liver transglutaminase. Role of GTP and calcium ions in modulating activity, *J. Biol. Chem. 262*, 1901–1906.

40. Bergamini, C. M., Signorini, M., Caselli, L., and Melandri, P. (1993) Regulation of transglutaminase activity by GTP in digitonin permeabilized Yoshida tumor cells, *Biochem. Mol. Biol. Int. 30*, 727–732.

41. Smethurst, P. A. and Griffin, M. (1996) Measurement of tissue transglutaminase activity in a permeabilized cell system: its regulation by Ca^{2+} and nucleotides, *Biochem. J. 313*, 803–808.

42. Zhang, J., Lesort, M., Guttmann, R. P., and Johnson, G. V. (1998) Modulation of the in situ activity of tissue transglutaminase by calcium and GTP, *J. Biol. Chem. 273*, 2288–2295.

43. Bergamini, C. M., Dondi, A., Lanzara, V., Squerzanti, M., Cervellati, C., Montin, K., Mischiati, C., Tasco, G., Collighan, R., Griffin, M., and Casadio, R. (2010) Thermodynamics of binding of regulatory ligands to tissue transglutaminase, *Amino Acids 39*, 297–304.

44. Monsonego, A., Shani, Y., Friedmann, I., Paas, Y., Eizenberg, O., and Schwartz, M. (1997) Expression of GTP-dependent and GTP-independent tissue-type transglutaminase in cytokine-treated rat brain astrocytes, *J. Biol. Chem. 272*, 7565–7565.

45. Lai, T. S., Liu, Y., Li, W., and Greenberg, C. S. (2007) Identification of two GTP-independent alternatively spliced forms of tissue transglutaminase in human leukocytes, vascular smooth muscle, and endothelial cells, *FASEB J. 21*, 4131–4143.

46. Lai, T. S., Bielawska, A., Peoples, K. A., Hannun, Y. A., and Greenberg, C. S. (1997) Sphingosylphosphocholine reduces the calcium ion requirement for activating tissue transglutaminase, *J. Biol. Chem. 272*, 16295–16300.

47. Ito, M., Kurita, T., and Kita, K. (1995) A novel enzyme that cleaves the *N*-acyl linkage of ceramides in various glycosphingolipids as well as sphingomyelin to produce their lyso forms, *J. Biol. Chem. 270*, 24370–24374.

48. Hess, D. T., Matsumoto, A., Kim, S. O., Marshall, H. E., and Stamler, J. S. (2005) Protein S-nitrosylation: Purview and parameters, *Nat. Rev. Mol. Cell Biol. 6*, 150–166.
49. Ischiropoulos, H. (2009) Protein tyrosine nitration—an update, *Arch. Biochem. Biophys. 484*, 117–121.
50. Melino, G., Bernassola, F., Knight, R. A., Corasaniti, M. T., Nistico, G., and Finazzi-Agro, A. (1997) S-nitrosylation regulates apoptosis, *Nature 388*, 432–433.
51. Catani, M. V., Bernassola, F., Rossi, A., and Melino, G. (1998) Inhibition of clotting factor XIII activity by nitric oxide, *Biochem. Biophys. Res. Commun. 249*, 275–278.
52. Lai, T. S., Hausladen, A., Slaughter, T. F., Eu, J. P., Stamler, J. S., and Greenberg, C. S. (2001) Calcium regulates S-nitrosylation, denitrosylation, and activity of tissue transglutaminase, *Biochemistry 40*, 4904–4910.
53. Telci, D., Collighan, R. J., Basaga, H., and Griffin, M. (2009) Increased TG2 expression can result in induction of transforming growth factor beta 1, causing increased synthesis and deposition of matrix proteins, which can be regulated by nitric oxide, *J. Biol. Chem. 284*, 29547–29558.
54. Skill, N. J., Griffin, M., El Nahas, A. M., Sanai, T., Haylor, J. L., Fisher, M., Jamie, M. F., Mould, N. N., and Johnson, T. S. (2001) Increases in renal epsilon-(gamma-glutamyl)-lysine crosslinks result from compartment-specific changes in tissue transglutaminase in early experimental diabetic nephropathy: pathologic implications, *Lab. Invest. 81*, 705–716.
55. Fesus, L. and Szondy, Z. (2005) Transglutaminase 2 in the balance of cell death and survival, *FEBS Lett. 579*, 3297–3302.
56. Filiano, A. J., Bailey, C. D. C., Tucholski, J., Gundemir, S., and Johnson, G. V. W. (2008) Transglutaminase 2 protects against ischemic insult, interacts with HIF1 beta, and attenuates HIF1 signaling, *FASEB J. 22*, 2662–2675.
57. Iismaa, S. E. and Graham, R. M. (2003) Dissecting cardiac hypertrophy and signaling pathways—evidence for an interaction between multifunctional G proteins and prostanoids, *Circ. Res. 92*, 1059–1061.
58. Dupuis, M., Levy, A., and Mhaouty-Kodja, S. (2004) Functional coupling of rat myometrial alpha 1-adrenergic receptors to Gh alpha/tissue transglutaminase 2 during pregnancy, *J. Biol. Chem. 279*, 19257–19263.
59. AbdAlla, S., Lother, H., el Missiry, A., Sergeev, P., Langer, A., el Faramawy, Y., and Quitterer, U. (2009) Dominant negative AT(2) receptor oligomers induce G-protein arrest and symptoms of neurodegeneration, *J. Biol. Chem. 284*, 6566–6574.
60. Fraij, B. M. and Gonzales, R. A. (1997) Organization and structure of the human tissue transglutaminase gene, *BBA-Gene Struct. Expr. 1354*, 65–71.
61. Nagy, L., Thomazy, V. A., Saydak, M. M., Stein, J. P., and Davies, P. J. (1997) The promoter of the mouse tissue transglutaminase gene directs tissue-specific, retinoid-regulated and apoptosis-linked expression, *Cell Death Differ. 4*, 534–547.
62. Lu, S. and Davies, P. J. (1997) Regulation of the expression of the tissue transglutaminase gene by DNA methylation, *Proc. Natl. Acad. Sci. U.S.A. 94*, 4692–4697.
63. Kuncio, G. S., Tsyganskaya, M., Zhu, J. L., Liu, S. L., Nagy, L., Thomazy, V., Davies, P. J. A., and Zern, M. A. (1998) TNF-alpha modulates expression of the tissue

transglutaminase gene in liver cells, *Am. J. Physiol. –Gastrointest. Liver Physiol. 37*, G240–G245.

64. Chiocca, E. A., Davies, P. J., and Stein, J. P. (1988) The molecular basis of retinoic acid action. Transcriptional regulation of tissue transglutaminase gene expression in macrophages, *J. Biol. Chem. 263*, 11584–11589.

65. Soehnlein, O., Eskafi, S., Schmeisser, A., Kloos, H., Daniel, W. G., and Garlichs, C. D. (2004) Atorvastatin induces tissue transglutaminase in human endothelial cells, *Biochem. Biophys. Res. Commun. 322*, 105–109.

66. Benedetti, L., Grignani, F., Scicchitano, B. M., Jetten, A. M., Diverio, D., Lo Coco, F., Avvisati, G., Gambacorti-Passerini, C., Adamo, S., Levin, A. A., Pelicci, P. G., and Nervi, C. (1996) Retinoid-induced differentiation of acute promyelocytic leukemia involves PML-RARalpha-mediated increase of type II transglutaminase, *Blood 87*, 1939–1950.

67. Kojima, S., Okuno, M., Matsushima-Nishiwaki, R., Friedman, S. L., and Moriwaki, H. (2004) Acyclic retinoid in the chemoprevention of hepatocellular carcinoma (Review), *Int. J. Oncol. 24*, 797–805.

68. Mehta, K., Fok, J., Miller, F. R., Koul, D., and Sahin, A. A. (2004) Prognostic significance of tissue transglutaminase in drug resistant and metastatic breast cancer, *Clin. Cancer Res. 10*, 8068–8076.

69. Jones, R. A., Nicholas, B., Mian, S., Davies, P. J., and Griffin, M. (1997) Reduced expression of tissue transglutaminase in a human endothelial cell line leads to changes in cell spreading, cell adhesion and reduced polymerisation of fibronectin, *J. Cell Sci. 110*, 2461–2472.

70. Gaudry, C. A., Verderio, E., Aeschlimann, D., Cox, A., Smith, C., and Griffin, M. (1999) Cell surface localization of tissue transglutaminase is dependent on a fibronectin-binding site in its N-terminal beta-sandwich domain, *J. Biol. Chem. 274*, 30707–30714.

71. Johnson, K. A. and Terkeltaub, R. A. (2005) External GTP-bound transglutaminase 2 is a molecular switch for chondrocyte hypertrophic differentiation and calcification, *J. Biol. Chem. 280*, 15004–15012.

72. Kawai, Y., Wada, F., Sugimura, Y., Maki, M., and Hitomi, K. (2008) Transglutaminase 2 activity promotes membrane resealing after mechanical damage in the lung cancer cell line A549, *Cell Biol. Int. 32*, 928–934.

73. Nicholas, B., Smethurst, P., Verderio, E., Jones, R., and Griffin, M. (2003) Cross-linking of cellular proteins by tissue transglutaminase during necrotic cell death: a mechanism for maintaining tissue integrity, *Biochem. J. 371*, 413–422.

74. Upchurch, H. F., Conway, E., Patterson, M. K., Jr., and Maxwell, M. D. (1991) Localization of cellular transglutaminase on the extracellular matrix after wounding: characteristics of the matrix bound enzyme, *J Cell Physiol. 149*, 375–382.

75. Akimov, S. S. and Belkin, A. M. (2001) Cell-surface transglutaminase promotes fibronectin assembly via interaction with the gelatin-binding domain of fibronectin: a role in TGF beta-dependent matrix deposition, *J. Cell Sci. 114*, 2989–3000.

76. Ichinose, A., Bottenus, R. E., and Davie, E. W. (1990) Structure of transglutaminases, *J. Biol. Chem. 265*, 13411–13414.

77. Folk, J. and Finlayson, J. (1977) The epsilon-(gamma-glutamyl)lysine crosslink and the catalytic role of transglutaminases, *Adv. Protein Chem. 31*, 1–133.
78. Akimov, S. S., Krylov, D., Fleischman, L. F., and Belkin, A. M. (2000) Tissue transglutaminase is an integrin-binding adhesion coreceptor for fibronectin, *J. Cell Biol. 148*, 825–838.
79. Scarpellini, A., Germack, R., Lortat-Jacob, H., Muramtsu, T., Johnson, T. S., Billett, E., and Verderio, E. A. (2009) Heparan sulphate proteoglycans are receptors for the cell-surface trafficking and biological activity of transglutaminase-2, *J. Biol. Chem. 284*, 18411–18423.
80. Gambetti, S., Dondi, A., Cervellati, C., Squerzanti, M., Pansini, F. S., and Bergamini, C. M. (2005) Interaction with heparin protects tissue transglutaminase against inactivation by heating and by proteolysis, *Biochimie 87*, 551–555.
81. Kuo, W. N., Kocis, J. M., and Mewar, M. (2002) Nitration/S-nitrosation of proteins by peroxynitrite-treatment and subsequent modification by glutathione S-transferase and glutathione peroxidase, *Mol. Cell. Biochem. 233*, 57–63.
82. Antonyak, M. A., Li, B., Regan, A. D., Feng, Q. Y., Dusaban, S. S., and Cerione, R. A. (2009) Tissue transglutaminase is an essential participant in the epidermal growth factor-stimulated signaling pathway leading to cancer cell migration and invasion, *J. Biol. Chem. 284*, 17914–17925.
83. Yi, S. J., Groffen, J., and Heisterkamp, N. (2009) Transglutaminase 2 regulates the GTPase-activating activity of Bcr, *J. Biol. Chem. 284*, 35645–35651.
84. Greenberg, C. S., Birckbichler, P. J., and Rice, R. H. (1991) Transglutaminases: multifunctional cross-linking enzymes that stabilize tissues, *FASEB J. 5*, 3071–30777.
85. Parameswaran, K. N., Cheng, X. F., Chen, E. C., Velasco, P. T., Wilson, J. H., and Lorand, L. (1997) Hydrolysis of gamma:epsilon isopeptides by cytosolic transglutaminases and by coagulation factor XIII(a), *J. Biol. Chem. 272*, 10311–10317.
86. Lee, J. M., Kim, Y. S., Choi, D. H., Bang, M. S., Han, T. R., Joh, T. H., and Kim, S. Y. (2004) Transglutaminase 2 induces nuclear factor-kappa B activation via a novel pathway in BV-2 microglia, *J. Biol. Chem. 279*, 53725–53735.
87. Verma, A. and Mehta, K. (2007) Tissue transglutaminase-mediated chemoresistance in cancer cells, *Drug Resist. Updat. 10*, 144–151.
88. Maiuri, L., Luciani, A., Giardino, I., Raia, V., Villella, V. R., D'Apolito, M., Pettoello-Mantovani, M., Guido, S., Ciacci, C., Cimmino, M., Cexus, O. N., Londei, M., and Quaratino, S. (2008) Tissue transglutaminase activation modulates inflammation in cystic fibrosis via PPAR gamma down-regulation, *J. Immunol. 180*, 7697–7705.
89. Verma, A., Guha, S., Wang, H. M., Fok, J. Y., Koul, D., Abbruzzese, J., and Mehta, K. (2008) Tissue transglutaminase regulates focal adhesion kinase/AKT activation by modulating PTEN expression in pancreatic cancer cells, *Clin. Cancer Res. 14*, 1997–2005.
90. Satpathy, M., Shao, M. H., Emerson, R., Donner, D. B., and Matei, D. (2009) Tissue transglutaminase regulates matrix metalloproteinase-2 in ovarian cancer by modulating cAMP-response element-binding protein activity, *J. Biol. Chem. 284*, 15390–15399.
91. Caccamo, D., Campisi, A., Curro, M., Li Volti, G., Vanella, A., and Ientile, R. (2004) Excitotoxic and post-ischemic neurodegeneration: involvement of transglutaminases, *Amino Acids 27*, 373–379.

92. De Laurenzi, V. and Melino, G. (2001) Gene disruption of tissue transglutaminase, *Mol. Cell. Biol. 21*, 148–155.

93. Nanda, N., Iismaa, S. E., Owens, W. A., Husain, A., Mackay, F., and Graham, R. M. (2001) Targeted inactivation of G(h)/tissue transglutaminase II, *J. Biol. Chem. 276*, 20673–20678.

94. Oliverio, S., Amendola, A., Rodolfo, C., Spinedi, A., and Piacentini, M. (1999) Inhibition of "tissue" transglutaminase increases cell survival by preventing apoptosis, *J. Biol. Chem. 274*, 34123–34128.

95. Robitaille, K., Daviau, A., Tucholski, J., Johnson, G. V., Rancourt, C., and Blouin, R. (2004) Tissue transglutaminase triggers oligomerization and activation of dual leucine zipper-bearing kinase in calphostin C-treated cells to facilitate apoptosis, *Cell Death Differ. 11*, 542–549.

96. Citron, B. A., Suo, Z., SantaCruz, K., Davies, P. J., Qin, F., and Festoff, B. W. (2002) Protein crosslinking, tissue transglutaminase, alternative splicing and neurodegeneration, *Neurochem. Int. 40*, 69–78.

97. Festoff, B. W., SantaCruz, K., Arnold, P. M., Sebastian, C. T., Davies, P. J., and Citron, B. A. (2002) Injury-induced "switch" from GTP-regulated to novel GTP-independent isoform of tissue transglutaminase in the rat spinal cord, *J. Neurochem. 81*, 708–718.

98. Fraij, B. M., Birckbichler, P. J., Patterson, M. K., Jr., Lee, K. N., and Gonzales, R. A. (1992) A retinoic acid-inducible mRNA from human erythroleukemia cells encodes a novel tissue transglutaminase homologue, *J. Biol. Chem. 267*, 22616–22623.

99. Antonyak, M. A., Jansen, J. M., Miller, A. M., Ly, T. K., Endo, M., and Cerione, R. A. (2006) Two isoforms of tissue transglutaminase mediate opposing cellular fates, *Proc. Natl. Acad. Sci. U.S.A. 103*, 18609–18614.

100. Chae, K. Y., Kim, J. H., Park, W. J., Kim, Y. G., Yun, H. Y., Kwon, N. S., Im, M. J., and Baek, K. J. (2005) Distinct pH modulation for dual function of Galphah (transglutaminase II), *J. Biochem. (Tokyo) 137*, 407–413.

101. Mishra, S. and Murphy, L. J. (2004) Tissue transglutaminase has intrinsic kinase activity: identification of transglutaminase 2 as an insulin-like growth factor-binding protein-3 kinase, *J. Biol. Chem. 279*, 23863–23868.

102. Malorni, W., Farrace, M. G., Matarrese, P., Tinari, A., Ciarlo, L., Mousavi-Shafaei, P., D'Eletto, M., Di Giacomo, G., Melino, G., Palmieri, L., Rodolfo, C., and Piacentini, M. (2009) The adenine nucleotide translocator 1 acts as a type 2 transglutaminase substrate: implications for mitochondrial-dependent apoptosis, *Cell Death Differ. 16*, 1480–1492.

103. Nakaoka, H., Perez, D. M., Baek, K. J., Das, T., Husain, A., Misono, K., Im, M. J., and Graham, R. M. (1994) Gh: a GTP-binding protein with transglutaminase activity and receptor signaling function, *Science 264*, 1593–1596.

104. Lee, K. H., Lee, N., Lim, S., Jung, H., Ko, Y. G., Park, H. Y., Jang, Y., Lee, H., and Hwang, K. C. (2003) Calreticulin inhibits the MEK1,2-ERK1,2 pathway in alpha(1)-adrenergic receptor/G(h)-stimulated hypertrophy of neonatal rat cardiomyocytes, *J. Steroid Biochem. Mol. Biol. 84*, 101–107.

105. Iismaa, S. E., Begg, G. E., and Graham, R. M. (2006) Cross-linking transglutaminases with G protein-coupled receptor signaling, *Sci. STKE 2006*, pe34.

106. Lee, M. Y., Chung, S., Bang, H. W., Baek, K. J., and Uhm, D. (1997) Modulation of large conductance Ca^{2+}-activated K^{+} channel by Galphah (transglutaminase II) in the vascular smooth muscle cell, *Pflugers Arch. 433*, 671–673.

107. AbdAlla, S., Lother, H., Langer, A., el Faramawy, Y., and Quitterer, U. (2004) Factor XIIIA transglutaminase crosslinks AT(1) receptor dimers of monocytes at the onset of atherosclerosis, *Cell 119*, 343–354.

108. Xu, L., Begum, S., Hearn, J. D., and Hynes, R. O. (2006) GPR56, an atypical G protein-coupled receptor, binds tissue transglutaminase, TG2, and inhibits melanoma tumor growth and metastasis, *Proc. Natl. Acad. Sci. U.S.A. 103*, 9023–9028.

109. Jurgensen, K., Aeschlimann, D., Cavin, V., Genge, M., and Hunziker, E. B. (1997) A new biological glue for cartilage–cartilage interfaces: tissue transglutaminase, *J. Bone Joint Surg. Am. 79*, 185–193.

110. Summey, B. T., Graff, R. D., Lai, T. S., Greenberg, C. S., and Lee, G. M. (2002) Tissue transglutaminase localization and activity regulation in the extracellular matrix of articular cartilage, *J. Orthop. Res. 20*, 76–82.

111. Sohn, J., Kim, T. I., Yoon, Y. H., Kim, J. Y., and Kim, S. Y. (2003) Novel transglutaminase inhibitors reverse the inflammation of allergic conjunctivitis, *J. Clin. Invest. 111*, 121–128.

112. Johnson, K. A., van Etten, D., Nanda, N., Graham, R. M., and Terkeltaub, R. A. (2003) Distinct transglutaminase 2-independent and transglutaminase 2-dependent pathways mediate articular chondrocyte hypertrophy, *J. Biol. Chem. 278*, 18824–18832.

113. Zemskov, E. A., Loukinova, E., Mikhailenko, I., Coleman, R. A., Strickland, D. K., and Belkin, A. M. (2009) Regulation of platelet-derived growth factor receptor function by integrin-associated cell surface transglutaminase, *J. Biol. Chem. 284*, 16693–16703.

114. Telci, D., Wang, Z., Li, X., Verderio, E. A. M., Humphries, M. J., Baccarini, M., Basaga, H., and Griffin, M. (2008) Fibronectin-tissue transglutaminase matrix rescues RGD-impaired cell adhesion through syndecan-4 and beta(1) integrin co-signaling, *J. Biol. Chem. 283*, 20937–20947.

115. Stephens, P., Grenard, P., Aeschlimann, P., Langley, M., Blain, E., Errington, R., Kipling, D., Thomas, D., and Aeschlimann, D. (2004) Crosslinking and G-protein functions of transglutaminase 2 contribute differentially to fibroblast wound healing responses, *J. Cell Sci. 117*, 3389–3403.

116. Chau, D. Y. S., Collighan, R. J., Verderio, E. A. M., Addy, V. L., and Griffin, M. (2005) The cellular response to transglutaminase-cross-linked collagen, *Biomaterials 26*, 6518–6529.

117. Belkin, A. M., Akimov, S. S., Zaritskaya, L. S., Ratnikov, B. I., Deryugina, E. I., and Strongin, A. Y. (2001) Matrix-dependent proteolysis of surface transglutaminase by membrane-type metalloproteinase regulates cancer cell adhesion and locomotion, *J. Biol. Chem. 276*, 18415–18422.

118. Telci, D. and Griffin, M. (2006) Tissue transglutaminase (TG2)—a wound response enzyme, *Front. Biosci. 11*, 867–882.

119. Verderio, E. A., Johnson, T., and Griffin, M. (2004) Tissue transglutaminase in normal and abnormal wound healing: review article, *Amino Acids 26*, 387–404.

120. Heath, D. J., Christian, P., and Griffin, M. (2002) Involvement of tissue transglutaminase in the stabilisation of biomaterial/tissue interfaces important in medical devices, *Biomaterials 23*, 1519–1526.

121. Al-Jallad, H. F., Nakano, Y., Chen, J. L. Y., McMillan, E., Lefebvre, C., and Kaartinen, M. T. (2006) Transglutaminase activity regulates osteoblast differentiation and matrix mineralization in MOT3-E1 osteoblast cultures, *Matrix Biol. 25*, 135–148.

122. Kojima, S., Nara, K., and Rifkin, D. B. (1993) Requirement for transglutaminase in the activation of latent transforming growth factor-beta in bovine endothelial cells, *J. Cell. Biol. 121*, 439–48.

123. Nunes, I., Gleizes, P. E., Metz, C. N., and Rifkin, D. B. (1997) Latent transforming growth factor-beta binding protein domains involved in activation and transglutaminase-dependent cross-linking of latent transforming growth factor-beta, *J. Cell Biol. 136*, 1151–1163.

124. Skill, N. J., Johnson, T. S., Coutts, I. G., Saint, R. E., Fisher, M., Huang, L., El Nahas, A. M., Collighan, R. J., and Griffin, M. (2004) Inhibition of transglutaminase activity reduces extracellular matrix accumulation induced by high glucose levels in proximal tubular epithelial cells, *J. Biol. Chem. 279*, 47754–47762.

125. Fisher, M., Jones, R. A., Huang, L. H., Haylor, J. L., El Nahas, M., Griffin, M., and Johnson, T. S. (2009) Modulation of tissue transglutaminase in tubular epithelial cells alters extracellular matrix levels: a potential mechanism of tissue scarring, *Matrix Biol. 28*, 20–31.

126. Spurlin, T. A., Bhadriraju, K., Chung, K. H., Tona, A., and Plant, A. L. (2009) The treatment of collagen fibrils by tissue transglutaminase to promote vascular smooth muscle cell contractile signaling, *Biomaterials 30*, 5486–5496.

127. Jones, R. A., Kotsakis, P., Johnson, T. S., Chau, D. Y. S., Ali, S., Melino, G., and Griffin, M. (2006) Matrix changes induced by transglutaminase 2 lead to inhibition of angiogenesis and tumor growth, *Cell Death Differ. 13*, 1442–1453.

128. Myrsky, E., Caja, S., Simon-Vecsei, Z., Korponay-Szabo, I. R., Nadalutti, C., Collighan, R., Mongeot, A., Griffin, M., Maki, M., Kaukinen, K., and Lindfors, K. (2009) Celiac disease IgA modulates vascular permeability in vitro through the activity of transglutaminase 2 and RhoA, *Cell. Mol. Life Sci. 66*, 3375–3385.

129. Tatsukawa, H., Fukaya, Y., Frampton, G., Martinez-Fuentes, A., Suzuki, K., Kuo, T. F., Nagatsuma, K., Shimokado, K., Okuno, M., Wu, J., Iismaa, S., Matsuura, T., Tsukamoto, H., Zern, M. A., Graham, R. M., and Kojima, S. (2009) Role of transglutaminase 2 in liver injury via cross-linking and silencing of transcription factor Sp1, *Gastroenterology 136*, 1783–1795.

130. Ientile, R., Caccamo, D., and Griffin, M. (2007) Tissue transglutaminase and the stress response, *Amino Acids 33*, 385–394.

131. Mirza, A., Liu, S. L., Frizell, E., Zhu, J. L., Maddukuri, S., Martinez, J., Davies, P., Schwarting, R., Norton, P., and Zern, M. A. (1997) A role for tissue transglutaminase in hepatic injury and fibrogenesis, and its regulation by NF-kappa B, *Am. J. Physiol. Gastrointest. Liver Physiol. 35*, G281–G288.

132. Griffin, M., Smith, L., and Wynne, J. (1979) Changes in transglutaminase activity in an experimental model of pulmonary fibrosis induced by paraquat, *Br. J. Exp. Pathol. 60*, 653–661.

133. Johnson, T. S., El-Koraie, A. F., Skill, N. J., Baddour, N. M., El Nahas, A. M., Njloma, M., Adam, A. G., and Griffin, M. (2003) Tissue transglutaminase and the progression of human renal scarring, *J. Am. Soc. Nephrol. 14*, 2052–2062.

134. Citron, B. A., SantaCruz, K. S., Davies, P. J., and Festoff, B. W. (2001) Intron-exon swapping of transglutaminase mRNA and neuronal Tau aggregation in Alzheimer's disease, *J. Biol. Chem. 276*, 3295–3301.

135. Muma, N. A. (2007) Transglutaminase is linked to neurodegenerative diseases, *J. Neuropathol. Exp. Neurol. 66*, 258–263.

136. Mehta, K. (2009) Biological and therapeutic significance of tissue transglutaminase in pancreatic cancer, *Amino Acids 36*, 709–716.

137. Kotsakis, P. and Griffin, M. (2007) Tissue transglutaminase in tumour progression: friend or foe? *Amino Acids 33*, 373–384.

138. Di Giacomo, G., Lentini, A., Beninati, S., Piacentini, M., and Rodolfo, C. (2009) In vivo evaluation of type 2 transglutaminase contribution to the metastasis formation in melanoma, *Amino Acids 36*, 717–724.

139. Luciani, A., Villella, V. R., Vasaturo, A., Giardino, I., Raia, V., Pettoello-Mantovani, M., D'Apolito, M., Guido, S., Leal, T., Quaratino, S., and Maiuri, L. (2009) SUMOylation of tissue transglutaminase as link between oxidative stress and inflammation, *J. Immunol. 183*, 2775–2784.

140. Periolo, N. and Chernavsky, A. C. (2006) Coeliac disease, *Autoimmun. Rev. 5*, 202–208.

141. Reif, S. and Lerner, A. (2004) Tissue transglutaminase—the key player in celiac disease: a review, *Autoimmun. Rev. 3*, 40–45.

142. Esposito, C., Paparo, F., Caputo, I., Rossi, M., Maglio, M., Sblattero, D., Not, T., Porta, R., Auricchio, S., Marzari, R., and Troncone, R. (2002) Anti-tissue transglutaminase antibodies from coeliac patients inhibit transglutaminase activity both in vitro and in situ, *Gut 51*, 177–181.

143. Caja, S., Myrsky, E., Korponay-Szabo, I. R., Nadalutti, C., Sulic, A. M., Lavric, M., Sblattero, D., Marzari, R., Collighan, R., Mongeot, A., Griffin, M., Maki, M., Kaukinen, K., and Lindfors, K. (2010) Inhibition of transglutaminase 2 enzymatic activity ameliorates the anti-angiogenic effects of coeliac disease autoantibodies, *Scand. J. Gastroenterol 45*, 421–427.

144. Chhabra, A., Verma, A., and Mehta, K. (2009) Tissue transglutaminase promotes or suppresses tumors depending on cell context, *Anticancer Res. 29*, 1909–1919.

145. Mann, A. P., Verma, A., Sethi, G., Manavathi, B., Wang, H. M., Fok, J. Y., Kunnumakkara, A. B., Kumar, R., Aggarwal, B. B., and Mehta, K. (2006) Overexpression of tissue transglutaminase leads to constitutive activation of nuclear factor-kappa B in cancer cells: delineation of a novel pathway, *Cancer Res. 66*, 8788–8795.

146. Kim, D. S., Park, K. S., and Kim, S. Y. (2009) Silencing of TGase 2 sensitizes breast cancer cells to apoptosis by regulation of survival factors, *Front. Biosci. 14*, 2514–2521.

147. Karpuj, M. V., Becher, M. W., Springer, J. E., Chabas, D., Pedotti, R., Youssef, S., Mitchell, D., and Steinman, L. (2002) Prolonged survival and decreased abnormal movements in transgenic model of Huntington disease, with administration of the transglutaminase inhibitor cystamine, *Nat. Med. 8*, 143–149.

148. Jeitner, T. M., Pinto, J. T., Krasnikov, B. F., Horswill, M., and Cooper, A. J. L. (2009) Transglutaminases and neurodegeneration, *J. Neurochem. 109*, 160–166.

149. Lesort, M., Lee, M., Tucholski, J., and Johnson, G. V. (2003) Cystamine inhibits caspase activity. Implications for the treatment of polyglutamine disorders, *J. Biol. Chem. 278*, 3825–3830.

150. Bailey, C. D. C. and Johnson, G. V. W. (2006) The protective effects of cystamine in the R6/2 Huntington's disease mouse involve mechanisms other than the inhibition of tissue transglutaminase, *Neurobiol. Aging 27*, 871–879.

151. Siegel, M. and Khosla, C. (2007) Transglutaminase 2 inhibitors and their therapeutic role in disease states, *Pharmacol. Ther. 115*, 232–245.

152. Kim, S. Y., Park, W. M., Jung, S. W., and Lee, J. (1997) Novel transglutaminase inhibitors reduce the cornified cell envelope formation, *Biochem. Biophys. Res. Commun. 233*, 39–44.

153. Killackey, J. J., Bonaventura, B. J., Castelhano, A. L., Billedeau, R. J., Farmer, W., DeYoung, L., Krantz, A., and Pliura, D. H. (1989) A new class of mechanism-based inhibitors of transglutaminase enzymes inhibits the formation of cross-linked envelopes by human malignant keratinocytes, *Mol. Pharmacol. 35*, 701–706.

154. Freund, K. F., Doshi, K. P., Gaul, S. L., Claremon, D. A., Remy, D. C., Baldwin, J. J., Pitzenberger, S. M., and Stern, A. M. (1994) Transglutaminase inhibition by 2-[(2-oxopropyl)thio]imidazolium derivatives: mechanism of factor XIIIa inactivation, *Biochemistry 33*, 10109–10119.

155. Pardin, C., Gillet, S. M. F. G., and Keillor, J. W. (2006) Synthesis and evaluation of peptidic irreversible inhibitors of tissue transglutaminase, *Bioorg. Med. Chem. 14*, 8379–8385.

156. de Macedo, P., Marrano, C., and Keillor, J. W. (2002) Synthesis of dipeptide-bound epoxides and alpha,beta-unsaturated amides as potential irreversible transglutaminase inhibitors, *Bioorg. Med. Chem. 10*, 355–360.

157. Hausch, F., Halttunen, T., Maki, M., and Khosla, C. (2003) Design, synthesis, and evaluation of gluten peptide analogs as selective inhibitors of human tissue transglutaminase, *Chem. Biol. 10*, 225–231.

158. Griffin, M., Mongeot, A., Collighan, R., Saint, R. E., Jones, R. A., Coutts, I. G. C., and Rathbone, D. L. (2008) Synthesis of potent water-soluble tissue transglutaminase inhibitors, *Bioorg. Med. Chem. Lett. 18*, 5559–5562.

159. Dolynchuk, K. N., Ziesmann, M., and Serletti, J. M. (1996) Topical putrescine (Fibrostat) in treatment of hypertrophic scars: phase II study, *Plast. Rec. Surg. 97*, 117–123.

160. Esposito, C., Caputo, I., and Troncone, R. (2007) New therapeutic strategies for coeliac disease: tissue transglutaminase as a target, *Curr. Med. Chem. 14*, 2572–2580.

161. Huang, L. H., Haylor, J. L., Hau, Z., Jones, R. A., Vickers, M. E., Wagner, B., Griffin, M., Saint, R. E., Coutts, I. G. C., El Nahas, A. M., and Johnson, T. S. (2009) Transglutaminase inhibition ameliorates experimental diabetic nephropathy, *Kidney Int. 76*, 383–394.

162. Johnson, T. S., Fisher, M., Haylor, J. L., Hau, Z., Skill, N. J., Jones, R., Saint, R., Coutts, I., Vickers, M. E., El Nahas, A. M., and Griffin, M. (2007) Transglutaminase

inhibition reduces fibrosis and preserves function in experimental chronic kidney disease, *J. Am. Soc. Nephrol. 18*, 3078–3088.

163. Yuan, L., Choi, K. H., Khosla, C., Zheng, X., Higashikubo, R., Chicoine, M. R., and Rich, K. M. (2005) Tissue transglutaminase 2 inhibition promotes cell death and chemosensitivity in glioblastomas, *Mol. Cancer Ther. 4*, 1293–1302.
164. Mor, O., Faerman, S., Wecsler, R., Ashush, H., Ferling, A., Spivak, I., Skaliter, R., Marzari, R., Sblattero, D., and Chajut, A. (2005) *8th International Conference on Protein Crosslinking and Transglutaminases (PCL8)*. Lubeck, Germany.
165. Myrsky, E., Kaukinen, K., Syrjanen, M., Korponay-Szabo, I. R., Maki, M., and Lindfors, K. (2008) Coeliac disease-specific autoantibodies targeted against transglutaminase 2 disturb angiogenesis, *Clin. Exp. Immunol. 152*, 111–119.
166. Di Niro, R., Ziller, F., Florian, F., Crovella, S., Stebel, M., Bestagno, M., Burrone, O., Bradbury, A. R., Secco, P., Marzari, R., and Sblattero, D. (2007) Construction of miniantibodies for the in vivo study of human autoimmune diseases in animal models, *BMC Biotechnol. 7*, 46.
167. Folk, J. E. (1980) Transglutaminases, *Ann. Rev. Biochem. 49*, 517–531.
168. Lorand, L., Hsu, L. K. H., Siefring, G. E., and Rafferty, N. S. (1981) Lens transglutaminase and cataract formation, *Proc. Natl. Acad. Sci. U.S.A. Biol. Sci. 78*, 1356–1360.
169. Lorand, L. (2007) Crosslinks in blood: transglutaminase and beyond, *FASEB J. 21*, 1627–1632.
170. Dieterich, W., Ehnis, T., Bauer, M., Donner, P., Volta, U., Riecken, E. O., and Schuppan, D. (1997) Identification of tissue transglutaminase as the autoantigen of celiac disease, *Nat. Med. 3*, 797–801.

Since preparation of this review, several relevant and important publications have been made that should be noted. The mechanisms involved in TG2 signalling via heparan sulphate proteoglycan binding have been thoroughly described by Wang et al. (J. Biol. Chem. 2010 285:40212–40229; Exp. Cell Res. 2011 317:367–381). A mechanism of TG2 secretion has been recently described by Zemskov et al. (Plos ONE 2011 6:e19414) involving interaction with integrin beta 1 inside recycling endosomes. In addition another role for TG2 in metastatic cancer has been described by Kumar et al. (Plos ONE 2010 5:e13390; Plos ONE 2011 6:e20701) and where it promotes epithelial-mesenchymal transition and cancer stem cell characteristics in mammary epithelial cells.

PHYSIOPATHOLOGICAL ROLES OF HUMAN TRANSGLUTAMINASE 2

VITTORIO GENTILE

CONTENTS

Advances in Enzymology and Related Areas of Molecular Biology, Volume 78.
Edited by Eric J. Toone.

I. INTRODUCTION

Transglutaminase 2 (TG2, "tissue" TGase) belongs to the family of the TGase enzymes that catalyze posttranslational modifications of proteins, such as the cross-linking of a glutaminyl residue of a protein/peptide substrate to a lysyl residue of a protein/peptide cosubstrate with the formation of an *N*-gamma-(epsilon-*L*-glutamyl)-*L*-lysine [GGEL] cross-link (isopeptidic bond) and the concomitant release of ammonia. The TGase family includes, to date, at least eight zymogens/enzymes (Table 1), which can also catalyze other important reactions for cell life. Recently, several findings concerning the relationships between the biochemical activities of the TGs and the basic molecular mechanisms responsible for some human diseases have been reported. For example, several neurodegenerative diseases, such as Alzheimer's disease (AD), Huntington's disease (HD), Parkinson's disease (PD), supranuclear palsy, etc., are characterized in part by aberrant cerebral TGs activity and by increased cross-linked proteins in affected brains. In this review, we describe the biochemistry, the molecular biology, and the physiopathological roles of the TG2 enzyme, with particular reference to human pathologies in which the molecular mechanism of disease can be due to its biochemical activities, such as in a very common human disease, celiac disease (CD), and also in some physiopathological states, including neuropsychiatric disorders.

II. BIOCHEMISTRY OF TG2

A. POSTTRANSLATIONAL PROTEIN-MODIFYING REACTIONS

TG2 ("tissue TG," tTG) belongs to the family of TGase enzymes (TGase; E.C. 2.3.2.13), which catalyze irreversible posttranslational modifications of

TABLE 1
TGase Enzymes and Their Major Biological Functions When Known

TGase	Physiological Role	Gene Map Location	Reference
Factor XIIIA (FXIIIA)	Blood clotting	6p24-25	[1]
TG1 (Keratinocyte TGase, kTG)	Skin differentiation	14q11.2	[2]
TG2 (Tissue TGase, tTG, cTG)	Apoptosis, cell adhesion, signal transduction	20q11-12	[3]
TG3 (Epidermal TGase, eTG)	Hair follicle differentiation	20p11.2	[4]
TG4 (Prostate TGase, pTG)	Suppression of sperm immunogenicity	3q21-2	[5]
TG5 (TGase X)	Epidermal differentiation	15q15.2	[6]
TG6 (TGase Y)	Unknown function	20p13	[6]
TG7 (TGase Z)	Unknown function	15q15.2	[6]

proteins. Examples of TGase-catalyzed reactions include (1) acyl transfer between the γ-carboxamide group of a protein/polypeptide glutaminyl residue and the ε-amino group of a protein/polypeptide lysyl residue, (2) attachment of a polyamine to the carboxamide of a glutaminyl residue, and (3) deamidation of the γ-carboxamide group of a protein/polypeptide glutaminyl residue (Figure 1) [7, 8]. These reactions catalyzed by TGs occur by a two-step mechanism (Figure 2). The transamidating activity of TGs is activated by the binding of Ca^{2+}, which exposes an active-site cysteine residue. This cysteine residue reacts with the γ-carboxamide group of an incoming glutaminyl residue of a protein/peptide substrate to yield a thioacyl-enzyme intermediate and ammonia (see Figure 2, Step 1). The thioacyl-enzyme intermediate then reacts with a nucleophilic primary amine substrate, resulting in the covalent attachment of the amine-containing donor to the substrate glutaminyl acceptor and regeneration of the cysteinyl residue at the active site (see Figure 2, Step 2). If the primary amine is donated by the ε-amino group of a lysyl residue in a protein/polypeptide, a N^{ε}-(γ-*L*-glutamyl)-*L*-lysine (GGEL) isopeptide bond is formed (see Figure 1, example I). On the other hand, if a polyamine or another primary amine (e.g., histamine) acts as the amine donor, a γ-glutamylpolyamine (or γ-glutamylamine) residue is formed (see Figure 1, example II). It is also possible for a polyamine to act as an *N*,*N*-bis- (γ-*L*-glutamyl)polyamine bridge between two glutaminyl acceptor residues either on the same protein/polypeptide or between two proteins/polypeptides [9]. If there is no primary amine present, water may act as the attacking nucleophile, resulting in the deamidation of glutaminyl residues to glutamyl residues (see Figure 1, example III). It is worthwhile noting that two of these reactions, in particular the deamidation of peptides obtained from the digestion of the gliadin, a protein present in wheat, and the GGEL isopeptide formation between these peptides and TG2, have been shown to be responsible for the formation of new antigenic epitopes responsible for the Celiac sprue, one of the most common human autoimmune diseases [10, 11]. The reactions catalyzed by TGs occur with little change in free energy and hence should theoretically be reversible. However, under physiological conditions, the cross-linking reactions catalyzed by TGs are usually irreversible. This irreversibility partly results from the metabolic removal of ammonia from the system and from thermodynamic considerations resulting from altered protein conformation. Literature data suggest that TGs may be able to catalyze the hydrolysis of GGEL cross-links isopeptide bonds in some soluble cross-linked proteins, the exchange of polyamines onto proteins, and esterification reactions, which result in fatty acid modification of an acceptor-protein glutamine side chain [12]. In

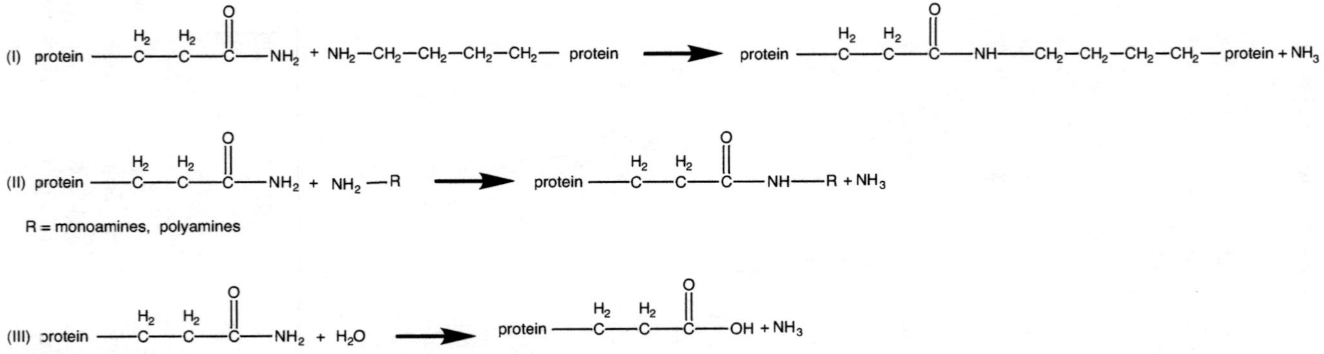

(I) $\text{protein}-CH_2-CH_2-C(=O)-NH_2 + NH_2-CH_2-CH_2-CH_2-CH_2-\text{protein} \longrightarrow \text{protein}-CH_2-CH_2-C(=O)-NH-CH_2-CH_2-CH_2-CH_2-\text{protein} + NH_3$

(II) $\text{protein}-CH_2-CH_2-C(=O)-NH_2 + NH_2-R \longrightarrow \text{protein}-CH_2-CH_2-C(=O)-NH-R + NH_3$

R = monoamines, polyamines

(III) $\text{protein}-CH_2-CH_2-C(=O)-NH_2 + H_2O \longrightarrow \text{protein}-CH_2-CH_2-C(=O)-OH + NH_3$

FIGURE 1. Examples of reactions catalyzed by TG2.

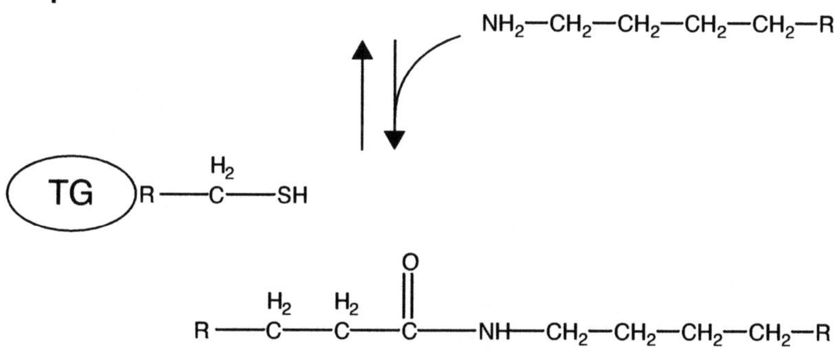

FIGURE 2. Mechanism of the reactions catalyzed by TG2.

TG2, like in some other TGs, additional catalytic activities, such as the ability to hydrolyze GTP (or ATP) into GDP (or ADP) and inorganic phosphate, as well as a protein disulfide isomerase activity, are present [13–15].

B. OTHER REACTIONS: TG2 IS A MULTIFUNCTIONAL ENZYME

Emerging evidence indicates that TG2 is a multifunctional enzyme with distinct and regulated catalytic activities. In fact, in physiological conditions, the transamidation activity of TG2 is latent [16]. By contrast, in some physiopathological states when the concentration of Ca^{2+} increases, the cross-linking activity of TG2 may contribute to important biological processes. As previously described, one of the most intriguing properties of TG2 is the ability to bind and hydrolyze GTP and furthermore, to bind to GTP and Ca^{2+}, which regulate its enzymatic activities, including protein cross-linking, in a reciprocal manner [13]. In practice, the binding of Ca^{2+} inhibits GTP-binding, and GTP-binding inhibits the TGase cross-linking activity of the TG2. Interestingly, TG2 shows no sequence homology with heterotrimeric or low-molecular-weight G-proteins, but there is evidence that TG2 (TG2/Ghα) [17] is involved in signal transduction, and, therefore, TG2/Ghα should also be classified as a high-molecular-weight G-protein. Other studies, along with ours, have shown that TG2/Ghα can mediate the activation of phospholipase C (PLC) by the α_{1b}-adrenergic receptor [17] and can modulate adenylyl cyclase activity [18]. TG2/Ghα can also mediate the activation of the δ1 isoform of PLC and of maxi-K channels [19]. Interestingly, the signaling function of TG2/Ghα is preserved even with the mutagenic inactivation of its cross-linking activity by the mutation of the active site cysteine [20]. However, evidence of a physiopathological role of the TG2 in cell signaling activity is still insufficient.

TG2 has also been reported to present serine/threonine kinase activity with substrates such as insulin-like growth factor binding protein-3, p53 tumor suppressor protein, or histones [21–23], as well as a protein disulfide isomerase activity [15]. However, the physiopathological role of TG2 in these last activities is still unknown.

III. MOLECULAR BIOLOGY OF TG2

A. GENE STRUCTURE AND REGULATION

Human TG2 is encoded by a gene that maps onto the chromosomal locus 20q11-12 [3], where other members of the TGs family have been mapped, TG3 (epidermal TGase, eTG) [4] and TG6 [6]. TG2 is ubiquitously expressed in the human body [24], and retinoids are major regulators of its gene expression [25], although there are also response elements for specificity protein-1, nuclear factor (NK)-1, interleukin (IL)-6, activator protein-1, activator protein-2, and glucocorticoids within the human *TG2* gene promoter [25]. The gene promoter of mouse *TG2* has been analyzed to gain information into control of murine *TG2* expression by using transgenic mouse lines expressing the *Escherichia coli* β-galactosidase (*lacZ*) reporter gene, under the control of a 3.8 kb 5′-flanking promoter region of the mouse

TG2 gene [26]. Correlation of transgene expression in embryos at 13.5 dpc with that of the endogenous TG2 protein demonstrated that the 3.8 kb gene promoter region of murine *TG2* was sufficient to confer specific patterns of expression during limb development. However, in contrast to expression of endogenous TG2, no β-galactosidase activity was detected in the liver or heart of transgenic embryos. Thus, the 3.8 kb 5′-flanking promoter region of the mouse *TG2* gene regulates some, but not all, of the physiological expression of *TG2*. Therefore, to date the characterization of the *TG2* gene regulation is still limited, and complex mechanisms, both at transcriptional and at translational levels, might be present for the expression of this gene also during the development of the mouse nervous system [27].

On the basis of bioinformatic studies, TG2 shows some homologies with other members of the TGase family. High sequence conservation and, therefore, a high degree of preservation of residue secondary structure indicate that TG2 (Figure 3) shares a four-domain tertiary structure that is similar to those of other TGs, such as Factor XIIIA and TG3, that have already been determined [12].

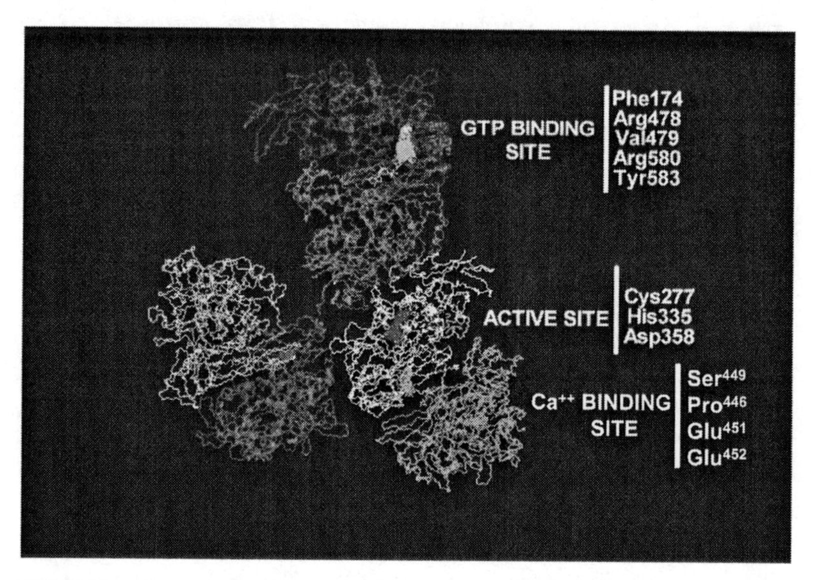

FIGURE 3. Representation of the tertiary structure of human TG2, with the amino acids responsible for the active site, the Ca^{++} binding site, and the GTP binding site. (See insert for color representation.)

B. PROTEIN EXPRESSION AND GENE VARIANTS

On the basis of the ubiquitous and abundant expression of TG2 and its developmental regulation [24], we may speculate that its absence would be lethal. However, this does not seem to be the case, since null mutants of TG2 (TG2$^{-/-}$) are phenotypically normal at birth [28, 29]. Two mixed strain (C57BL/6–129SVJ) knockout mouse models for TG2 were developed simultaneously by different groups to evaluate the in vivo function of TG2. One model was developed by the replacement of part of exon 5 and all exon 6 (exons that encode part of the catalytic core domain) with a neomycin resistance gene [28], while the other model was developed using the Cre/lox system with LoxP sites inserted into introns 5 and 8 for deletions of exons 6–8 of the catalytic core [29]. Both models showed an absence of TG2 protein in homozygote progeny. Although TG2 has been reported to either facilitate or attenuate apoptosis [30], TG2 null mutant mice showed no abnormal phenotype, with no abnormalities of developmental apoptosis and no difference in induced apoptosis of TG2 null mutant thymocytes and fibroblasts [28, 29]. These results were surprising, given the ubiquitous expression of TG2. Nevertheless, these mice have been useful in defining the role of TG2 in pathology and in conditions of stress.

Transgenic mice that specifically overexpress rat *TG2* cDNA in the heart under the control of the murine a-myosin heavy chain promoter [31–33] have been produced by two groups and have been used to investigate the role of TG2 in intracellular signaling pathways relevant to cardiac hypertrophy. A transgenic mouse model overexpressing human TG2 in neurons and heart has recently been produced using the murine prion promoter (MoPrP) from the MoPrP.XhoI expression transgenic vector [34]. This mouse model could help in the study of the role of TG2 in neurodegenerative diseases.

Several alternative splice variants (Figure 4) of the human *TG2* mRNA (full length 3.4 kb) [35], mostly in the 3′-end region, have been identified [34–37]. Interestingly, some of them are differently expressed in important human diseases, such as AD [38].

IV. PHYSIOPATHOLOGY OF TG2

Although many physiological findings of the TG2 have been obtained in the past by using biochemical and cell biology techniques, much has also been learned recently from studies of genetically engineered animal models,

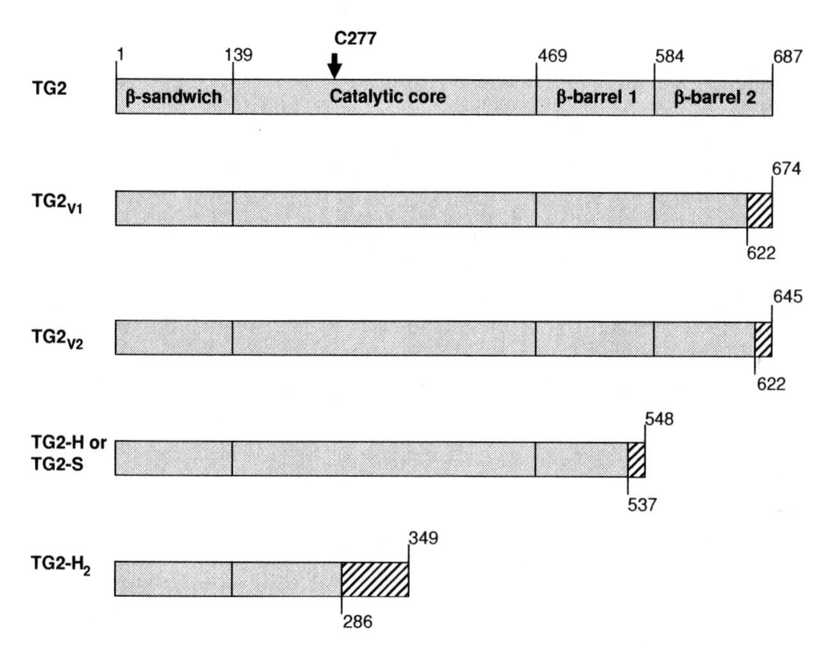

FIGURE 4. Schematic diagram of TG2 and its different isoforms. Gray boxes represent TG2 protein sequences, and shaded boxes represent alternate amino acid sequences due to changes in reading frames (see [35–38]).

as previously reported [39]. In contrast with in vitro experiments and cell culture models that are useful for dissecting protein functions, such studies allow the integrated actions of proteins to be evaluated in the setting of diverse cellular contexts and inputs and, while complicated, are the only ways to elucidate the full array of activities of a protein of interest. Moreover, if they model human diseases caused by an apparent inherited gene mutation, they can be a powerful tool for confirming disease causation, as well as for allowing the evaluation of disease mechanisms and the therapeutic potential of new drugs and therapeutic strategies. Furthermore, when no inherited disorder has been identified, due, for example, to redundancy or mutations, causing an embryonic lethal phenotype, animal models involving conditional gene inactivation may be the only approach for determining the physiological role of a protein of interest. Previous works showed that TG2 is an important factor in a number of physiological processes (Table 1), such as apoptosis, cell proliferation and differentiation, and extracellular matrix organization [40]. Strong evidences suggested that TG2 is involved in programed cell death or apoptosis [41,42]. TG2 is a predominantly cytosolic protein (80%), but is also localized in plasma (10–15%) and in nuclear membranes [39]. TG2 is not present in the mitochondria [43]. While the transamidation catalytic activity of TG2 seems to be silent in physiological conditions at the intracellular level [16], important functions are played by the enzyme at the extracellular level. The enzyme, in fact, is secreted by an unknown mechanism, from the cell, where it localizes to the cell surface and to the extracellular matrix. TG2 contributes significantly to the organization of the extracellular matrix by mediating cell–matrix interactions that affect cell spreading and migration and are crucial for wound healing [41,44], by promoting tissue mineralization [45,46] and by stabilizing dermo-epidermal junctions [47–49]. It is worthwhile noting that, with reference to the role played by the enzyme in the organization of the extracellular matrix, an increase of the TGase activity is observed in various inflammatory and fibrotic conditions (such as rheumatoid arthritis, renal and pulmonary fibrosis, erosive gastritis, Chron's disease, ulcerative colitis, etc.), leading to the hypothesis of its possible involvement in the pathogenesis of these fibrotic disorders [50], and it will be discussed later. The enzyme is expressed at low levels in the healthy small intestine, mainly in the submucosa, but its activity is greatly increased in several pathological conditions, including CD [51].

A. CELIAC DISEASE

CD, which affects 1–2% of the general population, is an autoimmune disorder triggered by dietary exposure to gluten present in wheat, barley, and rye. Preliminary studies showed that TGase activity is increased in CD and it was suggested that this activity could be responsible for CD [51]. Subsequently, other studies were carried out to explain the biochemical mechanisms by which TGase activity could cause CD. Studies on the ability of the gliadin to act as a substrate for TG2 were carried out more than a decade ago [52]. These studies carried out on TG2 and gliadin showed that gliadin and its proteolytic fragments were capable of acting in vitro as substrates for guinea pig liver TG2 ("tissue" TG). This enzyme was able to use gliadin and its proteolytic fragments for cross-linking reactions between Gln endoresidues and either Lys endoresidues or free polyamines. It is worthwhile noting that gliadin possesses a very large number of Gln residues (up to 36%) and that these residues are responsible for both Gln-Lys isodipeptide or polyamine-derivative formation. Sodium dodecyl sulfate-polyacrylamide gel electrophoresis (SDS-PAGE) analyses, followed by autoradiography detection, showed that gliadin and other food proteins, known to be responsible for CD and other food intolerances, were able to form large molecular weight aggregates in the presence of labeled polyamines and calcium-activated TG2. The ability of TGase activity to catalyze the formation of large molecular weight aggregates with polyglutamine peptides and polyglutamine rich proteins has also been reported as a possible mechanism of neurodegenerative diseases [53], and it will be discussed later. Indirect support for the hypothesis that TG2 activity may play a role in the pathogenesis of CD was obtained by biochemical studies [54], which showed that small molecular weight amines, mostly the polyamines spermidine and spermine, protect the celiac small intestine from the damaging activity of gliadin peptides in vitro. At present, it is not yet known whether polyamines exert their effect as competitors of the TG2 reactions or through some other molecular mechanisms. As previously reported, the direct involvement of TG2 in the pathogenesis of CD was shown when this enzyme was identified as one of the main antigens toward which specific immunoglobulins A, produced during CD and previously named antiendomysial, were directed [55]. The molecular mechanisms leading to the formation of an immune response toward TG2 in CD patients are today understood. Studies have shown that the enzyme can form immunoreactive complexes after cross-linking itself with gliadin [11]. The TG2–gliadin complex could then act as a strong hapten-carrier that initially stimulates the T-cell and subsequently the B-cell responses. The possible direct involvement of TG2 catalytic activity in the pathogenesis of CD was also shown by the demonstration of the ability of the enzyme to catalyze the deamidation of certain Gln residues present in the peptide sequences of gliadin. The deamidation of the amino acid residue Q 65 activates the immunological response responsible for CD (Figure 5) [10]. This reaction can be catalyzed in vitro when no amino donor groups, such as amino groups of Lys endoresidues or polyamines, are available to the enzyme. In order to demonstrate whether the TG2 deamidating activity, previously identified in noneukariotic TGs [56], was capable of increasing the immunogenic activity of A-gliadin peptides, primary T-cell lines and T-cell clones from CD patients and controls were used as an in vivo model to measure the binding activity of several peptides to the immune cells [57]. The peptides were obtained either as recombinant products or after digestion of crude gliadin with either pepsin or chimotrypsin. However, the deamidating reaction of Q residues in gliadin occurs in vitro at a slower rate than the

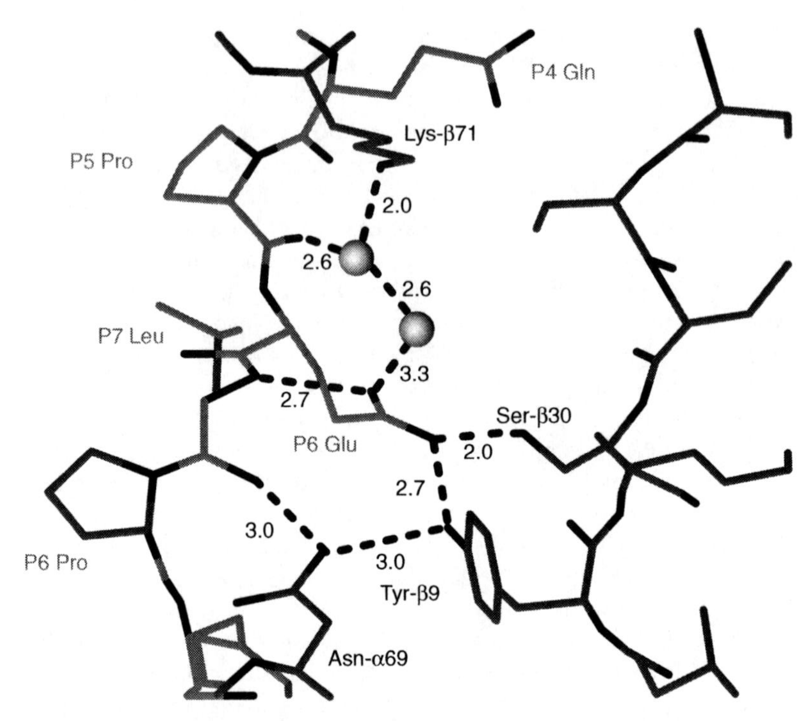

FIGURE 5. Picture of the structural binding between HLA DQ2 and deamidated gliadin epitope (see [10]). (See insert for color representation.)

reaction with donor amines, and further studies should be done to confirm this catalytic mechanism in the intestinal environment. A possible scenario showing the physiopathological mechanisms responsible for CD is reported in Figure 6.

B. BONE DISEASES

In spite of considerable evidence of the involvement of TG2 in bone development, no gross skeletal abnormalities are found in TG2 knockout mouse models [28, 29]. Thus, although the functions of TG2 may be coordinated in developing bone, with the enzyme potentially promoting cell adhesion, matrix assembly and mineralization, and matrix-mediated cell differentiation, TG2 alone is not critical for bone development and other TGs can perhaps compensate for the loss of TG2.

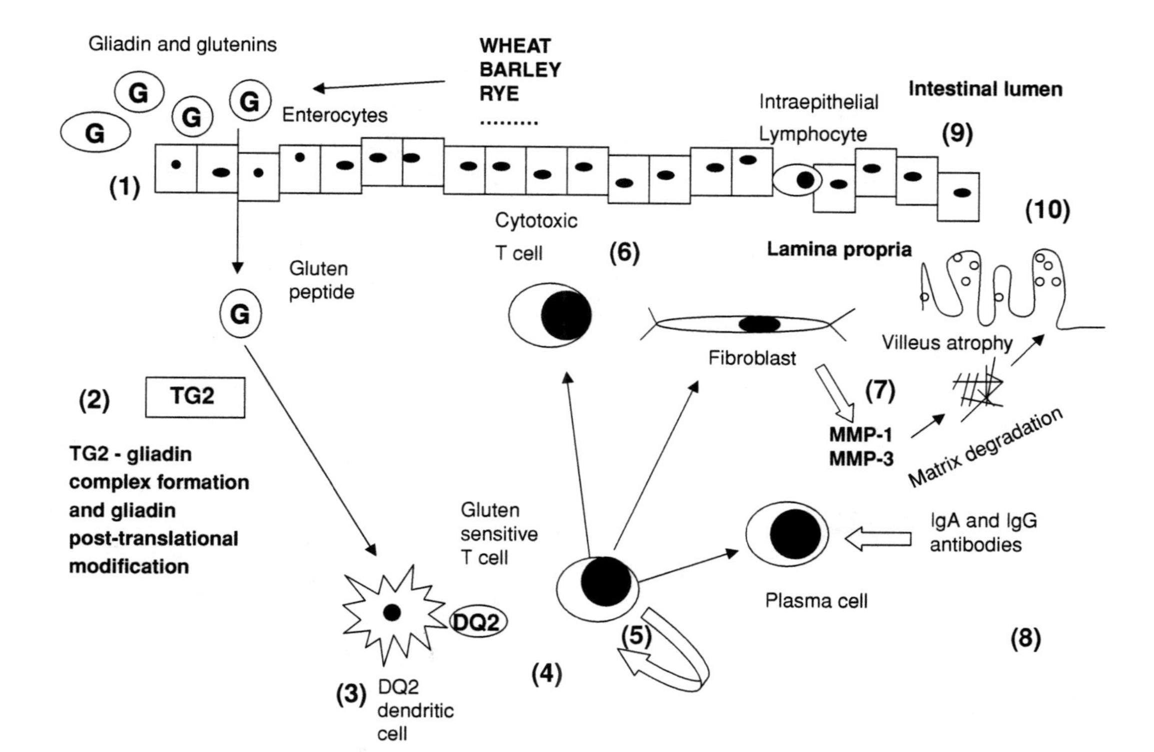

FIGURE 6. A probable scenario of the immunostimulatory pathway responsible for CD. Steps in the process: (1) gluten is digested to yield peptides, which are transported into mucosa; (2) key glutamines are deamidated by TG2; (3) epitope is processed and presented to dendrite cells by DQ2; (4) gluten-sensitive T cells recognize epitope and are stimulated; (5) lamina propria lymphocytes proliferate and recruit cellular infiltrate; (6) CD8 T cells with cytosolic markers increase in mucosa; (7) fibroblasts are activated and produce metalloproteinase to degrade the matrix; (8) plasma cells produce disease-specific celiac antibodies; (9) the role of primitive intra-epithelial lymphocytes remains unclear; (10) inflammatory reaction of intestinal tissue.

Osteoarthritis, a very frequent condition in western countries, is associated with regions of articular chondrocyte hypertrophy, formation of osteophytes (bony outgrowths), and degeneration of cartilage. Articular cartilage (hyaline cartilage without a perichondrium), which cushions the ends of bones in joints by elaborating synovial fluid in response to pressure, normally contains resting chondrocytes that do not differentiate or mineralize their surrounding matrix. A number of proinflammatory factors known to stimulate chondrocyte hypertrophy are upregulated in osteoarthritic cartilage, such as IL-1β [58] and the CXC chemokine subfamily members IL-8 (CXCL8) and growth-related oncogene α (GROα; CXCL1) [59]. IL-8 signals through CXCR1 and CXCR2, whereas GROα activates CXCR2, but not CXCR1 [60]. Both IL-8- and GROα-induced hypertrophy of wild-type murine chondrocytes is associated with increased transamidase activity [61]. Unlike lysates prepared from murine knee chondrocytes isolated from mixed-strain (SVJ129-C57BL/6) wild type mice, those from $TG2^{-/-}$ littermates showed no increase in transamidase catalytic activity or matrix calcification in response to stimulation with GROα, or the all-trans form of retinoic acid (ATRA). They also failed to exhibit features characteristic of chondrocyte hypertrophy, such as collagen X secretion [61, 62] and expression of the transcription core binding factor a1 Cbfal [61]. In contrast, stimulation of both wild-type mixed-strain (SVJ129-057BL/6) and $TG2^{-/-}$ littermate knee chondrocytes with C-type natriuretic peptide (CNP), an essential factor in endochondral development, resulted in comparable Cbfal expression and chondrocyte hypertrophy without associated mineralization or modulation of transamidase activity [62]. These studies thus support a direct contribution of TG2 to IL-1β-, GROα-, and ATRA-induced transamidase activity and mineralization, and demonstrate that there are distinct TG2-dependent (IL-1β-, GROα-, or ATRA-induced) and TG2-independent (CNP-induced) pathways of chondrocyte hypertrophy. The transamidase activity of TG2 has recently been implicated in promoting osteoarthritis progression. Studies using murine femoral head articular cartilage explants and knee chondrocytes from C57BL/6 $TG2^{-/-}$ mice have shown that the transamidase activity of TG2 is required to convert S100A11 (a calcium-binding, EF-hand S100/calgranulin protein secreted in response to IL-1β during low-grade chronic inflammation) into a covalent homodimer. This accelerates chondrocyte hypertrophy and matrix catabolism via receptor for advanced glycation end products signaling [63]. In a surgically induced knee joint instability model of osteoarthritis, mixed strain (SVJ129-C57BL/6) $TG2^{-/-}$ mice had reduced cartilage destruction and increased osteophyte formation relative to wild-type littermates [64]. This phenotype was associated with increased expression of TGF-β1, which promotes osteophyte formation [65], in the osteoarthritic tissues of the $TG2^{-/-}$ mice compared with $TG2^{+/+}$ mice [64]. Together, these studies point to a role for TG2 in modulating cartilage repair and stability in response to certain pathological stressors, and identify TG2 as a potential target for therapeutic intervention in osteoarthritis.

C. CANCER

A large number of conflicting reports implicate either TG2 up- or downregulation in this process [66]. TG2 expression is often downregulated in primary tumors and during tumor progression, whereas upregulation of TG2 is often associated with secondary metastatic tumors and resistance to chemotherapy [67]. Recent studies demonstrating hypermethylation of a CpG island that overlaps both the transcriptional and translational start site of the *TGM2* gene suggest that epigenetic gene silencing may mediate the reduction in TG2 expression in primary tumors [68]. In N-myc-overexpressing neuroblastomas, *N*-myc has been shown to repress TG2 expression by recruiting the transcription factor Sp1 and histone deacetylase (HDAC) 1 to Sp1 binding sites in the *TGM2* core promoter [69]. Moreover, upregulation of TG2 by *N*-myc siRNA or HDAC inhibitors correlated with marked reduction in tumor growth [69]. Similarly, intratumor TG2 injections inhibited CT26 colon carcinoma tumor growth in mice [70]. The host ECM plays an important role in the regulation of tumor growth, in metastatic spread, and in tumor angiogenesis. Thus, degradation of cell-surface TG2 by membrane type 1 matrix metalloproteinase and matrix metalloproteinase-2 at the boundary between normal and tumor-tissue-suppressed cell adhesion and migration on fibronectin, but stimulated migration on collagen [71,72]. Exogenous addition of TG2 increased ECM accumulation and reduced ECM turnover, conferring resistance to ECM degradation by matrix metalloproteinase-1 and blocking capillary growth in in vitro angiogenesis culture studies [70]. Moreover, studies with mixed strain (SVJ129–C57BL/6) [70] or C57BL/6 [73] $TG2^{-/-}$ mice have shown that subcutaneous implantation of B16F1 [70] or B16-F10 [71] melanoma cells resulted in increased tumor size [70, 73], and reduced survival of $TG2^{-/-}$ mice, relative to wild-type mice [70]. Hematogenous metastasis to lung and other organs of intravenously injected tumor cells was increased, although individual metastasis size was smaller, in $TG2^{-/-}$ compared with wild-type mice [73]. Inhibition of TG2, by treatment with TG2 inhibitors [73–75], antisense or ribozyme [76], or RNAi [77–80] technologies, sensitized cancer cells to chemotherapeutic agents by disrupting fibronectin assembly in the ECM and promoting cell death or autophagy [74, 80]. TG2 has also been shown to interact in the ECM with the novel G-protein-coupled receptor GPR56, a suppressor of melanoma tumor growth and metastasis that is markedly downregulated in highly metastatic cells [81]. Tumor cell growth and metastasis may thus involve localized interactions of TG2 with GPR56, integrins, and other matrix proteins such as fibronectin. However, the functional consequences of TG2 on tumor growth, metastasis, and angiogenesis still deserve further investigation.

D. CARDIAC DISEASES

Studies of various parameters of cardiac function (heart rate, systolic and diastolic blood pressure, maximum rates of contractility and relaxation) measured on TG2 knockout animal models, such as mixed-strain (SVJ129–C57BL/6) TG2$^{-/-}$ mice at rest, showed that they were not different from those of wild-type littermates [29]. In contrast, a recent study reported that TG2$^{-/-}$ mice (the background strain was not indicated) had reduced heart rate, coronary flow, aortic flow, and aortic pressure as well as reduced myocardial ATP levels, compared with C57BL/6 wild-type controls [82, 83]. Moreover, in response to an ischemia/reperfusion challenge, infarct size was reported to be increased in TG2$^{-/-}$ hearts, as was the incidence of reperfusion-induced ventricular fibrillation, and ATP levels were even further reduced [83]. Nevertheless, given that the preischemic coronary flow was already substantially reduced in these TG2$^{-/-}$ hearts, the validity of determining energy substrates, in this context, is questionable. Surprisingly, however, two earlier studies from the same group reported no significant difference in ATP levels in heart, liver, skeletal muscle [84], or brain [85] in the TG2$^{-/-}$ mice at rest. Only one of these studies [85] specified the background strain of the TG2$^{-/-}$ mice used, which was C57BL/6 (i.e., mixed SVJ129–C57BL/6 backcrossed for 10 generations to C57BL/6). Upon exercise [84] or in response to a toxin challenge [85], ATP content in brain and skeletal and cardiac muscle appeared to be depleted due to reduced activity of mitochondrial complex I, and increased activity of mitochondria complex II. Given the absence of precise animal husbandry details, the discrepancies in cardiac function and ATP levels of TG2$^{-/-}$ mice at rest [83 versus 29, 84, 85] may merely reflect different husbandry practices and/or the use of nonlittermate controls. Cardiac hypertrophy is an adaptive response to increased work load induced by pressure- or volume-overload, and in response to chronic overload, cardiac function deteriorates, resulting in decompensated heart failure. Steady-state *TG2* mRNA levels are increased in both volume- and pressure-overload-induced hypertrophied rat ventricles and are further increased during the transition to heart failure [86]. Increased TG2 expression has also been documented in failing hearts from patients with dilated but not ischemic cardiomyopathy, although both the GTP-binding and transamidase activities of TG2 were decreased in both [87]. The hypertrophic effect of transgenic cardiac-specific TG2 overexpression has been investigated in the mouse by two independent groups [31, 33]. Animals between 3 and 4 months of age developed a unique type of cardiac hypertrophy with a mild elevation in some markers of hypertrophy (α myosin heavy chain and α-skeletal actin) but not others (atrial naturietic factor) [31]. Older animals (7–10 months old) developed cardiac decompensation characterized by cardiomyocyte apoptosis and fibrosis, with an increase in mortality evident by 15 months [33]. The primary and initiating event underlying the cardiac failure observed in these mice was increased biosynthesis of the prostanoid thromboxane (TxA2) and enhanced signaling from its G-protein-coupled receptor, TP (tromboxane receptor) α. This response leads to ERK1/2 activation, which results in a feed forward loop that accounts for the hypertrophy, as well as for sustained activation of phospholipase A2 and induction of the prostanoid biosynthetic enzyme COX-2 and, as a consequence of the latter, enhanced TxA2 production [33, 88]. Coincident increased oxidant stress leads to oxidative modification of arachidonic acid and formation of free radical-catalyzed products called isoprostanes. Formed initially in situ in the phospholipid domain of cell membranes, they are then cleaved by phospholipases, circulate in esterified and unesterified forms, and importantly, act as incidental ligands for prostanoid receptors [89]. Oxidant stress, thus, contributes to the hypertrophy phenotype via TP stimulation [33, 88]. Apoptosis may be a direct result of enhanced TPα signaling [90] or may be secondary to TP-mediated vasoconstriction causing myocardial ischemia. Although intracellular TG2 levels were elevated in the cardiac TG2-overexpressors relative to control mice, intracellular transamidase activity was not measured in vivo or in intact cardiomyocytes. It is thus not clear if the transamidase activity of TG2 contributes to the regulation of COX-2 expression. Nevertheless, these studies provide compelling evidence for a potential link between TG2 and prostanoids in cardiac dysfunction due to hypertrophy disorders in humans.

E. VASCULAR DISEASES

Atherosclerosis is characterized by the deposition of fatty material in arterial walls. The subsequent migration of inflammatory cells (macrophages and T lymphocytes) and vascular smooth muscle cells into the lesion contributes to form a plaque, which results in a narrowing of the lumen that may eventually impair blood flow, either directly or as a result of plaque rupture. This last event can lead to the rapid formation of an occlusive thrombus. TG2 expression in intimal and medial smooth muscle cells, and in the extracellular matrix, increases concomitantly with the progression of atherosclerosis [91, 92], with leukocytes being the major source of TG2 in atherosclerotic lesions [93]. Apoptosis is observed in developing plaques, with endothelial cell and macrophage apoptosis increasing early in atherogenesis, and vascular smooth muscle cell and macrophage apoptosis contributing to plaque rupture [94]. Recent work indicates that macrophage-expressed TG2 limits expansion of atherosclerotic plaques [93]. Thus, following a high fat diet to induce atherosclerosis in irradiated C57BL/6 low-density lipoprotein receptor knockout ($LDLR^{-/-}$) mice that had been transplanted with mixed-strain (94% C57BL/6–6% SVJ-129) wild-type or $TG2^{-/-}$ bone marrow, significantly larger aortic valve atherosclerotic lesions were observed in those animals that received the $TG2^{-/-}$ marrow [93]. Lesions in recipients of $TG2^{-/-}$ bone marrow appeared to have more expanded necrotic cores, indicative of plaques that are vulnerable to rupture, with macrophages localized deeper in the intima than in those that received the wild-type bone marrow. The function of TG2 as an integrin-associated adhesion receptor involved in monocyte migration [44] may account for the lack of macrophages at or near the endothelia of knockout bone marrow recipients, while decreased extracellular matrix cross-links [93] and a decrease in the efficiency of apoptotic cell clearance [93, 95] by $TG2^{-/-}$ macrophages may contribute to atherosclerotic expansion in the same animals [93]. In addition, decreased macrophage activation of TGF-β [94], which is thought to stabilize plaques by stimulation of collagen synthesis [93], may promote plaque instability in knockout recipient. Interestingly, irradiated LDLR knockout mice transplanted with $TG2^{-/-}$ bone marrow had significantly lower triglyceride levels after 4 and 8 weeks on a high-fat diet [93]. However, the cause and potential implications of this finding remain unclear.

Atherosclerotic lesions calcify nonlinearly over time as a result of osteogenic and chondrogenic differentiation of vascular smooth muscle cells in a process akin to bone formation. Although minor differences were observed between freshly isolated (99.6% C57BL/6–0.4% SVJ-129) $TG2^{+/+}$ and $TG2^{-/-}$ mouse aortic smooth muscle cells and aortas, stimulation of smooth muscle cells (by an inorganic phosphate donor or by bone morphogenetic protein-2) to drive intra-arterial chondro-osseous differentiation and calcification, failed to induce either differentiation or calcification of $TG2^{-/-}$ smooth muscle cells [96]. This indicates a central role for TG2 in these processes. Exogenous addition of TG2 to mouse vascular smooth muscle cells has been shown to enhance calcification via binding to low-density lipoprotein receptor-related protein 5 (LRP5) and subsequent activation of β-catenin signaling [97]. TG2 released during artery wall repair may thus promote calcification.

F. INFLAMMATION AND TISSUE REPAIR/REMODELING

Tissue remodeling/repair is a dynamic process involving an initial inflammatory phase followed by tissue formation/stabilization and remodeling. Many different cell types, such as keratinocytes, platelets, macrophages, fibroblasts, and endothelial cells, interact to facilitate remodeling [98]. TG2 appears to be involved in tissue injury/remodeling, although its precise contribution to the various different stages of injury/remodeling has not yet been fully characterized.

1. Expression of TG2 in Inflammation

Increased TG2 expression and transamidation activity is a common feature of many inflammatory diseases and events. Cytokines and growth factors secreted during the initial phase of cell injury regulate TG2 expression. For example, transforming growth factor (TGF)-β acts to increase TG2 in keratinocytes [99] and on the surface of dermal fibroblasts [100] via a TGF-β1 response element in the *TGM2* gene promoter [101]. Tumor necrosis factor-α enhances TG2 synthesis in liver cells via activation of inhibitor of κB (IκB)α phosphorylation. This causes the dissociation of IκB from NFκB, allowing the nuclear translocation of NFκB, which then binds to the *TGM2* promoter [102]. TG2 enhances NFκB activation and, thus, inflammation in microglial cells, by cross-linking IκBα, which also results in NFκB dissociation and nuclear translocation [103]. Increased TG2 expression, resulting from increased binding of NFκB to the *TGM2* promoter, enhances extracellular TG2 cross-linking activity, a response observed in experimentally induced hepatic fibrogenesis in rats [104] and in patients with hepatic disease, who have increased TG2 levels [105, 106]. TG2 expression is also enhanced in cartilage tissue by IL-1 [107] and in liver cells by IL-6 [108]. Mallory body inclusions consisting primarily of tranglutaminase cross-linked keratins 8 and 18 are characteristic of several liver disorders, although it is not yet clear whether these bodies protect from, or promote, injury [109]. In response to cutaneous injury, TG2 expression and activity are increased at sites of neovascularization in the provisional fibrin matrix, endothelial cells, skeletal muscle cells, and macrophages infiltrating wounds in the border zone between normal and injured tissue and, later, in the granulation tissue matrix where it is thought to cross-link ECM substrates [110, 111]. Recent studies using lipopolysaccaride to induce murine endotoxic shock demonstrated increased survival of C57BL/6 TG2-deficent mice relative to their wild-type counterparts [112]. This was associated with decreased NFkB activation, decreased neutrophil recruitment into kidneys and peritoneum, and reduced renal and myocardial damage. This indicates that TG2 promotes the pathogenesis of endotoxic shock. Further delineation of the mechanism involved is required. Consistent with a stimulatory role for TG2 in inflammation is a recent work using airway epithelial cells from cystic fibrosis patients defective in the cystic fibrosis transmembrane conductance regulator gene. A marked upregulation of TG2 in these cells leads to cross-linking and subsequent proteasomal degradation of peroxisome proliferator-activated receptor-γ, a nuclear hormone receptor that negatively regulates inflammatory gene expression. As a result, these patients display an enhanced inflammatory response [113].

2. *Phagocytosis*

Macrophages that infiltrate a wound after injury contribute to the resolution of inflammation by generating TGF-β. In turn, activation of TGF-β stimulates phagocytosis of apoptotic cells and suppresses further proinflammatory mediator production [114]. During monocyte differentiation into macrophages, high levels of integrin-bound surface TG2 are induced [44, 115] concomitant with a decrease in FXIIIA expression [116]. This correlates with macrophage phagocytotic capacity [116, 117] as well as macrophage adhesion and migration [44]. TG2$^{-/-}$ macrophage adhesion and migration have not been investigated directly. However, TG2$^{-/-}$ mice have been reported to exhibit a defect in macrophage clearance of necrotic or apoptotic cells. Systems that have been studied include clearance of necrotic [105] or apoptotic [118] C57BL/6 TG2$^{+/+}$ and TG2$^{-/-}$ cells after hepatotoxin exposure, clearance of apoptotic cells (background strains of $TG2^{+/+}$ and $TG2^{-/-}$ mice not stated) during induced thymic involution [95], and clearance of apoptotic (94% C57BL/6–6% SVJ129) $TG2^{+/+}$ and $TG2^{-/-}$ neutrophils during gout-like peritoneal inflammation [119]. This defect was associated with the development of autoimmunity in aged mice [95]. However, ingestion of bacteria, yeast, or opsonized nonapoptotic thymocytes by $TG2^{-/-}$ macrophages was normal [95]. This finding is in contrast with monocytes of patients with FXIIIA deficiency, which have been reported to display impaired phagocytosis of sensitized red blood cells (FcγR-mediated) and of complement-coated (complement receptor-mediated) and uncoated (lectin-like receptor-mediated) yeast particles (120]. However, the clinical significance of these findings is unclear, since patients with FXIII deficiency do not have an increased susceptibility to infectious diseases resulting from impaired phagocytic elimination of microbes [121]. The impaired engulfment of apoptotic cells [118, 119] by $TG2^{-/-}$ macrophages appears to be due to a deficiency in activation of latent TGF-β1 [95, 119]. Although the transamidase activity of TG2 has been implicated in TGF-β release from latent matrix storage complexes [122, 123], rescue of the defective phagocytosis by $TG2^{-/-}$ macrophages appears not to require the transamidase activity of TG2, since it was observed even with the exogenous addition of a catalytically inactive mutant form of TG2 [119]. Furthermore, rescue of phagocytosis appeared to be dependent on exogenous TG2 binding to GDP or adenine nucleotides ADP and ATP, but not to GTP. Indeed, exogenous GTP-bound TG2-inhibited phagocytosis of apoptotic cells by wild-type macrophages [119]. Together, these results indicate that a particular conformation of TG2 is required for TGF-β activation during the resolution phase of inflammation. The mechanism by which exogenous TG2 activates TGF-β is not yet clear and requires investigation.

The phagocytic capacity of neutrophils stimulated by intraperitoneal administration of yeast extract was not compromised in mixed strain (SVJ129-C57BL/6) $TG2^{-/-}$ mice relative to those from wild-type mixed strain (SVJ129-C57BL/6) animals [124]. However, $TG2^{-/-}$ neutrophils were deficient in superoxide generation [124] due to downregulation of GP91PHOX [124], the major subunit of the NADPH oxidase system that generates reactive oxygen species in response to invading microrganisms as part of a phagosomal killing mechanism. $TG2^{-/-}$ mice have not yet been tested for their ability to resolve microbial infections. With the use of adhesion-dependent assays, migration of neutrophils into the peritoneum in response to yeast extract and chemotaxis of neutrophils towards f-Met-Leu-Phe have also been reported to be defective in TG2$^{-/-}$ mice [124]. However, this study did not rule out potential impairment of TG2$^{-/-}$ neutrophil adhesion as a contributing factor. Given that cell surface TG2 is involved in integrin-dependent fibroblast [125] and monocyte [44] adhesion and migration (see below), and that neutrophil migration is also integrin-dependent, the observed neutrophil migration defect may in part be related to an adhesion defect.

3. Cell Adhesion and Spreading

Interaction of cells with the surrounding ECM is critical for cell adhesion and migration. The role of TG2 in this process is being delineated primarily from cultured cell studies. TG2 overexpression in fibroblasts increased cell attachment [41, 126] and spreading [41]. In contrast, fibroblasts deficient in TG2 that were either isolated from mixed-strain (SVJ129–C57BL/6) mice [29] or generated by stable transfection of antisense constructs [127, 128] exhibited decreased adhesion [29, 127] and spreading [127, 128]. Available evidence indicates that TG2 contributes to cell adhesion in two ways. First, TG2 stabilizes ECM proteins by enhancing the initial phase of fibronectin matrix formation [129] and by cross-linking ECM proteins [111, 130, 131]. Second, cell-surface TG2 promotes fibroblast adhesion to fibronectin [125, 132–135] by interacting directly with β1/β3/β5 integrins [125, 135, 136], with heparan sulfate chains of the heparan sulfate proteoglycan receptor syndecan-4 [133, 134] and with the orphan G-protein-coupled cell-adhesion receptor GPR56 [81]. Enhancement of integrin- [125] or syndecan-4-mediated [133, 134] fibroblast adhesion and spreading on fibronectin, by cell surface TG2, is not mediated by its transamidase activity. Rather, it involves modulation of common targets of receptor-dependent outside-in signaling, such as PKC-α activation [133, 134]. During tissue injury and/or remodeling, activated matrix metalloproteinases degrade fibronectin and other matrix proteins to generate Arg-Gly-Asp (RGD)-containing peptides. These peptides compete with, and thus disrupt, integrin binding to the RGD-containing domains of matrix components, such as fibronectin, and, consequently, impair adhesion. Independent of its transamidase activity, extracellular TG2 restores fibroblast adhesion in the presence of RGD-containing peptides [134] by direct interaction, not with integrin, but with the heparan sulfate chains of syndecan-4. This results in intracellular cross-talk between syndecan-4 and β1 integrin via syndecan-4 activation of PKC-α and PKC-α-induced β1-integrin activation [133]. Thus, TG2 may have a physiological role in protecting from anoikis (detachment induced apoptosis) induced by inhibition of integrin-dependent adhesion following injury. Again, independent of its transamidase activity, intracellular TG2 may also regulate fibroblast spreading on fibronectin through PKC-α activation and FAK phosphorylation [128], as well as inhibiting smooth muscle cell motility by direct association with the cytoplasmic tail of α-integrin subunits [137]. Further studies should confirm and extend these findings.

4. *Arterial Remodeling*

In order to investigate the role of TG2 in arterial remodeling, preliminary studies have been carried out using TG2-knockout mice as animal model. Compared with wild-type littermates, arteries of mixed strain (SVJ129–C57BL/6) $TG2^{-/-}$ mice showed a delayed, but not suppressed, capacity to undergo inward remodeling (that is, a decrease in lumen diameter) in response to surgical reduction in blood flow [138] or nitric oxide inhibitor (N^{ω}-nitro-*L*-arginine methyl ester hydrochloride, *L*-NAME)-induced hypertension [139]. Remodeling was accompanied by increased arterial expression of only *FXIII-A* mRNA in $TG2^{-/-}$ mice and of *TG2* and *FXIII-A* mRNA in wild-type mice upon surgical reduction of blood flow [138], but no differences in *FXIII-A* or *TG2* mRNA levels were observed in wild-type or $TG2^{-/-}$ mice with *L*-NAME-induced hypertension [139]. Depletion of accumulated adventitial monocytes/macrophages by phagocytosis of the liposome-encapsulated toxic drug clodronate inhibited both arterial FXIIIA expression and surgically induced inward remodeling in $TG2^{-/-}$ mice [138], indicating potential compensation for TG2 in low flow artery remodeling by monocyte/macrophage-derived FXIIIA. Interestingly, *L*-NAME treatment resulted in stiffening of $TG2^{+/+}$ but not $TG2^{-/-}$ erythrocytes, indicating a role for TG2 in erythrocyte deformability [139], and it will be discussed later. The reduction in inward remodeling observed in $TG2^{-/-}$ mice in response to reduced blood flow is phenocopied in pure strain (SV129) vimentin$^{-/-}$ mice [140]. Given that vimentin is a major TGase cross-linking substrate in carotid atherosclerotic plaque [141], TG2-mediated dimerization of this integral cytoskeletal protein may regulate inward remodeling. Consistent with TG2 involvement in this process, inward remodeling in rats was inhibited by cystamine, a broad-spectrum inhibitor of sulfhydryl enzymes [142]. Outward remodeling (increase in vessel size), on the other hand, involves vimentin [140], but does not appear to involve TG2, as this was unaffected in $TG2^{-/-}$ mice [138], despite cystamine inhibition of outward remodeling in rats [142]. Clearly, further work is required to clarify the mechanisms responsible for vascular remodeling in these various models.

G. ERYTHROCYTE AGEING

The biological model in which an increase in intracellular Ca^{2+} concentration activates latent transamidase activity was first established for TG2 in human red blood cells (RBCs), by the demonstration that TG2-catalyzed remodeling of membrane skeletal proteins via N^{ε}-(γ-glutamyl)lysine cross-links was the immediate cause for the apoptosis-like morphological and physical alterations in these cells [143, 144]. Competitive or direct inhibitors of cross-linking activity offered effective protection against the permanent abnormal shape changes and loss of membrane deformability [145]. SDS-PAGE of ghosts from Ca^{2+}-treated RBCs revealed large polymeric clusters ranging in molecular weight from $\sim 1 \times 10^6$ to $\sim 6 \times 10^6$ Da; some polymers appeared to be situated peripherally to the inner leaflet (i.e., extractable by mild alkali), but a major fraction was firmly anchored into membrane and could not be stripped away by alkali treatment. Since neither chemical nor biochemical procedures are available at present to separate the N^{e}-(γ-glutamyl)lysine cross-linked structures into their polypeptide building blocks, indirect methods have been employed for assessing polymer composition, with good agreement between different modes of analyses. For example, immunological approaches have utilized cross-reactive antibodies for recognizing accessible epitopes both in monomers and polymers [146]. Membrane skeletal as well as cytoplasmic proteins are incorporated into the polymers (e.g., spectrins, ankyrin, band 4.1, hemoglobin, among several others), and the anion transporter (band 3) and possibly other transmembrane proteins serve to anchor the polymers firmly into the membrane.

High-molecular-weight polymers cross-linked by N^{ε}-(γ-glutamyl)lysine bonds have also been isolated from RBCs of an Hb-Koln [147]. A causal relationship is suggested between polymer formation and the shortened life span of these cells, whereby elevated Ca^{2+} levels in the pathological RBCs activate TG2 transamidase activity and the cells become susceptible to survival hazards in the circulation through changes in their physical properties (i.e., cell membrane stiffening and irreversible structural fixation of abnormal cell shape e.g., echinocyte, sickle cell), in addition to the known translocation of phosphatidylserine to the cell surface, which serves to attract phagocytes.

Elevated intracellular Ca^{2+} also activates membrane proteases that attack mainly glycophorin and the band 3 anion transporter [148]. TG2-mediated cross-linking can be uncoupled from proteolysis by differential inhibitors [149] or by a few days of blood bank storage, which greatly diminishes Ca^{2+}-induced proteolysis without impairing TG2-catalyzed modification of membrane skeleton [150]. In contrast to human RBCs, the prime response to Ca^{2+} overload of rat RBCs is proteolysis of membrane proteins (mostly band 4.1, band 3, and band 2.1 proteins) by cytosolic proteases (mainly calpain), rather than TG2-dependent cross-linking. This, incidentally, seems to explain the greater case of fusion of rodent compared with human RBC membranes and is attributed to the higher ratio protease to protease inhibitor (calpain to calpastatin) in the rodent RBC [151, 152]. Given this and other fundamental species differences, it is appropriate to question the relevance to human RBCs of observations made with rodent RBCs. RBCs from mixed strain (SWJ129-C57BL/6) $TG2^{-/-}$ mice had no transamidating activity, as expected, and RBC and reticulocyte counts, hemoglobin concentration, hematocrit, mean corpuscular volume, and mean corpuscular hemoglobin contents showed only minor abnormalities relative to wild-type mice, with a trend toward a reduced hematocrit (possibly due to anemia) in the $TG2^{-/-}$ mice [153]. Although no significant differences were noted by SDS-PAGE between $TG2^{-/-}$ and $TG2^{+/+}$ RBC membrane fractions, high-density ("old") but not low-density ("young"), $TG2^{-/-}$ RBCs were more resistant to osmotic lysis than wild-type RBCs, both in the absence of, or following, pretreatment with Ca^{2+}/ionophore to induce morphological changes [153].

As with the rat, the prime response of mouse RBCs to $Ca2^{+}$ influx is activation of calpain and membrane protein degradation, with TG2-mediated cross-linking of the small to the large subunits of μ-calpain-enhancing protease activity [154]. Following ionomycin addition to induce cell death, cell-surface phosphatidylserine exposure was delayed in mixed strain (SVJ129-C57BL/6) $TG2^{-/-}$ relative to wild-type RBCs, which accords with the notion that TG2, through calpain activation, promotes cell surface migration of phosphatidylserine [154]. This notion was strengthened by the observation that biotinylcadaverine, a competitive inhibitor of TG2 transamidase activity, delayed phosphatidylserine exposure in wild-type, but not $TG2^{-/-}$, RBCs. To investigate whether TG2 affects RBC survival, wild-type and $TG2^{-/-}$ RBCs were stained with a fluorescent dye, injected intraperitoneally into wild-type mice (rather than directly reinjecting them into the circulations of the individual donor mice), and their time-dependent disappearance from the circulation was followed. An unusual early increase in the life-span profile obtained for the $TG2^{-/-}$ cells was interpreted to indicate a delay, relative to wild-type cells, in $TG2^{-/-}$ cell uptake from the peritoneal cavity into the bloodstream, but otherwise, RBC survival rate was judged not to be significantly different [154].

H. CATARACTS

Age-related cataracts are characterized by aggregation of lens proteins with subsequent lens clouding. It has long been accepted that TG2-mediated cross-linking of some of the β-crystallins [155–157], αB-crystallin [158, 159], and the intermediate filament protein vimentin [160] is likely to be involved. Unphysiological levels of oxidative stress were shown to stimulate the in situ transamidase activity of TG2 in human lens epithelial cells [161] via TGF- β2 activation of the Smad3 signaling pathway, resulting in TG2-mediated cross-linking and aggregation of lens proteins [162]. Lenses from mixed-strain (C57BL/6–SV129) TG2-knockout mice cultured for 10 days with TGF-β2 showed no signs of cortical opacification, whereas lenses from wild-type, C57BL/6 mice opacified within 5 days of TGF-β2 treatment and were completely opaque by day 10 [162]. Thus, the transamidating activity of TG2 likely plays a critical role in the formation of lens protein aggregates induced by oxidative stress and may contribute to the pathogenesis of other diseases involving TGF-β-mediated responses.

I. DIABETES

Maturity-onset diabetes of the young (MODY) is a heritable variant of type 2 or noninsulin-dependent diabetes mellitus (NIDDM) that occurs in nonobese individuals and presents in adolescence or when subjects are in their early 20s. Mutations in several genes (e.g., glucokinase, transcription factors) that impair insulin secretion have been implicated in its pathogenesis [163]. Another potential candidate gene for MODY is *TGM2*, with heterozygous missense mutations that impair transamidase activity (M330R, I331N, or N333S) having been identified in one or more affected individuals of three different families [164, 165]. TG2 is the predominant TGase expressed in pancreatic islets [165], and it has been suggested that its transamidase activity modulates the exocytosis of glucose-stimulated insulin secretion that accompanies calcium influx into pancreatic β-cells [166]. Consistent with an involvement in insulin release but not synthesis, pancreatic islets from mixed strain (SVJ129-C57BL/6) $TG2^{-/-}$ mice were unchanged from wild-type islets with respect to total insulin content, islet size, or morphology [164]. However, less insulin was secreted relative to wild-type islets in response to a high glucose challenge. This was reflected in lower blood insulin and higher blood glucose levels in $TG2^{-/-}$ mice following intraperitoneal glucose loading [164]. Paradoxically, $TG2^{-/-}$ mice showed a hypoglycemic response to exogenous insulin, indicating increased insulin sensitivity [164], a phenomenon that has not been reported in MODY patients. The defect in $TG2^{-/-}$ glucose-stimulated insulin release has been attributed to mitochondrial TG2 functioning as a protein disulfide isomerase to reduce mitochondria respiratory chain activity and, consequently, ATP production, thereby resulting in a dramatic decrease in the general motility of $TG2^{-/-}$ mice [84]. However, such a pathogenetic mechanism is not generally associated with MODY. Moreover, a detailed study failed to find TG2 in mitochondria [43], and reduced motility due to low ATP levels (the basis for postulating the above pathogenetic mechanism) is not observed in all $TG2^{-/-}$ animal colonies (as discussed in Section "Cardiac Diseases"). Thus, the precise role, if any, of TG2 and *TGM2* mutations in MODY pathogenesis remains unclear.

J. HEPATIC AND RENAL DISEASES

TG2-knockout mice have been used as animal model in an attempt to investigate whether accumulation of TG2 during hepatic or renal injury plays a protective role. Consistent with a protective role for the TG2 in liver damage, treatment of C57BL/6 TG2-knockout mice with the hepatotoxin carbon tetrachloride resulted in an increased incidence of death postinjury, relative to control C57BL/6 mice, an effect associated with an increased ECM accumulation [105]. In contrast, consistent with a proapoptotic role for TG2 is the finding that, compared with their wild-type mixed-strain (SVJ129-C57BL/6) littermates, apoptosis of hepatocytes from $TG2^{-/-}$ mice was markedly reduced following treatment of isolated hepatocytes with ethanol or treatment of animals with a low dose (0.1 μg/g body wt) of the anti-Fas antibody Jo2, which is a known, potent inducer of endothelial cell and hepatocyte apoptosis that contributes to alcohol-induced liver damage [167]. This finding contrasts with that of an earlier study reporting increased death of mixed-strain (SVJ129–C57BL/6) TG2-knockout mice relative to wild-type FVB strain mice, when treated with a high dose (1 μg/g body wt.) of Jo2 [168]. The problem with this earlier study is that a Jo2 dose of 1 μg/g body wt. is sublethal in FVB mice, whereas C57BL/6 mice are more sensitive to Jo2, with lethality being observed at doses as low as 0.3 μg/g body wt. [168]; thus, it is not clear if the increased sensitivity of the TG2-knockout mice was due to the influence of their C57BL/6 genetic background or the lack of TG2. Indeed, when using wild-type SVJ129–C57BL/6 littermates from heterozygous $TG2^{+/-}$ crosses, the high dose of Jo2 used by Sarang et al. [168] caused massive hepatic necrosis both in $TG2^{+/+}$ and $TG2^{-/-}$ mice [167], indicating that the different genetic backgrounds of the wild-type and TG2-knockout mice used by Sarang et al. [168] probably accounts for the discrepancy between the two studies. The study of Tatsukawa et al. [167] also highlights a role for TG2 in promoting hepatic apoptosis caused by relatively low doses of Jo2, but not by higher doses. The proapototic effect of TG2 in the pathophysiology of liver cell death involves inactivation of the transcription factor Spl by TG2-mediated cross-linking with resultant inhibition of the expression of genes required for cell viability [167]. In a third type of liver injury model, C57BL/6 $TG2^{-/-}$ mice were found to be highly resistant to drug-induced Mallory body formation and associated liver hypertrophy relative to C57BL/6 $TG2^{+/+}$ mice [169]. However, hepatocellular damage was similar between wild type and $TG2^{-/-}$ mice, although $TG2^{-/-}$ mice had marked hepatic cholestasis, characterized by more gallstones, jaundice, and ductal proliferation than wild-type mice [169]. Thus, the amelioration of Mallory body formation and injury-mediated hypertrophy attributable to TG2 is associated not with an improvement in liver cell damage but, rather, with increased cholestasis.

Studies using C57BL/6 $TG2^{-/-}$ tubular epithelial cells, in conjunction with transfected opossum kidney proximal tubular epithelial cells, have shown that TG2 contributes to renal extracellular matrix accumulation primarily by accelerating the deposition of soluble collagens III and IV into the matrix [170]. Involvement of TG2 in interstitial renal fibrosis has also been demonstrated in C57BL/6 TG2-knockout mice subjected to unilateral ureteral obstruction [171]. Although apoptosis was similar, fibrosis development was substantially reduced in $TG2^{-/-}$ mice relative to $TG2^{+/+}$ mice. This was associated with reduced macrophage and myofibroblast infiltration and decreased collagen I synthesis due to decreased TGF-β activation.

K. NEURODEGENERATIVE DISEASES

An ever-growing number of scientific findings suggest that TGase activity is involved in the pathogenesis of neurodegenerative diseases. However, to date, only indirect evidence has been obtained about the involvement of these enzymes in the physiopathology of these neurological diseases. Protein aggregates in affected brain regions are histopathological hallmarks of many neurodegenerative diseases [172]. It was suggested more than twenty years ago that TGase activity might contribute to the formation of protein aggregates in AD brain [173]. Since then, tau protein has been shown to be an excellent in vitro substrate of TG2 [174], and GGEL cross-links have been found in the neurofibrillary tangles and paired helical filaments of AD brains [175]. More recently, TG2 activity has been shown to induce in vitro amyloid β (Aβ)-protein oligomerization and aggregation at physiologic levels [176]. By these molecular mechanisms, TGase activity could contribute to AD symptoms and progression [176]. In AD, also, symptom severity correlates with impaired $G\alpha_{q/11}$-coupling of the M1 muscarinic receptor [177]. Recently, studies carried out by using TG2-encoding lentivirus injection in a transgenic model of AD (APP^{Sw}) indicate that this is due to the formation of angiotensin II type 2 (AT_2) receptor oligomers in the brain that sequester $G\alpha_{q/11}$ [178]. AT_2 oligomers are triggered in a dose-dependent way by aggregated-Aβ in a two-step process involving oxidative cross-linking to form dissociable AT_2 receptors that do not sequester $G\alpha_{q/11}$ [178], followed by TG2-dependent cross-linking of residues in the receptor COOH terminus to form nondissociable AT2 receptor oligomers [179]. Moreover, there is evidence that TGase activity also contribute to the formation of proteinaceous deposits in PD [180, 181], in supranuclear palsy [182, 183] and in HD, a neurodegenerative disease caused by a CAG expansion (codifying for a polyglutamine domain) in the affected gene [184]. For example, expanded polyglutamine domains have been reported to be substrates of TGase activity [185–187] and therefore aberrant TGase activity could contribute to CAG-expansion diseases. However, although all these studies suggest the involvement of the TGase activity in the formation of protein aggregates, they do not indicate whether aberrant TGase activity per se directly determines the disease's progression. For example, several experimental findings reported that TG2 activity in vitro leads to the formation of soluble aggregates of α-synuclein [188] or polyQ proteins [189, 190]. However, studies carried out with genetically engineered animal models showed that TG2 ablation both in R6/1 [191] and R6/2 [192] mice models of HD is associated with increased lifespan and improved motor function, but in the same R6/1 and R6/2 animals an increase in aggregate formation also resulted [191, 192]. Therefore, it is possible that TG2 contributes the pathogenesis of HD by other mechanisms still not completely understood, and it must be

TABLE 2
List of Polyglutamine (CAG-Expansion) Diseases

Disease	Sites of Neuropathology	CAG Triplet Number		Gene Product (Intracellular Localization of Protein Deposits)	Reference
		Normal	Disease		
Corea major or Huntington's disease (HD)	Striatum (medium spiny neurons and cortex in late stage)	6–35	36–121	Huntingtin (n, c)	[196]
Spinocerebellar Ataxia type 1 (SCA1)	Cerebellar cortex (Purkinje cells), dentate nucleus and brain stem	6–39	40–81	Ataxin-1 (n, c)	[197]
Spinocerebellar ataxia type 2 (SCA2)	Cerebellum, pontine nuclei, substantia nigra	15–29	35–64	Ataxin-2 (c)	[198]
Spinocerebellar ataxia type 3 (SCA3) or Machado-Joseph disease (MJD)	Substantia nigra, globus pallidus, pontine nucleus, cerebellar cortex	13–42	61–84	Ataxin-3 (c)	[199]
Spinocerebellar ataxia type 6 (SCA6)	Cerebellar and mild brainstem atrophy	4–18	21–30	Calcium channel, subunit (α1A) (m)	[200]
Spinocerebellar ataxia type 7 (SCA7)	Photoreceptor and bipolar cells, cerebellar cortex, brainstem	7–17	37–130	Ataxin-7 (n)	[201]
Spinocerebellar ataxia type 12 (SCA12)	Cortical, cerebellar atrophy	7–32	41–78	Brain specific regulatory subunit of protein phosphatase PP2A (?)	[202]
Spinocerebellar ataxia type 17 (SCA17)	Gliosis and neuronal loss in the Purkinje cell layer	29–42	46–63	TATA-binding protein (TBP) (n)	[203]
Spinobulbar muscular atrophy (SBMA) or Kennedy disease	Motor neurons (anterior horn cells, bulbar neurons, and dorsal root ganglia)	11–34	40–62	Androgen receptor (n,c)	[204]
Dentatorubral–pallidoluysian atrophy (DRPLA)	Globus pallidus, dentato-rubral and subthalamic nucleus	7–35	49–88	Atrophin (n,c)	[205]

Cellular localization: c, cytosolic; m, transmembrane; n, nuclear.

considered that the human brain contains at least four TGs, including TG 1, 2, 3 [193], and possibly TG6 [194]. Conversely, to strengthen the possible central role of the TGase activity in neurodegenerative diseases, a study showed that anti-TG2 IgA antibodies are present in the gut and brain of patients with gluten ataxia, a nongenetic sporadic cerebellar ataxia, but not in ataxia control patients [195]. This study suggests that the involvement of a brain TG could represent a common denominator in several neurodegenerative diseases, which can lead to the determination of physiopathological consequences through different molecular mechanisms (biochemical, immunological, etc.). Moreover, as previously reported, at least ten human CAG-expansion diseases have been described to date (Table 2), and in at least nine of them, their neuropathology is caused by the expansion in the number of residues in the polyglutamine domain to a value beyond 35–40. Remarkably, the mutated proteins have no obvious similarities except for the expanded polyglutamine domain. Most of the mutated proteins are widely expressed both within the brain and elsewhere in the body. A major challenge then is to understand why the brain is primarily affected and why different regions within the brain are affected in the different CAG-expansion diseases, i.e., what accounts for the neurotoxic gain of function of each protein and for a selective vulnerability of each cell type. Possibly, the selective vulnerability [206] may be explained in part by the susceptibility of the expanded polyglutamine domains in the various CAG-expansion diseases to act as cosubstrates for a brain TG (as shown in Figure 7).

V. MEDICAL PERSPECTIVES AND FUTURE DIRECTIONS

A large number of scientific reports implicate aberrant TG2 activity in CD and in other human diseases, including neurodegenerative diseases, in which TG2 might also be involved through its pleiotropic functions. Since up to now there have been no long-term effective treatments in particular for human neurodegenerative diseases, then the possibility that selective TGase inhibitors may be of clinical benefit has been seriously considered. In this respect, some encouraging results have been obtained with TGase inhibitors in preliminary studies with different biological models of CAG-expansion diseases. As example, cystamine (Figure 8) is a potent in vitro inhibitor of enzymes that require an unmodified cysteine at the active site [207]. Inasmuch as TGs contain a crucial active-site cysteine, cystamine has the potential to inhibit these enzymes by disulfide interchange reactions. A disulfide interchange reaction results in the formation of cysteamine and a

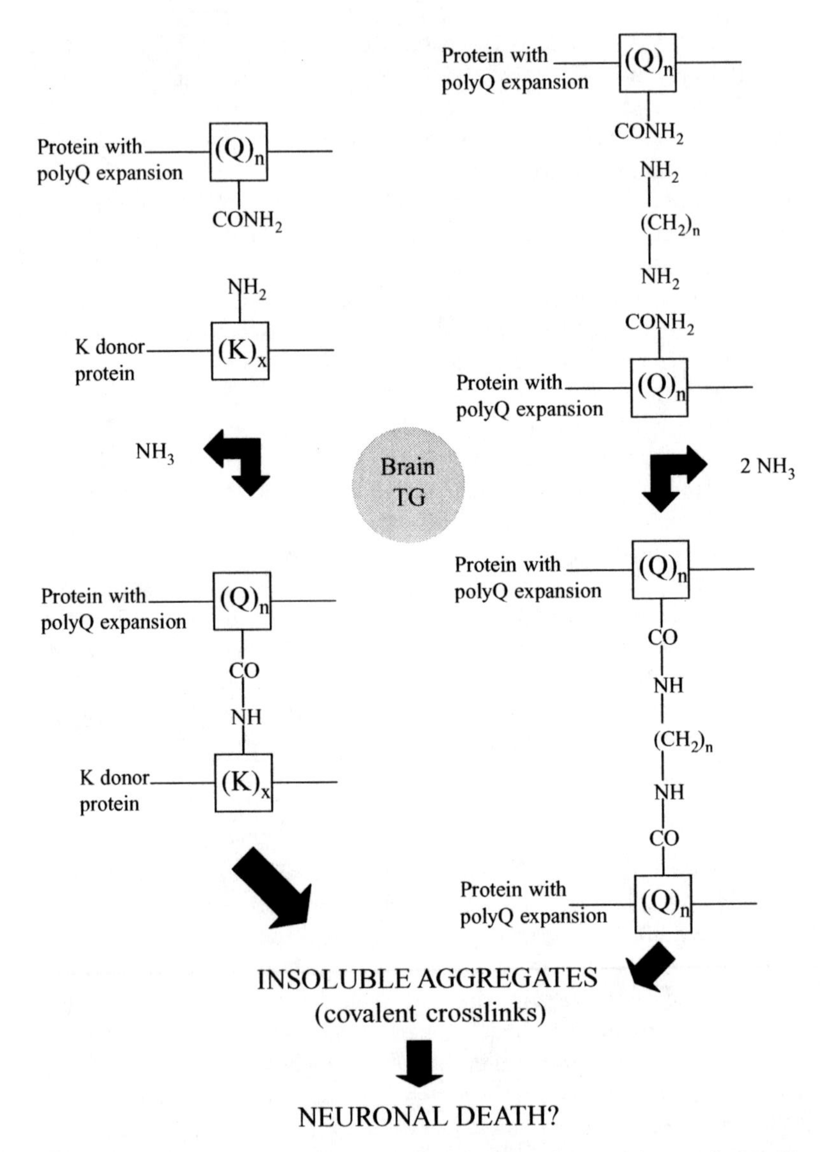

FIGURE 7. Biochemical reactions catalyzed by transglutaminase activity on polyglutamine proteins, suggested as pathogenetic mechanisms of polyglutamine diseases.

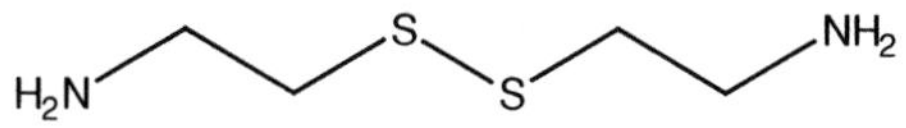

Cystamine

FIGURE 8. Chemical structure of cystamine.

cysteamine–cysteine-mixed disulfide residue at the active site. Recent studies have shown that cystamine decreases the number of protein inclusions in transfected cells expressing the atrophin (DRPLA) protein containing a pathological-length polyglutamine domain [208]. In other studies, cystamine administration to HD-transgenic mice resulted in an increase in life expectancy and amelioration of neurological symptoms [209, 210]. Neuronal inclusions were decreased in one of these studies [209]. Although all these data seem to support the hypothesis of a direct role of TGase activity in the pathogenesis of the polyglutamine diseases, cystamine is also found to act in the HD-transgenic mice by mechanisms other than the inhibition of TGs, such as the inhibition of Caspases [211]. Therefore, the pharmacodynamics and the pharmacokinetics of cystamine should be carefully investigated in order to confirm the same effectiveness in patients with HD and possibly in patients with other neurodegenerative diseases. Another critical problem in the use of TGase inhibitors in treating human diseases, including CD, neurological diseases, and others, relates to the fact that, as previously reported, human organs, like the human brain, contain several TGs. The human brain, for example, contains at least four TGs, including TG 1, 2, 3 [193], and possibly TG6 [194], and a strong nonselective inhibitor of TGs might also inhibit the plasma Factor XIII, causing a bleeding disorder. Therefore, from a number of standpoints, it would seem that a selective inhibitor that discriminates between TGs would be preferable to an indiscriminate TGase inhibitor. Interestingly, a recent study carried out on animal models affected by experimental diabetic nephropathy showed the efficacy of a TG2 inhibitor in reducing the clinical symptoms at renal level [212]. Thus, the determination of transamidase activity products and the use of inhibitors of TGs may be important diagnostic and therapeutic tools and, consequently, may be of clinical benefit. However, to minimize the possible side effects derived from indiscriminate inhibition of all TGs, selective inhibitors of each TG should be required. Progress in this area may be achieved in the near future with crystallographic analysis of other known TG or by pharmaco-genetic engineering.

ACKNOWLEDGMENTS

I like to thank all the colleagues whose work on TG2 is cited in the References and to apologize to all the other authors whose contributions have not been cited due to the focus of the chapter.

REFERENCES

1. Olaisen, B., Gedde-Dahl, T. Jr., Teisberg, P., Thorsby, E., Siverts, A., Jonassen, R., and Wilhelmy, M. C. (1985) A structural locus for coagulation factor XIIIA (F13A) is located distal to the HLA region on chromosome 6p in man, *Am. J. Hum. Genet. 37*, 215–220.
2. Yamanishi, K., Inazawa, J., Liew, F.-M., Nonomura, K., Ariyama, T., Yasuno, H., Abe, T., Doi, H., Hirano, J., and Fukushima, S. (1992) Structure of the gene for human transglutaminase 1, *J. Biol. Chem. 267*, 17858–17863.
3. Gentile, V., Davies, P. J. A., and Baldini, A. (1994) The human tissue transglutaminase gene maps on chromosome 20q12 by in situ fluorescence hybridisation, *Genomics 20*, 295–297.
4. Wang, M., Kim, I. G., Steinert, P. M., and McBride, O. W. (1994) Assignment of the human transglutaminase 2 (TGM2) and transglutaminase 3 (TGM3) to chromosome 20q11.2, *Genomics 23*, 721–722.
5. Gentile, V., Grant, F., Porta, R., and Baldini, A. (1995) Human prostate transglutaminase is localized on chromosome 3p21.33-p22 by in situ fluorescence hybridisation, *Genomics 27*, 219–220.
6. Grenard, P., Bates, M. K., and Aeschlimann, D. (2001) Evolution of transglutaminase genes: identification of a transglutaminase gene cluster on human chromosome 15q15. Structure of the gene encoding transglutaminase X and a novel gene family member, transglutaminase Z, *J. Biol. Chem. 276*, 33066–33078.
7. Folk, J. E. (1983) Mechanism and basis for specificity of transglutaminase-catalyzed ε-(γ-glutamyl)lysine bond formation, *Adv. Enzymol. Relat. Areas Mol. Biol. 54*, 1–56.
8. Lorand, L. and Conrad, S. M. (1984) Transglutaminases, *Mol. Cell. Biochem. 58*, 9–35.
9. Piacentini, M., Martinet, N., Beninati, S., and Folk, J. E. (1988) Free and protein conjugated-polyamines in mouse epidermal cells. Effect of high calcium and retinoic acid, *J. Biol. Chem. 263*, 3790–3794.
10. Kim, C. Y., Quarsten, H., Bergseng, E., Khosla, C., and Sollid, L. M. (2004) Structural basis for HLA-DQ2-mediated presentation of gluten epitopes in celiac disease, *Proc. Natl. Acad. Sci. U.S.A. 101*, 4175–4179.
11. Fleckenstein, B., Qiao, S. W., Larsen, M. R., Jung, G., Roepstorff, P., and Sollid, L. M. (2004) Molecular characterization of covalent complexes between tissue transglutaminase and gliadin peptides, *J. Biol. Chem. 279*, 17607–17616.
12. Lorand, L. and Graham, R. M. (2003). Transglutaminases: crosslinking enzymes with pleiotropic functions, *Nat. Mol. Cell Biol. 4*, 140–156.

13. Achyuthan, K. E. and Greenberg, C. S. (1987) Identification of a guanosine triphosphate-binding site on guinea pig liver transglutaminase. Role of GTP and calcium ions in modulating activity, *J. Biol. Chem. 262*, 1901–1906.

14. Hasegawa, G., Suwa, M., Ichikawa, Y., Ohtsuka, T., Kumagai, S., Kikuchi, M., Sato, Y., and Saito, Y. (2003) A novel function of tissue-type transglutaminase: protein disulfide isomerase, *Biochem. J. 373*, 793–803.

15. Lahav, J., Karniel, E., Bagoly, Z., Sheptovitsky, V., Dardik, R., and Inbal, A. (2009) Coagulation factor XIII serves as protein disulfide isomerase, *Thromb. Haemost. 101*, 840–844.

16. Smethurst, P. A. and Griffin, M. (1996) Measurement of tissue transglutaminase activity in a permeabilized cell system: its regulation by calcium and nucleotides, *Biochem. J. 313*, 803–808.

17. Nakaoka, H., Perez, D. M., Baek, K. J., Das, T., Husain, A., Misono, K., Im, M. J., and Graham R. M. (1994) Gh: a GTP-binding protein with transglutaminase activity and receptor signalling function, *Science 264*, 1593–1596.

18. Gentile, V., Porta, R., Chiosi, E., Spina, A., Caputo, I., Valente, F., Pezone, R., Davies, P. J. A., and Illiano, G. (1997) Tissue transglutaminase and adenylate cyclase interactions in Balb-C 3T3 fibroblast membranes, *Biochim. Biophys. Acta 1357*, 115–122.

19. Lee, M. Y., Chung, S., Bang, H. W., Baek, K. J., and Uhm, D. (1997) Modulation of large conductance Ca^{2+}-activated K^{+} channel by G alpha h (transglutaminase II) in the vascular smooth muscle cell, *Pflugers Arch. 433*, 761–763.

20. Mian, S., El Alaoui, S., Lawry, J., Gentile, V., Davies, P. J. A., and M. Griffin (1995) The importance of the GTP binding protein tissue transglutaminase in the regulation of cell cycle progression, *FEBS Lett. 370*, 27–31.

21. Mishra, S. and Murphy, L. J. (2004) Tissue transglutaminase has intrinsic kinase activity: identification of transglutaminase 2 as an insulin-like growth factor-binding protein-3 kinase, *J. Biol. Chem. 279*, 23863–23868.

22. Mishra, S. and Murphy, L. J. (2006) The p53 oncoprotein is a substrate for tissue transglutaminase kinase activiy, *Biochem. Biophys. Res. Commun. 339*, 726–730.

23. Mishra, S., Saleh, A., Espino, P. S., Davie, J. R., and Murphy, L. J. (2006) Phosphorylation of histones by tissue transglutaminase, *J. Biol. Chem. 281*, 5532–5538.

24. Thomazy, V. and Fesus, L. (1989) Differential distribution of tissue transglutaminase in human cells: an immunohistochemical study, *Cell Tissue Res. 255*, 215–224.

25. Lu, S., Saydak, M., Gentile, V., Stein, J. P., and Davies P. J. A. (1995) Isolation and characterization of the human tissue transglutaminase promoter, *J. Biol. Chem. 270*, 9748–9755.

26. Nagy, L., Thomazy, V., Saydak, M., Stein, J. P., and Davies, P. J. (1997) The promoter of the mouse transglutaminase directs tissue specific, retinoid-regulated, and apoptosis-linked expression, *Cell Death Differ. 4*, 534–547.

27. Bailey, C. D. C. and Johnson, G. V. W. (2004) Developmental regulation of tissue transglutaminase in the mouse forebrain, *J. Neurochem. 91*, 1369–1379.

28. De Laurenzi, V. and Melino, G. (2001) Gene disruption of tissue transglutaminase, *Mol. Cell. Biol. 21*, 148–155.

29. Nanda, N., Iismaa, S. E., Owens, W. A., Husain, A., Mackay, F., and Graham, R. M. (2001) Targeted inactivation of Gh/tissue transglutaminase II, *J. Biol. Chem. 276*, 20673–20678.

30. Fesus, L. and Szondy, Z. (2005) Transglutaminase 2 in the balance of cell death and survival, *FEBS Lett. 579*, 3297–3302.

31. Small, K., Feng, J. F., Lorenz, J., Donnelly, E. T., Yu, A., Im, M. J., Dorn, G. W. 2nd, and Liggett, S. B. (1999) Cardiac-specific overexpression of transglutaminase II (G_h) results in a unique hypertrophy phenotype independent of phospholipase C activation, *J. Biol. Chem. 274*, 21291–21296.

32. Vezza, R., Habib, A., and FitzGerald, G. A. (1999) Differential signaling by the thromboxane receptor isoforms via the novel GTP-binding protein, Gh, *J. Biol. Chem. 274*, 12774–12779.

33. Zhang, Z., Vezza, R., Plappert, T., McNamara, P., Lawson, J.A., Austin, S., Praticò, D., Sutton, M. S., and FitzGerald, G. A. (2003) COX-2-dependent cardiac failure in Gh/tTG transgenic mice, *Circ. Res. 92*, 1153–1161.

34. Tucholski, J., Roth, K. A., and Johnson, G. V. (2006) Tissue transglutaminase overexpression in the brain potentiates calcium-induced hippocampal damage, *J. Neurochem. 97*, 582–594.

35. Gentile, V., Saydak, M., Chiocca, E. A., Akande, N., Birchbickler, P. J., Lee, K. N., Stein, J. P., and Davies, P. J. (1991) Isolation and characterization of cDNA clones to mouse macrophage and human endothelial cell tissue transglutaminase, *J. Biol. Chem. 266*, 478–483.

36. Monsonego, A., Shani, Y., Friedmann, I., Paas, Y., Eizenberg, O., and Schwartz, M. (1997) Expression of GTP-dependent and GTP-independent tissue-type transglutaminase in cytokine-treated rat brain astrocytes, *J. Biol. Chem. 272*, 3724–3732.

37. Lai, T.-S., Liu, Y., Weidong, L., and Greenberg, C. (2007) Identification of two GTP-independent alternatively spliced forms of tissue transglutaminase in human leukocytes, vascular smooth muscle, and endothelial cells, *FASEB J. 21*, 4131–4134.

38. Citron, B. A., Santa Cruz, K. S., Davies, P. J., and Festoff, B. W. (2001) Intron–exon swapping of transglutaminase mRNA and neuronal tau aggregation in Alzheimer's disease, *J. Biol. Chem. 276*, 3295–3301.

39. Ismaa, S. E., Mearns, B. M., Lorand, L., and Graham, R. M. (2009) Transglutaminases and disaese: lessons from genetically engineered mouse models and inherited disorders, *Physiol. Rev. 89*, 991–1023.

40. Griffin, M., Casadio, R., and Bergamini, C. (2002) Transglutaminases: nature's biological glues, *Biochem. J. 368*, 377–396.

41. Gentile, V., Thomazy, V., Piacentini, M., Fesus, L., and Davies, P. J. A. (1992) Expression of tissue transglutaminase in Balb-C 3T3 fibroblasts: effects on cellular morphology and adhesion, *J. Cell Biol. 119*, 463–474.

42. Melino, G., Annichiarico-Petruzzelli, M., Piredda, L., Candi, E., Gentile, V., Davies, P. J. A., and Piacentini, M. (1994) Tissue transglutaminase and apoptosis: sense and antisense transfection studies with human neuroblastoma cells, *Mol. Cell. Biol. 14*, 6584–6596.

43. Krasnikov, B. F., Kim S. Y., McConoughey, S. J., Ryu, H., Xu, H., Stavrovskaya, I., Iismaa, S. E., Mearns, B. M., Ratan, R. R., Blass, J. P., Gibson, G. E., and Cooper, A. J. (2005) Transglutaminase activity is present in highly purified nonsynaptosomal mouse brain and liver mithochondria, *Biochemistry 44*, 7830–7843.

44. Akimov, S. S. and Belkin, A. M. (2001) Cell surface tissue transglutaminase is involved in adhesion and migration of monocytic cells on fibronectin, *Blood 98*, 1567–1576.

45. Aeschlimann, D., Kaupp, O., and Paulsson, M. (1995) Transglutaminase-catalyzed matrix cross-linking in differentiating cartilage: identification of osteonectin as a major glutaminyl substrate, *J. Cell Biol. 129*, 881–892.

46. Wozniak, M., Fausto, A., Carron, C. P., Meyer, D. M., and Hruska, K. A. (2000) Mechanically strained cells of the osteoblast lineage organize their extracellular matrix through unique sites of alpha(v)beta(3)-integrin expression, *J. Bone Miner. Res. 15*, 1731–1745.

47. Aeschlimann, D. and Paulsson, M. (1991) Cross-linking of laminin-nidogen complexes by tissue transglutaminase. A novel mechanism for basement membrane stabilization, *J. Biol. Chem. 266*, 15308–15317.

48. Qian, R. Q. and Glanville, R. W. (1997) Alignment of fibrillin molecules in elastic microfibrils is defined by transglutaminase-derived cross-links, *Biochemistry 36*, 15841–15847.

49. Trask, B. C., Broekelmann, T., Ritty, T. M., Trask, T. M., Tisdale, C., and Mecham, R. P. (2001) Posttranslational modifications of microfibril associated glycoprotein-1 (MAGP-1), *Biochemistry 40*, 4372–4380.

50. Kim, S.Y., Jeitner, T.M., and Steinert, P.M. (2002) Transglutaminases in disease, *Neurochem. Int. 40*, 85–103.

51. Bruce, S. E., Bjarnason I., and Peters, T. J. (1985) Human jejunal transglutaminase: demonstration of activity, enzyme kinetics and substrate specificity with special relation to gliadin and Coeliac disease, *Clin. Sci. (Colch) 68*, 573–579.

52. Porta, R., Gentile, V., Esposito, C., Mariniello, L., and Auricchio, S. (1990) Cereal dietary proteins with sites for cross-linking by transglutaminase, *Phytochemistry 29*, 2801–2804.

53. Green, H. (1993) Human genetic diseases due to codon reiteration: relationship to an evolutionary mechanism, *Cell 74*, 955–956.

54. Auricchio, S., De Ritis, G., De Vincenzi, M., Gentile, V., Maiuri, L., Mancini, E., Porta, R., and Raia, V. (1990) Amines protect in vitro the celiac small intestine from the damaging activity of gliadin peptides, *Gastroenterology 99*, 1668–1674.

55. Dietrich, W., Ehnis, T., Bauer, M., Donner, P., Volta, U., Riecken, E. O., and Schuppan, D. (1997) Identification of tissue transglutaminase as the autoantigen of Coeliac disease, *Nat. Med. 3*, 797–801.

56. Schmidt, G., Selzer, J., Lerm, M., and Aktories, K. (1998) The rho-deamidating cytotoxic necrotizing factor 1 from *Escherichia coli* possesses transglutaminase activity: cysteine 866 and histidine 881 are essential for enzyme activity, *J. Biol. Chem. 273*, 13669–13674.

57. Arentz-Hansen, H., Korner, R., Molberg, O., Quarsten, H., Vader, W., Kooy, Y. M. C., Lundin, K. E. A., Koning, F., Roepstorff, P., Sollid, L., and McAdam, S. N. (2000) The

intestinal T cell response to α-gliadin in adult Celiac disease is focused on a single deamidated glutamine targeted by tissue transglutaminase, *J. Exp. Med. 191*, 603–612.

58. Attur, M. G., Patel, I. R., Patel, R. N., Abramson, S. B., and Amin, A. R. (1998) Autocrine production of IL-1 beta by human osteoarthritis-affected cartilage and differential regulation of endogenous nitric oxide, IL-6, prostaglandin E2, and IL-8, *Proc. Assoc. Am. Physicians 110*, 65–72.

59. Borzi, R. M., Mazzetti, I., Macor, S., Silvestri, T., Bassi, A., Cattini, L., and Facchini., A. (1999) Flow cytometric analysis of intracellular chemokines in chondrocytes in vivo: constitutive expression and enhancement in osteoarthritis and rheumatoid arthritis, *FEBS Lett. 455*, 238–242.

60. Olson, T. S. and Ley, K. (2002) Chemokines and chemokine receptors in leukocyte trafficking, *Am. J. Physiol. Regul. Integr. Comp. Physiol. 283*, R7–R28.

61. Merz, D., Liu, R., Johnson, K., and Terkeltaub, R. (2003) IL-8/CXCL8 and growth-related oncogene alpha/CXCL1 induce chondrocyte hypertrophic differentiation, *J. Immunol. 171*, 4406–4415.

62. Johnson, K. A., van Etten, D., Nanda, N., Graham, R. M., and Terkeltaub, R. A. (2003) Distinct transglutaminase 2-independent and transglutaminase 2-dependent pathways mediate articular chondrocyte hypertrophy, *J. Biol. Chem. 278*, 18824–18832.

63. Cecil, D. L. and Terkeltaub, R. (2008) Transamidation by transglutaminase 2 transforms S100A11 calgranulin into a procatabolic cytokine for chondrocytes, *J. Immunol. 180*, 8378–8385.

64. Orlandi, A., Oliva, F., Taurisano, G., Candi, E., Di Lascio, A., Melino, G., Spagnoli, L. G., and Tarantino, U. (2009) Transglutaminase-2 differently regulates cartilage destruction and osteophyte formation in a surgical model of osteoarthritis, *Amino Acids 36*, 755–763.

65. Bakker, A. C., van de Loo, F. A., van Beuningen, H. M., Sime, P., van Lent, P. L., Van Der Kraan, P. M., Richards, C. D., and Van Den Berg, W. B. (2001) Overexpression of active TGF-beta-1 in the murine knee joint: evidence for synovial-layer-dependent chondro-osteophyte formation, *Osteoarthr. Cartilage 9*, 128–136.

66. Chhabra, A., Verma, A., and Mehta, K. (2009) Tissue transglutaminase promotes or suppresses tumors depending on cell context, *Anticancer Res. 29*, 1909–1920.

67. Kotsakis, P. and Griffin, M. (2007) Tissue transglutaminase in tumour progression: friend or foe? *Amino Acids 33*, 373–384.

68. Ai, L., Kim, W. J., Demircan, B., Dyer, L. M., Bray, K. J., Skehan, R. R., Massoll, N. A., and Brown, K. D. (2008) The transglutaminase 2 gene (TGM2), a potential molecular marker for chemotherapeutic drug sensitivity, is epigenetically silenced in breast cancer, *Carcinogenesis 29*, 510–518.

69. Liu, T., Tee, A. E., Porro, A., Smith, S. A., Dwarte, T., Liu, P. Y., Iraci, N., Sekyere, E., Haber, M., Norris, M. D., Diolaiti, D., Della Valle, G., Perini, G., and Marshall, G. M. (2007) Activation of tissue transglutaminase transcription by histone deacetylase inhibition as a therapeutic approach for Myc oncogenesis, *Proc. Natl. Acad. Sci. U.S.A. 104*, 18682–18787.

70. Jones, R. A., Kotsakis, P., Johnson, T. S., Chau, D. Y., Ali, S., Melino, G., and Griffin, M. (2006) Matrix changes induced by transglutaminase 2 lead to inhibition of angiogenesis and tumor growth, *Cell Death Differ. 13*, 1442–1453.

71. Belkin, A. M., Akimov, S. S., Zaritskaya, L. S., Ratnikov, B. I., Deryugina, E. I., and Strongin, A. Y. (2001) Matrix-dependent proteolysis of surface transglutaminase by membrane-type metalloproteinase regulates cancer cell adhesion and locomotion, *J. Biol. Chem. 276*, 18415–18422.

72. Belkin, A. M., Zemskov, E. A., Hang, J., Akimov, S. S., Sikora, S., and Strongin, A. Y. (2004) Cell-surface-associated tissue transglutaminase is a target of MMP-2 proteolysis, *Biochemistry 43*, 11760–11769.

73. Di Giacomo, G., Lentini, A., Beninati, S., Piacentini, M., and Rodolfo, C. (2009) In vivo evaluation of type 2 transglutaminase contribution to the metastasis formation in melanoma, *Amino Acids 36*, 717–724.

74. Yuan, L., Choi, K., Khosla, C., Zheng, X., Higashikubo, R., Chicoine, M. R., and Rich, K. M. (2005) Tissue transglutaminase 2 inhibition promotes cell death and chemosensitivity in glioblastomas, *Mol. Cancer Ther. 4*, 1293–1302.

75. Yuan, L., Behdad, A., Siegel, M., Khosla, C., Higashikubo, R., and Rich, K. M. (2008) Tissue transglutaminase 2 expression in meningiomas, *J. Neurooncol. 90*, 135–132.

76. Han, J. A. and Park, S. C. (1999) Reduction of transglutaminase 2 expression is associated with an induction of drug sensitivity in the PC-14 human lung cancer cell line, *J. Cancer Res. Clin. Oncol. 125*, 89–95.

77. Akar, U., Ozpolat, B., Mehta, K., Fok, J., Kondo, Y., and Lopez-Berestein, G. (2007) Tissue transglutaminase inhibits autophagy in pancreatic cancer cells, *Mol. Cancer Res. 5*, 241–249.

78. Herman, J. F., Mangala, L. S., and Mehta, K. (2006) Implications of increased tissue transglutaminase (TG2) expression in drug-resistant breast cancer (MCF-7) cells, *Oncogene 25*, 3049–3058.

79. Kim, D. S., Park, S. S., Nam, B. H., Kim, I. H., and Kim, S. Y. (2006) Reversal of drug resistance in breast cancer cells by transglutaminase 2 inhibition and nuclear factor-kappaB inactivation, *Cancer Res. 66*, 10936–10943.

80. Yuan, L., Siegel, M., Choi, K., Khosla, C., Miller, C. R., Jackson, E. N., Piwnica-Worms, D., and Rich, K. M. (2007) Transglutaminase 2 inhibitor, KCC009, disrupts fibronectin assembly in the extracellular matrix and sensitizes orthotopic glioblastomas to chemotherapy, *Oncogene 26*, 2563–2573.

81. Xu, L., Begum, S., Hearn, J. D., and Hynes, R. O. (2006) GPR56, an atypical G protein-coupled receptor, binds tissue transglutaminase, TG2, and inhibits melanoma tumor growth and metastasis, *Proc. Natl. Acad. Sci. U.S.A. 103*, 9023–9028.

82. Szondy, Z., Mastroberardino, P. G., Váradi, J., Farrace, M. G., Nagy, N., Bak, I., Viti, I., Wieckowski, M. R., Melino, G., Rizzuto, R., Tósaki, A., Fesus, L., and Piacentini, M. (2006) Tissue transglutaminase (TG2) protects cardiomyocytes against ischemia/reperfusion injury by regulating ATP synthesis, *Cell Death Differ. 13*, 1827–1829.

83. Sarang, Z., Tóth, B., Balajthy, Z., Köröskényi, K., Garabuczi, E., Fésüs, L., and Szondy, Z. (2009) Some lessons from the tissue transglutaminase knockout mouse, *Amino Acids 36*, 625–631.

84. Mastroberardino, P. G., Farrace, M. G., Viti, I., Pavone, F., Fimia, G. M., Melino, G., Rodolfo, C., and Piacentini, M. (2006) "Tissue" transglutaminase contributes to the

formation of disulphide bridges in proteins of mitochondrial respiratory complexes, *Biochim. Biophys. Acta 1757*, 1357–1365.

85. Battaglia, G., Farrace, M. G., Mastroberardino, P. G., Viti, I., Fimia, G. M., Van Beeumen, J., Devreese, B., Melino, G., Molinaro, G., Busceti, C. L., Biagioni, F., Nicoletti, F., and Piacentini, M. (2007) Transglutaminase 2 ablation leads to defective function of mitochondrial respiratory complex I affecting neuronal vulnerability in experimental models of extrapyramidal disorders, *J. Neurochem. 100*, 36–49.
86. Iwai, N., Shimoike, H., and Kinoshita, M. (1995) Genes up-regulated in hypertrophied ventricle, *Biochem. Biophys. Res. Commun. 209*, 527–534.
87. Hwang, K. C., Gray, C. D., Sweet, W. E., Moravec, C. S., and Im, M. J. (1996) Alpha 1-adrenergic receptor coupling with Gh in the failing human heart, *Circulation 94*, 718–726.
88. Iismaa, S. E. and Graham, R. M. (2003) Dissecting cardiac hypertrophy and signaling pathways: evidence for an interaction between multifunctional G proteins and prostanoids, *Circ. Res. 92*, 1059–1061.
89. Kunapuli, P., Lawson, J. A., Rokach, J. A., Meinkoth, J. L., and FitzGerald, G. A. (1998) Prostaglandin F2alpha (PGF2alpha) and the isoprostane, 8, 12-iso-isoprostane F2alpha-III, induce cardiomyocyte hypertrophy. Differential activation of downstream signaling pathways, *J. Biol. Chem. 273*, 22442–22452.
90. Gao, Y., Yokota, R., Tang, S., Ashton, A. W., and Ware, J. A. (2000) Reversal of angiogenesis in vitro, induction of apoptosis, and inhibition of AKT phosphorylation in endothelial cells by thromboxane A(2), *Circ. Res. 87*, 739–745.
91. Auld, G. C., Ritchie, H., Robbie, L. A., and Booth, N. A. (2001) Thrombin upregulates tissue transglutaminase in endothelial cells: a potential role for tissue transglutaminase in stability of atherosclerotic plaque, *Arterioscler. Thromb. Vasc. Biol. 21*, 1689–1694.
92. Sumi, Y., Inoue, N., Azumi, H., Seno, T., Okuda, M., Hirata, K., Kawashima, S., Hayashi, Y., Itoh, H., and Yokoyama, M. (2002) Expression of tissue transglutaminase and elafin in human coronary artery: implication for plaque instability, *Atherosclerosis 60*, 31–39.
93. Boisvert, W. A., Rose, D. M., Boullier, A., Quehenberger, O., Sydlaske, A., Johnson, K. A., Curtiss, L. K., and Terkeltaub, R. (2006) Leukocyte transglutaminase 2 expression limits atherosclerotic lesion size, *Arterioscler. Thromb. Vasc. Biol. 26*, 563–569.
94. Stoneman, V. E. and Bennett, M. R. (2004) Role of apoptosis in atherosclerosis and its therapeutic implications, *Clin Sci (Lond) 107*, 343–354.
95. Szondy, Z., Sarang, Z., Molnar, P., Nemeth, T., Piacentini, M., Mastroberardino, P. G., Falasca, L., Aeschlimann, D., Kovacs, J., Kiss, I., Szegezdi, E., Lakos, G., Rajnavolgyi, E., Birckbichler, P. J., Melino, G., and Fesus, L. (2003) Transglutaminase $2^{-/-}$ mice reveal a phagocytosis-associated crosstalk between macrophages and apoptotic cells, *Proc. Natl. Acad. Sci. U.S.A. 100*, 7812–7817.
96. Johnson, K. A., Polewski, M., and Terkeltaub, R. A. (2008) Transglutaminase 2 is central to induction of the arterial calcification program by smooth muscle cells, *Circ. Res. 102*, 529–537.
97. Faverman, L., Mikhaylova, L., Malmquist, J., and Nurminskaya, M. (2008) Extracellular transglutaminase 2 activates beta-catenin signaling in calcifying vascular smooth muscle cells, *FEBS Lett. 582*, 1552–1557.

98. Telci, D. and Griffin, M. (2006) Tissue transglutaminase (TG2)—a wound response enzyme, *Front. Biosci. 1*, 867–882.

99. George, M. D., Vollberg, T. M., Floyd, E. E., Stein, J. P., and Jetten, A. M. (1990) Regulation of transglutaminase type II by transforming growth factor-beta 1 in normal and transformed human epidermal keratinocytes, *J. Biol. Chem. 265*, 11098–11104.

100. Quan, G., Choi, J. Y., Lee, D. S., and Lee, S. C. (2005) TGF-beta1 up-regulates transglutaminase two and fibronectin in dermal fibroblasts: a possible mechanism for the stabilization of tissue inflammation, *Arch. Dermatol. Res. 297*, 84–90.

101. Ritter, S. J. and Davies, P. J. (1998) Identification of a transforming growth factor-beta1/bone morphogenetic protein 4 (TGF-beta1/BMP4) response element within the mouse tissue transglutaminase gene promoter, *J. Biol. Chem. 273*, 12798–12806.

102. Kuncio, G. S., Tsyganskaya, M., Zhu, J., Liu, S. L., Nagy, L., Thomazy, V., Davies, P. J., and Zern, M. A. (1998) TNF-alpha modulates expression of the tissue transglutaminase gene in liver cells, *Am. J. Physiol. 274*, G240–G245.

103. Lee, J., Kim, Y.S., Choi, D. H., Bang, M. S., Han, T. R., Joh, T. H., and Kim, S. Y. (2004) Transglutaminase 2 induces nuclear factor-kB activation via a novel pathway in BV-2 microglia, *J. Biol. Chem. 279*, 53725–53735.

104. Mirza, A., Liu, S. L., Frizell, E., Zhu, J., Maddukuri, S., Martinez, J., Davies, P., Schwarting, R., Norton, P., and Zern, M. A. (1997) A role for tissue transglutaminase in hepatic injury and fibrogenesis, and its regulation by NF-kappaB, *Am. J. Physiol. 272*, G281–G288.

105. Nardacci, R., Lo Iacono, O., Ciccosanti, F., Falasca, L., Addesso, M., Amendola, A., Antonucci, G., Craxì, A., Fimia, G. M., Iadevaia, V., Melino, G., Ruco, L., Tocci, G., Ippolito, G., and Piacentini, M. (2003) Transglutaminase type II plays a protective role in hepatic injury, *Am. J. Pathol. 162*, 1293–1303.

106. Piacentini, M., Farrace, M. G., Hassan, C., Serafini, B., and Autuori, F. (1999) "Tissue" transglutaminase release from apoptotic cells into extracellular matrix during human liver fibrogenesis, *J. Pathol. 189*, 92–98.

107. Johnson, K., Hashimoto, S., Lotz, M., Pritzker, K., and Terkeltaub, R. (2001) Interleukin-1 induces pro-mineralizing activity of cartilage tissue transglutaminase and factor XIIIa, *Am. J. Pathol. 159*, 149–163.

108. Suto, N., Ikura, K., and Sasaki, R. (1993) Expression induced by interleukin-6 of tissue-type transglutaminase in human hepatoblastoma HepG2 cells, *J. Biol. Chem. 268*, 7469–7473.

109. Zatloukal, K., French, S. W., Stumptner, C., Strnad, P., Harada, M., Toivola, D. M., Cadrin, M., and Omary, M. B. (2007) From Mallory to Mallory-Denk bodies: what, how and why? *Exp. Cell Res. 313*, 2033–20049.

110. Bowness, J. M., Tarr, A. H., and Wong, T. (1988) Increased transglutaminase activity during skin wound healing in rats, *Biochim. Biophys. Acta 967*, 234–240.

111. Haroon, Z. A., Hettasch, J. M., Lai, T. S., Dewhirst, M. W., and Greenberg, C. S. (1999) Tissue transglutaminase is expressed, active, and directly involved in rat dermal wound healing and angiogenesis, *FASEB J. 13*, 1787–1795.

112. Falasca, L., Farrace, M. G., Rinaldi, A., Tuosto, L., Melino, G., and Piacentini, M. (2008) Transglutaminase type II is involved in the pathogenesis of endotoxic shock, *J. Immunol. 180*, 2616–2624.

113. Maiuri, L., Luciani, A., Giardino, I., Raia, V., Villella, V. R., D'Apolito, M., Pettoello-Mantovani, M., Guido, S., Ciacci, C., Cimmino, M., Cexus, O. N., Londei, M., and Quaratino, S. (2008) Tissue transglutaminase activation modulates inflammation in cystic fibrosis via PPARgamma down-regulation, *J. Immunol. 180*, 7697–7705.

114. Fadok, V. A., Bratton, D. L., Konowal, A., Freed, P. W., Westcott, J. Y., and Henson, P. M. (1998) Macrophages that have ingested apoptotic cells in vitro inhibit proinflammatory cytokine production through autocrine/paracrine mechanisms involving TGF-beta, PGE2, and PAF, *J. Clin. Invest. 101*, 890–898.

115. Murtaugh, M. P., Mehta, K., Johnson, J., Myers, M., Julianok, R. L., and Davies, P. J. (1983) Induction of tissue transglutaminase in mouse peritoneal macrophages, *J. Biol. Chem. 258*, 11074–11081.

116. Seiving, B., Ohlsson, K., Linder, C., and Stenberg, P. (1991) Transglutaminase differentiation during maturation of human blood monocytes to macrophages, *Eur. J. Haematol. 46*, 263–271.

117. Schroff, G., Neumann, C., and Sorg, C. (1981) Transglutaminase as a marker for subsets of murine macrophages, *Eur. J. Immunol. 11*, 637–642.

118. Falasca, L., Iadevaia, V., Ciccosanti, F., Melino, G., Serafino, A., and Piacentini, M. (2005) Transglutaminase type II is a key element in the regulation of the anti-inflammatory response elicited by apoptotic cell engulfment, *J. Immunol. 174*, 7330–7340.

119. Rose, D. M., Sydlaske, A. D., Agha-Babakhani, A., Johnson, K., and Terkeltaub, R. (2006) Transglutaminase 2 limits murine peritoneal acute gout-like inflammation by regulating macrophage clearance of apoptotic neutrophils, *Arthritis Rheum. 54*, 3363–3371.

120. Sárváry, A., Szucs, S., Balogh, I., Becsky, A., Bárdos, H., Kávai, M., Seligsohn, U., Egbring, R., Lopaciuk, S., Muszbek, L., and Adány, R. (2004) Possible role of factor XIII subunit A in Fcgamma and complement receptor-mediated phagocytosis, *Cell. Immunol. 228*, 81–90.

121. Töröcsik, D., Bárdos, H., Nagy, L., and Adány, R. (2005) Identification of factor XIII-A as a marker of alternative macrophage activation, *Cell. Mol. Life Sci. 62*, 2132–2139.

122. Kojima, S., Nara, K., and Rifkin, D. B. (1993) Requirement for transglutaminase in the activation of latent transforming growth factor-beta in bovine endothelial cells, *J. Cell Biol. 121*, 439–448.

123. Nunes, I., Gleizes, P. E., Metz, C. N., and Rifkin, D. B. (1997) Latent transforming growth factor-beta binding protein domains involved in activation and transglutaminase-dependent cross-linking of latent transforming growth factor-beta, *J. Cell Biol. 136*, 1151–1163.

124. Balajthy, Z., Csomós, K., Vámosi, G., Szántó, A., Lanotte, M., and Fésüs, L. (2006) Tissue-transglutaminase contributes to neutrophil granulocyte differentiation and functions, *Blood 108*, 2045–2054.

125. Akimov, S. S., Krylov, D., Fleischman, L. F., and Belkin, A. M. (2000) Tissue transglutaminase is an integrin-binding adhesion coreceptor for fibronectin, *J. Cell Biol. 148*, 825–838.

126. Verderio, E., Nicholas, B., Gross, S., and Griffin, M. (1998) Regulated expression of tissue transglutaminase in Swiss 3T3 fibroblasts: effects on the processing of fibronectin, cell attachment, and cell death, *Exp. Cell Res. 239*, 119–138.

127. Jones, R. A., Nicholas, B., Mian, S., Davies, P. J., and Griffin, M. (1997) Reduced expression of tissue transglutaminase in a human endothelial cell line leads to changes in cell spreading, cell adhesion and reduced polymerisation of fibronectin, *J. Cell Sci. 110*, 2461–2472.

128. Stephens, P., Grenard, P., Aeschlimann, P., Langley, M., Blain, E., Errington, R., Kipling, D., Thomas, D., and Aeschlimann, D. (2004) Crosslinking and G-protein functions of transglutaminase 2 contribute differentially to fibroblast wound healing responses, *J. Cell Sci. 117*, 3389–3403.

129. Akimov, S. S. and Belkin, A. M. (2001) Cell-surface transglutaminase promotes fibronectin assembly via interaction with the gelatin-binding domain of fibronectin: a role in TGFbeta-dependent matrix deposition, *J. Cell Sci. 114*, 2989–3000.

130. Aeschlimann, D. and Thomazy, V. (2000) Protein crosslinking in assembly and remodelling of extracellular matrices: the role of transglutaminases, *Connect. Tissue Res. 41*, 1–27.

131. Forsprecher, J., Wang, Z., Nelea, V., and Kaartinen, M. T. (2009) Enhanced osteoblast adhesion on transglutaminase 2-crosslinked fibronectin, *Amino Acids 36*, 747–753.

132. Agah, A., Kyriakides, T. R., and Bornstein, P. (2005) Proteolysis of cell-surface tissue transglutaminase by matrix metalloproteinase-2 contributes to the adhesive defect and matrix abnormalities in thrombospondin-2-null fibroblasts and mice, *Am. J. Pathol. 167*, 81–88.

133. Telci, D., Wang, Z., Li, X., Verderio, E. A., Humphries, M. J., Baccarini, M., Basaga, H., and Griffin, M. (2008) Fibronectin-tissue transglutaminase matrix rescues RGD-impaired cell adhesion through syndecan-4 and beta1 integrin co-signaling, *J. Biol. Chem. 283*, 20937–20947.

134. Verderio, E. A., Telci, D., Okoye, A., Melino, G., and Griffin, M. (2003) A novel RGD-independent cel adhesion pathway mediated by fibronectin-bound tissue transglutaminase rescues cells from anoikis, *J. Biol. Chem. 278*, 42604–42614.

135. Zemskov, E. A., Janiak, A., Hang, J., Waghray, A., and Belkin, A. M. (2006) The role of tissue transglutaminase in cell-matrix interactions, *Front. Biosci. 11*, 1057–1076.

136. Gaudry, C. A., Verderio, E., Jones, R. A., Smith, C., and Griffin, M. (1999) Tissue transglutaminase is an important player at the surface of human endothelial cells: evidence for its externalization and its colocalization with the beta(1) integrin, *Exp. Cell Res. 252*, 104–113.

137. Kang, S. K., Yi, K. S., Kwon, N. S., Park, K. H., Kim, U. H., Baek, K. J., and Im, M. J. (2004) Alpha1B-adrenoceptor signaling and cell motility: GTPase function of Gh/transglutaminase 2 inhibits cell migration through interaction with cytoplasmic tail of integrin alpha subunits, *J. Biol. Chem. 279*, 36593–36600.

138. Bakker, E. N., Pistea, A., Spaan, J. A., Rolf, T., de Vries, C. J., van Rooijen, N., Candi, E., and VanBavel, E. (2006) Flow-dependent remodeling of small arteries in mice deficient for tissue-type transglutaminase: possible compensation by macrophage-derived factor XIII, *Circ. Res. 99*, 86–92.

139. Pistea, A., Bakker, E. N., Spaan, J. A., Hardeman, M. R., van Rooijen, N., and VanBavel, E. (2008) Small artery remodeling and erythrocyte deformability in L-NAME-induced hypertension: role of transglutaminases, *J. Vasc. Res. 45*, 10–18.

140. Schiffers, P. M., Henrion, D., Boulanger, C. M., Colucci-Guyon, E., Langa-Vuves, F., van Essen, H., Fazzi, G. E., Lévy, B. I., and De Mey, J. G. (2000) Altered flow-induced arterial remodeling in vimentin-deficient mice, *Arterioscler. Thromb. Vasc. Biol. 20*, 611–616.

141. Gupta, M., Greenberg, C. S., Eckman, D. M., and Sane, D. C. (2007) Arterial vimentin is a transglutaminase substrate: a link between vasomotor activity and remodeling? *J. Vasc. Res. 44*, 339–344.

142. Eftekhari, A., Rahman, A., Schaebel, L. H., Chen, H., Rasmussen, C. V., Aalkjaer, C., Buus, C. L., and Mulvany, M. J. (2007) Chronic cystamine treatment inhibits small artery remodelling in rats, *J. Vasc. Res. 44*, 471–482.

143. Lorand, L., Weissmann, L. B., Epel, D. L., and Bruner-Lorand, J. (1976) Role of the intrinsic transglutaminase in the Ca^{2+}-mediated crosslinking of erythrocyte proteins, *Proc. Natl. Acad. Sci. U.S.A. 73*, 4479–4481.

144. Siefring, G. E. Jr., Apostol, A. B., Velasco, P. T., and Lorand, L. (1978) Enzymatic basis for the Ca^{2+}-induced cross-linking of membrane proteins in intact human erythrocytes, *Biochemistry 17*, 2598–2604.

145. Smith, B. D., La Celle, P. L., Siefring, G. E. Jr., Lowe-Krentz, L., and Lorand, L. (1981) Effects of the calcium-mediated enzymatic cross-linking of membrane proteins on cellular deformability, *J. Membr. Biol. 61*, 75–80.

146. Bjerrum, O. J., Hawkins, M., Swanson, P., Griffin, M., and Lorand, L. (1981) An immunochemical approach for the analysis of membrane protein alterations in Ca^{2+}-loaded human erythrocytes, *J. Supramol. Struct. Cell Biochem. 16*, 289–301.

147. Lorand, L., Michalska, M., Murthy, S. N., Shohet, S. B., and Wilson, J. (1987) Cross-linked polymers in the red cell membranes of a patient with Hb-Koln disease, *Biochem. Biophys. Res. Commun. 147*, 602–607.

148. Lorand, L., Bjerrum, O. J., Hawkins, M., Lowe-Krentz, L., and Siefring, G. E. Jr. (1983) Degradation of transmembrane proteins in Ca^{2+}-enriched human erythrocytes. An immunochemical study, *J. Biol. Chem. 258*, 5300–5305.

149. Lorand, L., Barnes, N., Bruner-Lorand, J. A., Hawkins, M., and Michalska, M. (1987) Inhibition of protein cross-linking in Ca^{2+}-enriched human erythrocytes and activated platelets, *Biochemistry 26*, 308–313.

150. Lorand, L. and Michalska, M. (1985) Altered response of stored red cells to Ca^{2+} stress, *Blood 65*, 1025–1027.

151. Glaser, T. and Kosower, N. S. (1986) Calpain-calpastatin and fusion. Fusibility of erythrocytes is determined by a protease–protease inhibitor [calpain-calpastatin] balance, *FEBS Lett. 206*, 115–120.

152. Kosower, N. S., Glaser, T., and Kosower, E. M. (1983) Membrane-mobility agent-promoted fusion of erythrocytes: fusibility is correlated with attack by calcium-activated cytoplasmic proteases on membrane proteins, *Proc. Natl. Acad. Sci. U.S.A. 80*, 7542–7546.

153. Bernassola, F., Boumis, G., Corazzari, M., Bertini, G., Citro, G., Knight, R. A., Amiconi, G., and Melino, G. (2002) Osmotic resistance of high-density erythrocytes in transglutaminase 2-deficient mice, *Biochem. Biophys. Res. Commun. 291*, 1123–1127.

154. Sarang, Z., Mádi, A., Koy, C., Varga, S., Glocker, M. O., Ucker, D. S., Kuchay, S., Chishti, A. H., Melino, G., Fésüs, L., and Szondy, Z. (2007) Tissue transglutaminase (TG2) facilitates phosphatidylserine exposure and calpain activity in calcium-induced death of erythrocytes, *Cell Death Differ. 14*, 1842–1844.

155. Lorand, L., Hsu, L. K., Siefring, G. E. Jr., and Rafferty, N. S. (1981) Lens transglutaminase and cataract formation, *Proc. Natl. Acad. Sci. U.S.A. 78*, 1356–1360.

156. Mulders, J. W., Hoekman, W. A., Bloemendal, H., and de Jong, W. W. (1987) Beta B1 crystallin is an amine-donor substrate for tissue transglutaminase, *Exp. Cell Res. 171*, 296–305.

157. Velasco, P. T. and Lorand, L. (1987) Acceptor–donor relationships in the transglutaminase-mediated cross-linking of lens beta-crystallin subunits, *Biochemistry 26*, 4629–4634.

158. Groenen, P. J., Bloemendal, H., and de Jong, W. W. (1992) The carboxy-terminal lysine of alpha B-crystallin is an amine-donor substrate for tissue transglutaminase, *Eur. J. Biochem. 205*, 671–674.

159. Lorand, L., Velasco, P. T., Murthy, S. N., Wilson, J., and Parameswaran, K. N. (1992) Isolation of transglutaminase-reactive sequences from complex biological systems: a prominent lysine donor sequence in bovine lens, *Proc. Natl. Acad. Sci. U.S.A. 89*, 11161–11163.

160. Clément, S., Velasco, P. T., Murthy, S. N., Wilson, J. H., Lukas, T. J., Goldman, R. D., and Lorand, L. (1998) The intermediate filament protein, vimentin, in the lens is a target for cross-linking by transglutaminase, *J. Biol. Chem. 273*, 7604–7609.

161. Shin, D. M., Jeon, J. H., Kim, C. W., Cho, S. Y., Kwon, J. C., Lee, H. J., Choi, K. H., Park, S. C., and Kim, I. G. (2004) Cell type-specific activation of intracellular transglutaminase 2 by oxidative stress or ultraviolet irradiation: implications of transglutaminase 2 in age-related cataractogenesis, *J. Biol. Chem. 279*, 15032–15039.

162. Shin, D. M., Jeon, J. H., Kim, C. W., Cho, S. Y., Lee, H. J., Jang, G. Y., Jeong, E. M., Lee, D. S., Kang, J. H., Melino, G., Park, S. C., and Kim, I. G. (2008) TGFbeta mediates activation of transglutaminase 2 in response to oxidative stress that leads to protein aggregation, *FASEB J. 22*, 2498–2507.

163. Fajans, S. S., Bell, G. I., and Polonsky, K. S. (2001) Molecular mechanisms and clinical pathophysiology of maturity-onset diabetes of the young, *N. Engl. J. Med. 345*, 971–980.

164. Bernassola, F., Federici, M., Corazzari, M., Terrinoni, A., Hribal, M. L., De Laurenzi, V., Ranalli, M., Massa, O., Sesti, G., McLean, W. H., Citro, G., Barbetti, F., and Melino, G. (2002) Role of transglutaminase 2 in glucose tolerance: knockout mice studies and a putative mutation in a MODY patient, *FASEB J. 16*, 1371–1378.

165. Porzio, O., Massa, O., Cunsolo, V., Colombo, C., Malaponti, M., Bertuzzi, F., Hansen, T., Johansen, A., Pedersen, O., Meschi, F., Terrinoni, A., Melino, G., Federici, M., Decarlo, N., Menicagli, M., Campani, D., Marchetti, P., Ferdaoussi, M., Froguel, P., Federici, G., Vaxillaire, M., and Barbetti, F. (2007) Missense mutations in the TGM2 gene encoding transglutaminase 2 are found in patients with early-onset type 2 diabetes, *Hum. Mutat. 28*, 1150.

166. Bungay, P. J., Owen, R. A., Coutts, I. C., and Griffin, M. (1986) A role for transglutaminase in glucose-stimulated insulin release from the pancreatic beta-cell, *Biochem. J. 235*, 269–278.

167. Tatsukawa, H., Fukaya, Y., Frampton, G., Martinez-Fuentes, A., Suzuki, K., Kuo, T. F., Nagatsuma, K., Shimokado, K., Okuno, M., Wu, J., Iismaa, S., Matsuura, T., Tsukamoto, H., Zern, M. A., Graham, R. M., and Kojima, S. (2009) Role of transglutaminase 2 in liver injury via cross-linking and silencing of transcription factor Sp1, *Gastroenterology 136*, 1783–1795.

168. Sarang, Z., Molnár, P., Németh, T., Gomba, S., Kardon, T., Melino, G., Cotecchia, S., Fésüs, L., and Szondy, Z. (2005) Tissue transglutaminase (TG2) acting as G protein protects hepatocytes against Fas-mediated cell death in mice, *Hepatology 42*, 578–587.

169. Strnad, P., Harada, M., Siegel, M., Terkeltaub, R. A., Graham, R. M., Khosla, C., and Omary, M. B. (2007) Transglutaminase 2 regulates mallory body inclusion formation and injury-associated liver enlargement, *Gastroenterology 132*, 1515–1526.

170. Fisher, M., Jones, R. A., Huang, L., Haylor, J. L., El Nahas, M., Griffin, M., and Johnson, T. S. (2009) Modulation of tissue transglutaminase in tubular epithelial cells alters extracellular matrix levels: a potential mechanism of tissue scarring, *Matrix Biol. 28*, 20–31.

171. Shweke, N., Boulos, N., Jouanneau, C., Vandermeersch, S., Melino, G., Dussaule, J. C., Chatziantoniou, C., Ronco, P., and Boffa, J. J. (2008) Tissue transglutaminase contributes to interstitial renal fibrosis by favoring accumulation of fibrillar collagen through TGF-beta activation and cell infiltration, *Am. J. Pathol. 173*, 631–642.

172. Adams, R. D. and Victor, M. (1993) *Principles of Neurology*. McGraw-Hill, New York.

173. Selkoe, D. J., Abraham, C., and Ihara, Y. (1982) Brain transglutaminase: in vitro crosslinking of human neurofilament proteins into insoluble polymers, *Proc. Natl. Acad. Sci. U.S.A. 79*, 6070–6074.

174. Grierson, A. J., Johnson, G. V., and Miller, C. C. (2001) Three different human tau isoforms and rat neurofilament light, middle and heavy chain proteins are cellular substrates for transglutaminase, *Neurosci. Lett. 298*, 9–12.

175. Singer, S. M., Zainelli, G. M., Norlund, M. A., Lee, J. M., and Muma, N. A. (2002) Transglutaminase bonds in neurofibrillary tangles and paired helical filament tau early in Alzheimer's disease, *Neurochem. Int. 40*, 17–30.

176. Hartley, D. M., Zhao, C., Speier, A. C., Woodard, G. A., Li, S., Li, Z., and Walz, T. (2008) Transglutaminase induces protofibril-like amyloid beta-protein assemblies that are protease-resistant and inhibit long-term potentiation, *J. Biol. Chem. 283*, 16790–16800.

177. Tsang, S. W., Lai, M. K., Kirvell, S., Francis, P. T., Esiri, M. M., Hope, T., Chen, C. P., and Wong, P. T. (2006) Impaired coupling of muscarinic M1 receptors to G-proteins in the neocortex is associated with severity of dementia in Alzheimer's disease, *Neurobiol. Aging 27*, 1216–1223.

178. AbdAlla, S., Lother, H., el Missiry, A., Langer, A., Sergeev, P., el Faramawy, Y., and Quitterer, U. (2009) Angiotensin II AT2 receptor oligomers mediate G-protein dysfunction in an animal model of Alzheimer disease, *J. Biol. Chem. 284*, 6554–6565.

179. AbdAlla, S., Lother, H., el Missiry, A., Sergeev, P., Langer, A., el Faramawy, Y., and Quitterer, U. (2009) Dominant negative AT2 receptor oligomers induce G-protein arrest and symptoms of neurodegeneration, *J. Biol. Chem. 284*, 6566–6574.

180. Citron, B. A., Suo, Z., SantaCruz, K., Davies, P. J., Qin, F., and Festoff, B. W. (2002) Protein crosslinking, tissue transglutaminase, alternative splicing and neurodegeneration, *Neurochem. Int. 40*, 69–78.

181. Junn, E., Ronchetti, R. D., Quezado, M. M., Kim, S. Y., and Mouradian, M. M. (2003) Tissue transglutaminase-induced aggregation of alpha-synuclein: implications for Lewy body formation in Parkinson's disease and dementia with Lewy bodies, *Proc. Natl. Acad. Sci. U.S.A. 100*, 2047–2052.

182. Zemaitaitis, M. O., Lee, J. M., Troncoso, J. C., and Muma, N. A. (2000) Transglutaminase-induced cross-linking of tau proteins in progressive supranuclear palsy, *J. Neuropathol. Exp. Neurol. 59*, 983–989.

183. Zemaitaitis, M. O., Kim, S. Y., Halverson, R. A., Troncoso, J. C., Lee, J. M., and Muma, N. A. (2003) Transglutaminase activity, protein, and mRNA expression are increased in progressive supranuclear palsy, *J. Neuropathol. Exp. Neurol. 62*, 173–184.

184. Iuchi, S., Hoffner, G., Verbeke, P., Djian, P., and Green, H. (2003) Oligomeric and polymeric aggregates formed by proteins containing expanded polyglutamine, *Proc. Natl. Acad. Sci. U.S.A. 100*, 2409–2414.

185. Gentile, V., Sepe, C., Calvani, M., Melone, M. A. B., Cotrufo, R., Cooper, A. J. L., Blass, J. P., and Peluso, G. (1988) Tissue transglutaminase-catalyzed formation of high-molecular-weight aggregates *in vitro* is favored with long polyglutamine domains: a possible mechanism contributing to CAG-triplet diseases, *Arch. Biochem. Biophys. 352*, 314–321.

186. Kahlem, P., Green, H., and Djian, P. (1998) Transglutaminase action imitates Huntington's disease: selective polymerization of Huntingtin containing expanded polyglutamine, *Mol. Cell. 1*, 595–601.

187. Karpuj, M. V., Garren, H., Slunt, H., Price, D. L., Gusella, J., Becher, M. W., and Steinman, L. (1999) Transglutaminase aggregates huntingtin into nonamyloidogenic polymers, and its enzymatic activity increases in Huntington's disease brain nuclei, *Proc. Natl. Acad. Sci. U.S.A. 96*, 7388–7399.

188. Segers-Nolten, I. M., Wilhelmus, M. M., Veldhuis, G., van Rooijen, B. D., Drukarch, B., and Subramaniam, V. (2008) Tissue transglutaminase modulates alpha-synuclein oligomerization, *Protein Sci. 17*, 1395–1402.

189. Lai, T.-S., Tucker, T., Burke, J. R., Strittmatter, W. J., and Greenberg, C. S. (2004) Effect of tissue transglutaminase on the solubility of proteins containing expanded polyglutamine repeats, *J. Neurochem. 88*, 1253–1260.

190. Konno, T., Mori, T., Shimizu, H., Oiki, S., and Ikura, K. (2005) Paradoxical inhibition of protein aggregation and precipitation by transglutaminase-catalyzed intermolecular cross-linking, *J. Biol. Chem. 280*, 17520–17525.

191. Mastroberardino, P. G., Iannicola, C., Nardacci, R., Bernassola, F., De Laurenzi, V., Melino, G., Moreno, S., Pavone, F., Oliviero, S., Fesus, L., and Piacentini, M. (2002) “Tissue” transglutaminase ablation reduces neuronal death and prolongs survival in a mouse model of Huntington’s disease, *Cell Death Differ. 9*, 873–880.

192. Bailey, C. D. and Johnson, G. V. (2005) Tissue transglutaminase contributes to disease progression in the R6/2 Huntington’s disease mouse model via aggregate-independent mechanisms, *J. Neurochem. 92*, 83–92.

193. Kim, S.-Y., Grant, P., Lee, J. H. C., Pant, H. C., and Steinert, P. M. (1999) Differential expression of multiple transglutaminases in human brain, *J. Biol. Chem. 274*, 30715–30721.

194. Hadjivassiliou, M., Aeschlimann, P., Strigun, A., Sanders, D. S., Woodroofe, N., and Aeschlimann, D. (2008) Autoantibodies in gluten ataxia recognize a novel neuronal transglutaminase, *Ann. Neurol. 64*, 332–343.

195. Hadjivassiliou, M., Maki., M., Sanders, D. S., Williamson, C. A., Grunewald, R. A., Woodroofe, N. M., and Korponay-Szabo, I. R. (2006) Autoantibody targeting of brain and intestinal transglutaminase in gluten ataxia, *Neurology 66*, 373–377.

196. The Huntington’s Disease Collaborative Research Group (1993) A novel gene containing a trinucleotide repeat that is expanded and unstable on Huntington’s disease chromosome, *Cell 72*, 971–983.

197. Banfi, S., Chung, M. Y., Kwiatkowski, T. J. Jr., Ranum, L. P., McCall, A. E., Chinault, A. C., Orr, H. T., and Zoghbi, H. Y. (1993) Mapping and cloning of the critical region for the spinocerebellar ataxia type 1 gene (SCA1) in a yeast artificial chromosome contig spanning 1.2 Mb., *Genomics 18*, 627–635.

198. Sanpei, K., Takano, H., Igarashi, S., Sato, T., Oyake, M., Sasaki, H., Wakisaka, A., Tashiro, K., Ishida, Y., Ikeuchi, T., Koide, R., Saito, M., Sato, A., Tanaka, T., Hanyu, S., Takiyama, Y., Nishizawa, M., Shimizu, N., Nomura, Y., Segawa, M., Iwabuchi, K., Eguchi, I., Tanaka, H., Takahashi, H., and Tsuji, S. (1996) Identification of the spinocerebellar ataxia type 2 gene using a direct identification of repeat expansion and cloning technique, DIRECT, *Nat. Genet. 14*, 277–284.

199. Pujana, M. A., Volpini, V., and Estivill, X. (1998) Large CAG/CTG repeat templates produced by PCR, usefulness for the DIRECT method of cloning genes with CAG/CTG repeat expansions, *Nucleic Acids Res. 1*, 1352–1353.

200. Fletcher, C. F., Lutz, C. M., O’Sullivan, T. N., Shaughnessy, J. D. Jr., Hawkes, R., Frankel, W. N., Copeland, N. G., and Jenkins, N. A. (1996) Absence epilepsy in tottering mutant mice is associated with calcium channel defects, *Cell 87*, 607–617.

201. Vincent, J. B., Neves-Pereira, M. L., Paterson, A. D., Yamamoto, E., Parikh, S. V., Macciardi, F., Gurling, H. M., Potkin, S. G., Pato, C. N., Macedo, A., Kovacs, M., Davies, M., Lieberman, J. A., Meltzer, H. Y., Petronis, A., and Kennedy, J. L. (2000) An unstable trinucleotide-repeat region on chromosome 13 implicated in spinocerebellar ataxia: a common expansion locus, *Am. J. Hum. Genet. 66*, 819–829.

202. Holmes, S. E., O'Hearn, E., and Margolis, R. L. (2003) Why is SCA12 different from other SCAs? *Cytogenet. Genome Res. 100*, 189–197.

203. Imbert, G., Trottier, Y., Beckmann, J., and Mandel, J. L. (1994) The gene for the TATA binding protein (TBP) that contains a highly polymorphic protein coding CAG repeat maps to 6q27, *Genomics 21*, 667–668.

204. La Spada, A. R., Wilson, E. M., Lubahn, D. B., Harding, A. E., and Fischbeck, K. H. (1991) Androgen receptor gene mutations in X-linked spinal and bulbar muscular atrophy, *Nature 352*, 77–79.

205. Onodera, O., Oyake, M., Takano, H., Ikeuchi, T., Igarashi. S., and Tsuji, S. (1995) Molecular cloning of a full-length cDNA for dentatorubral-pallidoluysian atrophy and regional expressions of the expanded alleles in the CNS, *Am. J. Hum. Genet. 57*, 1050–1060.

206. Cooper, A. J. L., Sheu, K.-F. R., Burke, J. R., Strittmatter, W. J., Gentile, V., Peluso, G., and Blass, J. P. (1999) Pathogenesis of inclusion bodies in $(CAG)_n/Q_n$-expansion diseases with special reference to the role of tissue transglutaminase and to selective vulnerability, *J. Neurochem. 72*, 889–899.

207. Griffith, O. W., Larsson, A., and Meister, A. (1977) Inhibition of gamma-glutamylcysteine synthetase by cystamine: an approach to a therapy of 5-oxoprolinuria (pyroglutamic aciduria), *Biochem. Biophys. Res. Commun. 79*, 919–925.

208. Igarashi, S., Koide, R., Shimohata, T., Yamada, M., Hayashi, Y., Takano, H., Date, H., Oyake, M., Sato, A., Egawa, S., Ikeuchi, T., Tanaka, H., Nakanao, R., Tanaka, K., Hozumi, I., Inuzuka, T., Takahashi, H., and Tsuji, S. (1998) Suppression of aggregate formation and apoptosis by transglutaminase inhibitors in cells expressing truncated DRPLA protein with an expanded polyglutamine stretch, *Nat. Genet. 18*, 111–117.

209. Karpuj, M. V., Becher, M. W., Springer, J. E., Chabas, D., Youssef S., Pedotti, R., Mitchell, D., and Steinman L. (2002) Prolonged survival and decreased abnormal movements in transgenic model of Huntington disease, with administration of the transglutaminase inhibitor cystamine, *Nat. Med. 8*, 143–149.

210. Dedeoglu, A., Kubilus, J. K., Jeitner, T. M., Matson, S. A., Bogdanov, M., Kowall, N. W., Matson, W. R., Cooper, A. J., Ratan, R. R., Beal, M. F., Hersch, S. M., Ferrante, R. J. (2002) Therapeutic effects of cystamine in a murine model of Huntington's disease, *J. Neurosci. 22*, 8942–8950.

211. Lesort, M., Lee, M., Tucholski, J., and Johnson, G. V. W. (2003) Cystamine inhibits caspase activity. Implications for the treatment of polyglutamine disorders, *J. Biol. Chem. 278*, 3825–3830.

212. Huang, L., Haylor, J. L., Hau, Z., Jones, R. A., Vickers, M. E., Wagner, B., Griffin, M., Saint, R. E., Coutts, I. G., El Nahas, A. M., and Johnson, T. S. (2009) Transglutaminase inhibition ameliorates experimental diabetic nephropathy, *Kidney Int. 76*, 363–365.

TRANSGLUTAMINASE IN EPIDERMIS AND NEUROLOGICAL DISEASE OR WHAT MAKES A GOOD CROSS-LINKING SUBSTRATE

GUYLAINE HOFFNER
AMANDINE VANHOUTTEGHEM
WILLIAM ANDRÉ
PHILIPPE DJIAN

CONTENTS

Advances in Enzymology and Related Areas of Molecular Biology, Volume 78.
Edited by Eric J. Toone.

I. PHYSIOLOGICAL FUNCTIONS OF TRANSGLUTAMINASES

Transglutaminase catalyzes the formation of intermolecular isopeptide cross-links between glutamine and lysine residues of polypeptides [1, 2]. Because the cross-link is covalent and insensitive to proteolysis, polymers extensively cross-linked by transglutaminase are extremely resistant. Transglutaminase requires Ca^{2+} concentrations in the millimolar range in order to be active. Such concentrations are reached in extracellular compartments, but not inside the cell, where the concentration of free Ca^{2+} does not exceed a few micromolars. Therefore, cellular transglutaminase is latent and its activation requires either removal of the cell permeability barrier to extracellular Ca^{2+} or mobilization of the internal stores of Ca^{2+} present in organelles such as mitochondria. Both of these processes occur during cell death, whether programed or not. For instance, when keratinocytes reach the upper layers of the epidermis, they undergo a process of terminal differentiation in which the organelles are destroyed and the plasma membrane becomes permeable to Ca^{2+}. As a result, keratinocyte transglutaminase is activated [3]. As we shall see later, transient increases in Ca^{2+} concentration, sufficient for transglutaminase activation, can occur locally as part of the normal function of some cell types such as neurons. Transglutaminase may also be activated in cells subjected to forms of injury that are not lethal but that lead to an increase in Ca^{2+} concentration. In ethanol toxicity, hepatocytes acquire insoluble Mallory bodies consisting of cross-linked proteins [4, 5]. Hepatocytes containing Mallory bodies eventually succumb because of either persistence of the chemical injury or the effects of the cross-linked proteins.

There are eight transglutaminases known. Although erythrocyte band 4.2 is clearly paralogous to transglutaminases, it has lost all catalytic activity. Only three transglutaminases have a clear biological function: (1) factor XIII in blood coagulation, (2) transglutaminase-1 in the formation of the cornified envelope of epidermis, and (3) transglutaminase-4 in the synthesis of the copulatory plug of rodents. It is curious that the very abundant and widely distributed tissue transglutaminase (type 2) should have no obvious physiological relevance (Table 1). It has been proposed that the cross-linking activity of tissue transglutaminase was related to cell death, whether programed or due to toxic agents [6]. However, targeted inactivation of the mouse gene for transglutaminase-2 does not produce an obvious phenotype [7, 8]. Whether compensation by the other transglutaminases explains the lack of phenotype remains to be proven [8].

Factor XIII is a circulating transglutaminase that stabilizes the fibrin clot at the last step of coagulation [9]. The importance of factor XIII is shown by the fact that inactivating mutations cause severe bleeding in the human [10] and that disruption of the gene in mice causes death through massive hemorrhage [11, 12].

The coagulation of semen in a number of mammalian species results from the cross-linking action of prostatic transglutaminase on proteins secreted from the seminal vesicles [13]. The copulatory plug has been viewed as an "enzymatically fabricated chastity belt" [14], but this does not clarify its biological significance. It has been suggested that the purpose of the copulatory plug was to favor siring by dominant males, which, by producing the plug, would prevent subsequent mating of the female with other males. Although the prostate transglutaminase gene is expressed in the human, there is no appreciable cross-linking of seminal proteins in this species. In contrast to rodents, the function of transglutaminase-4 in the human, if any, is therefore unknown.

II. TRANSGLUTAMINASE AND THE FORMATION OF THE CROSS-LINKED ENVELOPE OF TERMINALLY DIFFERENTIATED KERATINOCYTES

A. THE CROSS-LINKED ENVELOPE

The plasma membrane of keratinocytes is unusual because on its cytoplasmic surface lies a rather uniform envelope about 12 nm in thickness [15, 16].

TABLE 1
Mammalian Transglutaminases and Erythrocyte Band 4.2

Gene	Chromosome (Human)	Protein Length (Amino Acids)	Tissue Distribution	Main Substrates	−/− Phenotype
F13A1	Chr6:5,989,514-6,189,722/	732	Platelets	Fibrin, fibronectin	Hemorrhage
Band 4-2	Chr15:41,276,718-41,300,615	691	Erythrocytes	No catalytic activity	Spherocytosis
TGM1	Chr14:23,788,160-23,802,256	817	Squamous epithelia	Envelope proteins	Lamellar ichthyosis
TGM2	Chr20:36,199,765-36,227,114	687	Ubiquitous	?	No obvious phenotype
TGM3	Chr20:2,224,613-2,269,725	693	Squamous epithelia	?	?
TGM4	Chr3:44,891,102-44,931,092	684	Prostate	?	?
TGM5	Chr15:41,312,667-41,346,347	720	Squamous epithelia	?	Peeling skin syndrome
TGM6	Chr20:2,309,554-2,361,399	706	?	?	?
TGM7	Chr15:41,355,771-41,381,745	710	?	?	?

This envelope, also known as the marginal band, is insoluble in very alkaline solutions and detergents, even in the presence of reducing agents [3, 15, 17]. It has been shown that the insoluble envelope of the upper layers of the epidermis and other stratified squamous epithelia was made of protein [18] assembled from soluble precursors [19] and cross-linked by transglutaminase [19–21].

B. TRANSGLUTAMINASES IN KERATINOCYTES

Keratinocytes contain transglutaminases-1, -2, -3, and -5, but only transglutaminases-1, -3, and -5 are thought to participate in envelope formation, because they are specific to the upper layers of the epidermis, whereas type 2 is found in all layers.

1. Transglutaminase-1 (Keratinocyte Transglutaminase)

In keratinocytes, nearly all transglutaminase activity is associated with the particulates, which include the insoluble proteins only extractable with detergents. This is in contrast to fibroblasts where transglutaminase activity is divided equally between the particulates and the cytosol [22]. Monoclonal antibodies raised to the particulate fraction have been shown to cross-react with one of the two transglutaminases of the cytosol of keratinocytes, but not with the other, presumably tissue transglutaminase [23]. Sequencing of the cDNA encoding particulate transglutaminase confirmed that it is distinct from tissue transglutaminase and the other transglutaminases [24]. The keratinocyte transglutaminase is bound to the plasma membrane, where it can presumably cross-link both cytosolic and membrane-bound proteins, and thereby catalyze the formation of the submembraneous cornified envelope. Binding of keratinocyte transglutaminase to the plasma membrane results from acylation of the enzyme by myristate and palmitate [25].

2. Transglutaminase-3 (Epidermal Transglutaminase)

The gene for transglutaminase-3 encodes a 77-kDa protein (Table 1). Cleavage by dispase of the 77-kDa enzyme greatly increases its specific activity by producing a very active 50-kDa *N*-terminal peptide containing the catalytic domain. It has been proposed that transglutaminase-3 exists as an inactive zymogen that becomes active upon proteolysis, and that the two peptides produced by proteolysis (50 and 27 kDa) associate noncovalently [26, 27]. However, in guinea pig skin, a specific antitransglutaminase-3 antibody stains a single protein with a molecular weight of 77 kDa and does not detect any proteolytic fragments [26]. Transglutaminase-3 is not specific to the skin since it is also found in the brain, the testes, the small intestine, and other tissues [28]. As of today, there is no publication on the knockout of the mouse transglutaminase-3 gene. However, one paper mentions that knockout of the transglutaminase-3 gene suppresses embryo implantation [29]. This would mean that the enzyme possesses an essential function unrelated to the epidermis. In the skin, transglutaminase-3 is specific to the upper layers of the epidermis and is cytoplasmic [30]. By producing recombinant human loricrin, a precursor of the cornified envelope (Table 2), Candi et al. showed that transglutaminase-1 and -3 treat the loricrin substrate differently: the cross-linking by transglutaminase-3 is mostly intrachain while that catalyzed by transglutaminase-1 is mostly interchain, thus resulting predominantly in the formation of multimers [48]. In vitro data show that recombinant small proline-rich repeat protein 1 or SPRR1, another envelope precursor (Table 2), is cross-linked first by transglutaminase-3 and then by transglutaminase-1 to form oligomers [42]. These studies suggest that transglutaminase-1 and -3 act in a complementary and sequential manner using different lysine and glutamine residues for cross-linking. Serum autoantibodies to transglutaminase-3 have been found in two diseases caused by gluten: (1) pediatric celiac disease [49] and (2) dermatitis herpetiformis. Dermopapillary deposits containing both IgA and transglutaminase-3 are characteristic of dermatitis herpetiformis and are likely to cause the disease [50–52].

3. Transglutaminase-5

The gene encoding transglutaminase-5 is located on chromosome 15q15.2 [53]. It encodes a protein of 720 amino acids (Table 1) with a molecular mass of 81 kDa [54]. Because the mRNA for transglutaminase-5 increases substantially in cultured keratinocytes induced to terminally differentiate, a function in the assembly of the envelope has been suggested [54]. In keeping with such a function, transglutaminase-5 is present in the upper layers of human epidermis, beginning in the spinous layer, and can cross-link efficiently recombinant loricrin, involucrin, and SPRR3, at least in vitro [55]. Treatment of cultured keratinocytes with high calcium results in a transient increase in transglutaminase-5 mRNA, in contrast to those encoding transglutaminase-1 and -3, which continue to accumulate. This shows that transglutaminases are differentially regulated during keratinocyte differentiation [56]. It has

TABLE 2
Envelope Precursors

Protein (Human)	Protein Length (Amino Acids)	Reactive Q(s)	Reactive K(s)	Location of Reactive Residue(s)	Reactive Qs in Q Repeat	Qs (%)	PolyQs in Protein	Ps (%)	PolyPs in Protein	Method of ID as Envelope Precursor	References
Cornifelin	112	?	?	?	?	**5.4**	None	**6.2**	None	a, b	[31, 32]
Desmoplakin	2871	Q_{2765}	K_{2778}, K_{2780}, K_{2786}	*C*-terminal	No	**7.6**	Q_3, 10xQ_2	1.9	None	b	[33, 34]
Elafin	57	Q_2, Q_{59}	K_6	*N*- & *C*-terminals	No	**5.1**	None	**9.4**	P_2	b	[33–35]
Envoplakin	2033	Q_{1970}, Q_{1973}	None	*C*-terminal	No	**9.7**	2xQ_3, 13xQ_2	4.3	4xP_2	b	[35–37]
Filaggrin	4061	Q_{2335}, Q_{2336}	none	Middle	Yes (Q_2)	**9.0**	19xQ_2	1.7	None	b	[33–35]
Involucrin	585	Q_{288}, Q_{465}, Q_{489}, Q_{495}, $\mathbf{Q_{496}}$	K_{468}, K_{485}, K_{508}	Toward *C*-terminal	Yes (Q_2 & Q_2)	**25.6**	5xQ_3, 34xQ_2	**7.5**	2xP_3, P_2	b, c	[34, 38, 39]
LCE1a	110	?	?	?	?	**8.2**	Q_2, Q_3	**13.6**	3xP_2, P_3	c	[40]
Loricrin	316	Q_4, Q_7, Q_{11}, Q_{154}, Q_{157}, $\mathbf{Q_{216}}$, $\mathbf{Q_{217}}$, Q_{220}, Q_{226}, Q_{306}, Q_{307}, Q_{309}	K_5, K_6, K_{89}, K_{308}, $\mathbf{K_{316}}$	In part *N*- & *C*-terminals	Yes (Q_2 & Q_2)	**4.5**	2xQ_2	2.9	P_2	b	[33, 41]
SPRR-1A	89	Q_4, Q_5, Q_6, Q_8, Q_{17}, Q_{18}, Q_{19}, Q_{20}, Q_{23}, Q_{26}, Q_{83}, Q_{84}, Q_{88}	K_7, K_{44}, K_{85}, K_{87}, K_{89}	In part *N*- & *C*-terminals	Yes (Q_2, Q_3 & Q_4)	**19.1**	Q_4, Q_3, Q_2	**30.3**	2xP_3	b, c	[33, 35, 42]

(Continued)

TABLE 2
(Continued)

Protein (Human)	Protein Length (Amino Acids)	Reactive Q(s)	Reactive K(s)	Location of Reactive Residue(s)	Reactive Qs in Q Repeat	Qs (%)	PolyQs in Protein	Ps (%)	PolyPs in Protein	Method of ID as Envelope Precursor	References
SPRR-2A	72	Q_4, Q_5, Q_6, Q_7, Q_{10}, Q_{13}, Q_{64}	K_9, K_{66}, K_{70}, K_{72}	*N*- & *C*-terminals	Yes (Q_4)	**16.7**	Q_4, $2xQ_2$	**38.0**	$3xP_3$, $4xP_2$	b, c	[33, 35, 43]
SPRR-3	169	Q_5, Q_6, Q_{17}, Q_{18}, Q_{19}, Q_{20}, Q_{168}	K_7, K_{22}, K_{165}, K_{167}, K_{169}	*N*- & *C*-terminals	Yes (Q_2 & Q_4)	**10.1**	Q_4, $2xQ_2$	**22.5**	$2xP_3$	c	[44]
Suprabasin	700	?	?	?	?	**9.6**	$2xQ_2$	1.0	None	c	[45]
S100A10	97	Q_4	None	*N*-terminal	No	**5.2**	None	3.1	None	c	[46]
S100A11	105	Q_{102}	K_3	*N*- & *C*-terminals	No	3.8	None	2.9	None	c	[47]

ID, identification.

Highly reactive Q and K residues are in bold type. Numbers in bold type indicate an above average content in either glutamine residues (Qs) or proline residues (Ps). Average content of mammalian proteins is 3.9% for Qs and 4.7% for Ps (see UniProtKB/Swiss-Prot protein knowledge base release 57.11).

a, coimmunoprecipitated with involucrin and loricrin; b, cross-linked peptides sequenced from digested cornified cell envelope; c, substrate of transglutaminase in vitro.

been suggested that, like transglutaminase-3, transglutaminase-5 existed as a zymogen, processed into a proteolytic fragment with increased specific activity [57].

4. Relative Importance of Transglutaminase-1, -3, and -5

Mutations inactivating human transglutaminase-1 produce lamellar ichthyosis [58, 59], due to the absence of cross-linked envelopes [60, 61]. Targeted disruption of the transglutaminase-1 gene in mice also causes the absence of envelopes, a severe disruption in barrier function and early lethality [62]. Therefore, it is clear that the cornified envelope is required for the protective function of the epidermis and that transglutaminase-1 is primarily responsible for the assembly of the cornified envelope since the other three transglutaminases present in keratinocytes (types 2, 3, and 5) do not compensate for its absence. Mice lacking transglutaminase-3 or -5 have not been reported and there is no description of a human disease due to loss-of-function mutations in transglutaminase-3. It is worth noting that cultured human keratinocytes can form cross-linked envelopes although they do not contain detectable transglutaminase-3 [27, 54]. Loss-of-function mutations of the human transglutaminase-5 gene produce acral peeling skin syndrome, a mild disease in which there is a shedding of the stratum corneum, limited to the back of the hands and feet [63]. Thus, the cross-linking activity of transglutaminase-5 is necessary to maintain the adhesion of the stratum corneum to the stratum granulosum. It is curious that lack of transglutaminase-5 activity should only affect the back of the hands and feet, and spare the rest of the epidermis, since the distribution of the enzyme is presumably similar in all body regions. Although cornified envelopes were not studied in patients with acral peeling skin syndrome, it appears unlikely that envelopes are significantly affected in view of the fact that this syndrome is much milder than lamellar ichthyosis and much more restricted in the body regions it affects.

III. SUBSTRATES OF KERATINOCYTE TRANSGLUTAMINASE

A. INVOLUCRIN: THE FIRST ENVELOPE PRECURSOR IDENTIFIED

Involucrin (from the Latin for envelope—involucrum) exists as a soluble precursor prior to the activation of the cross-linking and ultimately becomes incorporated into the cross-linked envelope [19, 21]. Involucrin is present only in enlarging cells undergoing terminal differentiation [64, 65]. Because human involucrin contains 38–42 repeats of a 10-amino acid sequence, over two-third of the protein is composed of repeats, each bearing three glutamine residues [66–69].

A large body of data supported the conclusion that involucrin was an essential component of the cross-linked envelope. The synthesis of involucrin is tightly regulated: it is restricted to terminally differentiated keratinocytes of stratified squamous epithelia. Human involucrin is a preferred substrate of transglutaminase-1, since after incubation of a crude extract of cultured epidermal cells in the presence of labeled putrescine (an amine-donor substrate of transglutaminase) and transglutaminase-1, involucrin is labeled at least 80 times more intensely than the average of the other cytosolic protein [22]. Involucrin has also been shown to promote the cross-linking of the particulate fraction, containing membrane proteins. Involucrin is rich in glutamine residues [66, 70], a property in keeping with its function as an amine acceptor in transglutaminase-catalyzed cross-linking; other envelope precursors such as loricrin, envoplakin, periplakin, or most of the SPRRs are not as glutamine-rich as involucrin (Table 2). For all these reasons, it is surprising that disruption of the gene causes no obvious phenotype and envelopes form normally in involucrinless mice [71]. The presence of these cross-linked envelopes in $inv^{-/-}$ mice clearly establishes that whatever properties might be conferred on envelopes by involucrin, this protein is not essential for the assembly of envelopes, which in its absence consist exclusively of other proteins. It remains to be seen whether more detailed examination of the skin of $inv^{-/-}$ mice will reveal a detectable difference from that of wild-type mice.

The use of the GenBank and Ensembl databases has made possible the identification of the involucrin genes of marsupials and birds, a discovery that could not have been made by previously available methods (Figure 1). The evolutionary persistence of involucrin and its continuing repeat additions are astonishing, in view of the fact that ablation of this gene in the mouse produces no detectable phenotype. This offers no support for an explanation of the evolution of involucrin based on natural selection [72].

B. OTHER SUBSTRATES OF KERATINOCYTE TRANSGLUTAMINASE

Since the amino acid composition of involucrin was shown to be different from that of cross-linked envelopes [19], it appeared likely that envelopes contained proteins other than involucrin. Indeed, a large number of envelope precursors have been discovered after involucrin. Their list and some of their

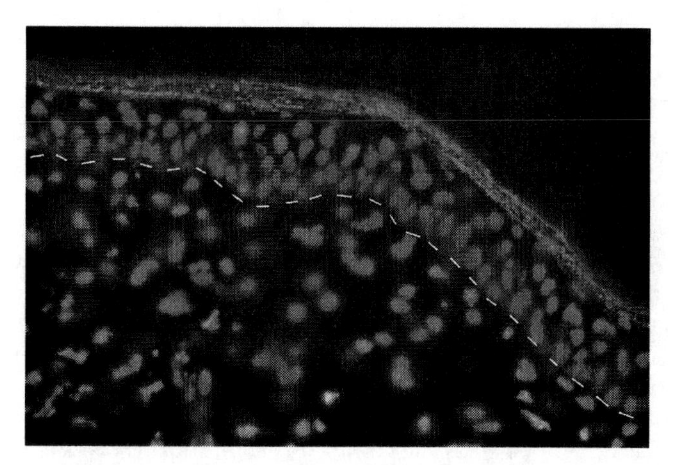

FIGURE 1. Detection of chicken involucrin with an antipeptide antibody. A section through *Gallus* skin was stained with an antiserum (red) prepared by Primm (Italy) and counterstained with DAPI (blue) for DNA. The cornified layer of the epidermis is strongly stained. The envelope precursor is concentrated peripherally. (See insert for color representation.)

properties are given in Table 2. Soon after involucrin was found, two other protein precursors of the cross-linked envelope were discovered by virtue of their ability to become nonextractable with detergents after transglutaminase activation. Both proteins were associated with the plasma membrane [36]. One had a molecular weight of 210 kDa and was later named envoplakin [37]. Another had a molecular weight of 195 kDa and was later named periplakin [73]. Lack of either envoplakin or periplakin causes no phenotype in mice [74, 75]. Triple knockout mice lacking involucrin, periplakin, and envoplakin can form cornified envelopes, although with reduced mechanical resistance [76].

Still other precursors were discovered, including filaggrin [77], SPRRs [78], loricrin [79], and late envelope proteins [40]. Profilaggrin is the major protein of keratohyalin granules present in keratinocytes of the granular layer. During terminal differentiation, profilaggrin is proteolytically cleaved into filaggrin peptides that aggregate keratin filaments [80]. Filaggrin has been shown to be an amine-acceptor substrate of transglutaminase and a component of the cornified envelope [81]. In 2006, Smith and colleagues identified two filaggrin mutations in the first repeat domain (nonsense and frameshift), both of which create stop codons and thereby prevent filaggrin synthesis. They showed that the semidominant alleles carrying these mutations are responsible for *ichthyosis vulgaris* [82, 83], and predispose to atopic dermatitis [84]. Although no filaggrin knockout mouse has been produced yet, a spontaneous filaggrin mutant mouse has been discovered. This loss-of-function mutant, called flaky tail, shows diffuse hyperkeratosis, but appears capable of forming cornified envelopes [85].

While studying the effect of UV irradiation on human keratinocyte gene expression, Kartasova and van de Putte identified cDNA clones encoding small proteins (8–10 kDa) rich in proline, glutamine, and cysteine [78], which they designated SPRR proteins. The cornifins, discovered later in the rabbit, are in fact the orthologs of the human SPRRs [86]. SPRRs are encoded by closely related members of a multigene family [87]. They are divided into four classes, composed in the human of two genes for SPRR1, seven genes for SPRR2, and one gene for SPRR3 and SPRR4. The SPRR proteins have a similar structure consisting of an *N*- and a *C*-terminal domain, both rich in glutamine and lysine residues, and a proline-rich central repetitive domain. They are specific to terminally differentiated keratinocytes [78, 87] and have been shown to be envelope precursors [88, 89]. Sequencing of peptides recovered from envelopes digested with proteases has shown that glutamine and lysine residues in the head and tail of SPRRs are cross-linked during the assembly of the envelope [33].

Hornerin is a 2496-residue protein first identified in the developing mouse skin [90]. The protein resembles profilaggrin since it comprises Ca^{2+} binding EF-hands at the *N*-terminus followed by a repetitive region rich in glycine (24%), serine (32%), and glutamine/glutamate (15%). Northern and western analyses have shown that the hornerin mRNA and protein appear at embryonic day 15.5 (E15.5) and increase towards E16.5. The expression profile is similar to that of profilaggrin. In the mouse, immunostaining of 3-day-old epidermis shows that hornerin is an abundant protein present in the keratohyalin granules of the granular and cornified layers [91]. This localization is similar in the other stratified squamous epithelia (oesophagus, tongue, forestomach). In the human, contradictory results have been obtained. Takaishi et al. reported that human hornerin was only present in regenerating and psoriatic skin but not in normal skin [92], whereas Wu et al. showed the presence of hornerin in the stratum granulosum and stratum corneum of normal human epidermis [93]. Because of its similarity with filaggrin and because its gene is closely linked with those encoding the envelope precursors, hornerin was initially considered as an envelope precursor. However, hornerin has never been shown to be a transglutaminase substrate.

Although loricrin has been thought to be a major component of the envelope [79], disruption of the loricrin gene does not prevent envelope assembly, but results only in transitory redness of skin immediately after birth [94, 95]. Mutations in the *C*-terminal part of human loricrin cause Vohwinkel syndrome, but the mutated alleles are dominant and the disease is likely to result from a gain of function [96, 97].

The late envelope proteins (LCEs) are encoded by a cluster of 18 genes. They are incorporated into the envelope after it is formed [40]. Deletion of the genes encoding LCE3B and LCE3C predisposes to psoriasis [98].

The genes encoding most envelope precursors (but not periplakin and envoplakin) are located in the region of human chromosome 1q21, the so-called EDC or epidermal differentiation complex [99]. This has been shown for involucrin [67], filaggrin [100], trichohyalin [101], loricrin [102], the SPRRs [78, 86, 88], the LCEs [40], and others [103]. Other proteins encoded in the EDC, such as cornulin [104], repetin [105], and hornerin [92] appear late in terminal differentiation, but are not likely to be envelope precursors. It has been proposed that since involucrin, loricrin, and the SPRRs have similar amino acid sequences in their *N*-terminal and *C*-terminal regions, they probably originated by successive duplications of a single ancestral gene, followed by divergence of each gene [106, 107].

It seems that while the keratinocyte transglutaminase is required in order to make the cross-linked envelope, none of the individual precursors is individually required. With the exception of the aforementioned triple involucrin/envoplakin/periplakin knockout [76], no mouse lacking multiple envelope precursors has been generated. The linkage of nearly all envelope precursor genes within the EDC, which is less than 2.5 Mb in length [108] precludes the generation of multiple knockouts from the crossing of individual knockout mice, since the frequency of recombination of the various null alleles would be too low to be of practical use.

Cystatin M/E is a 14-kDa inhibitor of cysteine proteases [109, 110] specific to the granular and cornified layers of epidermis. A null mutation in the mouse cystatin M/E gene causes harlequin ichthyosis, characterized by a disturbance in skin barrier function with hyperplastic, hyperkeratotic epidermis, and abnormal hair follicles. These mice die from dehydration a few days after birth [111, 112]. Although cystatin M/E acts in vitro as an acyl acceptor in transglutaminase-catalyzed cross-linking, cystatin M/E is unlikely to be an envelope precursor in view of the severity of the phenotype caused by its absence [113]. It has been suggested that lack of cystatin M/E increased the activity of epidermal proteases, such as legumain, thus leading to proteolytic cleavage of the relatively inactive transglutaminase-3 into its very active 50-kDa fragment. Unwanted cross-linking by transglutaminase-3 would explain the abnormal cornification of cystatin M/E-deficient mice [112]. This might be because transglutaminase-3 is diffused throughout the cytosol and therefore cannot form a regular envelope beneath the plasma membrane, as does the membrane-bound transglutaminase-1. Proteases and antiproteases appear to have important functions in epidermal differentiation since a number of human genetic diseases are caused by mutations therein. Such examples include the serine protease inhibitor Kazal-type 5 in Netherton syndrome [114], and cathepsins C and L in Papillon–Lefèvre syndrome [115] and the furless mouse [116], respectively. Human harlequin ichthyosis is not caused by lack of cystatin M/E [117], but by a null mutation in the lipid transporter ABCA12 [118].

Keratins have been found to be cross-linked in cornified envelopes [33, 119]. In contrast to keratins, which are found to some degree in other epithelial cell types, cross-linked envelopes are formed only in keratinocytes. The main function of keratins is in the formation of keratin filaments and not as precursors of the cornified envelope. Their abundance in keratinocytes might explain that they become incorporated to some degree in the envelope.

C. GLUTAMINE REPEATS: FROM INVOLUCRIN TO NEUROLOGICAL DISEASES

The amine-acceptor site of most transglutaminase substrates contains a segment of short glutamine repeats. Involucrin contains glutamine repeats, whose lengths vary in different mammalian lineages [70, 72, 120–122]. It has been proposed that the involucrin gene originated by repeated duplications of a single ancestral glutamine codon and therefore could function as a transglutaminase substrate since its origin [66]. The discovery that spinal and bulbar muscular atrophy [123] and Huntington's disease [124] both resulted from the expansion of a polyglutamine sequence in two unrelated proteins lead Howard Green to propose that the two proteins became substrates of the neuronal transglutaminase as a result of excessive glutamine reiteration [125]. It was later shown that all diseases of polyglutamine expansion were characteristically associated with the presence of insoluble neuronal aggregates [126, 127]. Since then a good deal of evidence has accumulated in support of the participation of transglutaminase in the stabilization of these aggregates. Transglutaminase has also been proposed to participate in the stabilization of the protein aggregates found in the brain of patients with Alzheimer's [128] and Parkinson's diseases [129]. However, the possibility of a new substrate is more difficult to consider in these two diseases, since the proteins that aggregate in Alzheimer's disease (tau and the β-amyloid peptide) and Parkinson's disease (α-synuclein) do not differ in their sequence from those of healthy persons. In these two diseases, one would have to postulate that the proteins are processed in a way that makes them better transglutaminase substrates or that activation of brain transglutaminase is part of the disease process.

Transglutaminase is found in all parts of the brain [130]. Neurons possess transglutaminases-1, -2, -3 [28], -6, and -7 [53], but transglutaminase-2 is by far the most abundant [131]. Although neuronal transglutaminase is mostly cytoplasmic, a significant fraction of the enzyme activity is nuclear [132].

The presence of glutamyl–lysine cross-links in normal brain shows that brain transglutaminase can be activated [131, 133, 134] and there might be several reasons for this. In the central nervous system, transient rises in Ca^{2+} concentration are physiological. Ca^{2+} concentration can increase locally more than two orders of magnitude above the micromolar concentration in the resting cytosol [135–137]. Other measurements have indicated that Ca^{2+} concentration in the vicinity of an open Ca^{2+} channel may be three orders of magnitude higher than normal cytosolic concentration, thus reaching optimal values for transglutaminase activation [138]. Because transglutaminase-2 and its regulator calmodulin have been found by immunostaining in nearly all the inclusions of Huntington's disease brain, it was suggested that calmodulin provides the Ca^{2+} necessary for in situ transglutaminase activation [139].

Alternative splicing of the transglutaminase-2 transcript results in the production of an enzyme isoform lacking the GTP binding site and therefore no longer negatively regulated by intracellular GTP. This short isoform has been hypothesized to participate in the protein aggregation observed in Alzheimer's disease and other neurological diseases; and indeed the mRNA encoding the short transglutaminase isoform has been found to be increased in the brain of patients with Alzheimer's disease [140–142].

IV. PROTEINS WITH ABNORMALLY LONG POLYGLUTAMINE AS SUBSTRATES OF NEURONAL TRANSGLUTAMINASES IN GENETIC DISEASES OF THE CENTRAL NERVOUS SYSTEM

A. COMMON PATHOGENIC MECHANISM OF DISEASES OF POLYGLUTAMINE EXPANSION

There are nine human diseases of the central nervous system, each associated with a different protein containing an expanded polyglutamine (Table 3): (1) spinal and bulbar muscular atrophy (SBMA) [123], (2) Huntington's disease [124], (3) dentatorubral–pallidoluysian atrophy (DRPLA) [159], (4) spinocerebellar ataxia (SCA) type 1 [160], (5) SCA type 2 [161–163], (6) SCA type 3 [164], (7) SCA type 6 [153], (8) SCA type 7, and (9) SCA type 17 [165]. The presence of a TATA-binding protein with an expanded polyQ is the cause of the very rare autosomal dominant cerebellar ataxia SCA17 [156, 166]. SBMA is caused by a polyglutamine expansion in the androgen receptor, SCA6 by an expansion in a calcium channel subunit and SCA7 by an expansion in a histone acetyl-transferase subunit. The functions of the other disease-causing proteins are still unclear and presumably unrelated to each other (Table 3). As we shall see below, it is probable that the normal functions of the disease-causing proteins have no direct bearing on the mechanism of the diseases.

The most frequent of the diseases of polyglutamine expansion is Huntington's disease: there are about 25,000 cases and 100,000 people at risk in the United States. In the causative proteins of all nine diseases, there is normally a polyglutamine sequence of about 20 glutamine residues. When the number exceeds some higher value (32 in some proteins, as high as 54 in others), the protein produces disease of the central nervous system (Table 3). Like expansion of microsatellites in intergenic DNA, expansion of coding polyCAG probably occurs by slipped strand mispairing/hairpin formation during DNA replication [167].

The fact that the diseases of expanded polyglutamine share many specific properties argues strongly for a common pathogenic mechanism. Common properties include the following: (1) autosomal dominant inheritance, with the exception of SBMA, which is linked to the X chromosome; (2) commonly late onset: the mean age of onset for Huntington's disease is 38 years; (3) all diseases tend to begin at an earlier age and be more severe with increasing polyglutamine length; (4) all diseases affect the central nervous

TABLE 3
Proteins Causing Neurological Disease When Containing an Abnormally Long Polyglutamine

Protein	Associated Disease	Normal Function of Protein	Protein Length (Amino Acids)	PolyQ Length	PolyP Adjacent to polyQ	Transglutaminase Substrate
Huntingtin	Huntington's disease	Essential but unknown [143]	3144	Normal range 6–38, mutated 38–180	P_{11} + P-rich region	Yes [139, 144–147]
Ataxin-1	SCA1	Learning and memory [148]	815	Normal range 6–44, mutated 39–83	No	Yes [149, 150]
Ataxin-2	SCA2	Lipid and glucose metabolism, fertility [151]	1313	Normal range 13–33, mutated 32–200	P_3 + P-rich region	Not tested
Ataxin-3	SCA3	Nonessential [152]	376	Normal range 12–41, mutated 54–89	No	Yes [149]
Ataxin-6	SCA6	Essential, Ca^{2+} channel [153, 154]	2505	Normal range 4–18, mutated 20–33	No	Not tested
Ataxin-7	SCA7	Histone acetyl-transferase [155]	892	Normal range 4–35, mutated 36–306	P_4 + P-rich region	Not tested
TATA-binding protein	SCA17	Essential, transcription [156]	339	Normal range 25–42, mutated 43-63	No	Not tested
Androgen receptor	SBMA	Transcription, sex determination [123]	919	Normal range 7–36, mutated 38-62	No	Yes [157]
Atrophin-1	DRPLA	Nonessential [158]	1185	Normal range 3–36, mutated 49–88	P_6 + P-rich region	Yes [149]

system, although the proteins are generally present in similar amounts in a large variety of unaffected tissues.

B. TOXIC GAIN OF FUNCTION RESULTING FROM ABNORMALLY LONG POLYGLUTAMINE

It appears very likely that the polyglutamine in itself, independently of its protein context, is toxic for neurons [168]. A polyglutamine inserted in a protein totally unrelated to any of the disease-causing proteins, the enzyme hypoxanthine phosphoribosyl transferase, produces disease of the nervous system in mice [169]. A transgene encoding an expanded polyglutamine surrounded by only 17 amino acids of huntingtin produces typical central nervous system disease in the mouse. It is very unlikely that such a fragment could have retained the function of a protein of over 3000 amino acids [170]. A toxic gain of function by the encoded protein would readily explain the dominance of the expanded allele.

Whether in human diseases or in their simulation in mice, the presence of the protein containing expanded polyglutamine is thought to be cell lethal and thus the cause of the extensive neuronal loss observed in the regions of the brain affected by the various diseases of expanded polyglutamine [171]. Progressive neuronal loss has also been observed in transgenic mice [172] and *Drosophila* [173] bearing an expanded polyglutamine.

What requires explanation in all of these diseases are the slowness of the lethality, which usually requires many years, and the selective toxicity of the disease-producing protein for neurons of certain regions of the central nervous system, when each of the protein is generally found in a variety of tissues. For instance, there is a large decrease in neuronal density within the caudate and cortex of patients with Huntington's disease [171], whereas no neuronal loss is observed in the cerebellum [174]. In transgenic animals, toxicity of the protein with expanded polyglutamine appears to be confined to neurons, in spite of a ubiquitous distribution of the proteins. The G-protein Rhes, which is largely specific to the striatum, has been shown to interact specifically with expanded huntingtin and to increase its toxicity [175].

C. NEURONAL AGGREGATES OF PROTEINS WITH EXPANDED POLYGLUTAMINE

Microscopic inclusions containing the protein with expanded polyglutamine are present in the neurons of the brain regions affected by each disease of polyglutamine expansion [126, 127, 176–183]. Inclusions can be either nuclear or cytoplasmic [127], but they tend to be predominantly nuclear in

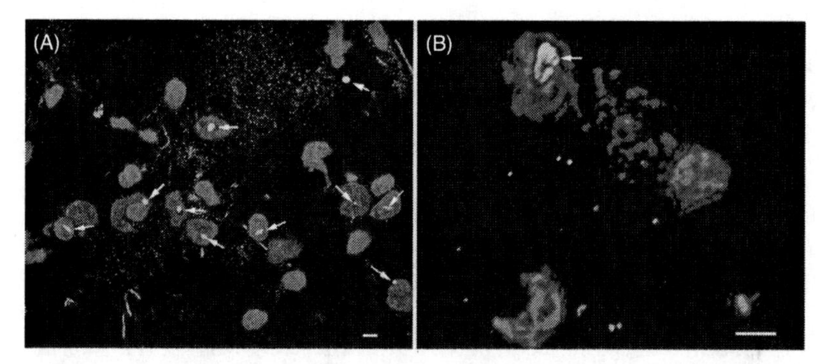

FIGURE 2. Neuronal inclusions in cerebral cortex of patient with juvenile Huntington's disease. Frozen sections were stained with an antibody directed against the *N*-terminus of huntingtin (green). (A) Nuclear and cytoplasmic inclusions (arrows) visualized with an LSM 510 confocal microsocope (Zeiss). DNA (red) was stained with TO-PRO-3. (B) Image of nuclear inclusion was obtained with a TE2000E microscope (Nikon). Spectral deconvolution of the image was performed using the ImageQuant software (GE Heathcare). DNA stained blue with DAPI. Bar: 5 μm. (See insert for color representation.)

the juvenile form of Huntington's disease, and mostly cytoplasmic in the adult form [184]. Presence of inclusions is linked to the toxic effect of the expanded protein, since inclusions occur in the most affected regions of the brain and are not found in less-affected regions, (Figure 2) and since the frequency of inclusions is correlated with polyglutamine length and severity of the disease [168, 185, 186]. Smaller huntingtin-containing aggregates are widespread in cortical dendrites and cortical dendritic spines of patients with Huntington's disease [187].

Because in Huntington's disease the expanded allele is dominant, patients are usually heterozygous and therefore should possess equal amounts of the normal and expanded protein. Western blot analysis has shown that in the most affected regions of the brain of patients with Huntington's disease, the discrete band of huntingtin containing the expanded polyglutamine is replaced by a broad band extending into the region of larger molecular weight, while the band of normal huntingtin is unchanged. In extracts of most cell types or of relatively spared regions of the central nervous system, the two forms of huntingtin are found in nearly equal amounts [144, 188]. Smearing of the expanded protein within the cortex does not result from mosaicism of the expanded allele and must therefore be the consequence of aggregation [189]. Filter retention assays on cellulose acetate membranes have demonstrated the presence of very large polymers of expanded huntingtin in the cortex, but not in the cerebellum of patients with Huntington's disease [189]. Large polymers of expanded polyglutamine are also found in transgenic mice bearing the first exon of the huntingtin gene with an expanded polyCAG [190].

D. PROTEIN AGGREGATION IN RELATION TO NEURONAL LETHALITY

Many experimental data support the view that aggregates of expanded polyglutamine are toxic. Aggregates of pure polyglutamine peptides formed in vitro and introduced into cultured cells produce cell death [191]. Neuritic aggregates in cultured striatal neurons containing expanded huntingtin block protein transport in neurites and cause neuritic degeneration [192]. It has been shown by various means that suppression of aggregate formation is correlated with suppression or alleviation of toxicity both in transgenic animals and in cell culture. Reduction of inclusion formation by overexpression of heat shock proteins, chaperones, IGF-1, ATP synthase, or components of the ubiquitin–proteasome system is associated with decreased cell death [193–197]. Monoclonal antibodies that inhibit aggregation of expanded huntingtin in cultured cells also inhibit cell death, whereas monoclonal antibodies that stimulate aggregation increase cell death [198]. A similar correlation between aggregation and cell death has been found in cultured cells, using either suppressor peptides or Congo red [199]. A strong correlation between formation of aggregates and cell death has been observed in *Drosophila* [200].

It has been argued that inclusions are not the cause of cell lethality and may even be protective [201, 202]. By various means, these investigators found absence of correlation between the presence of nuclear inclusions and cell death. Histological studies of the brain of patients with SCA1 and SCA3 have also provided arguments supporting a beneficial function of inclusions [203, 204]. All this evidence applies only to microscopically visible aggregates and does not disprove the toxicity of oligomeric or small polymeric forms of expanded huntingtin [168, 205, 206].

E. MECHANISM OF AGGREGATION OF PROTEINS CONTAINING EXPANDED POLYGLUTAMINE

1. *Polar Zippers*

Polyglutamine sequences of more than a few residues in length are highly insoluble because they form polar zippers stabilized by hydrogen bonds between their main chain and side chain amides [207, 208]. A short polyglutamine sequence (Q_{10}) incorporated into a small protein results in the formation of dimers and trimers [209]. A purified fragment of huntingtin containing 51 glutamines and a total of 120 amino acids forms insoluble filamentous aggregates in vitro [190].

The first evidence for polar-zipper formation by polyglutamine in cultured cells came from Hazeki et al. They showed that COS cells expressing the exon 1 of huntingtin, encoding an expanded polyQ, accumulated large aggregates, which could be reduced to monomer by concentrated formic acid [210]. Concentrated formic acid is an extremely effective solvent for otherwise insoluble proteins. It was later shown that the perinuclear inclusions formed in PC12 cells by an expanded polyQ flanked by the residues adjacent to the polyQ-containing region of the androgen receptor were mostly reduced to monomer after treatment with 96% formic acid [188].

Use of concentrated formic acid has provided direct evidence for the existence of noncovalent bonds (presumably polar zippers) in transgenic mice and the brain of patients. Formic acid releases a smear of monomeric *N*-terminal huntingtin fragments extending over a broad range of molecular weights from the brain of knockin mice bearing the entire huntingtin protein with Q_{150} [211, 212]. A smear of *N*-terminal fragments is also released from inclusions purified from the brain of patients with either juvenile or adult Huntington's disease [184].

2. *Transglutaminase-Catalyzed Cross-linking*

PolyQ-containing peptides as substrates for transglutaminase in vitro. As a model for polyglutamine-induced diseases, it was shown that a polyglutamine conferred excellent substrate properties on any peptide, as long as the peptides were rendered sufficiently soluble by the flanking residues. Lengthening the polyglutamine increased not only the reactivity of the peptide, but also the reactivity of each glutamine residue. Under optimal conditions, virtually all the glutamine residues acted as amine acceptors. Brain extract possessed sufficient enzyme to form the aggregates [149]. In a transglutaminase-catalyzed reaction, Q_{10} and Q_{62} fused to GST were later shown to accept putrescine much more efficiently than did dimethyl casein, a commonly used transglutaminase substrate. Under appropriate conditions, GST-Q_{52} incorporates nearly three times more putrescine than GST-Q_{33} [213]. Pure polyQs in dilute solution incorporate spermine in proportion to polyQ length [214]. These results established that a glutamine repeat, if exposed on the surface of a protein, should form cross-linked aggregates in the presence of any transglutaminase activated by Ca^{2+}.

These experiments were performed on model substrates containing polyglutamine but lacking most of the sequence of the disease-producing protein. It was later shown that full-length huntingtin was a substrate of transglutaminase in vitro and that the rate constant of the reaction increased strongly with the number of consecutive glutamine residues. As a result, huntingtin with expanded polyglutamine was preferentially incorporated into polymers [144]. Transglutaminase has been shown to polymerize to a greater extent an N-terminal fragment of huntingtin with Q_{67} than one with Q_{23} [145]. N-terminal fragments with Q_{148} have been found to be substrates of the three transglutaminases (types 1–3) present in brain [146].

Effect of overexpression or inhibition of transglutaminase in cultured cells and transgenic mice containing an expanded polyQ.

The treatment of cultured cells bearing an expanded polyglutamine with inhibitors of transglutaminase, such as cystamine, reduces aggregate formation and apoptotic cell death [215, 216]. When cultured cells are cotransfected with constructs expressing expanded polyglutamine and transglutaminase-2, respectively, the frequency of intracellular aggregates is increased, and this frequency is somewhat reduced by treating the cells with cystamine [215].

When transglutaminase-2 knockout mice are crossed with mice bearing a transgene consisting of the exon 1 of huntingtin encoding Q_{116} (R6/1 and R6/2 mice), the double mutants exhibit amelioration of the course of the disease, compared with the control animals bearing the transgene alone [217, 218]. The amount of brain isodipeptide, as measured by phenylisothiocyanate derivatization and high-pressure liquid chromatography, is greatly reduced in the double mutant [217]. In both studies, the double mutant shows a paradoxical (modest) increase in the frequency of inclusions. By using an in vitro solubility assay, it has been demonstrated that transglutaminase improves the solubility of a thioredoxine-Q_{62} fusion protein, presumably because the intermolecular cross-links introduced by the enzyme prevent polar-zipper formation [219]. Taken together, these results would mean that inclusions largely result from polar-zipper formation, but that the toxic species is represented by oligomers of expanded polyglutamine, whose formation is catalyzed by transglutaminase.

Three papers have reported that cystamine improves survival of the R6/2 mice, but with conflicting results on the inclusions [133, 220, 221]. Karpuj et al. did not observe a reduction of inclusions, but Dedeoglu et al. did.

The latter suggested that their earlier beginning of cystamine treatment (at 2 weeks of age instead of 7) was responsible for better alleviation of the disease and clear reduction in size and number of inclusions. Wang et al. observed a 50% decrease in the size of the striatal inclusions after 5 weeks of treatment with cystamine [221]. Oral administration of cystamine to symptomatic YAC128 mice, which express the entire huntingtin gene with Q_{128}, prevents most of the striatal neuronal loss, but does not improve motor function [222].

The role of transglutaminase in the aggregation of huntingtin with expanded polyglutamine has been challenged. Human neuroblastoma cells, stably transformed with a transglutaminase-2 antisense construct, form inclusions when they are transiently transfected with an expression vector encoding an N-terminal fragment of huntingtin containing Q_{82} [223]. Since these cells have lost all transglutaminase activity [224], their inclusions must be ascribed to polar-zipper formation. Bailey and Johnson have crossed R6/2 mice with transglutaminase knockout mice. The fact that cystamine extends lifespan and delays the onset of motor dysfunction to a similar extent in R6/2, transglutaminase-$2^{+/+}$ and in R6/2, transglutaminase-$2^{-/-}$ mice suggests that the inhibition of transglutaminase is not the primary mechanism of action of cystamine in R6/2 mice [225]. Nevertheless, it must be borne in mind that neurons contain transglutaminases other than transglutaminase-2.

Isodipeptide and transglutaminase in aggregates found in Huntington's disease's brain. After separation by liquid chromatography and electrochemical detection, the concentration of the glutamyl–lysyl isopeptide was found to be increased by about threefold in the cerebrospinal fluid of patients with Huntington's disease and by about fivefold in the cortex and caudate [226]. In this study, the cerebrospinal fluid of patients with Huntington's disease was compared with that of patients affected by stenosis of the spine. Using as controls patients with extensive neuronal loss resulting from causes unrelated to Huntington's disease would have clarified whether the increased level of isodipeptide was the consequence or the cause of the increased neuronal death. The neurons of the anterior horn of the spinal cord of transgenic mice carrying a full-length androgen receptor with Q_{112} were found to contain isopeptide cross-links by immunostaining [157]. Nearly complete colocalization of isopeptide cross-links with nuclear aggregates of huntingtin was observed in the frontal cortex of patients with Huntington's disease [147]. As the specificity of the monoclonal antibody used for detection has been questioned, this demonstration cannot be regarded as definitive [227].

Transglutaminase activity in Huntington's disease. A number of studies have found an increase in transglutaminase activity during the course of Huntington's disease. Both the mechanism of this increase and whether it is a consequence or a cause of the disease is still obscure. Since transglutaminase-3 and -5 are both thought to exist in cells as zymogens, it could be hypothesized that their cleavage by the proteolytic enzymes liberated during neuronal death may explain the increase in transglutaminase activity. Transglutaminase activity of cortex and striatum is increased two- to fourfold in Huntington's disease [145, 146, 228]. In one study, transglutaminase activity of affected brain regions increased with the grade of the disease [228]. The results in cerebellum were contradictory since Lesort et al. [228] found no increase in cerebellum, compared to controls, whereas Karpuj et al. [145] measured a fourfold increase. Results in lymphocytes were also contradictory [145, 229].

Tissue transglutaminase is predominantly cytoplasmic, but the inclusions associated with Huntington's disease are frequently nuclear [127, 230]. Although human neuroblastoma nuclei contain only 7% of the transglutaminase protein in the cell, the nuclear enzyme accounts for about half of the total enzyme activity. When intracellular Ca^{2+} levels are increased, some active transglutaminase appears to translocate to the nucleus [132].

In summary, it is clear that a substantial fraction of transglutaminase is located in the nucleus, where it can be activated by Ca^{2+} and thus potentially form nuclear aggregates. It is less clear whether the increased activity observed in cortex and striatum is related to the disease process, since in some cases this increase is also found in cerebellum and lymphocytes.

The aggregates characteristic of Huntington's disease may be stabilized by both hydrogen and isopeptide bonds. The use of concentrated formic acid has provided direct evidence as to the aggregation through polar zippers of *N*-terminal fragments of expanded huntingtin in the cerebral cortex of patients with Huntington's disease. The presence of covalent bonds in the inclusions is also suggested, albeit indirectly, by experiments with formic acid. These experiments have demonstrated that the cortical nuclei of patients with Huntington's disease contain a formic acid-resistant polymer that does not enter polyacrylamide gels and is too large to pass through a 0.2-μm filter. The resistance of the polymer to 96% formic acid suggests that it is stabilized by covalent bonds, presumably introduced by transglutaminase. Such a polymer is absent from cerebellum. We may conclude that both hydrogen and isopeptide bonds are likely to stabilize the cortical inclusions, but the relative abundance of the two types of bonds remains to be determined [184, 188].

In summary, it could be stated that although much evidence supports the role of transglutaminase in the aggregation of expanded polyglutamine, the evidence falls short of conclusive proof. What is lacking in order to prove unequivocally a role for transglutaminase is the demonstration of covalently cross-linked polymers containing isopeptide bridges. A very specific and sensitive method of detection of the glutamyl–lysyl cross-link has been recently described. This method relies on derivatization of the isopeptide, separation by liquid chromatography, and identification by multiple mass spectrometry [134, 231]. Analysis by this method of aggregates purified from Huntington's disease brain should provide definitive evidence of the participation of transglutaminase in polyglutamine aggregation.

V. TRANSGLUTAMINASE IN ALZHEIMER'S DISEASE

Alzheimer's disease is characterized by the deposition in the brain of extracellular senile plaques, the formation of intraneuronal neurofibrillary tangles, and neuronal loss. Selkoe et al. proposed that neurofibrillary tangles were cross-linked by transglutaminase. This hypothesis was based on the fact that paired helical filaments (PHF), the main structural component of neurofibrillary tangles, were insoluble in 70% formic acid and therefore presumably stabilized by covalent bonds [128], and that neurofilament proteins, then thought to be the main component of neurofibrillary tangles, could be cross-linked by transglutaminase in vitro [232]. However, it was later discovered that the main component of the neurofibrillary tangles was not neurofilament, but an abnormally phosphorylated form of the microtubule-associated protein tau [233, 234]. Tau has been shown to be an excellent substrate of tissue transglutaminase in vitro [235–239], and tissue transglutaminase colocalizes with hyperphosphorylated tau in the neurofibrillary tangles [240, 241].

The extracellular senile plaques, which are mostly composed of β-amyloid peptide, were initially presumed to be stabilized by noncovalent bonds since they are soluble in solution containing SDS [242, 243]. The implication of transglutaminase in the formation of the senile plaques became plausible when it was discovered that senile plaques contain immunodetectable transglutaminase, and that both the β-amyloid precursor protein and the β-amyloid peptide are good substrates of transglutaminase in vitro

[236, 244–246]. For self-aggregation to occur in vitro, a high concentration of the amyloid β-protein is required. Transglutaminase lowers the threshold of oligomerization of the amyloid β-protein from the high concentration required for self-assembly in vitro to concentrations approaching those found in brain. Interchain cross-linking could bring hydrophobic regions in close proximity and thus promote oligomerization [247].

Incubation of retinoic acid-differentiated neuronal SH-SY5Y cells with β-amyloid$_{1-42}$ induces cell death. The neurotoxicity is decreased by cotreatment of the cells with monodansylcadaverine, a competitive inhibitor, and increased by overexpression of catalytically active transglutaminase [248].

Transglutaminase activity of prefrontal cortex has been found to be increased by approximately threefold in Alzheimer's disease compared to age-matched controls and to Alzheimer's disease cerebellum, a region usually spared in Alzheimer's disease and where neurofibrillary tangles are not detected [249]. Other studies have found that transglutaminase activity and transglutaminase mRNA are two- to eightfold higher in Alzheimer's disease brain than in normal brain [131, 142].

In one study, liquid chromatography was used to quantify the free isodipeptide in the cerebrospinal fluid of patients with Alzheimer's disease and normal controls: a fivefold difference was found [250]. All other studies of cross-links in Alzheimer's disease brain relied on the same monoclonal anti-isopeptide antibody whose doubtful specificity has already been mentioned [227]. Isopeptide cross-links in PHF tau and some neurofilament proteins were found to be increased in Alzheimer's disease [251], even at the earliest stages [252]. Other studies showed the accumulation of cross-links in insoluble material [131] and in ubiquitinated particles found in the brain of patients with Alzheimer's disease. These particles contain both tau and neurofilament proteins [253]. Corpora amylacea are spherical bodies frequently observed in Alzheimer's and Parkinson's diseases. It has been proposed that transglutaminase-catalyzed cross-linking and consequent polymerization of cytoskeletal and cytoskeleton-associated proteins may underlie the formation of corpora amylacea [254].

Progressive supranuclear palsy resembles Alzheimer's disease by the deposition of aggregated tau protein in brain, and Parkinson's disease by its clinical presentation, which includes rigidity, akinesia, and postural instability. Transglutaminase activity, protein, and mRNA levels have been shown to be increased in affected brain regions [255]. Colocalization of glu–lys isodipeptide and PHF has been observed in the pons of patients with progressive supranuclear palsy [256].

VI. TRANSGLUTAMINASE IN PARKINSON'S DISEASE

Parkinson's disease is characterized by the loss of dopaminergic neurons in the substantia nigra and the aggregation of α-synuclein and other proteins in cytoplasmic Lewy bodies. The involvement of transglutaminase in the polymerization of α-synuclein and the formation of Lewy bodies has been suggested. Tissue transglutaminase catalyzes the formation of α-synuclein aggregates in vitro and in cultured cells. Treatment with cystamine of cultured cells overexpressing transglutaminase and α-synuclein reduces the formation of aggregates containing α-synuclein [129]. In older mice with a toxin-generated parkinsonian syndrome, low doses of cystamine are neuroprotective and prevent dopaminergic cell loss [257]. Transglutaminase mRNA, protein, and enzyme activity are higher in the substantia nigra of Parkinson's disease brain than in that of normal brain [258, 259]. The halo of Lewy bodies is stained by antibodies directed against the glu–lys isopeptide, transglutaminase, and α-synuclein [129].

The substantia nigra of patients with Parkinson's disease contains α-synuclein monomers and oligomers that can be immunoprecipitated with the antiglu–lys isopeptide antibody. The level of cross-linked monomer is correlated with disease progression, whereas that of the uncross-linked monomer is inversely correlated. It was concluded that intramolecular cross-linking of α-synuclein constitutes an early step in Parkinson's disease, preceding the aggregation of α-synuclein within the Lewy bodies. Therefore, transglutaminase might be involved in the early stages of the disease, not by promoting intermolecular polymerization of α-synuclein, but by promoting intramolecular cross-linking of the monomers, which may impair normal α-synuclein function [259]. Others have suggested that intramolecular cross-linking of α-synuclein prevents the formation of oligomeric toxic species and may therefore be beneficial [260]. The glutamine and lysine residues of α-synuclein that participate in intramolecular cross-linking have been identified [261].

VII. WHAT MAKES A GOOD TRANSGLUTAMINASE SUBSTRATE?

A. IMPORTANCE OF THE HIGHER ORDER STRUCTURE

Soon after transglutaminase was discovered, it was noted that some proteins might serve as amine acceptors, whereas others incorporated insignificant quantities of amine [1]. It was later found that native hemoglobin was not a transglutaminase substrate, but that heat denaturation of hemoglobin made it a good substrate, at least at certain glutamine positions. These results showed the importance of both the conformation and the primary sequence around the reactive glutamine [262]. Since then, it has been observed that transglutaminase is frequently quite specific with respect to the glutamine residues it will attack within a given protein [38, 263]. However, since no consensus sequence was identified either around the reactive glutamines or the reactive lysines, the selectivity of the active site was attributed to the higher order structure of the protein substrate, rather than to the influence of the residues neighboring the active residue [264]. More recent evidence has shown that the primary sequence around the reactive site is also important in the modulation of the reactivity of the substrate. One example of a natural substrate of transglutaminase whose reactivity is largely governed by its higher order structure is human involucrin. Although involucrin possesses nearly 150 glutamine residues, one (residue 496) is highly preferred by the enzyme as the amine acceptor. The primary sequence around the preferred glutamine residue is unlikely to explain the reactivity, since, owing to the repeated nature of the protein, glutamine residues with neighboring amino acids very similar to those neighboring residue 496 are present and do not react. When the *N*-terminal half of the molecule is deleted, an additional glutamine (residue 573) becomes as heavily labeled as residue 496. When the *N*-terminal half of involucrin is used as substrate, at least ten glutamine residues become reactive, none of which shows appreciable reactivity in intact involucrin. Remote amino acid residues could influence the reactivity of the molecule either by altering protein folding or by participating in binding of the enzyme at a location remote from the active site [38].

Fibrin has provided additional arguments as to the importance of the higher order structure of transglutaminase substrates. Mutagenesis of residues on either side of the active-site cysteine of factor XIII results in substantial loss of the enzyme activity without altering the ability of factor XIII to bind fibrin. This suggests that the sites used by the enzyme for substrate binding are different from those required for catalytic activity and therefore that the reactive residues of the substrate could also be different from those necessary for its binding to the enzyme [265]. In fibrin, the reactive glutamine and the reactive lysine are both located at the *C*-terminus, a generally exposed region [266]. A tendency of reactive glutamines and lysines to be located in intrinsically disordered regions has been recently suggested [267].

B. PRIMARY SEQUENCES AROUND REACTIVE GLUTAMINE RESIDUES

The first step in the transamidation reaction is the acylation of the cysteine at the active site of the enzyme by a protein-bound glutamine residue, resulting in the formation of a thioester intermediate between the transglutaminase and the reactive glutamine of the substrate. Because the glutamyl-enzyme intermediate could react with almost any nucleophilic amine, and in the absence of amine reacts with water to form glutamic acid, transglutaminase has been thought to have little selectivity toward amine donors, in contrast to amine acceptors [268, 269]. However, the use of labeled glutamine-containing oligopeptides has demonstrated the selectivity of transglutaminase for amine-donor peptides [270].

1. In Natural Substrates

It has long been known that in certain small proteins, most or all scattered glutamine residues may act as amine acceptors, at least in the absence of a higher order structure, preventing access of the enzyme [262, 271]; but in native proteins, the nature of the neighboring residues has appreciable influence on the reactivity of the glutamine residue and some glutamines are greatly preferred to others. A highly preferred glutamine has been found in natural substrates of transglutaminase-1 [38], transglutaminase-2 [272–275], and factor XIII [276]. Comparison of the reactive sites of these substrates shows that among preferred glutamine residues are ones adjacent to a second glutamine residue. This is observed in fibrin, in crystallin βA3, and in involucrin. The comparison of a substantial number of natural sites reactive vis-à-vis either transglutaminase-2 or factor XIII did not allow Coussons and collaborators to define a consensus sequence around the reactive glutamine. However, they could identify some features that were always absent from reactive sites and therefore presumably prevented the interaction of a protein with the enzyme. Among these are positively charged residues in close proximity to the glutamine, particularly *C*-terminal to it [277]. This conclusion does not apply to the tau protein since two of the eight amine-acceptor sites of tau are adjacent to residues with a basic side chain [278].

Studies of substance P, an 11-amino acid neuromediator and a substrate of transglutaminase-2 in vitro [279], have demonstrated the importance of the position of the glutamine residue. Substance P contains two consecutive glutamines in its central region. Deleting one of the glutamines and altering the position of the remaining glutamine do not affect the reactivity of the glutamine. Reactivity is suppressed only when the single glutamine is placed at the end of the peptide, whether it is the *N*- or *C*-terminus [280]. Reactive glutamine residues are very often located close to but not at the *N*- or *C*-terminus in natural substrates [267, 269].

2. *In Synthetic Peptides*

Systematic replacements or deletions of the amino acids surrounding the active glutamine residue have demonstrated that certain amino acid residues surrounding the active glutamine were more influential than others in directing the action of transglutaminase. However, no obvious consensus sequence around the glutamine donor sites has emerged [281–283]. It has been demonstrated that polyglutamine confers excellent substrate properties to any polypeptide containing it as long as the polyglutamine is rendered sufficiently soluble by the flanking residues. In these studies, polyglutamine flanked solely by arginine residues was an outstanding substrate for transglutaminase. It was concluded that any protein possessing a glutamine repeat exposed on its surface may act as a substrate for transglutaminase [149]. Heptapeptides composed of a glutamine residue flanked by three glycine residues are excellent substrates for transglutaminase-2 [284].

3. *In Gliadin Epitopes*

Much information has been obtained from the study of celiac disease, an autoimmune disease of the small intestine associated with the HLA-DQ2 allele. Celiac disease is caused by ingestion of the gliadin protein present in wheat gluten [285]. Transglutaminase-2, either released by intestinal cells or at the brush border, plays a central role in the pathogenesis by being the target of autoantibodies [286] and by generating new epitopes through deamidation of specific glutamine residues in gliadin peptides. Deamidated peptides are thought to be recognized by gliadin-specific, HLA-DQ2-restricted T-cells [287–289]. The propensity of transglutaminase-2 to catalyze deamidation of glutamine rather than its transamidation with a lysyl donor may be the result of the slightly acidic environment in the intestine: a pH below 6.5 favors water as acyl acceptor in the transglutaminase-catalyzed reaction [290]. It has also been proposed that transglutaminase-2 could hydrolyze isodipeptide bonds, thus resulting in an indirect deamidation [291–293]. Another contributor to celiac disease may be the microbial transglutaminase used by the food industry to improve the texture of protein-rich products, such as meat, fish, or pastries, since the microbial transglutaminase can generate deamidated gluten peptides identical to those generated by transglutaminase-2 [294].

Gliadin is extremely glutamine-rich and has been shown to be an excellent transglutaminase substrate. Among gliadin peptides, the most inflammatory epitopes are also the preferred substrates. Immunogenic gliadin peptides of wheat, rye, and barley are high-affinity substrates, whereas their non-toxic homologs found in rice, corn, and oat are much poorer substrates [295–297]. An exceptionally immunoreactive 33-residue peptide, resistant to intestinal proteases, has the highest specificity of all gliadin peptides toward transglutaminase-2-catalyzed deamidation [288].

A study of the deamidated epitopes recognized by $CD4^+$ lymphocytes revealed that the epitopes are not scattered across the gliadin molecule but tend to be located in distinct proline-rich clusters [298]. In most cases, the spacing between the target glutamine and the proline was QxP (where x represents a variable amino acid), rather than QP or QxxP [296]. Systematic investigation of the effect of amino acid replacements on the patterns of synthetic gliadin peptides directly deamidated by transglutaminase at acidic pH confirmed the strong effect of the spacing between the reactive glutamine and the immediately *C*-terminal proline [290, 299]. Large hydrophobic residues at position +3 also increased reactivity [290].

Indirect deamidation through transglutaminase-catalyzed hydrolysis of the isopeptide bond seems also to be favored by the presence of a proline residue at position +2 [293]. Therefore, the presence of a proline residue at +2 may promote both direct and indirect deamidation by transglutaminase. Proline residues may also be important because they promote binding of gliadin peptides to HLA-DQ2 [292] and increase the resistance of the gliadin peptides to degradation by the gut proteolytic enzymes (see below).

4. *In Peptides Identified by Phage Display*

Recently, two groups have reported the use of phage-display peptide libraries to study the amino acid sequence around reactive glutamines [300, 301]. The findings of the two groups are summarized in Table 4. Sugimura et al. screened a random M13 library displaying 12-residue peptides on phage coat protein by incorporation of biotinylated cadaverine under the catalytic action of either transglutaminase-2 or factor XIII. The phage clones that had incorporated substantial amount of the amine were purified by virtue of their affinity for avidin. The amino acid sequence of the displayed peptides was deduced from the encoding nucleotide sequence.

TABLE 4
Consensus Sequences of Amine Acceptor Sites Identified by Phage Display

TGase Type	Preferred Sequence Around Reactive Q	Peptide with Highest Reactivity and Specificity	References
TG1	Qx(K,R)øxxxWP	YEQHKLPSSWPF	[302]
TG2	Qx(P,Y)øD(P)	HQSYVDPWMLDH	[267, 300, 301]
	pQx(P,T,S)l	GQQQTPY	
		GLQQASV	
		WQTPMNS	
Factor XIII	Qxxø(xWP)	DQMMLPWPAVKL	[300]
Microbial TGase	⊗xLQRP. . .Y		[303]

X, an amino acid residue; Ø, hydrophobic residue; ⊗, aromatic residue; p, polar residue; l, aliphatic residue.

Several features common to the reactive sites preferred by transglutaminase-2 were thus identified: (1) the reactive glutamine residue is often located close to the *N*-terminus and (2) half of the clones displayed sequences containing QxP (where x represents any amino acid). More precisely, four groups of clones were found: (1) QxPøD(P) (where ø represents a hydrophobic amino acid), (2) QxPø, (3) QxxøDP, and (4) sequences with no apparent motif. This result agrees with the above-mentioned results on gliadin peptides, whose preferred transglutaminase-catalyzed deamidation site is QxP [296]. Although factor XIII like transglutaminase-2 preferred glutamines close to the *N*-terminus, its sequence requirements proved different since QxP was rarely found. However, as for transglutaminase-2, a hydrophobic residue was often present at +3 from the glutamine and the sequence preferred by factor XIII was QxxøxWP [300]. The validity of the results was confirmed by the demonstration that insertion of the most reactive sequences into a modified GST protein conferred very good substrate properties to the otherwise unreactive GST. The reactivity of the GST fusion peptides was found to be higher than that of peptides derived from known transglutaminase reactive sites such as those of fibrinogen and α2-plasmin inhibitor. It is worth noting that the site of cross-linking of human and bovine fibrin by factor XIII is QQHHL [276] and therefore does not conform to the sequences identified by phage display, since it does not possess a hydrophobic residue at position +3.

The same group has applied phage display to microbial transglutaminase [303] and transglutaminase-1 [302]. The sequences preferred by the microbial enzyme are quite different from those preferred by either transglutaminase-2 or factor XIII, since they contain: (1) an aromatic residue at position −3 relative to the glutamine; (2) leucine, arginine, and proline residues at positions −1, +1, and +2, respectively; and (3) a tyrosine at the *C*-terminal side [303]. In the case of transglutaminase-1, most of the sequences contained either QxK or QxR, and generally the consensus motif QxK/RøxxxWP. Closely related motifs lacking the hydrophobic residue are present in two natural substrates of transglutaminase-1: (1) loricrin (QQKQAPTWP) and (2) involucrin (QQKQEVQWP). The motif in loricrin does represent one of the cross-linking sites of loricrin [48] but that in involucrin does not, since the cross-linking site of intact involucrin is QQVGQPKHL [38].

In another study, Keresztessy et al. screened a phage library displaying random heptapeptides by using direct binding of the peptides to GST-immobilized transglutaminase-2 [301]. Peptides were then eluted with the synthetic amine-donor substrate 5-(biotinamido)pentylamine. After several rounds of amplification and selection, 26 peptides were obtained, and pQx(P, T, S [in order of frequency])l (where p represents a polar residue, x any amino acid residue, and l an aliphatic residue) emerged as the dominant sequence (Table 4). This sequence agrees with the QxPø consensus sequence identified by Sugimura et al. A protein database search using the three most highly reactive peptides (GQQQTPY, GLQQASV, and WQTPMNS) identified about 60 proteins, of which 14 were known transglutaminase-2 substrates. However, since virtually none of the reactive sites of these 14 proteins is known, the value of phage display for predicting reactive sites cannot be assessed. Because phage display relies solely on the primary sequence, it takes into account neither the necessary exposure of the reactive glutamine on the surface of the molecule nor the importance of residues remote from the reactive site for the binding to the enzyme. Natural cross-linking sites may not necessarily be the best possible sequences for catalysis. The cases of gliadin and huntingtin with an expanded polyglutamine both show that being an outstanding substrate for transglutaminase is not necessarily advantageous. Long glutamine repeats, which are excellent transglutaminase substrates, are absent from natural proteins, presumably because of their deleterious effect [304].

In summary, no obvious consensus around reactive glutamine residues can be defined. The presence of contiguous glutamines is often found in the active site of natural and artificial substrates [149, 301]. This feature has been used to synthesize highly reactive peptides of pharmacological use [305]. The main contribution of phage display appears to be in the identification of sequences that are excellent and selective substrates of transglutaminase and could be of practical use, as in the synthesis of inhibitors specific for a given transglutaminase.

C. PRIMARY SEQUENCES AROUND REACTIVE LYSINE RESIDUES

The initial work on the amine specificity of the enzyme showed that most compounds that possess a primary amino group can be incorporated into proteins, although to somewhat different degrees [306]. It was then demonstrated that the amine is bound to the acyl-enzyme intermediate and does not simply compete with water in a nucleophilic displacement reaction [307]. The presence of a specific amine-binding site on the enzyme implies that the residues neighboring the reactive lysine could influence the affinity of the peptide for the enzyme. Gross and Folk were the first to study the sequence requirements of peptide-bound lysyl donor sites. They showed the absence of appreciable differences in the substrate properties of an aliphatic amine and those of lysine residues in a polypeptide in which three glycine residues flank the lysine residue (N^{α}-acetyl-Gly_3-Lys-Gly_3). However, they noted that a leucine residue flanking the amino side of the reactive lysine increases the reactivity of the lysine residue [308].

Other studies have shown that not all peptide-bound lysines are equal as transglutaminase substrates. Bovine βA3-, βA4-, βB1-, and αB-crystallins are selective amine donors for transglutaminase [309, 310]. Although βA3- and αB-crystallins each possesses ten lysine residues, a single lysine residue acts as amine donor in each of the two proteins [309, 311]. The number of reactive lysines is 1/5 in β-endorphin [312], 4/20 in vimentin [313], 1/26 in keratin 1, 1/25 in keratin 2e [33], and 1/10 in both S100A11 and bovine pancreatic RNase A [46, 314]. Therefore, it appears that a single lysine is often highly preferred by transglutaminase and, in this respect, lysine does not differ from glutamine [311, 314].

Mutagenesis in recombinant proteins and amino acid replacements in synthetic peptides have been used to study the influence of the position of reactive lysine residues and of the primary sequence around them. The tendency of reactive lysines to be located in the *C*-terminal region has been noticed [267, 315, 316]. The introduction of a lysine at the *C*-terminal or penultimate position of the otherwise unreactive αA-crystallin is sufficient to make αA-crystallin an excellent substrate. Changing the amino acid around the reactive lysine decreases its reactivity without abolishing it [316, 317]. For instance, valine, phenylalanine, and arginine residues

N-terminal to the amine-donor lysine have a positive effect, whereas glycine and aspartic acid have adverse effects. The positive effect of arginine has also been demonstrated independently in the tau protein [278]. The *C*-terminal adjacent residue is also important: reactivity is somewhat increased by valine, more so by glycine, but decreased by proline. A recent publication has demonstrated that the presence of positively charged residues (Arg at −4 and +3, Lys at −2 and +2), the presence of Pro (at −2 and −3), and the absence of Lys at +1 increase lysine reactivity [267]. These studies show that accessibility is the primary requirement for a lysine residue to act as amine-donor substrate and that the enzyme has a relatively broad tolerance towards the amino acid residues surrounding the reactive lysine. Therefore, although there is no consensus sequence around reactive lysine residues, the neighboring amino acids influence reactivity. Our own experiments confirm the idea that only a minority of proteins can act as amine donors. Electrophoresis of the mouse brain proteins that have incorporated either glutamine- or lysine-donor probes under the catalytic action of endogenous transglutaminase demonstrates that, in brain, amine-donor proteins are fewer than amine-acceptor proteins (Figure 3).

D. SPECIFICITY OF SUBSTRATES WITH RESPECT TO TRANSGLUTAMINASE TYPE

Transglutaminases differ in their substrate preferences. Several substrates of factor XIII can also serve as substrates for transglutaminase-2, albeit with a much lower affinity [318]. β-casein is an excellent substrate for factor XIII and a relatively poor one for transglutaminase-2 [281, 283]. Factor XIII, transglutaminase-2, and hair follicle transglutaminase (presumably type 3) all stabilize the blood clot in vitro, but generate different fibrin polymers. Transglutaminase-2 polymerizes α-monomers first and then γ-monomers. Factor XIII does the reverse: formation of γ–γ dimers followed by α-polymers. Hair follicle transglutaminase polymerizes α- and γ-monomers simultaneously [319, 320]. In addition to their preference for a given fibrin subunit, factor XIII and transglutaminase-2 prefer different glutamine residues within the α-subunit [321]. Small proline-rich proteins and trichohyalin are other such examples since their in vitro cross-linking patterns differ according to whether transglutaminase-1, -2, or -3 is used [44, 322].

Experiments with synthetic peptides have also demonstrated that neighboring residues could greatly affect the reactivity for a given transglutaminase, while having little effect on reactivity for another transglutaminase. For

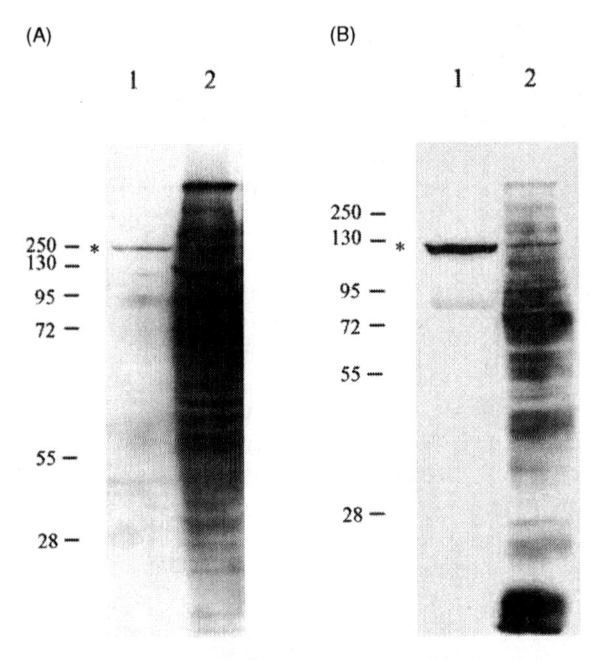

FIGURE 3. Western blot showing mouse brain proteins that have incorporated a biotinylated probe under the action of endogenous transglutaminase. A mouse brain homogenate was incubated for 2 hours in the presence of the biotinylated probes, whose incorporation resulted from the action of the transglutaminase present in the brain extracts since no exogenous enzyme was used. At the end of the incubation, proteins (80 μg/lane in (A) and 60 μg/lane in (B)) were resolved by 10% SDS-PAGE, transferred to nitrocellulose, and stained with horseradish peroxidase-conjugated streptavidin. (A) Visualization of amine-acceptor proteins by incorporation of the amine donor 5-biotinamidopentylamine ($BPNH_2$). A large number of amine-acceptor proteins with a wide distribution in molecular weights are detected. Lane 1, no probe; lane 2, with probe. (B) Detection of amine-donor proteins by incorporation of the amine acceptor peptide biotin-TVQQEL. Several amine donors are detected, but their number appears much smaller than that of glutamine-donors. A few proteins were stained in the absence of the probes (asterisks); they probably correspond to endogenous streptavidin-binding proteins. Intensity of nonspecific bands was higher in (B) than in (A) because of longer exposure of the film in the latter.

instance, the lysine residue present at the second next position *C*-terminal to the reactive glutamine is essential for recognition by factor XIII, but not for recognition by transglutaminase-2, which has a preference for the absence of a positive charge at this position. In general, factor XIII appears to have more stringent specificity requirements than transglutaminase-2 [282, 283].

E. SUBSTRATE REACTIVE SITES AS DRUGS AND FOR PROTEIN LABELING

The screening of phage displayed peptide libraries appears to be the most promising technique for defining the amine-acceptor and amine-donor sequences preferred by each transglutaminase. This will make possible the synthesis of tagged peptides (biotinylated or coupled with a fluorochrome for instance) that could be employed as baits in order to identify the natural substrates preferred by each transglutaminase. This information will be useful for elucidating the physiological function of transglutaminases, particularly the poorly characterized ones, such as transglutaminases-4, -5, -6, and -7. Bait peptides (oligoglutamine, TVQQEL, 5-(biotinamido)pentylamine) have already been used successfully to identify general transglutaminase substrates in human placenta [323], in rat liver extracts [324], and in human intestinal epithelial cell lines [325]. Identifying substrates may also shed light on the mechanisms by which transglutaminases participate in disease.

Determining the amino acid sequences that confer specificity for a given transglutaminase may open the way to the development of competitive inhibitors specific for each transglutaminase. Such inhibitors have potential applications in neurodegenerative diseases [326, 327], celiac disease [328, 329], dermatitis herpetiformis, gluten ataxia, inflammatory diseases [330–335], and cancer [336–343]. For instance, cystamine, a nonspecific inhibitor, has been shown to have neuroprotective effects in animal models of Huntington's and Parkinson's diseases [133, 220, 222, 257], and cysteamine, the reduced form of cystamine, is currently under evaluation in a phase II clinical trial for treatment of patients with Huntington's disease [327]. Cystamine has also been shown to block the proliferation of gluten-responsive T-cells in patient-derived small intestine biopsies exposed ex vivo to gluten [329, 344]. Another transglutaminase inhibitor, L682777 (a 2-[(2-oxopropyl)thio]imidazolium derivative), prevents gluten peptide deamidation and reduces T-cell activation in biopsies of celiac small intestine [345]. In spite of their promising effects in vitro and in animals, the development of transglutaminase inhibitors as drugs for humans may be hampered by the undesirable inhibition of nontargeted transglutaminases, especially factor XIII with its deleterious effects on blood coagulation. Some inhibitors showing specificity for transglutaminase-2 in vitro have been recently described: thieno[2,3-d]pyrimidin-4-one acylhydrazide derivatives [346], substituted cinnamoyl benzotriazolyl amides [347], and peptidomimetic inhibitors with a gliadin peptide backbone [348, 349]. It remains to be demonstrated whether these molecules possess the ability to inhibit selectively and efficiently transglutaminase-2 in vivo.

Defining substrate sequences with high affinity for transglutaminase may also be useful in the development of specific tags for the cross-linking of functional proteins. Such tags have been described for fluorescent labeling [350] and site-specific glycosidation of target proteins [351]. The transglutaminase-mediated site-specific immobilization method (TRANSIM), based on the use of a peptide identified by phage display [300], is a novel system for the site-specific conjugation of proteins with nonprotein material [352]. A Transglutaminase Substrate Database (TRANSDAB at http://genomics.dote.hu/wiki/index.php/Main_Page) has been initiated recently in order to provide a list of regularly updated substrate proteins for each transglutaminase as well as the location of the glutamine- and lysine-cross-linking sites, when possible [353]. This database can be freely updated by the scientific community. Another database, called TRANSIT (TRANsglutamination SITes at http://bioinformatica.isa.cnr.it/TRANSIT), contains known cross-linking sites for six transglutaminases [354].

VIII. FEATURES SHARED BY TRANSGLUTAMINASE SUBSTRATES AND PROTEINS ASSOCIATED WITH DISEASES OF POLYGLUTAMINE EXPANSION

While polyQ sequences have been shown to be sufficient to confer to various peptides and proteins, the ability to act as transglutaminase substrates [144, 149], many proteins associated with neurological disease and many precursors of the cornified cell envelope share features similar to those found in synthetic peptides highly reactive toward transglutaminase. First, in a large fraction of envelope precursors, the amine-acceptor site is located close to one end of the molecule, the *N*-terminal end or the *C*-terminal end or sometimes both (Table 2). The expanded polyglutamine is also located toward the end of the molecule in all disease-causing proteins but atrophin-1.

Second, a number of transglutaminase substrates and disease-causing proteins are rich in glutamine residues and/or contain stretches of consecutive glutamine residues and nearly half of reactive glutamine residues are adjacent to at least one glutamine residue; this is true in envelope precursors (Table 2) and probable in the disease-causing proteins. These data are in agreement with those derived from the screening of phage-displayed peptide libraries, in which the identified glutamine-donor sites often contain repeated glutamines [301].

Third, transglutaminase substrates and disease-causing proteins tend to contain an unusually large number of proline residues either scattered or repeated. The polyQs in huntingtin, ataxin-2, ataxin-7, and atrophin are all followed by either a polyP or a proline-rich region (Table 3). Although it has been shown that a proline directly *C*-terminal to the reactive glutamine decreases the reactivity of the glutamine, it is unlikely that the polyP has an impact on the reactivity of the entire expanded polyQ, as virtually all the glutamine residues of the repeat can act as amine acceptors [149]. It could be suggested that the polyP adjacent to the expanded polyQ of huntingtin protects the polyQ-bearing peptides (whether aggregated or not) from intracellular degradation, just like the proline-rich region of gliadin protects gliadin peptides from intestinal digestion [298]. The inclusions found in the cortex of patients with Huntington's disease have indeed been shown to be largely composed of *N*-terminal fragments of huntingtin including the polyP [184]. However, none of ataxins-1, -3, -6, TBP, or atrophin contains a proline repeat in the vicinity of the expanded polyQ. Proline and glutamine codons in adjacent polyQ and polyP repeats generally differ by only one nucleotide (CAG for glutamine and CCG for proline) and therefore are likely to have originated from the duplication of an ancestral codon (either CAG or CCG), followed by divergence of the duplicate into a CCG and a CAG codon and further independent reiterations of the two diverged codons.

IX. PERSPECTIVES

An increasing number of studies support the causative role of transglutaminase in the protein aggregation associated with diseases of the central nervous system, particularly those caused by a polyQ expansion. However, none of these studies provides incontrovertible proof of the role of transglutaminase in the disease process. This is because the demonstration is either indirect, for instance when it relies on formic acid, or suffers from methodological shortcomings. Indisputable evidence of cross-linking by isopeptide will be needed to provide firm support for therapeutic or prophylactic measures based on the role of transglutaminase.

ACKNOWLEDGEMENTS

The research of the authors is aided by the CNRS.

REFERENCES

1. Mycek, M., Clatke, D., Neidle, A., and Waelsch, H. (1959) Amine incorporation into insulin as catalyzed by transglutaminase, *Arch. Biochem. Biophys. 84*, 528–540.
2. Pisano, J., Finlayson, J., and Peyton, M. (1968) Cross-link in fibrin polymerized by factor 13: ε-(γ-glutamyl)lysine, *Science 160*, 892–893.
3. Green, H. (1979) The keratinocyte as differentiated cell type, *Harvey Lect. 74*, 101–139.
4. Zatloukal, K., Fesus, L., Denk, H., Tarcsa, E., Spurej, G., and Böck, G. (1992) High amount of ε-(γ-glutamyl)lysine cross-links in Mallory bodies, *Lab. Invest. 66*, 774–777.
5. Strnad, P., Harada, M., Siegel, M., Terkeltaub, R., Graham, R., Khosla, C., and Omary, M. (2007) Transglutaminase 2 regulates mallory body inclusion formation and injury-associated liver enlargement, *Gastroenterology 132*, 1515–1526.
6. Fesus, L., Thomazy, V., and Falus, A. (1987) Induction and activation of tissue transglutaminase during programmed cell death, *FEBS Lett. 224*, 104–108.
7. De Laurenzi, V. and Melino, G. (2001) Gene disruption of tissue transglutaminase, *Mol. Cell Biol. 21*, 148–155.
8. Nanda, N., Iismaa, S., Owens, W., Husain, A., Mackay, F., and Graham, R. (2001) Targeted inactivation of Gh/tissue transglutaminase II, *J. Biol. Chem. 276*, 20673–20678.
9. Laki, K. and Lóránd, L. (1948) On the solubility of fibrin clots, *Science 108*, 280.
10. Fisher, S., Rikover, M., and Naor, S. (1966) Factor 13 deficiency with severe hemorrhagic diathesis, *Blood 28*, 34–39.
11. Souri, M., Koseki-Kuno, S., Takeda, N., Yamakawa, M., Takeishi, Y., Degen, J., and Ichinose, A. (2008) Male-specific cardiac pathologies in mice lacking either the A or B subunit of factor XIII, *Thromb. Haemost. 99*, 401–408.
12. Koseki-Kuno, S., Yamakawa, M., Dickneite, G., and Ichinose, A. (2003) Factor XIII A subunit-deficient mice developed severe uterine bleeding events and subsequent spontaneous miscarriages, *Blood 102*, 4410–4412.
13. Williams-Ashman, H., Notides, A., Pabalan, S., and Lorand, L. (1972) Transamidase reactions involved in the enzymic coagulation of semen: isolation of -glutamyl- -lysine dipeptide from clotted secretion protein of guinea pig seminal vesicle, *Proc. Natl. Acad. Sci. U.S.A. 69*, 2322–2325.
14. Williams-Ashman, H. G. (1984) Transglutaminases and the clotting of mammalian seminal fluids, *Mol. Cell Biochem. 58*, 51.
15. Matoltsy, A. and Balsamo, C. (1955) A study of the components of the cornified epithelium of human skin, *J. Biophys. Biochem. Cytol. 1*, 339–360.
16. Brody, I. (1959) An ultrastructural study on the role of the keratohyalin granules in the keratinization process, *J. Ultrastruct. Res. 3*, 84–104.
17. Matoltsy, A. and Matoltsy, M. (1966) The membrane protein of horny cells, *J. Invest. Dermatol. 46*, 127–129.
18. Sun, T. and Green, H. (1976) Differentiation of the epidermal keratinocyte in cell culture: formation of the cornified envelope, *Cell 9*, 511–521.

19. Rice, R. and Green, H. (1979) Presence in human epidermal cells of a soluble protein precursor of the cross-linked envelope: activation of the cross-linking by calcium ions, *Cell 18*, 681–694.
20. Rice, R. H. and Green, H. (1977) The cornified envelope of terminally differentiated human epidermal keratinocytes consists of cross-linked protein, *Cell 11*, 417–422.
21. Rice, R. and Green, H. (1978) Relation of protein synthesis and transglutaminase activity to formation of the cross-linked envelope during terminal differentiation of the cultured human epidermal keratinocyte, *J. Cell Biol. 76*, 705–711.
22. Simon, M. and Green, H. (1985) Enzymatic cross-linking of involucrin and other proteins by keratinocyte particulates in vitro, *Cell 40*, 677–683.
23. Thacher, S. and Rice, R. (1985) Keratinocyte-specific transglutaminase of cultured human epidermal cells: relation to cross-linked envelope formation and terminal differentiation, *Cell 40*, 685–695.
24. Phillips, M., Stewart, B., Qin, Q., Chakravarty, R., Floyd, E., Jetten, A., and Rice, R. (1990) Primary structure of keratinocyte transglutaminase, *Proc. Natl. Acad. Sci. U.S.A. 87*, 9333–9337.
25. Chakravarty, R. and Rice, R. (1989) Acylation of keratinocyte transglutaminase by palmitic and myristic acids in the membrane Anchorage region, *J. Biol. Chem. 264*, 625–629.
26. Kim, H., Lewis, M., Gorman, J., Park, S., Girard, J., Folk, J., and Chung, S. (1990) Protransglutaminase E from guinea pig skin. Isolation and partial characterization, *J. Biol. Chem. 265*, 21971–21978.
27. Kim, I., Gorman, J., Park, S., Chung, S., and Steinert, P. (1993) The deduced sequence of the novel protransglutaminase E (TGase3) of human and mouse, *J. Biol. Chem. 268*, 12682–12690.
28. Hitomi, K., Horio, Y., Ikura, K., Yamanishi, K., and Maki, M. (2001) Analysis of epidermal-type transglutaminase (TGase 3) expression in mouse tissues and cell lines, *Int. J. Biochem. Cell Biol. 33*, 491–498.
29. Ahvazi, B., Boeshans, K., and Rastinejad, F. (2004) The emerging structural understanding of transglutaminase 3, *J. Struct. Biol. 147*, 200–207.
30. Hitomi, K., Presland, R., Nakayama, T., Fleckman, P., Dale, B., and Maki, M. (2003) Analysis of epidermal-type transglutaminase (transglutaminase 3) in human stratified epithelia and cultured keratinocytes using monoclonal antibodies, *J. Dermatol. Sci. 32*, 95–103.
31. Robinson, N., Lapic, S., Welter, J., and Eckert, R. (1997) S100A11, S100A10, annexin I, desmosomal proteins, small proline-rich proteins, plasminogen activator inhibitor-2, and involucrin are components of the cornified envelope of cultured human epidermal keratinocytes, *J. Biol. Chem. 272*, 12035–12046.
32. Michibata, H., Chiba, H., Wakimoto, K., Seishima, M., Kawasaki, S., Okubo, K., Mitsui, H., Torii, H., and Imai, Y. (2004) Identification and characterization of a novel component of the cornified envelope, cornifelin, *Biochem. Biophys. Res. Commun. 318*, 803–813.
33. Steinert, P. and Marekov, L. (1995) The proteins elafin, filaggrin, keratin intermediate filaments, loricrin, and small proline-rich proteins 1 and 2 are isodipeptide cross-linked

components of the human epidermal cornified cell envelope, *J. Biol. Chem. 270*, 17702–17711.

34. Steinert, P. and Marekov, L. (1997) Direct evidence that involucrin is a major early isopeptide cross-linked component of the keratinocyte cornified cell envelope, *J. Biol. Chem. 272*, 2021–2030.
35. Steinert, P., Kartasova, T., and Marekov, L. (1998) Biochemical evidence that small proline-rich proteins and trichohyalin function in epithelia by modulation of the biomechanical properties of their cornified cell envelopes, *J. Biol. Chem. 273*, 11758–11769.
36. Simon, M. and Green, H. (1984) Participation of membrane-associated proteins in the formation of the cross-linked envelope of the keratinocyte, *Cell 36*, 827–834.
37. Ruhrberg, C., Hajibagheri, M., Simon, M., Dooley, T., and Watt, F. (1996) Envoplakin, a novel precursor of the cornified envelope that has homology to desmoplakin, *J. Cell Biol. 134*, 715–729.
38. Simon, M. and Green, H. (1988) The glutamine residues reactive in transglutaminase-catalyzed cross-linking of involucrin, *J. Biol. Chem. 263*, 18093–18098.
39. Nemes, Z., Marekov, L., and Steinert, P. (1999) Involucrin cross-linking by transglutaminase 1. Binding to membranes directs residue specificity, *J. Biol. Chem. 274*, 11013–11021.
40. Marshall, D., Hardman, M., Nield, K., and Byrne, C. (2001) Differentially expressed late constituents of the epidermal cornified envelope, *Proc. Natl. Acad. Sci. U.S.A. 98*, 13031–13036.
41. Hohl, D., Mehrel, T., Lichti, U., Turner, M., Roop, D., and Steinert, P. (1991) Characterization of human loricrin. Structure and function of a new class of epidermal cell envelope proteins, *J. Biol. Chem. 266*, 6626–6636.
42. Candi, E., Tarcsa, E., Idler, W., Kartasova, T., Marekov, L., and Steinert, P. (1999) Transglutaminase cross-linking properties of the small proline-rich 1 family of cornified cell envelope proteins. Integration with loricrin, *J. Biol. Chem. 274*, 7226–7237.
43. Tarcsa, E., Candi, E., Kartasova, T., Idler, W., Marekov, L., and Steinert, P. (1998) Structural and transglutaminase substrate properties of the small proline-rich 2 family of cornified cell envelope proteins, *J. Biol. Chem. 273*, 23297–23303.
44. Steinert, P., Candi, E., Tarcsa, E., Marekov, L., Sette, M., Paci, M., Ciani, B., Guerrieri, P., and Melino, G. (1999) Transglutaminase crosslinking and structural studies of the human small proline rich 3 protein, *Cell Death Differ. 6*, 916–930.
45. Park, G., Lim, S., Jang, S., and Morasso, M. (2002) Suprabasin, a novel epidermal differentiation marker and potential cornified envelope precursor, *J. Biol. Chem. 277*, 45195–45202.
46. Ruse, M., Lambert, A., Robinson, N., Ryan, D., Shon, K., and Eckert, R. (2001) S100A7, S100A10, and S100A11 are transglutaminase substrates, *Biochemistry 40*, 3167–3173.
47. Robinson, N. and Eckert, R. (1998) Identification of transglutaminase-reactive residues in S100A11, *J. Biol. Chem. 273*, 2721–2728.
48. Candi, E., Melino, G., Mei, G., Tarcsa, E., Chung, S., Marekov, L., and Steinert, P. (1995) Biochemical, structural, and transglutaminase substrate properties of human loricrin, the major epidermal cornified cell envelope protein, *J. Biol. Chem. 270*, 26382–26390.

49. Jaskowski, T., Hamblin, T., Wilson, A., Hill, H., Book, L., Meyer, L., Zone, J., and Hull, C. (2009) IgA anti-epidermal transglutaminase antibodies in dermatitis herpetiformis and pediatric celiac disease, *J. Invest. Dermatol. 129*, 2728–2730.

50. Sardy, M., Kárpáti, S., Merkl, B., Paulsson, M., and Smyth, N. (2002) Epidermal transglutaminase (TGase 3) is the autoantigen of dermatitis herpetiformis, *J. Exp. Med. 195*, 747–757.

51. Donaldson, M., Zone, J., Schmidt, L., Taylor, T., Neuhausen, S., Hull, C., and Meyer, L. (2007) Epidermal transglutaminase deposits in perilesional and uninvolved skin in patients with dermatitis herpetiformis, *J. Invest. Dermatol. 127*, 1268–1271.

52. Hull, C., Liddle, M., Hansen, N., Meyer, L., Schmidt, L., Taylor, T., Jaskowski, T., Hill, H., and Zone, J. (2008) Elevation of IgA anti-epidermal transglutaminase antibodies in dermatitis herpetiformis, *Br. J. Dermatol. 159*, 120–124.

53. Grenard, P., Bates, M., and Aeschlimann, D. (2001) Evolution of transglutaminase genes: identification of a transglutaminase gene cluster on human chromosome 15q15. Structure of the gene encoding transglutaminase X and a novel gene family member, transglutaminase Z, *J. Biol. Chem.276*, 33066–33078.

54. Aeschlimann, D., Koeller, M., Allen-Hoffmann, B., and Mosher, D. (1998) Isolation of a cDNA encoding a novel member of the transglutaminase gene family from human keratinocytes. Detection and identification of transglutaminase gene products based on reverse transcription-polymerase chain reaction with degenerate primers, *J. Biol. Chem. 273*, 3452–3460.

55. Candi, E., Oddi, S., Paradisi, A., Terrinoni, A., Ranalli, M., Teofoli, P., Citro, G., Scarpato, S., Puddu, P., and Melino, G. (2002) Expression of transglutaminase 5 in normal and pathologic human epidermis, *J. Invest. Dermatol. 119*, 670–677.

56. Candi, E., Oddi, S., Terrinoni, A., Paradisi, A., Ranalli, M., Finazzi-Agró, A., and Melino, G. (2001) Transglutaminase 5 cross-links loricrin, involucrin, and small proline-rich proteins in vitro, *J. Biol. Chem. 276*, 35014–35023.

57. Pietroni, V., Di Giorgi, S., Paradisi, A., Ahvazi, B., Candi, E., and Melino, G. (2008) Inactive and highly active, proteolytically processed transglutaminase-5 in epithelial cells, *J. Invest. Dermatol.128*, 2760–2766.

58. Huber, M., Rettler, I., Bernasconi, K., Frenk, E., Lavrijsen, S., Ponec, M., Bon, A., Lautenschlager, S., Schorderet, D., and Hohl, D. (1995) Mutations of keratinocyte transglutaminase in lamellar ichthyosis, *Science 267*, 525–528.

59. Russell, L., DiGiovanna, J., Rogers, G., Steinert, P., Hashem, N., Compton, J., and Bale, S. (1995) Mutations in the gene for transglutaminase 1 in autosomal recessive lamellar ichthyosis, *Nat. Genet. 9*, 279–283.

60. Jeon, S., Djian, P., and Green, H. (1998) Inability of keratinocytes lacking their specific transglutaminase to form cross-linked envelopes: absence of envelopes as a simple diagnostic test for lamellar ichthyosis, *Proc. Natl. Acad. Sci. U.S.A. 95*, 687–690.

61. Rice, R., Crumrine, D., Hohl, D., Munro, C., and Elias, P. (2003) Cross-linked envelopes in nail plate in lamellar ichthyosis, *Br. J. Dermatol. 149*, 1050–1054.

62. Matsuki, M., Yamashita, F., Ishida-Yamamoto, A., Yamada, K., Kinoshita, C., Fushiki, S., Ueda, E., Morishima, Y., Tabata, K., Yasuno, H., Hashida, M., Iizuka, H., Ikawa, M., Okabe, M., Kondoh, G., Kinoshita, T., Takeda, J., and Yamanishi, K. (1998) Defective

stratum corneum and early neonatal death in mice lacking the gene for transglutaminase 1 (keratinocyte transglutaminase). *Proc. Natl. Acad. Sci. U.S.A. 95*, 1044–1049.

63. Cassidy, A., van Steensel, M., Steijlen, P., van Geel, M., Van Der Velden, J., Morley, S., Terrinoni, A., Melino, G., Candi, E., and McLean, W. (2005) A homozygous missense mutation in TGM5 abolishes epidermal transglutaminase 5 activity and causes acral peeling skin syndrome, *Am. J. Hum. Genet. 77*, 909–917.
64. Watt, F. and Green, H. (1981) Involucrin synthesis is correlated with cell size in human epidermal cultures, *J. Cell Biol. 90*, 738–742.
65. Watt, F. and Green, H. (1982) Stratification and terminal differentiation of cultured epidermal cells, *Nature 295*, 434–436.
66. Eckert, R. and Green, H. (1986) Structure and evolution of the human involucrin gene, *Cell 46*, 583–589.
67. Simon, M., Phillips, M., Green, H., Stroh, H., Glatt, K., Burns, G., and Latt, S. A. (1989) Absence of a single repeat from the coding region of the human involucrin gene leading to RFLP, *Am. J. Hum. Genet. 45*, 910–916.
68. Simon, M., Phillips, M., and Green, H. (1991) Polymorphism due to variable number of repeats in the human involucrin gene, *Genomics 9*, 576–580.
69. Djian, P., Delhomme, B., and Green, H. (1995) Origin of the polymorphism of the involucrin gene in Asians, *Am. J. Hum. Genet. 56*, 1367–1372.
70. Djian, P., Phillips, M., Easley, K., Huang, E., Simon, M., Rice, R., and Green, H. (1993) The involucrin genes of the mouse and the rat: study of their shared repeats, *Mol. Biol. Evol. 10*, 1136–1149.
71. Djian, P., Easley, K., and Green, H. (2000) Targeted ablation of the murine involucrin gene, *J. Cell Biol. 151*, 381–388.
72. Vanhoutteghem, A., Djian, P., and Green, H. (2008) Ancient origin of the gene encoding involucrin, a precursor of the cross-linked envelope of epidermis and related epithelia, *Proc. Natl. Acad. Sci. U.S.A. 105*, 15481–15486.
73. Ruhrberg, C., Hajibagheri, M., Parry, D., and Watt, F. (1997) Periplakin, a novel component of cornified envelopes and desmosomes that belongs to the plakin family and forms complexes with envoplakin, *J. Cell Biol. 139*, 1835–1849.
74. Määttä, A., DiColandrea, T., Groot, K., and Watt, F. (2001) Gene targeting of envoplakin, a cytoskeletal linker protein and precursor of the epidermal cornified envelope, *Mol. Cell Biol. 21*, 7047–7053.
75. Aho, S., Li, K., Ryoo, Y., McGee, C., Ishida-Yamamoto, A., Uitto, J., and Klement, J. (2004) Periplakin gene targeting reveals a constituent of the cornified cell envelope dispensable for normal mouse development, *Mol. Cell Biol. 24*, 6410–6418.
76. Sevilla, L., Nachat, R., Groot, K., Klement, J., Uitto, J., Djian, P., Määttä, A., and Watt, F. (2007) Mice deficient in involucrin, envoplakin, and periplakin have a defective epidermal barrier, *J. Cell Biol. 179*, 1599–1612.
77. Dale, B., Holbrook, K., and Steinert, P. (1978) Assembly of stratum corneum basic protein and keratin filaments in macrofibrils, *Nature 276*, 729–731.
78. Kartasova, T. and van de Putte, P. (1988) Isolation, characterization, and UV-stimulated expression of two families of genes encoding polypeptides of related structure in human epidermal keratinocytes, *Mol. Cell Biol. 8*, 2195–2203.

79. Mehrel, T., Hohl, D., Rothnagel, J., Longley, M., Bundman, D., Cheng, C., Lichti, U., Bisher, M., Steven, A., and Steinert, P. (1990) Identification of a major keratinocyte cell envelope protein, loricrin, *Cell 61*, 1103–1112.
80. Gan, S., McBride, O., Idler, W., Markova, N., and Steinert, P. (1990) Organization, structure, and polymorphisms of the human profilaggrin gene, *Biochemistry 29*, 9432–9440.
81. Richards, S., Scott, I., Harding, C., Liddell, J., Powell, G., and Curtis, C. (1988) Evidence for filaggrin as a component of the cell envelope of the newborn rat, *Biochem. J. 253*, 153–160.
82. Smith, F., Irvine, A., Terron-Kwiatkowski, A., Sandilands, A., Campbell, L., Zhao, Y., Liao, H., Evans, A., Goudie, D., Lewis-Jones, S., Arseculeratne, G., Munro, C., Sergeant, A., O'Regan, G., Bale, S., Compton, J., DiGiovanna, J., Presland, R., Fleckman, P., and McLean, W. (2006) Loss-of-function mutations in the gene encoding filaggrin cause ichthyosis vulgaris, *Nat. Genet. 38*, 337–342.
83. Sandilands, A., Sutherland, C., Irvine, A., and McLean, W. (2009) Filaggrin in the frontline: role in skin barrier function and disease, *J. Cell Sci. 122*, 1285–1294.
84. Palmer, C., Irvine, A., Terron-Kwiatkowski, A., Zhao, Y., Liao, H., Lee, S., Goudie, D., Sandilands, A., Campbell, L., Smith, F., O'Regan, G., Watson, R., Cecil, J., Bale, S., Compton, J., DiGiovanna, J., Fleckman, P., Lewis-Jones, S., Arseculeratne, G., Sergeant, A., Munro, C., El Houate, B., McElreavey, K., Halkjaer, L., Bisgaard, H., Mukhopadhyay, S., and McLean, W. (2006) Common loss-of-function variants of the epidermal barrier protein filaggrin are a major predisposing factor for atopic dermatitis, *Nat. Genet. 38*, 441–446.
85. Fallon, P., Sasaki, T., Sandilands, A., Campbell, L., Saunders, S., Mangan, N., Callanan, J., Kawasaki, H., Shiohama, A., Kubo, A., Sundberg, J., Presland, R., Fleckman, P., Shimizu, N., Kudoh, J., Irvine, A., Amagai, M., and McLean, W. (2009) A homozygous frameshift mutation in the mouse Flg gene facilitates enhanced percutaneous allergen priming, *Nat. Genet. 41*, 602–608.
86. Marvin, K., George, M., Fujimoto, W., Saunders, N., Bernacki, S., and Jetten, A. (1992) Cornifin, a cross-linked envelope precursor in keratinocytes that is down-regulated by retinoids, *Proc. Natl. Acad. Sci. U.S.A. 89*, 11026–11030.
87. Gibbs, S., Fijneman, R., Wiegant, J., van Kessel, A., van De Putte, P., and Backendorf, C. (1993) Molecular characterization and evolution of the SPRR family of keratinocyte differentiation markers encoding small proline-rich proteins, *Genomics 16*, 630–637.
88. Hohl, D., de Viragh, P., Amiguet-Barras, F., Gibbs, S., Backendorf, C., and Huber, M. (1995) The small proline-rich proteins constitute a multigene family of differentially regulated cornified cell envelope precursor proteins, *J. Invest. Dermatol. 104*, 902–909.
89. Cabral, A., Sayin, A., de Winter, S., Fischer, D., Pavel, S., and Backendorf, C. (2001) SPRR4, a novel cornified envelope precursor: UV-dependent epidermal expression and selective incorporation into fragile envelopes, *J. Cell Sci. 114*, 3837–3843.
90. Makino, T., Takaishi, M., Morohashi, M., and Huh, N. (2001) Hornerin, a novel profilaggrin-like protein and differentiation-specific marker isolated from mouse skin, *J. Biol. Chem. 276*, 47445–47452.
91. Makino, T., Takaishi, M., Toyoda, M., Morohashi, M., and Huh, N. (2003) Expression of hornerin in stratified squamous epithelium in the mouse: a comparative analysis with profilaggrin, *J. Histochem. Cytochem. 51*, 485–492.

92. Takaishi, M., Makino, T., Morohashi, M., and Huh, N. (2005) Identification of human hornerin and its expression in regenerating and psoriatic skin, *J. Biol. Chem. 280*, 4696–4703.

93. Wu, Z., Meyer-Hoffert, U., Reithmayer, K., Paus, R., Hansmann, B., He, Y., Bartels, J., Gläser, R., Harder, J., and Schröder, J. (2009) Highly complex peptide aggregates of the S100 fused-type protein hornerin are present in human skin, *J. Invest. Dermatol. 129*, 1446–1458.

94. Koch, P., de Viragh, P., Scharer, E., Bundman, D., Longley, M., Bickenbach, J., Kawachi, Y., Suga, Y., Zhou, Z., Huber, M., Hohl, D., Kartasova, T., Jarnik, M., Steven, A., and Roop, D. (2000) Lessons from loricrin-deficient mice: compensatory mechanisms maintaining skin barrier function in the absence of a major cornified envelope protein, *J. Cell Biol. 151*, 389–400.

95. Jarnik, M., de Viragh, P., Schärer, E., Bundman, D., Simon, M., Roop, D., and Steven, A. (2002) Quasi-normal cornified cell envelopes in loricrin knockout mice imply the existence of a loricrin backup system, *J. Invest. Dermatol. 118*, 102–109.

96. Maestrini, E., Monaco, A., McGrath, J., Ishida-Yamamoto, A., Camisa, C., Hovnanian, A., Weeks, D., Lathrop, M., Uitto, J., and Christiano, A. (1996) A molecular defect in loricrin, the major component of the cornified cell envelope, underlies Vohwinkel's syndrome, *Nat. Genet. 13*, 70–77.

97. Suga, Y., Jarnik, M., Attar, P., Longley, M., Bundman, D., Steven, A., Koch, P., and Roop, D. (2000) Transgenic mice expressing a mutant form of loricrin reveal the molecular basis of the skin diseases, Vohwinkel syndrome and progressive symmetric erythrokeratoderma, *J. Cell Biol. 151*, 401–412.

98. de Cid, R., Riveira-Munoz, E., Zeeuwen, P., Robarge, J., Liao, W., Dannhauser, E., Giardina, E., Stuart, P., Nair, R., Helms, C., Escaramís, G., Ballana, E., Martín-Ezquerra, G., den Heijer, M., Kamsteeg, M., Joosten, I., Eichler, E., Lázaro, C., Pujol, R., Armengol, L., Abecasis, G., Elder, J., Novelli, G., Armour, J., Kwok, P., Bowcock, A., Schalkwijk, J., and Estivill, X. (2009) Deletion of the late cornified envelope LCE3B and LCE3C genes as a susceptibility factor for psoriasis, *Nat. Genet. 41*, 211–215.

99. Mischke, D., Korge, B., Marenholz, I., Volz, A., and Ziegler, A. (1996) Genes encoding structural proteins of epidermal cornification and S100 calcium-binding proteins form a gene complex ("epidermal differentiation complex") on human chromosome 1q21, *J. Invest. Dermatol. 106*, 989–992.

100. McKinley-Grant, L., Idler, W., Bernstein, I., Parry, D., Cannizzaro, L., Croce, C., Huebner, K., Lessin, S., and Steinert, P. (1989) Characterization of a cDNA clone encoding human filaggrin and localization of the gene to chromosome region 1q21, *Proc. Natl. Acad. Sci. U.S.A. 86*, 4848–4852.

101. Rothnagel, J. and Rogers, G. (1986) Trichohyalin, an intermediate filament-associated protein of the hair follicle, *J. Cell Biol. 102*, 1419–1429.

102. Yoneda, K., Hohl, D., McBride, O., Wang, M., Cehrs, K., Idler, W., and Steinert, P. (1992) The human loricrin gene, *J. Biol. Chem. 267*, 18060–18066.

103. Volz, A., Korge, B., Compton, J., Ziegler, A., Steinert, P., and Mischke, D. (1993) Physical mapping of a functional cluster of epidermal differentiation genes on chromosome 1q21, *Genomics 18*, 92–99.

104. Contzler, R., Favre, B., Huber, M., and Hohl, D. (2005) Cornulin, a new member of the "fused gene" family, is expressed during epidermal differentiation, *J. Invest. Dermatol.* *124*, 990–997.

105. Huber, M., Siegenthaler, G., Mirancea, N., Marenholz, I., Nizetic, D., Breitkreutz, D., Mischke, D., and Hohl, D. (2005) Isolation and characterization of human repetin, a member of the fused gene family of the epidermal differentiation complex, *J. Invest. Dermatol.* *124*, 998–1007.

106. Green, H., and Djian, P. (1992) Consecutive actions of different gene-altering mechanisms in the evolution of involucrin, *Mol. Biol. Evol.* *9*, 977–1017.

107. Backendorf, C., and Hohl, D. (1992) A common origin for cornified envelope proteins? *Nat. Genet.* *2*, 91.

108. South, A., Cabral, A., Ives, J., James, C., Mirza, G., Marenholz, I., Mischke, D., Backendorf, C., Ragoussis, J., and Nizetic, D. (1999) Human epidermal differentiation complex in a single 2.5 Mbp long continuum of overlapping DNA cloned in bacteria integrating physical and transcript maps, *J. Invest. Dermatol.* *112*, 910–918.

109. Sotiropoulou, G., Anisowicz, A., and Sager, R. (1997) Identification, cloning, and characterization of cystatin M, a novel cysteine proteinase inhibitor, down-regulated in breast cancer, *J. Biol. Chem.* *272*, 903–910.

110. Ni, J., Abrahamson, M., Zhang, M., Fernandez, M., Grubb, A., Su, J., Yu, G., Li, Y., Parmelee, D., Xing, L., Coleman, T., Gentz, S., Thotakura, R., Nguyen, N., Hesselberg, M., and Gentz, R. (1997) Cystatin E is a novel human cysteine proteinase inhibitor with structural resemblance to family 2 cystatins, *J. Biol. Chem.* *272*, 10853–10858.

111. Zeeuwen, P., van Vlijmen-Willems, I., Hendriks, W., Merkx, G., and Schalkwijk, J. (2002) A null mutation in the cystatin M/E gene of ichq mice causes juvenile lethality and defects in epidermal cornification, *Hum. Mol. Genet.* *11*, 2867–2875.

112. Zeeuwen, P., van Vlijmen-Willems, I., Olthuis, D., Johansen, H., Hitomi, K., Hara-Nishimura, I., Powers, J., James, K., op den Camp, H., Lemmens, R., and Schalkwijk, J. (2004) Evidence that unrestricted legumain activity is involved in disturbed epidermal cornification in cystatin M/E deficient mice, *Hum. Mol. Genet.* *13*, 1069–1079.

113. Zeeuwen, P., Van Vlijmen-Willems, I., Jansen, B., Sotiropoulou, G., Curfs, J., Meis, J., Janssen, J., Van Ruissen, F., and Schalkwijk, J. (2001) Cystatin M/E expression is restricted to differentiated epidermal keratinocytes and sweat glands: a new skin-specific proteinase inhibitor that is a target for cross-linking by transglutaminase, *J. Invest. Dermatol.* *116*, 693–701.

114. Chavanas, S., Bodemer, C., Rochat, A., Hamel-Teillac, D., Ali, M., Irvine, A., Bonafé, J., Wilkinson, J., Taïeb, A., Barrandon, Y., Harper, J., de Prost, Y., and Hovnanian, A. (2000) Mutations in SPINK5, encoding a serine protease inhibitor, cause Netherton syndrome, *Nat. Genet.* *25*, 141–142.

115. Toomes, C., James, J., Wood, A., Wu, C., McCormick, D., Lench, N., Hewitt, C., Moynihan, L., Roberts, E., Woods, C., Markham, A., Wong, M., Widmer, R., Ghaffar, K., Pemberton, M., Hussein, I., Temtamy, S., Davies, R., Read, A., Sloan, P., Dixon, M., and Thakker, N. (1999) Loss-of-function mutations in the cathepsin C gene result in periodontal disease and palmoplantar keratosis, *Nat. Genet.* *23*, 421–424.

116. Roth, W., Deussing, J., Botchkarev, V., Pauly-Evers, M., Saftig, P., Hafner, A., Schmidt, P., Schmahl, W., Scherer, J., Anton-Lamprecht, I., Von Figura, K., Paus, R., and Peters, C. (2000) Cathepsin L deficiency as molecular defect of furless: hyperproliferation of keratinocytes and pertubation of hair follicle cycling, *FASEB J. 14*, 2075–2086.

117. Zeeuwen, P., Dale, B., de Jongh, G., van Vlijmen-Willems, I., Fleckman, P., Kimball, J., Stephens, K., and Schalkwijk, J. (2003) The human cystatin M/E gene (CST6): exclusion candidate gene for harlequin ichthyosis, *J. Invest. Dermatol. 121*, 65–68.

118. Thomas, A., Cullup, T., Norgett, E., Hill, T., Barton, S., Dale, B., Sprecher, E., Sheridan, E., Taylor, A., Wilroy, R., DeLozier, C., Burrows, N., Goodyear, H., Fleckman, P., Stephens, K., Mehta, L., Watson, R., Graham, R., Wolf, R., Slavotinek, A., Martin, M., Bourn, D., Mein, C., O'Toole, E., and Kelsell, D. (2006) ABCA12 is the major harlequin ichthyosis gene, *J. Invest. Dermatol. 126*, 2408–2413.

119. Candi, E., Tarcsa, E., Digiovanna, J., Compton, J., Elias, P., Marekov, L., and Steinert, P. (1998) A highly conserved lysine residue on the head domain of type II keratins is essential for the attachment of keratin intermediate filaments to the cornified cell envelope through isopeptide crosslinking by transglutaminases, *Proc. Natl. Acad. Sci. U.S.A. 95*, 2067–2072.

120. Djian, P. and Green, H. (1989) The involucrin gene of the orangutan: generation of the late region as an evolutionary trend in the hominoids, *Mol. Biol. Evol. 6*, 469–477.

121. Phillips, M., Djian, P., and Green, H. (1990) The involucrin gene of the galago. Existence of a correction process acting on its segment of repeats, *J. Biol. Chem. 265*, 7804–7807.

122. Djian, P. and Green, H. (1991) Involucrin gene of tarsioids and other primates: alternatives in evolution of the segment of repeats, *Proc. Natl. Acad. Sci. U.S.A. 88*, 5321–5325.

123. La Spada, A., Wilson, E., Lubahn, D., Harding, A., and Fischbeck, K. (1991) Androgen receptor gene mutations in X-linked spinal and bulbar muscular atrophy, *Nature 352*, 77–79.

124. The Huntington's Disease Collaborative Research Group. (1993) A novel gene containing a trinucleotide repeat that is expanded and unstable on Huntington's disease chromosomes, *Cell 72*, 971–983.

125. Green, H. (1993) Human genetic diseases due to codon reiteration: relationship to an evolutionary mechanism, *Cell 74*, 955–956.

126. Davies, S., Turmaine, M., Cozens, B., DiFiglia, M., Sharp, A., Ross, C., Scherzinger, E., Wanker, E., Mangiarini, L., and Bates, G. (1997) Formation of neuronal intranuclear inclusions underlies the neurological dysfunction in mice transgenic for the HD mutation, *Cell 90*, 537–548.

127. DiFiglia, M., Sapp, E., Chase, K., Davies, S., Bates, G., Vonsattel, J., and Aronin, N. (1997) Aggregation of huntingtin in neuronal intranuclear inclusions and dystrophic neurites in brain, *Science 277*, 1990–1993.

128. Selkoe, D., Ihara, Y., and Salazar, F. (1982) Alzheimer's disease: insolubility of partially purified paired helical filaments in sodium dodecyl sulfate and urea, *Science 215*, 1243–1245.

129. Junn, E., Ronchetti, R., Quezado, M., Kim, S., and Mouradian, M. (2003) Tissue transglutaminase-induced aggregation of α-synuclein: implications for Lewy body

formation in Parkinson's disease and dementia with Lewy bodies, *Proc. Natl. Acad. Sci. U.S.A. 100*, 2047–2052.

130. Gilad, G. and Varon, L. (1985) Transglutaminase activity in rat brain: characterization, distribution, and changes with age, *J. Neurochem. 45*, 1522–1526.
131. Kim, S., Grant, P., Lee, J., Pant, H., and Steinert, P. (1999) Differential expression of multiple transglutaminases in human brain. Increased expression and cross-linking by transglutaminases 1 and 2 in Alzheimer's disease, *J. Biol. Chem. 274*, 30715–30721.
132. Lesort, M., Attanavanich, K., Zhang, J., and Johnson, G. (1998) Distinct nuclear localization and activity of tissue transglutaminase, *J. Biol. Chem. 273*, 11991–11994.
133. Dedeoglu, A., Kubilus, J., Jeitner, T., Matson, S., Bogdanov, M., Kowall, N., Matson, W., Cooper, A., Ratan, R., Beal, M., Hersch, S., and Ferrante, R. (2002) Therapeutic effects of cystamine in a murine model of Huntington's disease, *J. Neurosci. 22*, 8942–8950.
134. Hoffner, G., Van Der Rest, G., Dansette, P., and Djian, P. (2009) The end product of transglutaminase crosslinking: simultaneous quantitation of [Nε-(γ-glutamyl) lysine] and lysine by HPLC-MS3, *Anal. Biochem. 384*, 296–304.
135. Simon, S. and Llinás, R. (1985) Compartmentalization of the submembrane calcium activity during calcium influx and its significance in transmitter release, *Biophys. J. 48*, 485–498.
136. Yamada, W. and Zucker, R. (1992) Time course of transmitter release calculated from simulations of a calcium diffusion model, *Biophys. J. 61*, 671–682.
137. Neher, E. (1998) Vesicle pools and Ca2+ microdomains: new tools for understanding their roles in neurotransmitter release, *Neuron 20*, 389–399.
138. Stern, M. (1992) Buffering of calcium in the vicinity of a channel pore, *Cell Calcium 13*, 183–192.
139. Zainelli, G., Ross, C., Troncoso, J., Fitzgerald, J., and Muma, N. (2004) Calmodulin regulates transglutaminase 2 cross-linking of huntingtin, *J. Neurosci. 24*, 1954–1961.
140. Monsonego, A., Shani, Y., Friedmann, I., Paas, Y., Eizenberg, O., and Schwartz, M. (1997) Expression of GTP-dependent and GTP-independent tissue-type transglutaminase in cytokine-treated rat brain astrocytes, *J. Biol. Chem. 272*, 3724–3732.
141. Monsonego, A., Friedmann, I., Shani, Y., Eisenstein, M., and Schwartz, M. (1998) GTP-dependent conformational changes associated with the functional switch between Gα and cross-linking activities in brain-derived tissue transglutaminase, *J. Mol. Biol. 282*, 713–720.
142. Citron, B., SantaCruz, K., Davies, P., and Festoff, B. (2001) Intron-exon swapping of transglutaminase mRNA and neuronal Tau aggregation in Alzheimer's disease, *J. Biol. Chem. 276*, 3295–3301.
143. Nasir, J., Floresco, S., O'Kusky, J., Diewert, V., Richman, J., Zeisler, J., Borowski, A., Marth, J., Phillips, A., and Hayden, M. (1995) Targeted disruption of the Huntington's disease gene results in embryonic lethality and behavioral and morphological changes in heterozygotes, *Cell 81*, 811–823.
144. Kahlem, P., Green, H., and Djian, P. (1998) Transglutaminase action imitates Huntington's disease: selective polymerization of Huntingtin containing expanded polyglutamine, *Mol. Cell 1*, 595–601.

145. Karpuj, M., Garren, H., Slunt, H., Price, D., Gusella, J., Becher, M., and Steinman, L. (1999) Transglutaminase aggregates huntingtin into nonamyloidogenic polymers, and its enzymatic activity increases in Huntington's disease brain nuclei, *Proc. Natl. Acad. Sci. U.S.A. 96*, 7388–7393.

146. Zainelli, G., Dudek, N., Ross, C., Kim, S., and Muma, N. (2005) Mutant huntingtin protein: a substrate for transglutaminase 1, 2, and 3, *J. Neuropathol. Exp. Neurol. 64*, 58–65.

147. Zainelli, G., Ross, C., Troncoso, J., and Muma, N. (2003) Transglutaminase cross-links in intranuclear inclusions in Huntington disease, *J. Neuropathol. Exp. Neurol. 62*, 14–24.

148. Matilla, A., Roberson, E., Banfi, S., Morales, J., Armstrong, D., Burright, E., Orr, H., Sweatt, J., Zoghbi, H., and Matzuk, M. (1998) Mice lacking ataxin-1 display learning deficits and decreased hippocampal paired-pulse facilitation, *J. Neurosci. 18*, 5508–5516.

149. Kahlem, P., Terré, C., Green, H., and Djian, P. (1996) Peptides containing glutamine repeats as substrates for transglutaminase-catalyzed cross-linking: relevance to diseases of the nervous system, *Proc. Natl. Acad. Sci. U.S.A. 93*, 14580–14585.

150. D'Souza, D., Wei, J., Shao, Q., Hebert, M., Subramony, S., and Vig, P. (2006) Tissue transglutaminase crosslinks ataxin-1: possible role in SCA1 pathogenesis, *Neurosci. Lett. 409*, 5–9.

151. Kiehl, T., Nechiporuk, A., Figueroa, K., Keating, M., Huynh, D., and Pulst, S. (2006) Generation and characterization of Sca2 (ataxin-2) knockout mice, *Biochem. Biophys. Res. Commun. 339*, 17–24.

152. Schmitt, I., Linden, M., Khazneh, H., Evert, B., Breuer, P., Klockgether, T., and Wuellner, U. (2007) Inactivation of the mouse Atxn3 (ataxin-3) gene increases protein ubiquitination, *Biochem. Biophys. Res. Commun. 362*, 734–739.

153. Zhuchenko, O., Bailey, J., Bonnen, P., Ashizawa, T., Stockton, D., Amos, C., Dobyns, W., Subramony, S., Zoghbi, H., and Lee, C. (1997) Autosomal dominant cerebellar ataxia (SCA6) associated with small polyglutamine expansions in the α 1A-voltage-dependent calcium channel, *Nat. Genet. 15*, 62–69.

154. Jun, K., Piedras-Rentería, E., Smith, S., Wheeler, D., Lee, S., Lee, T., Chin, H., Adams, M., Scheller, R., Tsien, R., and Shin, H. (1999) Ablation of P/Q-type Ca(2+) channel currents, altered synaptic transmission, and progressive ataxia in mice lacking the α(1A)-subunit, *Proc. Natl. Acad. Sci. U.S.A. 96*, 15245–15250.

155. Helmlinger, D., Hardy, S., Sasorith, S., Klein, F., Robert, F., Weber, C., Miguet, L., Potier, N., Van-Dorsselaer, A., Wurtz, J., Mandel, J., Tora, L., and Devys, D. (2004) Ataxin-7 is a subunit of GCN5 histone acetyltransferase-containing complexes, *Hum. Mol. Genet. 13*, 1257–1265.

156. Koide, R., Kobayashi, S., Shimohata, T., Ikeuchi, T., Maruyama, M., Saito, M., Yamada, M., Takahashi, H., and Tsuji, S. (1999) A neurological disease caused by an expanded CAG trinucleotide repeat in the TATA-binding protein gene: a new polyglutamine disease? *Hum. Mol. Genet. 8*, 2047–2053.

157. Mandrusiak, L., Beitel, L., Wang, X., Scanlon, T., Chevalier-Larsen, E., Merry, D., and Trifiro, M. (2003) Transglutaminase potentiates ligand-dependent proteasome dysfunction induced by polyglutamine-expanded androgen receptor, *Hum. Mol. Genet. 12*, 1497–1506.

158. Shen, Y., Lee, G., Choe, Y., Zoltewicz, J., and Peterson, A. (2007) Functional architecture of atrophins, *J. Biol. Chem. 282*, 5037–5044.

159. Koide, R., Ikeuchi, T., Onodera, O., Tanaka, H., Igarashi, S., Endo, K., Takahashi, H., Kondo, R., Ishikawa, A., and Hayashi, T. (1994) Unstable expansion of CAG repeat in hereditary dentatorubral-pallidoluysian atrophy (DRPLA). *Nat. Genet. 6*, 9–13.

160. Orr, H., Chung, M., Banfi, S., Kwiatkowski, T. J., Servadio, A., Beaudet, A., McCall, A., Duvick, L., Ranum, L., and Zoghbi, H. (1993) Expansion of an unstable trinucleotide CAG repeat in spinocerebellar ataxia type 1, *Nat. Genet. 4*, 221–226.

161. Sanpei, K., Takano, H., Igarashi, S., Sato, T., Oyake, M., Sasaki, H., Wakisaka, A., Tashiro, K., Ishida, Y., Ikeuchi, T., Koide, R., Saito, M., Sato, A., Tanaka, T., Hanyu, S., Takiyama, Y., Nishizawa, M., Shimizu, N., Nomura, Y., Segawa, M., Iwabuchi, K., Eguchi, I., Tanaka, H., Takahashi, H., and Tsuji, S. (1996) Identification of the spinocerebellar ataxia type 2 gene using a direct identification of repeat expansion and cloning technique, DIRECT, *Nat. Genet. 14*, 277–284.

162. Pulst, S., Nechiporuk, A., Nechiporuk, T., Gispert, S., Chen, X., Lopes-Cendes, I., Pearlman, S., Starkman, S., Orozco-Diaz, G., Lunkes, A., DeJong, P., Rouleau, G., Auburger, G., Korenberg, J., Figueroa, C., and Sahba, S. (1996) Moderate expansion of a normally biallelic trinucleotide repeat in spinocerebellar ataxia type 2, *Nat. Genet. 14*, 269–276.

163. Imbert, G., Saudou, F., Yvert, G., Devys, D., Trottier, Y., Garnier, J., Weber, C., Mandel, J., Cancel, G., Abbas, N., Dürr, A., Didierjean, O., Stevanin, G., Agid, Y., and Brice, A. (1996) Cloning of the gene for spinocerebellar ataxia 2 reveals a locus with high sensitivity to expanded CAG/glutamine repeats, *Nat. Genet. 14*, 285–291.

164. Kawaguchi, Y., Okamoto, T., Taniwaki, M., Aizawa, M., Inoue, M., Katayama, S., Kawakami, H., Nakamura, S., Nishimura, M., and Akiguchi, I. (1994) CAG expansions in a novel gene for Machado-Joseph disease at chromosome 14q32.1, *Nat. Genet. 8*, 221–228.

165. David, G., Abbas, N., Stevanin, G., Dürr, A., Yvert, G., Cancel, G., Weber, C., Imbert, G., Saudou, F., Antoniou, E., Drabkin, H., Gemmill, R., Giunti, P., Benomar, A., Wood, N., Ruberg, M., Agid, Y., Mandel, J., and Brice, A. (1997) Cloning of the SCA7 gene reveals a highly unstable CAG repeat expansion, *Nat. Genet. 17*, 65–70.

166. Zuhlke, C., Hellenbroich, Y., Dalski, A., Kononowa, N., Hagenah, J., Vieregge, P., Riess, O., Klein, C., and Schwinger, E. (2001) Different types of repeat expansion in the TATA-binding protein gene are associated with a new form of inherited ataxia, *Eur. J. Hum. Genet. 9*, 160–164.

167. Djian, P. (1998) Evolution of simple repeats in DNA and their relation to human disease, *Cell 94*, 155–160.

168. Hoffner, G. and Djian, P. (2005) Transglutaminase and diseases of the central nervous system, *Front. Biosci. 10*, 3078–3092.

169. Ordway, J., Tallaksen-Greene, S., Gutekunst, C., Bernstein, E., Cearley, J., Wiener, H., Dure, L. T., Lindsey, R., Hersch, S., Jope, R., Albin, R., and Detloff, P. (1997) Ectopically expressed CAG repeats cause intranuclear inclusions and a progressive late onset neurological phenotype in the mouse, *Cell 91*, 753–763.

170. Mangiarini, L., Sathasivam, K., Seller, M., Cozens, B., Harper, A., Hetherington, C., Lawton, M., Trottier, Y., Lehrach, H., Davies, S., and Bates, G. (1996) Exon 1 of the HD gene with an expanded CAG repeat is sufficient to cause a progressive neurological phenotype in transgenic mice, *Cell 87*, 493–506.

171. Vonsattel, J., Myers, R., Stevens, T., Ferrante, R., Bird, E., and Richardson, E. J. (1985) Neuropathological classification of Huntington's disease, *J. Neuropathol. Exp. Neurol. 44*, 559–577.

172. Ikeda, H., Yamaguchi, M., Sugai, S., Aze, Y., Narumiya, S., and Kakizuka, A. (1996) Expanded polyglutamine in the Machado-Joseph disease protein induces cell death in vitro and in vivo, *Nat. Genet. 13*, 196–202.

173. Warrick, J., Paulson, H., Gray-Board, G., Bui, Q., Fischbeck, K., Pittman, R., and Bonini, N. (1998) Expanded polyglutamine protein forms nuclear inclusions and causes neural degeneration in Drosophila, *Cell 93*, 939–949.

174. Tellez-Nagel, I., Johnson, A., and Terry, R. (1974) Studies on brain biopsies of patients with Huntington's chorea, *J. Neuropathol. Exp. Neurol. 33*, 308–332.

175. Subramaniam, S., Sixt, K., Barrow, R., and Snyder, S. (2009) Rhes, a striatal specific protein, mediates mutant-huntingtin cytotoxicity, *Science 324*, 1327–1330.

176. Paulson, H., Perez, M., Trottier, Y., Trojanowski, J., Subramony, S., Das, S., Vig, P., Mandel, J., Fischbeck, K., and Pittman, R. (1997) Intranuclear inclusions of expanded polyglutamine protein in spinocerebellar ataxia type 3, *Neuron 19*, 333–344.

177. Skinner, P., Koshy, B., Cummings, C., Klement, I., Helin, K., Servadio, A., Zoghbi, H., and Orr, H. (1997) Ataxin-1 with an expanded glutamine tract alters nuclear matrix-associated structures, *Nature 389*, 971–974.

178. Hayashi, Y., Kakita, A., Yamada, M., Koide, R., Igarashi, S., Takano, H., Ikeuchi, T., Wakabayashi, K., Egawa, S., Tsuji, S., and Takahashi, H. (1998) Hereditary dentatorubral-pallidoluysian atrophy: detection of widespread ubiquitinated neuronal and glial intranuclear inclusions in the brain, *Acta Neuropathol. 96*, 547–552.

179. Holmberg, M., Duyckaerts, C., Dürr, A., Cancel, G., Gourfinkel-An, I., Damier, P., Faucheux, B., Trottier, Y., Hirsch, E., Agid, Y., and Brice, A. (1998) Spinocerebellar ataxia type 7 (SCA7): a neurodegenerative disorder with neuronal intranuclear inclusions, *Hum. Mol. Genet. 7*, 913–918.

180. Ishikawa, K., Fujigasaki, H., Saegusa, H., Ohwada, K., Fujita, T., Iwamoto, H., Komatsuzaki, Y., Toru, S., Toriyama, H., Watanabe, M., Ohkoshi, N., Shoji, S., Kanazawa, I., Tanabe, T., and Mizusawa, H. (1999) Abundant expression and cytoplasmic aggregations of [α]1A voltage-dependent calcium channel protein associated with neurodegeneration in spinocerebellar ataxia type 6, *Hum. Mol. Genet. 8*, 1185–1193.

181. Koyano, S., Uchihara, T., Fujigasaki, H., Nakamura, A., Yagishita, S., and Iwabuchi, K. (1999) Neuronal intranuclear inclusions in spinocerebellar ataxia type 2: triple-labeling immunofluorescent study, *Neurosci. Lett. 273*, 117–120.

182. Nakamura, K., Jeong, S., Uchihara, T., Anno, M., Nagashima, K., Nagashima, T., Ikeda, S., Tsuji, S., and Kanazawa, I. (2001) SCA17, a novel autosomal dominant cerebellar ataxia caused by an expanded polyglutamine in TATA-binding protein, *Hum. Mol. Genet. 10*, 1441–1448.

183. Fujigasaki, H., Martin, J., De Deyn, P., Camuzat, A., Deffond, D., Stevanin, G., Dermaut, B., Van Broeckhoven, C., Dürr, A., and Brice, A. (2001) CAG repeat expansion in the TATA box-binding protein gene causes autosomal dominant cerebellar ataxia, *Brain 124*, 1939–1947.

184. Hoffner, G., Island, M., and Djian, P. (2005) Purification of neuronal inclusions of patients with Huntington's disease reveals a broad range of N-terminal fragments of expanded huntingtin and insoluble polymers, *J. Neurochem. 95*, 125–136.

185. Becher, M., Kotzuk, J., Sharp, A., Davies, S., Bates, G., Price, D., and Ross, C. (1998) Intranuclear neuronal inclusions in Huntington's disease and dentatorubral and pallidoluysian atrophy: correlation between the density of inclusions and IT15 CAG triplet repeat length, *Neurobiol. Dis. 4*, 387–397.

186. Sieradzan, K., Mechan, A., Jones, L., Wanker, E., Nukina, N., and Mann, D. (1999) Huntington's disease intranuclear inclusions contain truncated, ubiquitinated huntingtin protein, *Exp. Neurol. 156*, 92–99.

187. Gutekunst, C., Li, S., Yi, H., Mulroy, J., Kuemmerle, S., Jones, R., Rye, D., Ferrante, R., Hersch, S., and Li, X. (1999) Nuclear and neuropil aggregates in Huntington's disease: relationship to neuropathology, *J. Neurosci. 19*, 2522–2534.

188. Iuchi, S., Hoffner, G., Verbeke, P., Djian, P., and Green, H. (2003) Oligomeric and polymeric aggregates formed by proteins containing expanded polyglutamine, *Proc. Natl. Acad. Sci. U.S.A. 100*, 2409–2414.

189. Kahlem, P. and Djian, P. (2000) The expanded CAG repeat associated with juvenile Huntington disease shows a common origin of most or all neurons and glia in human cerebrum, *Neurosci. Lett. 286*, 203–207.

190. Scherzinger, E., Lurz, R., Turmaine, M., Mangiarini, L., Hollenbach, B., Hasenbank, R., Bates, G., Davies, S., Lehrach, H., and Wanker, E. (1997) Huntingtin-encoded polyglutamine expansions form amyloid-like protein aggregates in vitro and in vivo, *Cell 90*, 549–558.

191. Yang, W., Dunlap, J., Andrews, R., and Wetzel, R. (2002) Aggregated polyglutamine peptides delivered to nuclei are toxic to mammalian cells, *Hum. Mol. Genet. 11*, 2905–2917.

192. Li, H., Li, S., Yu, Z., Shelbourne, P., and Li, X. (2001) Huntingtin aggregate-associated axonal degeneration is an early pathological event in Huntington's disease mice, *J. Neurosci. 21*, 8473–8481.

193. Carmichael, J., Chatellier, J., Woolfson, A., Milstein, C., Fersht, A., and Rubinsztein, D. (2000) Bacterial and yeast chaperones reduce both aggregate formation and cell death in mammalian cell models of Huntington's disease, *Proc. Natl. Acad. Sci. U.S.A. 97*, 9701–9705.

194. Kitamura, A., Kubota, H., Pack, C., Matsumoto, G., Hirayama, S., Takahashi, Y., Kimura, H., Kinjo, M., Morimoto, R., and Nagata, K. (2006) Cytosolic chaperonin prevents polyglutamine toxicity with altering the aggregation state, *Nat. Cell Biol. 8*, 1163–1170.

195. Iwata, A., Nagashima, Y., Matsumoto, L., Suzuki, T., Yamanaka, T., Date, H., Deoka, K., Nukina, N., and Tsuji, S. (2009) Intranuclear degradation of polyglutamine aggregates by the ubiquitin-proteasome system, *J. Biol. Chem. 284*, 9796–9803.

196. Palazzolo, I., Stack, C., Kong, L., Musaro, A., Adachi, H., Katsuno, M., Sobue, G., Taylor, J., Sumner, C., Fischbeck, K., and Pennuto, M. (2009) Overexpression of IGF-1

in muscle attenuates disease in a mouse model of spinal and bulbar muscular atrophy, *Neuron 63*, 316–328.

197. Wang, H., Xu, Y., Zhao, X., Zhao, H., Yan, J., Sun, X., Guo, J., and Zhu, C. (2009) Overexpression of F(0)F(1)-ATP synthase α suppresses mutant huntingtin aggregation and toxicity in vitro, *Biochem. Biophys. Res. Commun. 390*, 1294–1298.
198. Wang, C., Zhou, H., McGuire, J., Cerullo, V., Lee, B., Li, S., and Li, X. (2008) Suppression of neuropil aggregates and neurological symptoms by an intracellular antibody implicates the cytoplasmic toxicity of mutant huntingtin, *J. Cell Biol. 181*, 803–816.
199. Sanchez, I., Mahlke, C., and Yuan, J. (2003) Pivotal role of oligomerization in expanded polyglutamine neurodegenerative disorders, *Nature 421*, 373–379.
200. Kazantsev, A., Walker, H., Slepko, N., Bear, J., Preisinger, E., Steffan, J., Zhu, Y., Gertler, F., Housman, D., Marsh, J., and Thompson, L. (2002) A bivalent Huntingtin binding peptide suppresses polyglutamine aggregation and pathogenesis in Drosophila, *Nat. Genet. 30*, 367–376.
201. Saudou, F., Finkbeiner, S., Devys, D., and Greenberg, M. (1998) Huntingtin acts in the nucleus to induce apoptosis but death does not correlate with the formation of intranuclear inclusions, *Cell 95*, 55–66.
202. Arrasate, M., Mitra, S., Schweitzer, E., Segal, M., and Finkbeiner, S. (2004) Inclusion body formation reduces levels of mutant huntingtin and the risk of neuronal death, *Nature 431*, 805–810.
203. Uchihara, T., Iwabuchi, K., Funata, N., and Yagishita, S. (2002) Attenuated nuclear shrinkage in neurons with nuclear aggregates–a morphometric study on pontine neurons of Machado-Joseph disease brains, *Exp. Neurol. 178*, 124–128.
204. Nagaoka, U., Uchihara, T., Iwabuchi, K., Konno, H., Tobita, M., Funata, N., Yagishita, S., and Kato, T. (2003) Attenuated nuclear shrinkage in neurones with nuclear inclusions of SCA1 brains, *J. Neurol. Neurosurg. Psychiatry 74*, 597–601.
205. Ross, C., and Poirier, M. (2004) Protein aggregation and neurodegenerative disease, *Nat. Med. 10(Suppl)*, S10–S17.
206. Steffan, J., Agrawal, N., Pallos, J., Rockabrand, E., Trotman, L., Slepko, N., Illes, K., Lukacsovich, T., Zhu, Y., Cattaneo, E., Pandolfi, P., Thompson, L., and Marsh, J. (2004) SUMO modification of Huntingtin and Huntington's disease pathology, *Science 304*, 100–104.
207. Perutz, M., Johnson, T., Suzuki, M., and Finch, J. (1994) Glutamine repeats as polar zippers: their possible role in inherited neurodegenerative diseases, *Proc. Natl. Acad. Sci. U.S.A. 91*, 5355–5358.
208. Perutz, M. (1996) Glutamine repeats and inherited neurodegenerative diseases: molecular aspects, *Curr. Opin. Struct. Biol. 6*, 848–858.
209. Stott, K., Blackburn, J., Butler, P., and Perutz, M. (1995) Incorporation of glutamine repeats makes protein oligomerize: implications for neurodegenerative diseases, *Proc. Natl. Acad. Sci. U.S.A. 92*, 6509–6513.
210. Hazeki, N., Tukamoto, T., Goto, J., and Kanazawa, I. (2000) Formic acid dissolves aggregates of an N-terminal huntingtin fragment containing an expanded polyglutamine tract: applying to quantification of protein components of the aggregates, *Biochem. Biophys. Res. Commun. 277*, 386–393.

211. Lin, C., Tallaksen-Greene, S., Chien, W., Cearley, J., Jackson, W., Crouse, A., Ren, S., Li, X., Albin, R., and Detloff, P. (2001) Neurological abnormalities in a knock-in mouse model of Huntington's disease, *Hum. Mol. Genet. 10*, 137–144.

212. Zhou, H., Cao, F., Wang, Z., Yu, Z., Nguyen, H., Evans, J., Li, S., and Li, X. (2003) Huntingtin forms toxic NH2-terminal fragment complexes that are promoted by the age-dependent decrease in proteasome activity, *J. Cell Biol. 163*, 109–118.

213. Cooper, A. J., Sheu, K. F., Burke, J. R., Onodera, O., Strittmatter, W. J., Roses, A. D., and Blass, J. P. (1997) Polyglutamine domains are substrates of tissue transglutaminase: does transglutaminase play a role in expanded CAG/poly-Q neurodegenerative diseases? *J. Neurochem. 69*, 431–434.

214. Gentile, V., Sepe, C., Calvani, M., Melone, M. A., Cotrufo, R., Cooper, A. J., Blass, J. P., and Peluso, G. (1998) Tissue transglutaminase-catalyzed formation of high-molecular-weight aggregates in vitro is favored with long polyglutamine domains: a possible mechanism contributing to CAG-triplet diseases, *Arch. Biochem. Biophys. 352*, 314–321.

215. Igarashi, S., Koide, R., Shimohata, T., Yamada, M., Hayashi, Y., Takano, H., Date, H., Oyake, M., Sato, T., Sato, A., Egawa, S., Ikeuchi, T., Tanaka, H., Nakano, R., Tanaka, K., Hozumi, I., Inuzuka, T., Takahashi, H., and Tsuji, S. (1998) Suppression of aggregate formation and apoptosis by transglutaminase inhibitors in cells expressing truncated DRPLA protein with an expanded polyglutamine stretch, *Nat. Genet. 18*, 111–117.

216. de Cristofaro, T., Affaitati, A., Cariello, L., Avvedimento, E., and Varrone, S. (1999) The length of polyglutamine tract, its level of expression, the rate of degradation, and the transglutaminase activity influence the formation of intracellular aggregates, *Biochem. Biophys. Res. Commun. 260*, 150–158.

217. Mastroberardino, P., Iannicola, C., Nardacci, R., Bernassola, F., De Laurenzi, V., Melino, G., Moreno, S., Pavone, F., Oliverio, S., Fesus, L., and Piacentini, M. (2002) 'Tissue' transglutaminase ablation reduces neuronal death and prolongs survival in a mouse model of Huntington's disease, *Cell Death Differ. 9*, 873–880.

218. Bailey, C. and Johnson, G. (2005) Tissue transglutaminase contributes to disease progression in the R6/2 Huntington's disease mouse model via aggregate-independent mechanisms, *J. Neurochem. 92*, 83–92.

219. Lai, T., Tucker, T., Burke, J., Strittmatter, W., and Greenberg, C. (2004) Effect of tissue transglutaminase on the solubility of proteins containing expanded polyglutamine repeats, *J. Neurochem. 88*, 1253–1260.

220. Karpuj, M., Becher, M., Springer, J., Chabas, D., Youssef, S., Pedotti, R., Mitchell, D., and Steinman, L. (2002) Prolonged survival and decreased abnormal movements in transgenic model of Huntington disease, with administration of the transglutaminase inhibitor cystamine, *Nat. Med. 8*, 143–149.

221. Wang, X., Sarkar, A., Cicchetti, F., Yu, M., Zhu, A., Jokivarsi, K., Saint-Pierre, M., and Brownell, A. (2005) Cerebral PET imaging and histological evidence of transglutaminase inhibitor cystamine induced neuroprotection in transgenic R6/2 mouse model of Huntington's disease, *J. Neurol. Sci. 231*, 57–66.

222. Van Raamsdonk, J., Pearson, J., Bailey, C., Rogers, D., Johnson, G., Hayden, M., and Leavitt, B. (2005) Cystamine treatment is neuroprotective in the YAC128 mouse model of Huntington disease, *J. Neurochem. 95*, 210–220.

223. Chun, W., Lesort, M., Tucholski, J., Ross, C., and Johnson, G. (2001) Tissue transglutaminase does not contribute to the formation of mutant huntingtin aggregates, *J. Cell Biol. 153*, 25–34.
224. Tucholski, J., Lesort, M., and Johnson, G. (2001) Tissue transglutaminase is essential for neurite outgrowth in human neuroblastoma SH-SY5Y cells, *Neuroscience 102*, 481–491.
225. Bailey, C. and Johnson, G. (2006) The protective effects of cystamine in the R6/2 Huntington's disease mouse involve mechanisms other than the inhibition of tissue transglutaminase, *Neurobiol. Aging 27*, 871–879.
226. Jeitner, T., Bogdanov, M., Matson, W., Daikhin, Y., Yudkoff, M., Folk, J., Steinman, L., Browne, S., Beal, M., Blass, J., and Cooper, A. (2001) N(ε)-(γ-L-glutamyl)-L-lysine (GGEL) is increased in cerebrospinal fluid of patients with Huntington's disease, *J. Neurochem. 79*, 1109–1112.
227. Johnson, G. and LeShoure, R. J. (2004) Immunoblot analysis reveals that isopeptide antibodies do not specifically recognize the ε-(γ-glutamyl)lysine bonds formed by transglutaminase activity, *J. Neurosci. Methods 134*, 151–158.
228. Lesort, M., Chun, W., Johnson, G., and Ferrante, R. (1999) Tissue transglutaminase is increased in Huntington's disease brain, *J. Neurochem. 73*, 2018–2027.
229. Cariello, L., de Cristofaro, T., Zanetti, L., Cuomo, T., Di Maio, L., Campanella, G., Rinaldi, S., Zanetti, P., Di Lauro, R., and Varrone, S. (1996) Transglutaminase activity is related to CAG repeat length in patients with Huntington's disease, *Hum. Genet. 98*, 633–635.
230. Hoffner, G. and Djian, P. (2002) Protein aggregation in Huntington's disease, *Biochimie 84*, 273–278.
231. Hoffner, G., Hoppilliard, Y., Van Der Rest, G., Dansette, P., Djian, P., and Ohanessian, G. (2008) [Nε-(γ-glutamyl) lysine] as a potential biomarker in neurological diseases: new detection method and fragmentation pathways, *J. Mass Spectrom. 43*, 456–469.
232. Selkoe, D., Abraham, C., and Ihara, Y. (1982) Brain transglutaminase: in vitro crosslinking of human neurofilament proteins into insoluble polymers, *Proc. Natl. Acad. Sci. U.S.A. 79*, 6070–6074.
233. Lee, V., Balin, B., Otvos, L. J., and Trojanowski, J. (1991) A68: a major subunit of paired helical filaments and derivatized forms of normal Tau, *Science 251*, 675–678.
234. Goedert, M., Spillantini, M., Cairns, N., and Crowther, R. (1992) Tau proteins of Alzheimer paired helical filaments: abnormal phosphorylation of all six brain isoforms, *Neuron 8*, 159–168.
235. Dudek, S. and Johnson, G. (1993) Transglutaminase catalyzes the formation of sodium dodecyl sulfate-insoluble, Alz-50-reactive polymers of tau, *J. Neurochem. 61*, 1159–1162.
236. Dudek, S. and Johnson, G. (1994) Transglutaminase facilitates the formation of polymers of the β-amyloid peptide, *Brain Res. 651*, 129–133.
237. Miller, M. and Johnson, G. (1995) Transglutaminase cross-linking of the tau protein, *J. Neurochem. 65*, 1760–1770.

238. Appelt, D. and Balin, B. (1997) The association of tissue transglutaminase with human recombinant tau results in the formation of insoluble filamentous structures, *Brain Res.* *745*, 21–31.

239. Halverson, R., Lewis, J., Frausto, S., Hutton, M., and Muma, N. (2005) Tau protein is cross-linked by transglutaminase in P301L tau transgenic mice, *J. Neurosci.* *25*, 1226–1233.

240. Appelt, D., Kopen, G., Boyne, L., and Balin, B. (1996) Localization of transglutaminase in hippocampal neurons: implications for Alzheimer's disease, *J. Histochem. Cytochem.* *44*, 1421–1427.

241. Wilhelmus, M., Grunberg, S., Bol, J., van Dam, A., Hoozemans, J., Rozemuller, A., and Drukarch, B. (2009) Transglutaminases and transglutaminase-catalyzed cross-links colocalize with the pathological lesions in Alzheimer's disease brain, *Brain Pathol.* *19*, 612–622.

242. Gorevic, P., Goñi, F., Pons-Estel, B., Alvarez, F., Peress, N., and Frangione, B. (1986) Isolation and partial characterization of neurofibrillary tangles and amyloid plaque core in Alzheimer's disease: immunohistological studies, *J. Neuropathol. Exp. Neurol.* *45*, 647–664.

243. Selkoe, D., Abraham, C., Podlisny, M., and Duffy, L. (1986) Isolation of low-molecular-weight proteins from amyloid plaque fibers in Alzheimer's disease, *J. Neurochem.* *46*, 1820–1834.

244. Ikura, K., Takahata, K., and Sasaki, R. (1993) Cross-linking of a synthetic partial-length (1–28) peptide of the Alzheimer β/A4 amyloid protein by transglutaminase, *FEBS Lett.* *326*, 109–111.

245. Ho, G., Gregory, E., Smirnova, I., Zoubine, M., and Festoff, B. (1994) Cross-linking of β-amyloid protein precursor catalyzed by tissue transglutaminase, *FEBS Lett.* *349*, 151–154.

246. Rasmussen, L., Sørensen, E., Petersen, T., Gliemann, J., and Jensen, P. (1994) Identification of glutamine and lysine residues in Alzheimer amyloid β A4 peptide responsible for transglutaminase-catalysed homopolymerization and cross-linking to α 2M receptor, *FEBS Lett.* *338*, 161–166.

247. Hartley, D., Zhao, C., Speier, A., Woodard, G., Li, S., Li, Z., and Walz, T. (2008) Transglutaminase induces protofibril-like amyloid β-protein assemblies that are protease-resistant and inhibit long-term potentiation, *J. Biol. Chem.* *283*, 16790–16800.

248. Wakshlag, J., Antonyak, M., Boehm, J., Boehm, K., and Cerione, R. (2006) Effects of tissue transglutaminase on β -amyloid1-42-induced apoptosis, *Protein J.* *25*, 83–94.

249. Johnson, G., Cox, T., Lockhart, J., Zinnerman, M., Miller, M., and Powers, R. (1997) Transglutaminase activity is increased in Alzheimer's disease brain, *Brain Res.* *751*, 323–329.

250. Nemes, Z., Fésüs, L., Egerházi, A., Keszthelyi, A., and Degrell, I. (2001) N(ε)(γ-glutamyl)lysine in cerebrospinal fluid marks Alzheimer type and vascular dementia, *Neurobiol. Aging* *22*, 403–406.

251. Norlund, M., Lee, J., Zainelli, G., and Muma, N. (1999) Elevated transglutaminase-induced bonds in PHF tau in Alzheimer's disease, *Brain Res.* *851*, 154–163.

252. Singer, S., Zainelli, G., Norlund, M., Lee, J., and Muma, N. (2002) Transglutaminase bonds in neurofibrillary tangles and paired helical filament tau early in Alzheimer's disease, *Neurochem. Int. 40*, 17–30.

253. Nemes, Z., Devreese, B., Steinert, P., Van Beeumen, J., and Fésüs, L. (2004) Cross-linking of ubiquitin, HSP27, parkin, and α-synuclein by γ-glutamyl-ε-lysine bonds in Alzheimer's neurofibrillary tangles, *FASEB J. 18*, 1135–1137.

254. Wilhelmus, M., Verhaar, R., Bol, J., van Dam, A., Hoozemans, J., Rozemuller, A., and Drukarch, B. (2011) Novel role of transglutaminase 1 in corpora amylacea formation? *Neurobiol. Aging. 32*, 845–856.

255. Zemaitaitis, M., Kim, S., Halverson, R., Troncoso, J., Lee, J., and Muma, N. (2003) Transglutaminase activity, protein, and mRNA expression are increased in progressive supranuclear palsy, *J. Neuropathol. Exp. Neurol. 62*, 173–184.

256. Zemaitaitis, M., Lee, J., Troncoso, J., and Muma, N. (2000) Transglutaminase-induced cross-linking of tau proteins in progressive supranuclear palsy, *J. Neuropathol. Exp. Neurol. 59*, 983–989.

257. Tremblay, M., Saint-Pierre, M., Bourhis, E., Lévesque, D., Rouillard, C., and Cicchetti, F. (2006) Neuroprotective effects of cystamine in aged parkinsonian mice, *Neurobiol. Aging 27*, 862–870.

258. Citron, B., Suo, Z., SantaCruz, K., Davies, P., Qin, F., and Festoff, B. (2002) Protein crosslinking, tissue transglutaminase, alternative splicing and neurodegeneration, *Neurochem. Int. 40*, 69–78.

259. Andringa, G., Lam, K., Chegary, M., Wang, X., Chase, T., and Bennett, M. (2004) Tissue transglutaminase catalyzes the formation of α-synuclein crosslinks in Parkinson's disease, *FASEB J. 18*, 932–934.

260. Segers-Nolten, I., Wilhelmus, M., Veldhuis, G., van Rooijen, B., Drukarch, B., and Subramaniam, V. (2008) Tissue transglutaminase modulates α-synuclein oligomerization, *Protein Sci. 17*, 1395–1402.

261. Schmid, A., Chiappe, D., Pignat, V., Grimminger, V., Hang, I., Moniatte, M., and Lashuel, H. (2009) Dissecting the mechanisms of tissue transglutaminase-induced cross-linking of α-synuclein: implications for the pathogenesis of Parkinson disease, *J. Biol. Chem. 284*, 13128–13142.

262. Pincus, J. and Waelsch, H. (1968) The specificity of transglutaminase. I. Human hemoglobin as a substrate for the enzyme, *Arch. Biochem. Biophys. 126*, 34–43.

263. Aeschlimann, D. and Paulsson, M. (1994) Transglutaminases: protein cross-linking enzymes in tissues and body fluids, *Thromb. Haemost. 71*, 402–415.

264. Lorand, L. (1996) Neurodegenerative diseases and transglutaminase, *Proc. Natl. Acad. Sci. U.S.A. 93*, 14310–14313.

265. Hettasch, J. and Greenberg, C. (1994) Analysis of the catalytic activity of human factor XIIIa by site-directed mutagenesis, *J. Biol. Chem. 269*, 28309–28313.

266. Spraggon, G., Everse, S., and Doolittle, R. (1997) Crystal structures of fragment D from human fibrinogen and its crosslinked counterpart from fibrin, *Nature 389*, 455–462.

267. Csosz, E., Bagossi, P., Nagy, Z., Dosztanyi, Z., Simon, I., and Fesus, L. (2008) Substrate preference of transglutaminase 2 revealed by logistic regression analysis and intrinsic disorder examination, *J. Mol. Biol. 383*, 390–402.

268. Greenberg, C., Birckbichler, P., and Rice, R. (1991) Transglutaminases: multifunctional cross-linking enzymes that stabilize tissues, *FASEB J. 5*, 3071–3077.

269. Aeschlimann, D., Paulsson, M., and Mann, K. (1992) Identification of Gln726 in nidogen as the amine acceptor in transglutaminase-catalyzed cross-linking of laminin-nidogen complexes, *J. Biol. Chem. 267*, 11316–11321.

270. Parameswaran, K., Velasco, P., Wilson, J., and Lorand, L. (1990) Labeling of ε-lysine crosslinking sites in proteins with peptide substrates of factor XIIIa and transglutaminase, *Proc. Natl. Acad. Sci. U.S.A. 87*, 8472–8475.

271. Toda, H. and Folk, J. (1969) Determination of protein-bound glutamine, *Biochim. Biophys. Acta 175*, 427–430.

272. Tamaki, T. and Aoki, N. (1982) Cross-linking of α 2-plasmin inhibitor to fibrin catalyzed by activated fibrin-stabilizing factor, *J. Biol. Chem. 257*, 14767–14772.

273. Berbers, G., Bentlage, H., Brans, A., Bloemendal, H., and de Jong, W. (1983) β-Crystallin: endogenous substrate of lens transglutaminase. Characterization of the acyl-donor site in the β Bp chain, *Eur. J. Biochem. 135*, 315–320.

274. Berbers, G., Feenstra, R., Van Den Bos, R., Hoekman, W., Bloemendal, H., and de Jong, W. (1984) Lens transglutaminase selects specific β-crystallin sequences as substrate, *Proc. Natl. Acad. Sci. U.S.A. 81*, 7017–7020.

275. Bowness, J., Folk, J., and Timpl, R. (1987) Identification of a substrate site for liver transglutaminase on the aminopropeptide of type III collagen, *J. Biol. Chem. 262*, 1022–1024.

276. Chen, R. and Doolittle, R. (1971) - cross-linking sites in human and bovine fibrin, *Biochemistry 10*, 4487–4491.

277. Coussons, P., Price, N., Kelly, S., Smith, B., and Sawyer, L. (1992) Factors that govern the specificity of transglutaminase-catalysed modification of proteins and peptides, *Biochem. J. 282(Pt 3)*, 929–930.

278. Murthy, S., Wilson, J., Lukas, T., Kuret, J., and Lorand, L. (1998) Cross-linking sites of the human tau protein, probed by reactions with human transglutaminase, *J. Neurochem. 71*, 2607–2614.

279. Ferrándiz, C., Pérez-Payá, E., Braco, L., and Abad, C. (1994) Gln5 selectively monodansylated substance P as a sensitive tool for interaction studies with membranes, *Biochem. Biophys. Res. Commun. 203*, 359–365.

280. Pastor, M., Diez, A., Pérez-Payá, E., and Abad, C. (1999) Addressing substrate glutamine requirements for tissue transglutaminase using substance P analogues, *FEBS Lett. 451*, 231–234.

281. Gorman, J. and Folk, J. (1980) Structural features of glutamine substrates for human plasma factor XIIIa (activated blood coagulation factor XIII). *J. Biol. Chem. 255*, 419–427.

282. Gorman, J. and Folk, J. (1981) Structural features of glutamine substrates for transglutaminases. Specificities of human plasma factor XIIIa and the guinea pig liver enzyme toward synthetic peptides, *J. Biol. Chem. 256*, 2712–2715.

283. Gorman, J. and Folk, J. (1984) Structural features of glutamine substrates for transglutaminases. Role of extended interactions in the specificity of human plasma factor XIIIa and of the guinea pig liver enzyme, *J. Biol. Chem. 259*, 9007–9010.

284. Gross, M., Whetzel, N., and Folk, J. (1975) The extended active site of guinea pig liver transglutaminase, *J. Biol. Chem. 250*, 4648–4655.

285. Sollid, L. (2002) Coeliac disease: dissecting a complex inflammatory disorder, *Nat. Rev. Immunol. 2*, 647–655.

286. Dieterich, W., Ehnis, T., Bauer, M., Donner, P., Volta, U., Riecken, E., and Schuppan, D. (1997) Identification of tissue transglutaminase as the autoantigen of celiac disease, *Nat. Med. 3*, 797–801.

287. Molberg, O., Mcadam, S., Körner, R., Quarsten, H., Kristiansen, C., Madsen, L., Fugger, L., Scott, H., Norén, O., Roepstorff, P., Lundin, K., Sjöström, H., and Sollid, L. (1998) Tissue transglutaminase selectively modifies gliadin peptides that are recognized by gut-derived T cells in celiac disease, *Nat. Med. 4*, 713–717.

288. Shan, L., Molberg, Ø., Parrot, I., Hausch, F., Filiz, F., Gray, G., Sollid, L., and Khosla, C. (2002) Structural basis for gluten intolerance in celiac sprue, *Science 297*, 2275–2279.

289. Mowat, A. (2003) Coeliac disease–a meeting point for genetics, immunology, and protein chemistry, *Lancet 361*, 1290–1292.

290. Fleckenstein, B., Molberg, Ø., Qiao, S., Schmid, D., von der Mülbe, F., Elgstøen, K., Jung, G., and Sollid, L. (2002) Gliadin T cell epitope selection by tissue transglutaminase in celiac disease. Role of enzyme specificity and pH influence on the transamidation versus deamidation process, *J. Biol. Chem. 277*, 34109–34116.

291. Parameswaran, K., Cheng, X., Chen, E., Velasco, P., Wilson, J., and Lorand, L. (1997) Hydrolysis of $\gamma:\varepsilon$ isopeptides by cytosolic transglutaminases and by coagulation factor XIIIa, *J. Biol. Chem. 272*, 10311–10317.

292. Qiao, S., Piper, J., Haraldsen, G., Oynebråten, I., Fleckenstein, B., Molberg, O., Khosla, C., and Sollid, L. (2005) Tissue transglutaminase-mediated formation and cleavage of histamine-gliadin complexes: biological effects and implications for celiac disease, *J. Immunol. 174*, 1657–1663.

293. Stamnaes, J., Fleckenstein, B., and Sollid, L. (2008) The propensity for deamidation and transamidation of peptides by transglutaminase 2 is dependent on substrate affinity and reaction conditions, *Biochim. Biophys. Acta 1784*, 1804–1811.

294. Dekking, E., Van Veelen, P., De Ru, A., Kooy-Winkelaar, E., Groneveld, T., Nieuwenhuizen, W., and Konig, F. (2008) Microbial transglutaminase generates T cell stimulatory epitopes involved in celiac disease, *J. Cereal Sci. 47*, 339–346.

295. Bruce, S., Bjarnason, I., and Peters, T. (1985) Human jejunal transglutaminase: demonstration of activity, enzyme kinetics and substrate specificity with special relation to gliadin and coeliac disease, *Clin. Sci. (Lond.) 68*, 573–579.

296. Piper, J., Gray, G., and Khosla, C. (2002) High selectivity of human tissue transglutaminase for immunoactive gliadin peptides: implications for celiac sprue, *Biochemistry 41*, 386–393.

297. Skovbjerg, H., Norén, O., Anthonsen, D., Moller, J., and Sjöström, H. (2002) Gliadin is a good substrate of several transglutaminases: possible implication in the pathogenesis of coeliac disease, *Scand. J. Gastroenterol. 37*, 812–817.

298. Arentz-Hansen, H., McAdam, S., Molberg, Ø., Fleckenstein, B., Lundin, K., Jørgensen, T., Jung, G., Roepstorff, P., and Sollid, L. (2002) Celiac lesion T cells recognize epitopes

that cluster in regions of gliadins rich in proline residues, *Gastroenterology 123*, 803–809.

299. Vader, W., Kooy, Y., Van Veelen, P., De Ru, A., Harris, D., Benckhuijsen, W., Peña, S., Mearin, L., Drijfhout, J., and Koning, F. (2002) The gluten response in children with celiac disease is directed toward multiple gliadin and glutenin peptides, *Gastroenterology 122*, 1729–1737.
300. Sugimura, Y., Hosono, M., Wada, F., Yoshimura, T., Maki, M., and Hitomi, K. (2006) Screening for the preferred substrate sequence of transglutaminase using a phage-displayed peptide library: identification of peptide substrates for TGASE 2 and Factor XIIIA, *J. Biol. Chem. 281*, 17699–17706.
301. Keresztessy, Z., Csosz, E., Hársfalvi, J., Csomós, K., Gray, J., Lightowlers, R., Lakey, J., Balajthy, Z., and Fésüs, L. (2006) Phage display selection of efficient glutamine-donor substrate peptides for transglutaminase 2, *Protein Sci. 15*, 2466–2480.
302. Sugimura, Y., Hosono, M., Kitamura, M., Tsuda, T., Yamanishi, K., Maki, M., and Hitomi, K. (2008) Identification of preferred substrate sequences for transglutaminase 1–development of a novel peptide that can efficiently detect cross-linking enzyme activity in the skin, *FEBS J. 275*, 5667–5677.
303. Sugimura, Y., Yokoyama, K., Nio, N., Maki, M., and Hitomi, K. (2008) Identification of preferred substrate sequences of microbial transglutaminase from Streptomyces mobaraensis using a phage-displayed peptide library, *Arch. Biochem. Biophys. 477*, 379–383.
304. Green, H. and Wang, N. (1994) Codon reiteration and the evolution of proteins, *Proc. Natl. Acad. Sci. U.S.A. 91*, 4298–4302.
305. Hu, B. and Messersmith, P. (2003) Rational design of transglutaminase substrate peptides for rapid enzymatic formation of hydrogels, *J. Am. Chem. Soc. 125*, 14298–14299.
306. Clarke, D., Mycek, M., Neidle, A., and Waelsch, H. (1959) The Incorporation of Amines into Protein, *Arch. Biochem. Biophys. 79*, 338–354.
307. Pincus, J. and Waelsch, H. (1968) The specificity of transglutaminase. II. Structural requirements of the amine substrate, *Arch. Biochem. Biophys. 126*, 44–52.
308. Gross, M., Whetzel, N., and Folk, J. (1977) Amine binding sites in acyl intermediates of transglutaminases. Human blood plasma enzyme (activated coagulation factor XIII) and guinea pig liver enzyme, *J. Biol. Chem. 252*, 3752–3759.
309. Groenen, P., Bloemendal, H., and de Jong, W. (1992) The carboxy-terminal lysine of α B-crystallin is an amine-donor substrate for tissue transglutaminase, *Eur. J. Biochem. 205*, 671–674.
310. Groenen, P., Seccia, M., Smulders, R., Gravela, E., Cheeseman, K., Bloemendal, H., and de Jong, W. (1993) Exposure of β H-crystallin to hydroxyl radicals enhances the transglutaminase-susceptibility of its existing amine-donor and amine-acceptor sites, *Biochem. J. 295(Pt 2)*, 399–404.
311. Lorand, L., Velasco, P., Murthy, S., Wilson, J., and Parameswaran, K. (1992) Isolation of transglutaminase-reactive sequences from complex biological systems: a prominent lysine donor sequence in bovine lens, *Proc. Natl. Acad. Sci. U.S.A. 89*, 11161–11163.

312. Pucci, P., Malorni, A., Marino, G., Metafora, S., Esposito, C., and Porta, R. (1988) B-endorphin modification by transglutaminase in vitro: identification by FAB/MS of glutamine-11 and lysine-29 as acyl donor and acceptor sites, *Biochem. Biophys. Res. Commun. 154*, 735–740.

313. Clement, S., Velasco, P., Murthy, S., Wilson, J., Lukas, T., Goldman, R., and Lorand, L. (1998) The intermediate filament protein, vimentin, in the lens is a target for cross-linking by transglutaminase, *J. Biol. Chem. 273*, 7604–7609.

314. Murthy, S., Lukas, T., Jardetzky, T., and Lorand, L. (2009) Selectivity in the post-translational, transglutaminase-dependent acylation of lysine residues, *Biochemistry 48*, 2654–2660.

315. Groenen, P., Grootjans, J., Lubsen, N., Bloemendal, H., and de Jong, W. (1994) Lys-17 is the amine-donor substrate site for transglutaminase in β A3-crystallin, *J. Biol. Chem. 269*, 831–833.

316. Grootjans, J., Groenen, P., and de Jong, W. (1995) Substrate requirements for transglutaminases. Influence of the amino acid residue preceding the amine donor lysine in a native protein, *J. Biol. Chem. 270*, 22855–22858.

317. Groenen, P., Smulders, R., Peters, R., Grootjans, J., Van Den Ijssel, P., Bloemendal, H., and de Jong, W. (1994) The amine-donor substrate specificity of tissue-type transglutaminase. Influence of amino acid residues flanking the amine-donor lysine residue, *Eur. J. Biochem. 220*, 795–799.

318. McDonagh, J. and Fukue, H. (1996) Determinants of substrate specificity for factor XIII, *Semin. Thromb. Hemost. 22*, 369–376.

319. Chung, S. (1972) Comparative studies on tissue transglutaminase and factor XIII, *Ann. N. Y. Acad. Sci. 202*, 240–255.

320. Hettasch, J., Peoples, K., and Greenberg, C. (1997) Analysis of factor XIII substrate specificity using recombinant human factor XIII and tissue transglutaminase chimeras, *J. Biol. Chem. 272*, 25149–25156.

321. Shainoff, J., Urbanic, D., and DiBello, P. (1991) Immunoelectrophoretic characterizations of the cross-linking of fibrinogen and fibrin by factor XIIIa and tissue transglutaminase. Identification of a rapid mode of hybrid α-/γ-chain cross-linking that is promoted by the γ-chain cross-linking, *J. Biol. Chem. 266*, 6429–6437.

322. Tarcsa, E., Marekov, L., Andreoli, J., Idler, W., Candi, E., Chung, S., and Steinert, P. (1997) The fate of trichohyalin. Sequential post-translational modifications by peptidyl-arginine deiminase and transglutaminases, *J. Biol. Chem. 272*, 27893–27901.

323. Robinson, N., Baker, P., Jones, C., and Aplin, J. (2007) A role for tissue transglutaminase in stabilization of membrane-cytoskeletal particles shed from the human placenta, *Biol. Reprod. 77*, 648–657.

324. Ichikawa, A., Ishizaki, J., Morita, M., Tanaka, K., and Ikura, K. (2008) Identification of new amine acceptor protein substrate candidates of transglutaminase in rat liver extract: use of 5-(biotinamido) pentylamine as a probe, *Biosci. Biotechnol. Biochem. 72*, 1056–1062.

325. Orru, S., Caputo, I., D'Amato, A., Ruoppolo, M., and Esposito, C. (2003) Proteomics identification of acyl-acceptor and acyl-donor substrates for transglutaminase in a human

intestinal epithelial cell line. Implications for celiac disease, *J. Biol. Chem. 278*, 31766–31773.

326. Wilhelmus, M., van Dam, A., and Drukarch, B. (2008) Tissue transglutaminase: a novel pharmacological target in preventing toxic protein aggregation in neurodegenerative diseases, *Eur. J. Pharmacol. 585*, 464–472.

327. Hoffner, G., André, W., Vanhoutteghem, A., Souès, S., and Djian, P. (2010) Transglutaminase-catalyzed cross linking in neurological disease: from experimental evidence to therapeutic inhibition, *CNS Neurol. Disord. Drug Targets 9*, 217–231.

328. Gentile, V. and Cooper, A. (2004) Transglutaminases - possible drug targets in human diseases, *Curr. Drug Targets CNS Neurol. Disord. 3*, 99–104.

329. Sollid, L. and Khosla, C. (2005) Future therapeutic options for celiac disease, *Nat. Clin. Pract. Gastroenterol. Hepatol. 2*, 140–147.

330. Siegel, M., Bethune, M., Gass, J., Ehren, J., Xia, J., Johannsen, A., Stuge, T., Gray, G., Lee, P., and Khosla, C. (2006) Rational design of combination enzyme therapy for celiac sprue, *Chem. Biol. 13*, 649–658.

331. Kim, S. Y. (2006) Transglutaminase 2 in inflammation, *Front. Biosci. 11*, 3026–3035.

332. Pinkas, D. M., Strop, P., Brunger, A. T., and Khosla, C. (2007) Transglutaminase 2 undergoes a large conformational change upon activation, *PLoS Biol. 5*, e327.

333. Falasca, L., Farrace, M. G., Rinaldi, A., Tuosto, L., Melino, G., and Piacentini, M. (2008) Transglutaminase type II is involved in the pathogenesis of endotoxic shock, *J. Immunol. 180*, 2616–2624.

334. Maiuri, L., Luciani, A., Giardino, I., Raia, V., Villella, V., D'Apolito, M., Pettoello-Mantovani, M., Guido, S., Ciacci, C., Cimmino, M., Cexus, O., Londei, M., and Quaratino, S. (2008) Tissue transglutaminase activation modulates inflammation in cystic fibrosis via PPARγ down-regulation, *J. Immunol. 180*, 7697–7705.

335. Elli, L., Bergamini, C., Bardella, M., and Schuppan, D. (2009) Transglutaminases in inflammation and fibrosis of the gastrointestinal tract and the liver, *Dig. Liver Dis. 41*, 541–550.

336. Siegel, M. and Khosla, C. (2007) Transglutaminase 2 inhibitors and their therapeutic role in disease states, *Pharmacol. Ther. 115*, 232–245.

337. Verma, A. and Mehta, K. (2007) Tissue transglutaminase-mediated chemoresistance in cancer cells, *Drug Resist Updat 10*, 144–151.

338. Kotsakis, P. and Griffin, M. (2007) Tissue transglutaminase in tumour progression: friend or foe? *Amino Acids 33*, 373–384.

339. Ai, L., Kim, W. J., Demircan, B., Dyer, L. M., Bray, K. J., Skehan, R. R., Massoll, N. A., and Brown, K. D. (2008) The transglutaminase 2 gene (TGM2), a potential molecular marker for chemotherapeutic drug sensitivity, is epigenetically silenced in breast cancer, *Carcinogenesis 29*, 510–518.

340. Jiang, W., Ablin, R., Kynaston, H., and Mason, M. (2009) The prostate transglutaminase (TGase-4, TGaseP) regulates the interaction of prostate cancer and vascular endothelial cells, a potential role for the ROCK pathway, *Microvasc. Res. 77*, 150–157.

341. Uemura, N., Nakanishi, Y., Kato, H., Saito, S., Nagino, M., Hirohashi, S., and Kondo, T. (2009) Transglutaminase 3 as a prognostic biomarker in esophageal cancer revealed by proteomics, *Int. J. Cancer 124*, 2106–2115.

342. Park, K., Kim, H., Lee, J., Choi, Y., Park, S., Yang, S., Kim, S., and Hong, K. (2010) Transglutaminase 2 as a cisplatin resistance marker in non-small cell lung cancer, *J. Cancer Res. Clin. Oncol. 136*, 493–502.

343. Kim, D., Park, K., Jeong, K., Lee, B., Lee, C., and Kim, S. (2009) Glucosamine is an effective chemo-sensitizer via transglutaminase 2 inhibition, *Cancer Lett. 273*, 243–249.

344. Molberg, Ø., Solheim Flaete, N., Jensen, T., Lundin, K., Arentz-Hansen, H., Anderson, O., Kjersti Uhlen, A., and Sollid, L. (2003) Intestinal T-cell responses to high-molecular-weight glutenins in celiac disease, *Gastroenterology 125*, 337–344.

345. Maiuri, L., Ciacci, C., Ricciardelli, I., Vacca, L., Raia, V., Rispo, A., Griffin, M., Issekutz, T., Quaratino, S., and Londei, M. (2005) Unexpected role of surface transglutaminase type II in celiac disease, *Gastroenterology 129*, 1400–1413.

346. Case, A. and Stein, R. (2007) Kinetic analysis of the interaction of tissue transglutaminase with a nonpeptidic slow-binding inhibitor, *Biochemistry 46*, 1106–1115.

347. Pardin, C., Pelletier, J., Lubell, W., and Keillor, J. (2008) Cinnamoyl inhibitors of tissue transglutaminase, *J. Org. Chem. 73*, 5766–5775.

348. Hausch, F., Halttunen, T., Mäki, M., and Khosla, C. (2003) Design, synthesis, and evaluation of gluten peptide analogs as selective inhibitors of human tissue transglutaminase, *Chem. Biol. 10*, 225–231.

349. Griffin, M., Mongeot, A., Collighan, R., Saint, R., Jones, R., Coutts, I., and Rathbone, D. (2008) Synthesis of potent water-soluble tissue transglutaminase inhibitors, *Bioorg. Med. Chem. Lett. 18*, 5559–5562.

350. Jager, M., Nir, E., and Weiss, S. (2006) Site-specific labeling of proteins for single-molecule FRET by combining chemical and enzymatic modification, *Protein Sci. 15*, 640–646.

351. Valdivia, A., Villalonga, R., Di Pierro, P., Pérez, Y., Mariniello, L., Gómez, L., and Porta, R. (2006) Transglutaminase-catalyzed site-specific glycosidation of catalase with aminated dextran, *J. Biotechnol. 122*, 326–333.

352. Sugimura, Y., Ueda, H., Maki, M., and Hitomi, K. (2007) Novel site-specific immobilization of a functional protein using a preferred substrate sequence for transglutaminase 2, *J. Biotechnol. 131*, 121–127.

353. Csosz, E., Meskó, B., and Fésüs, L. (2009) Transdab wiki: the interactive transglutaminase substrate database on web 2.0 surface, *Amino Acids 36*, 615–617.

354. Facchiano, A., Facchiano, A., and Facchiano, F. (2003) Active Sequences Collection (ASC) database: a new tool to assign functions to protein sequences, *Nucleic Acids Res. 31*, 379–382.

TRANSGLUTAMINASE 2: A NEW PARADIGM FOR NF-κB INVOLVEMENT IN DISEASE

SOO-YOUL KIM

CONTENTS

I. INTRODUCTION

A. DISCOVERY OF A NEW ROLE OF TGase 2 IN NF-κB ACTIVATION

For decades, my group and many others have observed that TGase 2 expression is often increased in inflammation (see review [1, 2]). The mechanisms

Advances in Enzymology and Related Areas of Molecular Biology, Volume 78.
Edited by Eric J. Toone.

underlying the beneficial effect of TGase 2 inhibition in neurodegenerative diseases were not known, although it was observed that TGase 2 inhibition had favorable outcomes in several disease models, including allergic conjunctivitis, hepatic injury, atherosclerosis, and cancer [1,2]. TGase 2 may act via diverse mechanisms because it is widely used in many biological systems for generic tissue stabilization or immediate defense in infection. It was unclear how TGase 2 contributed to the pathogenesis of inflammation prior to the identification of TGase 2-mediated NF-κB activation [3]. The knowledge that direct activation of NF-κB occurs via polymerization of I-κBα provides new insights into the involvement of TGase 2 in disease pathogenesis [2,4–6].

Two major findings have led to the discovery of TGase 2-mediated NF-κB activation. One was that TGase 2 expression is induced by various inflammatory stresses (see review [1, 2]). The other finding was the clear evidence that TGase 2 inhibition reduces inflammation in cells and in vivo [3, 7, 8]. Therefore, combining the knowledge of these two phenomena provided the key to understanding the role of TGase 2 in inflammation. Fox et al. reported that TGase 2 was the most highly induced gene in the brain following simian immunodeficiency virus (SIV) infection [9]. They showed increased TGase 2 expression localized in the activated microglia and the infiltrated macrophages in brain. This result inspired me to test whether LPS treatment in microglia induces TGase 2 and iNOS expression [10]. LPS-induced TGase 2 expression could occur through NF-κB activation, because the TGase 2 promoter region has an NF-κB binding domain [11]. We also found that TGase inhibition using both cystamine and the TGase-specific R2 (KVLDGQDP) peptide inhibitor dramatically reduced the level of iNOS, as well as secreted nitric oxide [10]. Therefore, we tested whether TGase 2 could affect the NF-κB signaling pathway, since iNOS was regulated by NF-κB. We found that TGase 2 directly activates NF-κB through I-κBα polymerization, rather than through the kinase–ubiquitin–proteasome pathway [3]. This system has been confirmed in microglia, neuroblastomas, human kidney cells, Hela cells, and human breast cancer cells [2, 3, 5]. Specific lysine (K22,23,177) and glutamine residues (Q266,267,315) in I-κBα have been shown to be used for polymerization [4, 5]. Point mutations of these residues reverse I-κBα polymerization [5]. Furthermore, we found that TGase 2 specifically targets I-κBα, but not I-κBβ, despite the fact that the homology between the two genes is over 60% [4]. We found that endogenous TGase 2 expression determines the duration of NF-κB activation. TNF-α treatment results in different NF-κB activation responses depending on the level of TGase 2 expression [163]. The duration of NF-κB activation was correlated with TGase 2 expression in a dose-dependent manner [163].

Many questions remain to be answered, including how TGase 2 expression is induced and what causes TGase 2 activation. Recent studies have provided clues that TGase 2 expression can be induced by demethylation of certain regions in the promoter, which results in constitutive NF-κB activation, contributing to drug resistance [12, 13]. Maiuri et al. showed that mutation of cystic fibrosis (CF) transmembrane conductance regulatory genes could cause TGase 2 activation via loss of ion transport, including calcium ions [14]. This study suggests that activated TGase 2 may induce TGase 2 expression through NF-κB activation, because TGase 2 contains an NF-κB-binding site in the promoter.

B. THE OUROBOROS THEORY OF NF-κB ACTIVATION

Ouroboros is an ancient symbol depicting a serpent swallowing its own tail and forming a circle. I suggest ouroboros theory of NF-κB activation that implies constitutive NF-κB activation through NF-κB induction together with TGase 2 induction (Figure 1). It is a very curious question why NF-κB activation induced by inflammatory signals continues for hours and days, despite the fact that I-κBα is restored to normal levels within 30 minutes. It has been proposed that another NF-κB activation system may exist, because NF-κB can be activated without kinase activation in an IKK1/2 double knockout cell line [9]. Furthermore, constitutive NF-κB activation was observed without I-κBα phosphorylation [9]. Therefore, an ouroboros system appears to be involved with that the target induced by NF-κB activation, or another factor may reactivate NF-κB. One of the targets induced by NF-κB activation is TGase 2, which can reactivate NF-κB via I-κBα polymerization. This vicious cycle may be balanced by homeostasis between the levels of TGase 2 and I-κBα.

TGase 2 expression is normally low in most tissues, if blood contamination is avoided during preparation. TGase 2 activity is also tightly regulated by GTP–calcium homeostasis [15, 16]. Recent studies showed that TGase 2 activation, caused by an increase in intracellular calcium, induced TGase 2 expression [17]. It is also well accepted that TGase 2 is induced by various attacks to tissues or cells. The major source of TGase 2 production remains in question, but sources include immune surveillance cells, endothelial cells, and fibroblasts. For example, in the wound healing process, TGase 2 is

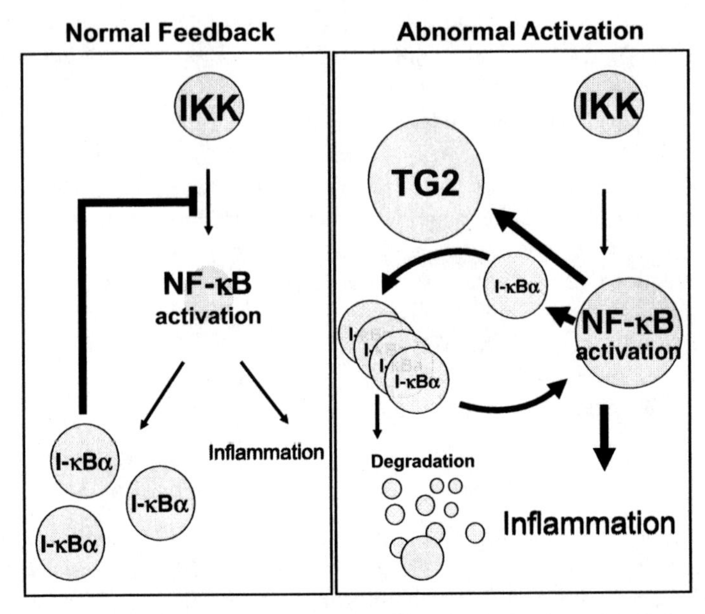

FIGURE 1. TGase 2 activation exacerbates NF-κB activation (Ouroboros theory). In normal stress, feedback mechanisms allow recovery from inflammation. However, under conditions of highly induced TGase 2 expression, inflammatory stress is increased by the depletion of I-κBα via TGase 2 polymerization. TGase 2 can be induced by various epigenetic mechanisms, including demethylation of the TGM2 promoter. TGase 2 expression can also be induced directly by NF-κB activation, because the TGase 2 promoter has an NF-κB binding motif. This is an intriguing finding because many stimuli in addition to LPS trigger NF-κB activation, including viral infection, UV, oxidative stress, and inflammatory cytokines. In the case of high TGase 2 expression, inflammation is maintained by a vicious cycle, which I have termed Ouroboros. (See insert for color representation.)

induced in endothelial cells, macrophages, and skeletal muscle cells. Increased TGase 2 expression is associated with the production of TGF-β, TNF-α, IL-6, and VEGF in wounds [18]. There are putative transcription regulation sites in the TGase 2 promoter, including a RARE (−1597), glucocorticoid response element (−1399), NF-κB binding site (−1338), and IL-6 response element (−1190), as well as AP2 (−634), AP1 (−183), and SP1 (−51, −40, +57) response elements [11]. Among them, only binding of SP1 [19] and NF-κB [20] to the TGase 2 promoter region have been confirmed. Following TGF-β treatment, binding of NF-κB to the TGase 2 promoter region was demonstrated in a liver cell line in which inflammatory stress directly induces TGase 2 expression [20]. Recently, it was demonstrated in porcine islets treated with an inflammatory cytokine cocktail, including TNF-α, IL-1β, and IFN-γ treatment, that the most highly induced gene was TGase 2 [21]. A number of inflammatory stresses trigger increased TGase 2 expression through direct or indirect signaling. These include biological stresses including microbial and viral infection, physical stresses including radiation and UV, and chemical stresses including inflammatory cytokines, glutamate, and anticancer drugs [11]. However, the stress adaptor molecules or signaling pathways involved in these responses are not clearly understood.

Interestingly, we found that inhibition of TGase 2 using cystamine and R2 inhibitor in LPS-induced microglia dramatically decreased the expression of TGase 2 (Figure 2). This explains how NF-κB-mediated TGase 2 induction

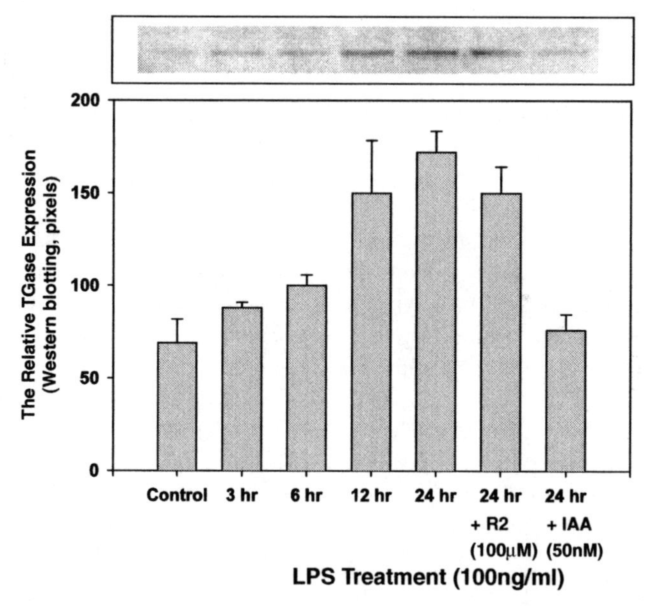

FIGURE 2. Effect of TGase inhibitors on TGase 2 expression in LPS-activated BV-2 microglia. BV-2 cells were pretreated with two TGase inhibitors, R2 peptide (100 μM), and IAA (50 nM) for 30 minutes prior to LPS treatment (100 ng/mL). After 24 hours of LPS treatment, the BV-2 cells were analyzed. The protein levels of TGase 2 were determined by Western analysis. Interestingly, TGase 2 inhibition reduced TGase 2 expression in LPS-induced microglia. At the same time, the level of LPS-induced iNOS was increased threefold at 24 hours, and TGase 2 inhibition reduced iNOS expression in the LPS-activated BV-2 cells [10].

may reactivate NF-κB after I-κBα feedback inhibition by NF-κB activation. This vicious cycle must be controlled in normal conditions by sumoylation of I-κBα, which prevents I-κBα polymerization by TGase 2 [22], or methylation of the TGase 2 promoter, which suppresses TGase 2 transcription [12, 13, 23]. Consistent with this hypothesis, in TGase 2-transfected cells, inflammatory stress results in a more dramatic increase in NF-κB activation than in control cells [163].

This theory leaves a major question remaining, namely why is TGase 2 important when the canonical NF-κB signaling pathway is clearly sufficient. Many people regard TGase 2 not as a major regulator, but rather as a complementary inducer. Although this situation may hold for the normal state, in inflammatory disease, TGase 2 may be the most important regulator in the recovery period after inflammation. In most inflammatory diseases, NF-κB is constitutively activated without activation of the canonical pathway, but with activation of TGase 2 expression.

II. REVISITING DISEASES IN RELATION TO THE TGase 2-NF-κB MECHANISM

A. CELIAC DISEASE

Celiac disease (CD), an adaptive immune disorder accompanying inflammation, is characterized by small intestinal damage with loss of absorptive villi and hyperplasia of the crypts, typically leading to malabsorption. In addition to nutrient deficiencies, prolonged CD is associated with an increased risk of malignancy, particularly intestinal T-cell lymphoma [24]. CD is precipitated by ingestion of the protein gliadin (a component of wheat gluten). Gliadin initiates mucosal damage, which involves an immunological process in individuals with a genetic predisposition to CD. IgA antibodies against gliadin and TGase 2 are valuable tools for identifying patients with CD and for judging the effects of therapy. The CD population comprises approximately 3.2% of patients with inflammatory bowel syndrome (IBS) [25]. Therefore, once we select the group of CD patients from IBS, TGase 2 inhibition may be a useful therapeutic approach [24].

TGase 2 is considered to play a role only as an immediate defense system. Therefore, research has focused on the immediate increases in TGase 2 expression in macrophages under innate immunity challenge. However, recent reports have shown that TGase 2 plays a central role in adaptive immune responses to CD. TGase 2 triggers a T-cell-mediated adaptive immune response by provoking the specific glutamine deamidation of gluten peptides [26–29] in the lamina propria of the intestine. These peptides are internalized and loaded onto the surface of antigen-presenting cells, which results in activation and clonal expansion of gluten-specific DQ2- and DQ8-restricted CD4-T cells in the lamina propria. This initiates a cell-mediated Th1 response, causing crypt cell hyperplasia and villous flattening. We do not yet know the details of how TGase 2 presents its antigenicity in people of certain HLA types. It may occur through the formation of hapten-carrier complexes in the form of covalent TGase 2-gluten peptide complexes in activated T cells, which stimulate B cells that express the TGase 2-specific antibody [30].

Although there is no direct evidence that TGase 2 induction plays a key role in constitutive NF-κB in CD, it appears to exacerbate inflammation associated with constitutive NF-κB activation. TGase 2 also plays a central role in the immediate activation of macrophages, as well as in the inflammation of intestinal epithelia cells, through NF-κB activation. TGase 2 expression is significantly elevated in jejunal biopsies of CD patients [31]. It is possible that this contributes to CD by increasing the frequency of gliadin-TGase 2 cross-linking and/or deamidation, as well as depleting free I-κBα for NF-κB activation. The increase in TGase 2 expression by proinflammatory cytokines is a possible novel mechanism for causing CD. The inflammatory infiltrates of jejunal tissues from patients with CD are rich in T cells [31]. T cells are not the only source of cytokines, as duodenal biopsies from CD patients also show increases in INF-γ of greater than 1000-fold relative to normal individuals. Elevations in IL-2, IL-4, IL-6, and TNF-α have also been reported in CD patients [32]. We have also shown that INF-γ can induce expression of TGase 2 in the rat IEC-6 small intestinal cells [33].

TGase activity was first reported in human peripheral lymphocytes almost three decades ago and was found to increase by up to 15-fold within 30 minutes of concanavalin A treatment of lymphocytes [34]. Because of its immediate induction, TGase 2 was identified as a new marker for macrophages in a certain differentiation or activation state [35]. TGase 2 enzyme was also reported to increase approximately 150-fold within 90 minutes of exposure of macrophages to a heat-labile constituent of serum or plasma [36]. The induction of TGase 2 is responsible for differentiation of cultured human monocytes into macrophages [37, 38].

Although increased TGase 2 causes CD only in people with specific HLA types, severe inflammation may occur in inflammatory bowel disease by inflammatory challenge. This may occur under conditions of high levels of TGase 2 expression in intestinal epithelium, via constitutive NF-κB activation. Therefore, exploration of TGase 2 induction in IBS will also be interesting.

B. SEPTIC SHOCK

Sepsis is characterized by the inability to regulate the inflammatory response, involving both cellular and humoral defense mechanisms [39]. It is the host response, rather than the infection itself, which determines the outcome of sepsis [39], and of particular relevance is the development of acute renal failure and heart failure [40].

Interestingly, TGase 2 expression has been reported to be induced by LPS in several tissues and organs [10, 41, 42]. However, the expression of hundreds of genes is changed in response to LPS treatment. Therefore, the role of TGase 2 was determined by testing whether TGase 2 inhibition reversed the inflammatory response. Indeed, the TGase 2 inhibitor R2 reversed LPS-induced microglia [10], septic lung injury [43], and septic brain injury [3]. A clue to the discovery of TGase 2-mediated NF-κB activation was the observation that iNOS was downregulated by TGase 2 inhibition [10]. It is known that NF-κB activation is responsible for the induction of iNOS. It has been demonstrated clearly that TGase 2 activity is responsible for NF-κB activation using microglia and neuroblastomas in vitro, as well as in vivo, in an LPS-induced brain injury model [3, 10].

Recently, it was demonstrated that TGase 2 knockout mice have increased survival compared to wild type following LPS-induced murine endotoxic shock [44]. This was associated with decreased neutrophil recruitment into the kidney and peritoneum, and reduced renal and myocardial tissue damage, due to defects in NF-κB activation [44]. The analysis of NF-κB in peritoneal macrophages, isolated from control and LPS-treated WT and TGase 2 knockout mice, revealed that LPS induced a significant increase in the p65 in the nucleus of WT cells, whereas no difference was observed in the absence of TGase 2 [44].

This phenotype indicates that TGase 2 promotes the pathogenesis of endotoxic shock through continuous activation of NF-κB [3, 43, 44]. As we discussed earlier, TGase 2 has an NF-κB binding site in its promoter region.

Therefore, TGase 2 expression is increased during endotoxic shock and is directly involved in the mechanisms of NF-κB activation. This may cause continuous activation of the inflammatory process, which would contribute to the development of the pathogenesis of sepsis.

C. ALLERGIC CONJUNCTIVITIS

Allergic conjunctivitis is caused by various allergens and ranges from being a mild to a vision-threatening condition [45]. Previous studies have proposed that inhibition of TGase 2 may be a new approach to the treatment of allergic conjunctivitis [3, 7, 43, 46]. We showed that recombinant peptides from lipocortin-1 and proelafin (cementoin) had a strong anti-inflammatory effect on allergic conjunctivitis in vivo [7]. Another study revealed that the anti-inflammatory effect of these antiflammins was due to inhibition of TGase 2 rather than PLA2 [47]. This strongly supports the proposal that TGase 2 plays a central role in the inflammatory process. Recently, we found that TGase 2 activates the transcriptional activator NF-κB, which upregulates the expression of inflammatory genes such as inducible nitric oxide synthase and TNF-α [3]. Therefore, TGase 2-mediated NF-κB activation may provide a new pathway for understanding the pathological progression of inflammation, since NF-κB is a key inducer of inflammation in immune cells. The induction of TGase 2 expression occurred in the same cells as those with NF-κB activation (Figure 3).

Blocking TGase 2 activity is a promising approach to ameliorating NF-κB-mediated inflammation during allergic conjunctivitis; however, to achieve this goal, safe TGase inhibitors are required. TGase 2 inhibitors have been developed based on the chemical backbone structure of dihydroisoxazole [48]. Although the efficacy of such drugs has been demonstrated in the micromolar range, drug safety and stability issues need to be addressed in vivo. As an alternative, we screened a pool of natural extracts for stable and safe TGase inhibitors and found that glucosamine has TGase 2 inhibitory activity [49]. Glucosamine bearing certain amino groups effectively inhibited TGase 2 in EcR293/TGase (human kidney cell line EcR293 stably transfected with TGase 2) cells. Furthermore, the glucosamine metabolite glucosamine-6-phosphate increased the TGase 2 inhibitory effect via binding to the GTP binding site [50]. We examined whether glucosamine effectively rescued inflammation via TGase 2 inhibition in a guinea pig model of allergic conjunctivitis. Interestingly, TGase 2 expression colocalized with

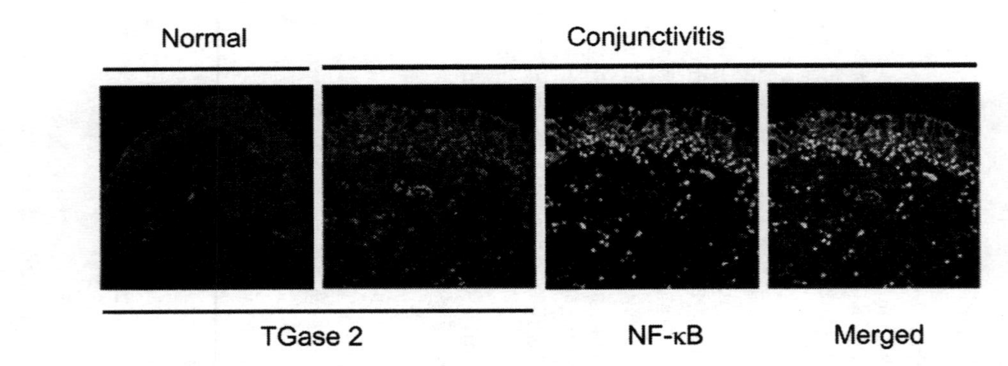

FIGURE 3. Colocalization of TGase 2 and NF-κB in conjunctivitis. An in vivo experimental model of allergic conjunctivitis was employed, as previously described [7]. Conjuctiva tissue was double-immunostained using FITC-conjugated anti-TGase 2 (red) and Alexafluor594-conjugated anti-NF-κB (green) antibodies. This model showed markedly increased colocalization of TGase 2 and NF-κB compared to that seen in normal conjuctiva. (See insert for color representation.)

NF-κB expression in this model. Glucosamine dramatically reduced NF-κB activation as effectively as conventional drugs (data not shown). TGase 2 inhibition may be sufficient to reverse serious allergic conjunctivitis because TGase 2 expression is required to prolong NF-κB activation.

D. CYSTIC FIBROSIS

A characteristic of CF is the exaggerated inflammation seen in response to infection. In some cases, inflammation may occur prior to other symptoms. Lung injuries occur during infection and inflammation, and may continue at a slower pace even when infection is clinically controlled. The injuries in turn make the lungs more susceptible to infection, and the very thick purulent mucus resulting from infection and inflammation further aggravates the prognosis. This vicious cycle compounds the severe structural alterations and functional deterioration of the lungs, eventually resulting in death [51]. This inflammatory process is reminiscent of the ouroboros theory of NF-κB activation that I mentioned at the beginning of this review.

The defective gene in CF patients has been identified [52–54]. It encodes an integral membrane protein termed the CF transmembrane conductance regulator (CFTR), which is a chloride ion-conducting channel. Using cells cultured directly from uninfected CF tissues, defects in CFTR localization have been found to contribute to endogenous activation of NF-κB and consequently to the exaggerated production of the proinflammatory cytokine IL-8, even in the absence of bacteria [55]. However, the mechanism causing constitutive NF-κB activation remains unknown.

Recent reports have shown that bronchial epithelial cells from CF patients can spontaneously initiate the inflammatory cascade. During the exploration of the molecular mechanism, a stimulatory role for TGase 2 in inflammation was found using airway epithelial cells from CF patients defective in the CFTR gene [14]. A marked upregulation of TGase 2 was found in these cells, leading to cross-linking and subsequent proteasomal degradation of peroxisome proliferator-activated receptor (PPAR)-γ, a nuclear hormone receptor that negatively regulates inflammatory gene expression through NF-κB inhibition [56]. The authors also showed that PPAR-γ levels were restored after treatment with a TGase 2 inhibitor, via reduction of PPAR-γ polymerization. As a result, CF patients display an enhanced inflammatory response via NF-κB activation, possibly through both I-κBα and PPAR-γ polymerization [14]. Therefore, TGase 2 inhibition appears to be an effective therapeutic strategy in CF.

E. HEPATIC INJURY

Increased TGase 2 expression has been proposed to be involved in liver pathologies and can lead to severe injury [20]. Various animal models of hepatic injury, including the hepatotoxin test (using carbon tetrachloride, CCl_4), hepatitis C infection, alcohol damage, and Mallory body (MB) formation test (3,5-diethoxycarbonyl-1,4-dihydrocollidine, DDC fed), have been used to investigate whether accumulation of TGase 2 during injury plays a protective or pathological role. There are conflicting results regarding the role of TGase 2 in different injury models. The effect may differ depending on the stage of damage or cell-type. As we know, NF-κB activation may contribute to protection, while its activation in immune cells has a negative impact on damaged tissue. The early stage of fibrogenesis is associated with inflammation and the late stage affects the extracellular matrix (ECM) that underlies liver architecture [57]. TGase 2 may play an important role in fibrogenesis, depending on the stage and cellular location of liver damage. However, it is clear that induction of TGase 2 expression is associated with liver damage, and there is no doubt that TGase 2 inhibition ameliorates liver damage.

1. Hepatotoxin Test (CCl_4)

Although the relationship between oxidative stress, cytotoxic cytokines, and liver cell injury has not been fully clarified, NF-κB is believed to have a pivotal function during cell injury [58]. NF-κB plays an essential and beneficial role in normal physiology; however, the upregulation of this factor has been implicated in the pathogenesis of several diseases, including inflammation and liver diseases [58]. Animal models of acute hepatic damage also support a role for NF-κB in the development and progression of cirrhosis [58]. Reactive oxygen species (ROS) have been suggested to be the triggering inducer of NF-κB activation. Consistent with this, antioxidants reduce NF-κB activation in a CCl_4-treatment liver damage model [59]. Liver damage may also be caused by TGase 2 in hepatic fibrogenesis, which has been shown to be associated with the fibrosis induced by CCl_4 administration. TGase 2 is suggested to play a role in fibrosis because it acts on ECM organization through cross-linking [60]. Interestingly, TGase 2 can also be activated by ROS signals [61]. ROS increase TGase 2 expression through NF-κB activation [62]. Therefore, it is also possible that TGase 2 can be induced by ROS through NF-κB activation, which participates in liver fibrogenesis. Continuously elevated TGase 2 expression by various cytokines including TGF-β or IL-6 [63] may potentially cause hepatic damage after hepatotoxin administration. Consistent with a protective role for TGase 2 in liver damage, treatment of TGase 2 knockout mice with CCl_4 resulted in an increased incidence of death postinjury, relative to control mice; the effect was associated with an increased inflammatory response and increased ECM accumulation [64].

2. *HCV Infection*

The NF-κB signaling pathway has particular relevance to several liver diseases, including hepatitis (HCV), liver fibrosis and cirrhosis, and hepatocellular carcinoma [65]. HCV activates NF-κB in hepatocytes through multiple mechanisms, including ER stress [66], ssRNA via TLR [67], and dsRNA via RIGI [68].

As previously stated, TGase 2 expression appears to be induced by NF-κB activation, since TGase 2 has an NF-κB binding site in its promoter region [11]. Evidence for the role of TGase 2 in hepatic fibrogenesis has been obtained by studying an animal model in which induction of fibrosis was achieved by HCV infection [64]. This report showed that increased TGase 2 levels, as well as changes in its distribution, are strictly related to the different stages of liver damage. TGase 2 increases early during liver injury and its intracellular/extracellular localization varies according to the stage of the disease. In HCV-infected patients, the enzyme level was increased 15-fold during periods of more active liver fibrogenesis and inflammation [64], while expression was decreased in stages characterized by prominent ECM deposition. Therefore, it appears certain that TGase 2 is involved in the early inflammatory process via NF-κB activation. However, Piacentini et al. have obtained contradictory results to those mentioned above in their TGase 2 knockout mouse model. It is interesting to note that, in the absence of TGase 2, mice develop an increased inflammatory response in association with alterations in liver architecture [64]. This suggests another role for TGase 2 as a general protective factor in hepatic homeostasis.

3. *Alcohol Damage*

Excessive alcohol intake can cause alcoholic hepatitis or liver cancer, given that alcohol-related liver disease is a leading cause of cirrhosis. Alcoholic liver injury is a complex process involving several injury mechanisms and multiple cellular targets. Although some associated factors have been found, the mechanism by which this inhibition of the factors occurs has not been delineated. Chen et al. demonstrated that both the ERK1/2 and PI3K/Akt pathways can mediate the effects of ethanol on NF-κB-dependent transcription, and that these signaling pathways appear to be involved in ethanol-mediated TGase 2 induction [69]. This induction includes upregulation of importin-a3, which facilitates TGase 2 transfer into the nucleus [19]. TGase 2 appears to play a causative role in ethanol- and Fas-induced hepatocyte apoptosis. Tatsukawa et al. identified a potential novel link between TGase 2 and ethanol-mediated liver injury [19]. The transcription factor Sp1 can be cross-linked, oligomerized, and inactivated by nuclear TGase 2, resulting in reduced expression of growth factor receptors such as c-Met [19, 70]. The TGase 2-induced impairment of Sp1-c-met signaling is of clear biological relevance. However, the question of whether depletion of Sp1 selectively affects c-met signaling remains to be answered, because the effects on other Sp1 target genes, including Bcl2, Bcl-XL, and p21, remain unknown.

In contrast to the above, and in agreement with a proapoptotic role for TGase 2, is the finding that apoptosis of isolated hepatocytes from TGase 2 knockout mice was markedly lower than in wild-type hepatocytes following ethanol treatment [19]. TGase 2 knockout mice also exhibited decreased hepatic apoptosis following treatment with a low dose of the anti-Fas antibody Jo2, which is a known, potent inducer of endothelial cell and hepatocyte apoptosis in alcohol-induced liver damage [19]. The proapoptotic effect of TGase 2 in the pathophysiology of liver cell death involves inactivation of the transcription factor Sp1 by TGase 2-mediated cross-linking, resulting in the inhibition of the expression of genes required for cell viability [19].

4. MB Formation

MB formation is a characteristic feature of common liver disorders, which shares similarities with cytoplasmic inclusions observed in neural diseases and myopathies. Cytoplasmic inclusions can be induced in mice by long-term feeding with chemicals such as griseofulvin or 3,5-diethoxycarbonyl-1,4-dihydrocollidine (DDC) [71, 72]. TGase 2 induction may be responsible for MB inclusion formation. This hypothesis is supported by the association between MB induction and upregulation of TGase 2 [73–75], by the fact that keratins K8/K18 are known TGase substrates, and by the presence of cross-linked keratins in MBs [76–78]. It has recently been demonstrated that DDC-fed TGase 2 knockout mice have decreased MB formation and liver hypertrophy compared to DDC-fed wild-type mice [79]. To achieve TGase 2-mediated inclusion formation, TGase 2 expression has to be induced in liver by inflammatory stress. As previously discussed, TGase 2 is increased by NF-κB activation. Interestingly, the level of NF-κB1/p105 mRNA significantly increased in DDC-primed hepatocytes after 24 hours of culture and in normal control hepatocytes after 48 hours of culture [80]. In DDC-primed hepatocytes, siRNA-mediated knockdown of NF-κB for 6 days significantly decreased the mRNA expression of Src, p105/NF-κB1, ERK1, MEKK1, and JNK1/2. Thus, the amelioration of MB formation and injury-mediated hypertrophy in DDC-fed TGase 2 knockout mice is not associated with an improvement in liver cell damage, but may be associated with reduced NF-κB activation.

F. RENAL INJURY

Worldwide, increasing numbers of patients are affected by chronic kidney diseases. Renal fibrosis is the final common pathway of a wide variety of chronic kidney diseases, irrespective of the initial cause of nephropathy. Renal fibrosis is defined as an excessive accumulation and deposit of ECM components, leading to complete destruction of kidney tissue and renal failure [81]. NF-κB participates in vascular and target-organ damage. It has been demonstrated that endothelial cell-specific NF-κB suppression ameliorates renal damage [82]. However, the key regulator leading to constitutive NF-κB activation in renal fibrosis is unclear because, normally, feedback inhibition of NF-κB activation limits its retention time in the nucleus. Therefore, another factor must disturb the feedback mechanism to increase the NF-κB nuclear retention in renal damage.

TGase 2 expression and the tissue levels of ε(γ-glutamyl)lysine cross-links are increased in subtotal nephrectomy and in a unilateral ureteral obstruction model [83, 84], and recent studies indicate that TGase inhibition by pharmacological agents reduces fibrosis and preserves function in a subtotal nephrectomy model [85]. In humans, a strong association was observed between the expression of TG2 and its cross-link products, and renal fibrosis and function, irrespective of the original cause of nephropathy [60, 87]. Diabetic nephropathy (DN) is the most common cause of end-stage renal disease [88, 89], and its increase is associated with the current prevalence of type 2 diabetes and the reduced mortality of DN patients, resulting from better management. The progression of DN is accompanied by the development of proteinuria and early glomerular hyperfiltration, followed by a decline of the glomerular filtration rate. Morphologically, it is characterized by excessive renal ECM accumulation in the glomeruli and tubulointerstitial space, causing glomerulosclerosis and tubulointerstitial fibrosis. A role for TGase 2 in the pathogenesis of DN has been reported in both the streptozotocin (STZ)-induced model of type 1 diabetes [90] and human diabetic kidney disease [91]. Additionally, increased TGase 2-mediated ε-(γ-glutamyl)lysine cross-link formation is tightly associated with both glomerular and tubulointerstitial ECM expansion in diabetic kidneys. TGase 2 inhibition preserves kidney function, reduces albuminuria, and ameliorates the progression of the histological changes associated with the formation of scar tissue in DN [92]. The involvement of TGase 2 in interstitial renal fibrosis has also been demonstrated in TGase 2 knockout mice subjected to unilateral ureteral obstruction [93]. Although apoptosis was unaffected, there was a substantial reduction in fibrosis development in TGase 2 knockout relative to wild-type mice. This was associated with reduced macrophage and myofibroblast infiltration and decreased collagen I synthesis. The degree of macrophage infiltration in the damaged kidney was also dramatically reduced in TGase 2 knockout mice [93]. The authors stated that the reduction in renal fibrosis in TGase 2 knockout mouse was not related to NF-κB activation [93]. However, the dramatic decrease in macrophage infiltration in renal tissue of TGase 2 knockout mouse cannot be explained without invoking NF-κB downregulation, because NF-κB activation is responsible for the activation and chemotactic migration of macrophages and monocytes [37, 38].

In summary, TGase 2 plays an important role in the development of renal fibrosis. During the early stages of fibrosis in obstructive nephropathy, TGase 2 plays key roles in immune cell infiltration and ECM accumulation, indicating that it is functional in two different types of cells. TGase 2 induction by injury is responsible for long-term macrophage activation, which is mediated through NF-κB activation. TGase 2 induction by injury in renal epithelial cell appears to increase ECM proteins through NF-κB activation. TGase 2 increases the resistance of ECM proteins to degradation. A group of studies provides evidence that the inhibition of TGase 2 limits renal scarring [93]. Targeting TGase 2 expression and/or activation can provide a novel and important additional treatment for combating renal fibrosis.

G. ATHEROSCLEROSIS

Inflammatory and innate immune mechanisms employing monocytes, innate receptors, and innate cytokines may be involved in atherogenesis. Early atherogenesis is characterized by leukocyte recruitment and expression of proinflammatory cytokines [94]. Also, studies on cytokine-deficient animals have provided direct evidence for a role of cytokines in atherosclerosis.

Atherosclerosis is characterized by proliferation and dedifferentiation of VSMCs. Studies in mice have also shown that leukocyte recruitment and expression of proinflammatory Th1 cytokines typically characterize early atherogenesis, and modulation of inflammatory mediators reduces atheroma formation [94, 95]. NF-κB regulates the functions of multiple genes that have been connected to atherogenesis [96, 97]. The products of those genes include VCAM-1, ICAM-1, and E-selectin [98]. Inflammatory cytokines and oxidants, such as oxidized LDL, also induce atherosclerosis with NF-κB activation [99]. Moreover, NF-κB activation occurs in vascular tissue during smooth muscle cells' (SMCs) proliferation after lumen injury [100]. There are many results suggesting that NF-κB plays a key role in the pathogenesis of atherosclerosis. It has been suggested that NF-κB activation is involved in the arterial response to balloon injury in rats [100–102]. Interruption of blood flow caused by ligation of the left common carotid just proximal to the left carotid bifurcation results in an 80% reduction in the lumen area, through a combination of intimal hyperplasia and decreased vessel diameter [103]. In this mouse model of vascular remodeling, neointimal lesion formation in p105 (a precursor of p50) knockout mice was greatly reduced compared to in wild-type mice [104]. A series of reports also demonstrated that NF-κB inhibition reversed atherogenesis induced by ischemia or angiotensin II- and isoproterenol-treatment [105–108].

During development of cardiac hypertrophy, TGase 2 expression was strongly increased [109]. It has been suggested that TGase 2 expression may be limited to macrophages in the atherosclerotic region, which may also participate in apoptotic cell clearance [110]. However, TGase 2 expression is also detected in other cell types. TGase 2 is essential for the SMC transition to chondro-osseous differentiation and plays a central role in the calcification of cultured SMCs and aortic ring explants in organ culture [111]. In TGase 2 knockout mice, the failure to deregulate several genes was associated with chondro-osseous differentiation and a sharp increase in the expression of physiological inhibitors of artery calcification, including Matrix Gla protein (a BMP-2 inhibitor), osteopontin, and osteoprotegerin, in response to Pi donor treatment [111]. It is very interesting that osteopontin is induced by NF-κB activation [112]. Therefore, TGase 2 induction may be responsible for NF-κB-induced osteopontin expression.

As atherosclerosis progresses to an advanced-stage atherosclerotic lesion, both the expression and activity of TGase 2 increases markedly in the vessel wall, mainly in neointimal SMCs [8]. TGase 2 colocalizes with activated NF-κB in human atherosclerotic lesions [8]. Furthermore, it has been demonstrated that NF-κB can be activated by TGase 2 in cell lines overexpressing TGase2.

H. OSTEOARTHRITIS

Osteoarthritis is associated with regions of articular chondrocyte hypertrophy, formation of osteophytes, and degeneration of cartilage. Articular cartilage normally contains resting chondrocytes that do not differentiate or mineralize their surrounding matrix. A number of proinflammatory factors known to stimulate chondrocyte hypertrophy are upregulated in osteoarthritic cartilage, such as IL-1β [113], and the CXC chemokine subfamily members IL-8 (CXCL8) and growth-related oncogene α (GRO α; CXCL1) [114]. IL-8 signals through CXCR1 and CXCR2, whereas GRO α activates CXCR2, but not CXCR1 [115]. Both IL-8- and GRO α-induced hypertrophy of wild-type murine chondrocytes is associated with increased transglutaminase activity [116].

Although NF-κB plays an essential role in normal physiology, inappropriate regulation of NF-κB activity has been implicated in the pathogenesis of rheumatoid arthritis and osteoarthritis [117]. In human articular chondrocytes, NF-κB plays key roles in mediating chondrocyte activation induced by the fibronectin fragment and in increasing the expression of proinflammatory cytokines, chemokines, and MMPs such as IL-6, IL-8, MCP-1, GRO-α, GRO-β, GRO-γ, and MMP-13 [118, 119]. Furthermore, NF-κB participates in the receptor for advanced glycation end (RAGE) signaling that induces increases in MMP-13 expression in monkey and human articular chondrocytes [120]. In addition, NF-κB production increased with donor age in IL-1β-stimulated human articular chondrocytes [119]. Although the normal regulation of NF-κB signaling is well understood, the mechanism underlying the constitutive activation of NF-κB in osteoarthritis remains unclear.

TGase 2 has recently been implicated in promoting osteoarthritis progression. TGase 2 is required for cultured chondrocyte maturation to terminal hypertrophic differentiation and for the calcification of the matrix by chondrocytes in response to retinoic acid and certain inflammatory cytokines [120, 121]. TGase 2, although lacking a signal peptide, is released by cells [121, 122], and nanomolar concentrations of exogenously added TGase 2 are sufficient to promote chondrocyte hypertrophy [121]. In addition, direct actions of extracellular TGase 2 on cell differentiation have been linked to the effects of TGase 2-catalyzed pericellular matrix protein cross-linking [122, 123].

Lysates prepared from murine knee chondrocytes isolated from wild-type mice have been shown to increase matrix calcification. However, lysates prepared from TGase 2 knockout mice showed no increase in matrix calcification in response to stimulation with IL-1β, GRO α, or the all-trans form of retinoic acid. TGase 2 knockout mice showed no characteristic features of chondrocyte hypertrophy, such as collagen X secretion [116, 124] and loss of expression of the transcription core binding factor α 1 [124]. Studies using murine femoral head articular cartilage explants and knee chondrocytes have shown that TGase 2 is required to convert S100A11 into a covalent homodimer. This accelerates chondrocyte hypertrophy and matrix catabolism via RAGE signaling [125]. In a surgically induced knee joint instability model of osteoarthritis, TGase 2 knockout mice had reduced cartilage destruction and increased osteophyte formation relative to wild-type mice [126]. This phenotype was associated with the increased expression of TGF-β1, which promoted osteophyte formation [127] more efficiently in the osteoarthritic tissues of wild-type mice than in TGase 2 knockout mice [126]. Together, these studies point to a role for TGase 2 in modulating cartilage repair and stability in response to certain pathological stressors, and identify TGase 2 as a potential target for therapeutic intervention in osteoarthritis.

Although there is no direct demonstration that TGase 2 expression induces NF-κB activation in chondrocyte or SMCs, the increased expression of TGase 2 in cells or the ECM has been shown to contribute to NF-κB activation or ECM resistance. Additionally, the therapeutic benefit of glucosamine in arthritis has been suggested to be due to its ability to inhibit NF-κB [128, 129]. As previously discussed, glucosamine possesses TGase 2 inhibitory activity, which results in decreased NF-κB activity via inhibition of TGase 2-mediated NF-κB activation [49, 50].

I. CANCER

An important signaling pathway that is often altered in human cancers is the I-κB kinase (IKK)/NF-κB pathway. Aberrant NF-κB regulation has been observed in many cancers, including both solid and hematopoietic malignancies [130]. NF-κB affects all hallmarks of cancer through the transcriptional activation of genes associated with self-sufficiency in proliferative growth, insensitivity to growth inhibitors, evasion of apoptosis, acquisition of limitless replicable potential, induction of angiogenesis, induction of invasion, and metastasis [131]. However, three controversies remain to be resolved. One is that NF-κB activation can be induced by anticancer drug treatment [132]. Most anticancer drugs trigger stresses such as DNA damage or intercellular cross talk via cytotoxic chemokines. The cancer cells that survive and adapt following constitutive anticancer drug treatment may grow colonies of untreatable drug-resistant cancer cells. The second issue is the demonstration that NF-κB activation by anticancer drug treatment occurred in an IKK1/2-deficient cell line [133]. Therefore, NF-κB activation caused by anticancer drugs must occur via a pathway other than the IKK signaling pathway. The third issue is that NF-κB activation in most cancer cases is due to defective I-κBα activity, rather than IKK mutations [134].

Interestingly, it has been reported that TGase 2 expression correlates with drug resistance [135–137]. For decades, the mechanism was unknown, until our group found that TGase 2 can activate NF-kB directly through I-kBα polymerization. Furthermore, TGase 2 inhibition resulted in increased sensitivity to chemotherapeutic drugs, which may represent a potentially useful treatment for certain cancers [6]. TGase 2 inhibition using KCC009 also demonstrated sensitization of drug-resistant cancer in a xenograft model [138, 139]. Interestingly, we found that glucosamine, which inhibits TGase 2, had a chemo-sensitization effect on drug-resistant cancer cells [140]. Cancer cell lines with high expression of TGase 2 show constitutive NF-κB activation, with depletion of free I-kBa protein (Figure 4). The

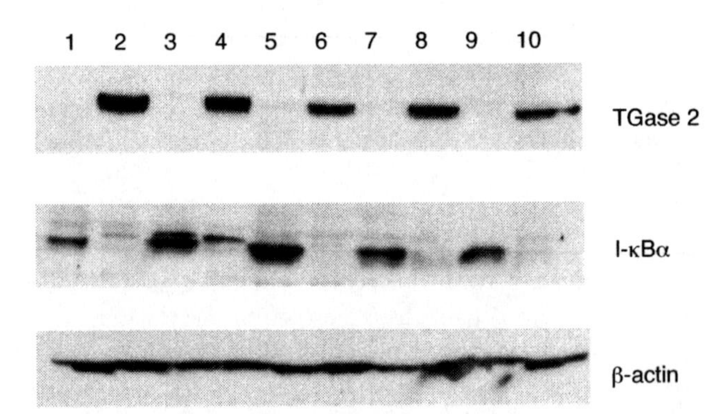

FIGURE 4. Western analysis of TGase 2 and I-κBα in cancer cell lines. Western analysis of cancer cell lines shows an inverse relationship between TGase 2 and I-κBα levels. Thyroid cancer cells (lane 1:WRO, lane 2:FRO), liver cancer cells (lane 3:HepG2, lane 4:SK-Hep1), breast cancer cells (lane 5:MCF7, lane 6:MDA-MB-231), lung cancer cells (lane 7:HCC95, lane 8:A549), and ovarian cancer cells (lane 9:NIH-OVCAR3, lane 10:NCI/ADR-RES).

involvement of TGase 2 in these mechanisms may help resolve the three controversies described above.

In relation to energy consumption, TGase 2-mediated NF-κB activation is a very cost-efficient system, since TGase 2 does not require ATP consumption, in contrast to the ubiquitin–proteasome system. Cancer cells appear to adopt this system to survive. Therefore, aberrant expression of TGase 2 contributes to the development of disease pathologies in cancers [4, 141, 142]. Recent studies have suggested that TGase 2 expression could be induced by demethylation of certain regions of the TGM2 promoter [12, 13, 23], which leads to constitutive NF-κB activation associated with drug resistance in breast, ovarian, and lung cancers [24]. Cell lines with TGM2 promoter-methylation, HCC-95 and HCC-1588, showed relatively higher sensitivity to cisplatin than TGM2-expressing cell lines (NCI-H1299 and HCC-1195). Downregulation and overexpression of TGM2 in NSCLC cells also suggested a positive correlation between cisplatin sensitivity and TGM2 inhibition [12]. Although increased expression levels of TGase 2 may contribute to multiple pathophysiologies, it is clear that increased TGase 2 activity triggers NF-κB activation without I-κB kinase signaling (Figure 1). NF-κB is responsible for many steps in cancer progression, including angiogenesis, metastasis, cancer cell proliferation, drug resistance, and so on [131].

Although many of the pathways responsible for bridging inflammation and cancer still remain to be discovered, the targeting of NF-κB through the inhibition of TGase 2 is certainly one pathway worthy of further exploration. Cancer cells are more likely to depend on TGase 2 compared to normal cells for constitutive NF-κB activation. The administration of a TGase 2 inhibitor in combination with other conventional anticancer drugs may turn out to be more effective than the administration of a TGase 2 inhibitor alone. Although multiple factors are involved in chronic inflammation, TGase 2 inhibition reduces inflammation. Therefore, we believe that combination therapy with a TGase 2 inhibitor may help attenuate the cytotoxic effects of anticancer drugs and/or other side effects of anti-inflammatory drugs.

J. NEURODEGENERATIVE DISEASES

Increased levels and activity of TGase 2 have been observed in many neurodegenerative diseases (reviewed in [143]). TGases are likely to contribute to the formation of protein aggregates. Isoforms of TGase colocalize with plaques and tangles in the brains of Alzheimer's disease patients [144–148], and the number of ε(γ-glutamyl)lysine linkages in insoluble proteins from Alzheimer's disease brains is 30–50 times greater than the number found in normal brain tissue [148]. TGase 2 is responsible for Lewy body formation in Parkinson's disease and dementia with Lewy bodies [149]. TGase 2 is also responsible for nuclear inclusion formation [150] as well as mitochondrial protein aggregates in Huntington's disease [151]. Generally, these protein aggregates are considered to be cytotoxic and to trigger apoptosis.

The above observations provide a strong argument for the participation of TGase 2 in the formation of insoluble protein aggregates including plaques and tangles in neurodegenerative disease brain. However, an opposing hypothesis relies on the observation that the amount of amyloid deposits at the time of death does not correlate with the cognitive ability of either Alzheimer's disease or Down's syndrome patients [152, 153]. Therefore, it is possible that TGase 2 may play other roles in neurodegenerative diseases, such as contributing to neuro-inflammation through NF-κB activation.

Multiple sclerosis and AIDS dementia complex (also called HIV-associated dementia) are two examples of CNS diseases with a strong inflammatory component [154]. In particular, macrophage/microglia activation in the deep white matter is a key feature of both diseases [154]. Cerebrospinal fluid from multiple sclerosis patients contains autoantibodies against myelin basic protein and myelin-associated glycoprotein [155]. It is not clear why and how autoantibodies are generated in multiple sclerosis. Interestingly, it has been demonstrated that glial filaments and myelin basic protein can serve as TGase 2 substrates in vitro [156]. Recently, we found that TGase 2 activation plays a key role in at least one component of microglial NF-κB activation [10]. These findings suggest that uncontrollable activation of TGase 2 in activated microglia may contribute to the development of inflammation in multiple sclerosis, either by continuous inflammatory signaling through NF-κB activation or by autoantibody generation through modification of myelin basic proteins. SIVE arises sporadically in infected rhesus macaques, similar to HIVE in humans. Using human GeneChips, researchers identified the gene transcripts in infiltrating and activated macrophages [9]. TGase 2 was significantly upregulated in the frontal lobe of brains with SIVE, where inflammation was highly advanced, compared to brains from uninfected animals [9]. Interestingly, nuclear staining for NF-κB also colocalized predominantly to perivascular microglia and macrophages in the basal ganglia and deep white matter together with TGase 2 induction, and correlated with the severity of the AIDS–dementia complex [157]. Following this discovery, we demonstrated that LPS induces TGase 2 together with iNOS in microglia [10]. Furthermore, we found that inhibition of TGase 2 reverses iNOS induction, together with decrease of TGase 2 expression itself (Figure 2). Ientile et al. found that NF-κB activation by glutamate treatment in astrocytes induces TGase 2 expression [158]. Therefore, TGase 2 expression can be induced by NF-κB activation via glutamate, which may extend NF-κB activation by I-κBα depletion via TGase 2-mediated protein polymerization. The extended NF-κB activation of immune surveillance cells, including microglia and astrocytes, may exacerbate neuronal apoptosis.

III. CONCLUSION

The complex roles of TGase 2 in cell homeostasis may become clearer if we examine the roles played by ancestors of the TGase 2 gene in invertebrates. For instance, the main function of the TGase 2 gene in the defense system might be preserved in a living fossil, such as the horseshoe crab, which belongs to the subclass Xiphosura that spans more than 500 million years of evolution [159]. There exists significant amino acid sequence similarity (32.3%) between the horseshoe crab TGase and the guinea pig liver TGase [160]. The horseshoe crab defense reaction consists of two major steps: (1) the immobilization of invaders by coagulation factors, including TGase, and (2) the degradation of the coagulated protein mass by proteases [161]. Interestingly, TGase 2 and matrix metalloproteases are often associated in the lesions of inflammatory diseases such as rheumatoid arthritis [162]. There is no doubt that TGase 2 plays a key role in the immediate defense system of the horseshoe crab, which mimics innate immunity in higher vertebrates. However, in higher vertebrates with canonical NF-κB activation, the TGase 2 system does not appear to be necessary in the primary defense system. It would seem that the host defense role played by TGase 2 in invertebrates was replaced by the more sophisticated kinase–ubiquitin–proteasome system involved in NF-κB activation in higher vertebrates. This theory would explain why the TGase 2 knockout mouse has a normal phenotype [1]. However, a role for TGase 2 in regulating the retention period of NF-κB activity was probably retained during evolution. At the beginning of this review, I proposed the ouroboros theory, which proposes that inflammatory stress induces TGase 2 expression through NF-κB activation, thereby regulating the retention period of NF-κB activity through I-κBα polymerization. Therefore, in tissues with TGase 2 induced by factors other than inflammation, such as epigenetic changes, inflammatory stress results in explosive

NF-κB activation [163], which may lead to many pathological problems, including CD, septic shock, conjunctivitis, CF, hepatic injury, renal injury, atherosclerosis, osteoarthritis, and cancer. Future studies directed toward defining the groups of patients that have TGase 2 induction will provide us with a detailed understanding of the disease pathology, as well as with a guide to more efficient treatments.

ACKNOWLEDGEMENTS

This work was supported by a research grant (NCC1110011-1) from the National Cancer Center in Korea.

REFERENCES

1. Iismaa, S. E., Mearns, B. M., Lorand, L., and Graham, R. M. (2009) Transglutaminases and disease: lessons from genetically engineered mouse models and inherited disorders, *Physiol. Rev. 89*, 991–1023.
2. Kim, S. Y. (2006) Transglutaminase 2 in inflammation, *Front. Biosci. 11*, 3026–3035.
3. Lee, J., Kim, Y. S., Choi, D. H., Bang, M. S., Han, T. R., Joh, T. H., and Kim, S. Y. (2004) Transglutaminase 2 induces nuclear factor-kappaB activation via a novel pathway in BV-2 microglia, *J. Biol. Chem. 279*, 53725–53735.
4. Kim, J. M., Voll, R. E., Ko, C., Kim, D. S., Park, K. S., and Kim, S. Y. (2008) A new regulatory mechanism of NF-kappaB activation by I-kappaBbeta in cancer cells, *J. Mol. Biol. 384*, 756–765.
5. Park, S. S., Kim, J. M., Kim, D. S., Kim, I. H., and Kim, S. Y. (2006) Transglutaminase 2 mediates polymer formation of I-kappaBalpha through C-terminal glutamine cluster, *J. Biol. Chem. 281*, 34965–34972.
6. Kim, D. S., Park, S. S., Nam, B. H., Kim, I. H., and Kim, S. Y. (2006) Reversal of drug resistance in breast cancer cells by transglutaminase 2 inhibition and nuclear factor-kappaB inactivation, *Cancer Res. 66*, 10936–10943.
7. Sohn, J., Kim, T. I., Yoon, Y. H., Kim, J. Y., and Kim, S. Y. (2003) Novel transglutaminase inhibitors reverse the inflammation of allergic conjunctivitis, *J. Clin. Invest. 111*, 121–128.
8. Cho, B. R., Kim, M. K., Suh, D. H., Hahn, J. H., Lee, B. G., Choi, Y. C., Kwon, T. J., Kim, S. Y., and Kim, D. J. (2008) Increased tissue transglutaminase expression in human atherosclerotic coronary arteries, *Coron. Artery Dis. 19*, 459–468.
9. Roberts, E. S., Zandonatti, M. A., Watry, D. D., Madden, L. J., Henriksen, S. J., Taffe, M. A., and Fox, H. S. (2003) Induction of pathogenic sets of genes in macrophages and neurons in NeuroAIDS, *Am. J. Pathol. 162*, 2041–2057.
10. Park, K. C., Chung, K. C., Kim, Y. S., Lee, J., Joh, T. H., and Kim, S. Y. (2004) Transglutaminase 2 induces nitric oxide synthesis in BV-2 microglia, *Biochem. Biophys. Res. Commun. 323*, 1055–1062.

11. Ientile, R., Caccamo, D., and Griffin, M. (2007) Tissue transglutaminase and the stress response, *Amino Acids 33*, 385–394.

12. Park, K. S., Kim, H. K., Lee, J. H., Choi, Y. B., Park, S. Y., Yang, S. H., Kim, S. Y., and Hong, K. M. (2009) Transglutaminase 2 as a cisplatin resistance marker in non-small cell lung cancer, *J. Cancer Res. Clin. Oncol. 136(4)*, 493–502.

13. Ai, L., Kim, W. J., Demircan, B., Dyer, L. M., Bray, K. J., Skehan, R. R., Massoll, N. A., and Brown, K. D. (2008) The transglutaminase 2 gene (TGM2), a potential molecular marker for chemotherapeutic drug sensitivity, is epigenetically silenced in breast cancer, *Carcinogenesis 29*, 510–518.

14. Maiuri, L., Luciani, A., Giardino, I., Raia, V., Villella, V. R., D'Apolito, M., Pettoello-Mantovani, M., Guido, S., Ciacci, C., Cimmino, M., Cexus, O. N., Londei, M., and Quaratino, S. (2008) Tissue transglutaminase activation modulates inflammation in cystic fibrosis via PPARgamma down-regulation, *J. Immunol. 180*, 7697–7705.

15. Liu, S., Cerione, R. A., and Clardy, J. (2002) Structural basis for the guanine nucleotide-binding activity of tissue transglutaminase and its regulation of transamidation activity, *Proc. Natl. Acad. Sci. U.S.A. 99*, 2743–2747.

16. Begg, G. E., Carrington, L., Stokes, P. H., Matthews, J. M., Wouters, M. A., Husain, A., Lorand, L., Iismaa, S. E., and Graham, R. M. (2006) Mechanism of allosteric regulation of transglutaminase 2 by GTP, *Proc. Natl. Acad. Sci. U.S.A. 103*, 19683–19688.

17. Siegel, M., Strnad, P., Watts, R. E., Choi, K., Jabri, B., Omary, M. B., and Khosla, C. (2008) Extracellular transglutaminase 2 is catalytically inactive, but is transiently activated upon tissue injury, *PLoS One 3*, e1861.

18. Haroon, Z. A., Hettasch, J. M., Lai, T. S., Dewhirst, M. W., and Greenberg, C. S. (1999) Tissue transglutaminase is expressed, active, and directly involved in rat dermal wound healing and angiogenesis, *FASEB J. 13*, 1787–1795.

19. Tatsukawa, H., Fukaya, Y., Frampton, G., Martinez-Fuentes, A., Suzuki, K., Kuo, T. F., Nagatsuma, K., Shimokado, K., Okuno, M., Wu, J., Iismaa, S., Matsuura, T., Tsukamoto, H., Zern, M. A., Graham, R. M., and Kojima, S. (2009) Role of transglutaminase 2 in liver injury via cross-linking and silencing of transcription factor Sp1, *Gastroenterology 136*, 1783–1795 e1710.

20. Mirza, A., Liu, S. L., Frizell, E., Zhu, J., Maddukuri, S., Martinez, J., Davies, P., Schwarting, R., Norton, P., and Zern, M. A. (1997) A role for tissue transglutaminase in hepatic injury and fibrogenesis, and its regulation by NF-kappaB, *Am. J. Physiol. 272*, G281–G288.

21. Dvorak, C. M., Hardstedt, M., Xie, H., Wang, M., Papas, K. K., Hering, B. J., Murtaugh, M. P., and Fahrenkrug, S. C. (2007) Transcriptional profiling of stress response in cultured porcine islets, *Biochem. Biophys. Res. Commun. 357*, 118–125.

22. Luciani, A., Villella, V. R., Vasaturo, A., Giardino, I., Raia, V., Pettoello-Mantovani, M., D'Apolito, M., Guido, S., Leal, T., Quaratino, S., and Maiuri, L. (2009) SUMOylation of tissue transglutaminase as link between oxidative stress and inflammation, *J. Immunol. 183*, 2775–2784.

23. Cacciamani, T., Virgili, S., Centurelli, M., Bertoli, E., Eremenko, T., and Volpe, P. (2002) Specific methylation of the CpG-rich domains in the promoter of the human tissue transglutaminase gene, *Gene 297*, 103–112.

24. Schuppan, D., Junker, Y., and Barisani, D. (2009) Celiac disease: from pathogenesis to novel therapies, *Gastroenterology 137(6)*, 1912–1933.
25. Leeds, J. S., Horoldt, B. S., Sidhu, R., Hopper, A. D., Robinson, K., Toulson, B., Dixon, L., Lobo, A. J., McAlindon, M. E., Hurlstone, D. P., and Sanders, D. S. (2007) Is there an association between coeliac disease and inflammatory bowel diseases? A study of relative prevalence in comparison with population controls, *Scand. J. Gastroenterol. 42*, 1214–1220.
26. Dieterich, W., Ehnis, T., Bauer, M., Donner, P., Volta, U., Riecken, E. O., and Schuppan, D. (1997) Identification of tissue transglutaminase as the autoantigen of celiac disease, *Nat. Med. 3*, 797–801.
27. Arentz-Hansen, H., Korner, R., Molberg, O., Quarsten, H., Vader, W., Kooy, Y. M., Lundin, K. E., Koning, F., Roepstorff, P., Sollid, L. M., and McAdam, S. N. (2000) The intestinal T cell response to alpha-gliadin in adult celiac disease is focused on a single deamidated glutamine targeted by tissue transglutaminase, *J. Exp. Med. 191*, 603–612.
28. Molberg, O., McAdam, S. N., Korner, R., Quarsten, H., Kristiansen, C., Madsen, L., Fugger, L., Scott, H., Noren, O., Roepstorff, P., Lundin, K. E., Sjostrom, H., and Sollid, L. M. (1998) Tissue transglutaminase selectively modifies gliadin peptides that are recognized by gut-derived T cells in celiac disease, *Nat. Med. 4*, 713–717.
29. van de Wal, Y., Kooy, Y., van Veelen, P., Pena, S., Mearin, L., Papadopoulos, G., and Koning, F. (1998) Selective deamidation by tissue transglutaminase strongly enhances gliadin-specific T cell reactivity, *J. Immunol. 161*, 1585–1588.
30. Sollid, L. M., Molberg, O., McAdam, S., and Lundin, K. E. (1997) Autoantibodies in coeliac disease: tissue transglutaminase–guilt by association?, *Gut 41*, 851–852.
31. Bruce, S. E., Bjarnason, I., and Peters, T. J. (1985) Human jejunal transglutaminase: demonstration of activity, enzyme kinetics and substrate specificity with special relation to gliadin and coeliac disease, *Clin. Sci. (Lond.) 68*, 573–579.
32. Nilsen, E. M., Jahnsen, F. L., Lundin, K. E., Johansen, F. E., Fausa, O., Sollid, L. M., Jahnsen, J., Scott, H., and Brandtzaeg, P. (1998) Gluten induces an intestinal cytokine response strongly dominated by interferon gamma in patients with celiac disease, *Gastroenterology 115*, 551–563.
33. Kim, S. Y., Jeong, E. J., and Steinert, P. M. (2002) IFN-gamma induces transglutaminase 2 expression in rat small intestinal cells, *J. Interferon Cytokine Res. 22*, 677–682.
34. Novogrodsky, A., Quittner, S., Rubin, A. L., and Stenzel, K. H. (1978) Transglutaminase activity in human lymphocytes: early activation by phytomitogens, *Proc. Natl. Acad. Sci. U.S.A. 75*, 1157–1161.
35. Schroff, G., Neumann, C., and Sorg, C. (1981) Transglutaminase as a marker for subsets of murine macrophages, *Eur. J. Immunol. 11*, 637–642.
36. Murtaugh, M. P., Mehta, K., Johnson, J., Myers, M., Juliano, R. L., and Davies, P. J. (1983) Induction of tissue transglutaminase in mouse peritoneal macrophages, *J. Biol. Chem. 258*, 11074–11081.
37. Murtaugh, M. P., Arend, W. P., and Davies, P. J. (1984) Induction of tissue transglutaminase in human peripheral blood monocytes, *J. Exp. Med. 159*, 114–125.

38. Seiving, B., Ohlsson, K., Linder, C., and Stenberg, P. (1991) Transglutaminase differentiation during maturation of human blood monocytes to macrophages, *Eur. J. Haematol. 46*, 263–271.
39. Cohen, J. (2002) The immunopathogenesis of sepsis, *Nature 420*, 885–891.
40. Hotchkiss, R. S. and Karl, I. E. (2003) The pathophysiology and treatment of sepsis, *N. Engl. J. Med. 348*, 138–150.
41. Leu, R. W., Herriott, M. J., Moore, P. E., Orr, G. R., and Birckbichler, P. J. (1982) Enhanced transglutaminase activity associated with macrophage activation. Possible role in Fc-mediated phagocytosis, *Exp. Cell Res. 141*, 191–199.
42. Bowness, J. M. and Tarr, A. H. (1997) Increase in transglutaminase and its extracellular products in response to an inflammatory stimulus by lipopolysaccharide, *Mol. Cell. Biochem. 169*, 157–163.
43. Suh, G. Y., Ham, H. S., Lee, S. H., Choi, J. C., Koh, W. J., Kim, S. Y., Lee, J., Han, J., Kim, H. P., Choi, A. M., and Kwon, O. J. (2006) A Peptide with anti-transglutaminase activity decreases lipopolysaccharide-induced lung inflammation in mice, *Exp. Lung Res. 32*, 43–53.
44. Falasca, L., Farrace, M. G., Rinaldi, A., Tuosto, L., Melino, G., and Piacentini, M. (2008) Transglutaminase type II is involved in the pathogenesis of endotoxic shock, *J. Immunol. 180*, 2616–2624.
45. Joneja, J. M. V. and Bielory, L. (1990) *Understanding Allergy, Sensitivity, and Immunity: A Comprehensive Guide*. Rutgers University Press, New Brunswick, pp. 126–136.
46. Chae, J. B., Lee, S., Kim, S. Y., Kim, J. G., and Yoon, Y. H. (2005) Novel transglutaminase inhibitors ameliorates endotoxininduced uveitis (EIU), *J. Korean Ophthalmol. Soc. (Suppl. 46)*, 96.
47. Moreno, J. J. (2006) Effects of antiflammins on transglutaminase and phospholipase A2 activation by transglutaminase, *Int. Immunopharmacol. 6*, 300–303.
48. Choi, K., Siegel, M., Piper, J. L., Yuan, L., Cho, E., Strnad, P., Omary, B., Rich, K. M., and Khosla, C. (2005) Chemistry and biology of dihydroisoxazole derivatives: selective inhibitors of human transglutaminase 2, *Chem. Biol. 12*, 469–475.
49. Kim, D. S., Park, K. S., Jeong, K. C., Lee, B. I., Lee, C. H., and Kim, S. Y. (2009) Glucosamine is an effective chemo-sensitizer via transglutaminase 2 inhibition, *Cancer Lett. 273*, 243–249.
50. Jeong, K. C., Ahn, K. O., Lee, B. I., Lee, C. H., and Kim, S. Y. (2009) The mechanism of transglutaminase 2 inhibition with glucosamine: implications of a possible anti-inflammatory effect through transglutaminase inhibition, *J. Cancer Res. Clin. Oncol. 136(1)*, 143–150.
51. Rottner, M., Freyssinet, J. M., and Martinez, M. C. (2009) Mechanisms of the noxious inflammatory cycle in cystic fibrosis, *Respir. Res. 10*, 23.
52. Rommens, J. M., et al. (1989) Identification of the cystic fibrosis gene: chromosome walking and jumping, *Science 245*, 1059–1065.
53. Kerem, B., Rommens, J. M., Buchanan, J. A., Markiewicz, D., Cox, T. K., Chakravarti, A., Buchwald, M., and Tsui, L. C. (1989) Identification of the cystic fibrosis gene: genetic analysis, *Science 245*, 1073–1080.

54. Riordan, J. R., et al. (1989) Identification of the cystic fibrosis gene: cloning and characterization of complementary DNA, *Science 245*, 1066–1073.

55. Weber, A. J., Soong, G., Bryan, R., Saba, S., and Prince, A. (2001) Activation of NF-kappaB in airway epithelial cells is dependent on CFTR trafficking and Cl- channel function, *Am. J. Physiol. Lung Cell. Mol. Physiol. 281*, L71–L78.

56. Bright, J. J., Kanakasabai, S., Chearwae, W., and Chakraborty, S. (2008) PPAR regulation of inflammatory signaling in CNS diseases, *PPAR Res. 2008*, 658520.

57. Grenard, P., Bresson-Hadni, S., El Alaoui, S., Chevallier, M., Vuitton, D. A., and Ricard-Blum, S. (2001) Transglutaminase-mediated cross-linking is involved in the stabilization of extracellular matrix in human liver fibrosis, *J. Hepatol. 35*, 367–375.

58. Heyninck, K., Wullaert, A., and Beyaert, R. (2003) Nuclear factor-kappa B plays a central role in tumour necrosis factor-mediated liver disease, *Biochem. Pharmacol. 66*, 1409–1415.

59. Campo, G. M., Avenoso, A., Campo, S., Nastasi, G., Traina, P., D'Ascola, A., Rugolo, C. A., and Calatroni, A. (2008) The antioxidant activity of chondroitin-4-sulphate, in carbon tetrachloride-induced acute hepatitis in mice, involves NF-kappaB and caspase activation, *Br. J. Pharmacol. 155*, 945–956.

60. Johnson, T. S., El-Koraie, A. F., Skill, N. J., Baddour, N. M., El Nahas, A. M., Njloma, M., Adam, A. G., and Griffin, M. (2003) Tissue transglutaminase and the progression of human renal scarring, *J. Am. Soc. Nephrol. 14*, 2052–2062.

61. Lee, Z. W., Kwon, S. M., Kim, S. W., Yi, S. J., Kim, Y. M., and Ha, K. S. (2003) Activation of in situ tissue transglutaminase by intracellular reactive oxygen species, *Biochem. Biophys. Res. Commun. 305*, 633–640.

62. Curro, M., Condello, S., Caccamo, D., Ferlazzo, N., Parisi, G., and Ientile, R. (2009) Homocysteine-induced toxicity increases TG2 expression in Neuro2a cells, *Amino Acids 36*, 725–730.

63. Aeschlimann, D. and Thomazy, V. (2000) Protein crosslinking in assembly and remodelling of extracellular matrices: the role of transglutaminases, *Connect. Tissue Res. 41*, 1–27.

64. Nardacci, R., Lo Iacono, O., Ciccosanti, F., Falasca, L., Addesso, M., Amendola, A., Antonucci, G., Craxi, A., Fimia, G. M., Iadevaia, V., Melino, G., Ruco, L., Tocci, G., Ippolito, G., and Piacentini, M. (2003) Transglutaminase type II plays a protective role in hepatic injury, *Am. J. Pathol. 162*, 1293–1303.

65. Sun, B. and Karin, M. (2008) NF-kappaB signaling, liver disease and hepatoprotective agents, *Oncogene 27*, 6228–6244.

66. Joyce, M. A., Walters, K. A., Lamb, S. E., Yeh, M. M., Zhu, L. F., Kneteman, N., Doyle, J. S., Katze, M. G., and Tyrrell, D. L. (2009) HCV induces oxidative and ER stress, and sensitizes infected cells to apoptosis in SCID/Alb-uPA mice, *PLoS Pathog. 5*, e1000291.

67. Zhang, Y. L., Guo, Y. J., Bin, L., and Sun, S. H. (2009) Hepatitis C virus single-stranded RNA induces innate immunity via Toll-like receptor 7, *J. Hepatol. 51*, 29–38.

68. Wang, Q., Nagarkar, D. R., Bowman, E. R., Schneider, D., Gosangi, B., Lei, J., Zhao, Y., McHenry, C. L., Burgens, R. V., Miller, D. J., Sajjan, U., and Hershenson, M. B.

(2009) Role of double-stranded RNA pattern recognition receptors in rhinovirus-induced airway epithelial cell responses, *J. Immunol. 183*, 6989–6997.

69. Chen, C. S., Wu, C. H., Lai, Y. C., Lee, W. S., Chen, H. M., Chen, R. J., Chen, L. C., Ho, Y. S., and Wang, Y. J. (2008) NF-kappaB-activated tissue transglutaminase is involved in ethanol-induced hepatic injury and the possible role of propolis in preventing fibrogenesis, *Toxicology 246*, 148–157.
70. Han, J. A. and Park, S. C. (2000) Transglutaminase-dependent modulation of transcription factor Sp1 activity, *Mol. Cells 10*, 612–618.
71. Denk, H., Gschnait, F., and Wolff, K. (1975) Hepatocellar hyalin (Mallory bodies) in long term griseofulvin-treated mice: a new experimental model for the study of hyalin formation, *Lab. Invest. 32*, 773–776.
72. Yokoo, H., Harwood, T. R., Racker, D., and Arak, S. (1982) Experimental production of Mallory bodies in mice by diet containing 3,5-diethoxycarbonyl-1,4-dihydrocollidine, *Gastroenterology 83*, 109–113.
73. Tazawa, J., Irie, T., and French, S. W. (1983) Mallory body formation runs parallel to gamma-glutamyl transferase induction in hepatocytes of griseofulvin-fed mice, *Hepatology 3*, 989–1001.
74. Denk, H., Bernklau, G., and Krepler, R. (1984) Effect of griseofulvin treatment and neoplastic transformation on transglutaminase activity in mouse liver, *Liver 4*, 208–213.
75. Riley, N. E., Li, J., Worrall, S., Rothnagel, J. A., Swagell, C., van Leeuwen, F. W., and French, S. W. (2002) The Mallory body as an aggresome: in vitro studies, *Exp. Mol. Pathol. 72*, 17–23.
76. Zatloukal, K., Denk, H., Lackinger, E., and Rainer, I. (1989) Hepatocellular cytokeratins as substrates of transglutaminases, *Lab. Invest. 61*, 603–608.
77. Cadrin, M., Marceau, N., and French, S. W. (1992) Cytokeratin of apparent high molecular weight in livers from griseofulvin-fed mice, *J. Hepatol. 14*, 226–231.
78. Strnad, P., Siegel, M., Toivola, D. M., Choi, K., Kosek, J. C., Khosla, C., and Omary, M. B. (2006) Pharmacologic transglutaminase inhibition attenuates drug-primed liver hypertrophy but not Mallory body formation, *FEBS Lett. 580*, 2351–2357.
79. Tatsukawa, H., Fukaya, Y., Frampton, G., Martinez-Fuentes, A., Suzuki, K., Kuo, T. F., Nagatsuma, K., Shimokado, K., Okuno, M., Wu, J., Iismaa, S., Matsuura, T., Tsukamoto, H., Zern, M. A., Graham, R. M., and Kojima, S. (2009) Role of transglutaminase 2 in liver injury via cross-linking and silencing of transcription factor Sp1, *Gastroenterology 136*, 1783–1795 e1710.
80. Nan, L., Wu, Y., Bardag-Gorce, F., Li, J., French, B. A., Wilson, L. T., and French, S. W. (2005) The p105/50 NF-kappaB pathway is essential for Mallory body formation, *Exp. Mol. Pathol. 78*, 198–206.
81. Meguid El Nahas, A., and Bello, A. K. (2005) Chronic kidney disease: the global challenge, *Lancet 365*, 331–340.
82. Henke, N., Schmidt-Ullrich, R., Dechend, R., Park, J. K., Qadri, F., Wellner, M., Obst, M., Gross, V., Dietz, R., Luft, F. C., Scheidereit, C., and Muller, D. N. (2007) Vascular endothelial cell-specific NF-kappaB suppression attenuates hypertension-induced renal damage, *Circ. Res. 101*, 268–276.

83. Johnson, T. S., Griffin, M., Thomas, G. L., Skill, J., Cox, A., Yang, B., Nicholas, B., Birckbichler, P. J., Muchaneta-Kubara, C., and Meguid El Nahas, A. (1997) The role of transglutaminase in the rat subtotal nephrectomy model of renal fibrosis, *J. Clin. Invest.* *99*, 2950–2960.

84. Chen, D., Huang, H. C., and Yu, L. (2005) Expression and implication of tissue transglutaminase and connective tissue growth factor at fibrotic tubulointerstitium in kidneys from UUO rats, *Beijing Da Xue Xue Bao 37*, 143–146.

85. Johnson, T. S., Fisher, M., Haylor, J. L., Hau, Z., Skill, N. J., Jones, R., Saint, R., Coutts, I., Vickers, M. E., El Nahas, A. M., and Griffin, M. (2007) Transglutaminase inhibition reduces fibrosis and preserves function in experimental chronic kidney disease, *J. Am. Soc. Nephrol. 18*, 3078–3088.

86. Johnson, T. S., Abo-Zenah, H., Skill, J. N., Bex, S., Wild, G., Brown, C. B., Griffin, M., and El Nahas, A. M. (2004) Tissue transglutaminase: a mediator and predictor of chronic allograft nephropathy? *Transplantation 77*, 1667–1675.

87. Ikee, R., Kobayashi, S., Hemmi, N., Saigusa, T., Namikoshi, T., Yamada, M., Imakiire, T., Kikuchi, Y., Suzuki, S., and Miura, S. (2007) Involvement of transglutaminase-2 in pathological changes in renal disease, *Nephron Clin Pract 105*, c139–c146.

88. Parving, H. H. (2001) Diabetic nephropathy: prevention and treatment, *Kidney Int 60*, 2041–2055.

89. Remuzzi, G., Schieppati, A., and Ruggenenti, P. (2002) Clinical practice. Nephropathy in patients with type 2 diabetes, *N. Engl. J. Med. 346*, 1145–1151.

90. Skill, N. J., Griffin, M., El Nahas, A. M., Sanai, T., Haylor, J. L., Fisher, M., Jamie, M. F., Mould, N. N., and Johnson, T. S. (2001) Increases in renal epsilon-(gamma-glutamyl)-lysine crosslinks result from compartment-specific changes in tissue transglutaminase in early experimental diabetic nephropathy: pathologic implications, *Lab. Invest. 81*, 705–716.

91. El Nahas, A. M., Abo-Zenah, H., Skill, N. J., Bex, S., Wild, G., Griffin, M., and Johnson, T. S. (2004) Elevated epsilon-(gamma-glutamyl)lysine in human diabetic nephropathy results from increased expression and cellular release of tissue transglutaminase, *Nephron Clin Pract 97*, c108–c117.

92. Huang, L., Haylor, J. L., Hau, Z., Jones, R. A., Vickers, M. E., Wagner, B., Griffin, M., Saint, R. E., Coutts, I. G., El Nahas, A. M., and Johnson, T. S. (2009) Transglutaminase inhibition ameliorates experimental diabetic nephropathy, *Kidney Int. 76*, 383–394.

93. Shweke, N., Boulos, N., Jouanneau, C., Vandermeersch, S., Melino, G., Dussaule, J. C., Chatziantoniou, C., Ronco, P., and Boffa, J. J. (2008) Tissue transglutaminase contributes to interstitial renal fibrosis by favoring accumulation of fibrillar collagen through TGF-beta activation and cell infiltration, *Am. J. Pathol. 173*, 631–642.

94. Shimizu, K., Mitchell, R. N., and Libby, P. (2006) Inflammation and cellular immune responses in abdominal aortic aneurysms, *Arterioscler. Thromb. Vasc. Biol. 26*, 987–994.

95. Barath, P., Fishbein, M. C., Cao, J., Berenson, J., Helfant, R. H., and Forrester, J. S. (1990) Detection and localization of tumor necrosis factor in human atheroma, *Am. J. Cardiol. 65*, 297–302.

96. Brand, K., Page, S., Walli, A. K., Neumeier, D., and Baeuerle, P. A. (1997) Role of nuclear factor-kappa B in atherogenesis, *Exp. Physiol. 82*, 297–304.

97. de Winther, M. P., Kanters, E., Kraal, G., and Hofker, M. H. (2005) Nuclear factor kappaB signaling in atherogenesis, *Arterioscler. Thromb. Vasc. Biol. 25*, 904–914.

98. Collins, T. and Cybulsky, M. I. (2001) NF-kappaB: pivotal mediator or innocent by-stander in atherogenesis? *J. Clin. Invest. 107*, 255–264.

99. Brand, K., Eisele, T., Kreusel, U., Page, M., Page, S., Haas, M., Gerling, A., Kaltschmidt, C., Neumann, F. J., Mackman, N., Baeurele, P. A., Walli, A. K., and Neumeier, D. (1997) Dysregulation of monocytic nuclear factor-kappa B by oxidized low-density lipoprotein, *Arterioscler. Thromb. Vasc. Biol. 17*, 1901–1909.

100. Cercek, B., Yamashita, M., Dimayuga, P., Zhu, J., Fishbein, M. C., Kaul, S., Shah, P. K., Nilsson, J., and Regnstrom, J. (1997) Nuclear factor-kappaB activity and arterial response to balloon injury, *Atherosclerosis 131*, 59–66.

101. Lindner, V. (1998) The NF-kappaB and IkappaB system in injured arteries, *Pathobiology 66*, 311–320.

102. Landry, D. B., Couper, L. L., Bryant, S. R., and Lindner, V. (1997) Activation of the NF-kappa B and I kappa B system in smooth muscle cells after rat arterial injury. Induction of vascular cell adhesion molecule-1 and monocyte chemoattractant protein-1, *Am. J. Pathol. 151*, 1085–1095.

103. Bryant, S. R., Bjercke, R. J., Erichsen, D. A., Rege, A., and Lindner, V. (1999) Vascular remodeling in response to altered blood flow is mediated by fibroblast growth factor-2, *Circ. Res. 84*, 323–328.

104. Squadrito, F., Deodato, B., Bova, A., Marini, H., Saporito, F., Calo, M., Giacca, M., Minutoli, L., Venuti, F. S., Caputi, A. P., and Altavilla, D. (2003) Crucial role of nuclear factor-kappaB in neointimal hyperplasia of the mouse carotid artery after interruption of blood flow, *Atherosclerosis 166*, 233–242.

105. Freund, C., Schmidt-Ullrich, R., Baurand, A., Dunger, S., Schneider, W., Loser, P., El-Jamali, A., Dietz, R., Scheidereit, C., and Bergmann, M. W. (2005) Requirement of nuclear factor-kappaB in angiotensin II- and isoproterenol-induced cardiac hypertrophy in vivo, *Circulation 111*, 2319–2325.

106. Herrmann, O., Baumann, B., de Lorenzi, R., Muhammad, S., Zhang, W., Kleesiek, J., Malfertheiner, M., Kohrmann, M., Potrovita, I., Maegele, I., Beyer, C., Burke, J. R., Hasan, M. T., Bujard, H., Wirth, T., Pasparakis, M., and Schwaninger, M. (2005) IKK mediates ischemia-induced neuronal death, *Nat. Med. 11*, 1322–1329.

107. Luedde, T., Assmus, U., Wustefeld, T., Meyer zu Vilsendorf, A., Roskams, T., Schmidt-Supprian, M., Rajewsky, K., Brenner, D. A., Manns, M. P., Pasparakis, M., and Trautwein, C. (2005) Deletion of IKK2 in hepatocytes does not sensitize these cells to TNF-induced apoptosis but protects from ischemia/reperfusion injury, *J. Clin. Invest. 115*, 849–859.

108. Muller, D. N., Heissmeyer, V., Dechend, R., Hampich, F., Park, J. K., Fiebeler, A., Shagdarsuren, E., Theuer, J., Elger, M., Pilz, B., Breu, V., Schroer, K., Ganten, D., Dietz, R., Haller, H., Scheidereit, C., and Luft, F. C. (2001) Aspirin inhibits NF-kappaB and protects from angiotensin II-induced organ damage, *FASEB J. 15*, 1822–1824.

109. Iwai, N., Shimoike, H., and Kinoshita, M. (1995) Genes up-regulated in hypertrophied ventricle, *Biochem. Biophys. Res. Commun. 209*, 527–534.

110. Boisvert, W. A., Rose, D. M., Boullier, A., Quehenberger, O., Sydlaske, A., Johnson, K. A., Curtiss, L. K., and Terkeltaub, R. (2006) Leukocyte transglutaminase 2 expression limits atherosclerotic lesion size, *Arterioscler. Thromb. Vasc. Biol. 26*, 563–569.

111. Johnson, K. A., Polewski, M., and Terkeltaub, R. A. (2008) Transglutaminase 2 is central to induction of the arterial calcification program by smooth muscle cells, *Circ. Res. 102*, 529–537.

112. Tozawa, K., Yasui, T., Okada, A., Hirose, M., Hamamoto, S., Itoh, Y., and Kohri, K. (2008) NF-kappaB activation in renal tubular epithelial cells by oxalate stimulation, *Int. J. Urol. 15*, 924–928.

113. Attur, M. G., Patel, I. R., Patel, R. N., Abramson, S. B., and Amin, A. R. (1998) Autocrine production of IL-1 beta by human osteoarthritis-affected cartilage and differential regulation of endogenous nitric oxide, IL-6, prostaglandin E2, and IL-8, *Proc. Assoc. Am. Physicians 110*, 65–72.

114. Borzi, R. M., Mazzetti, I., Macor, S., Silvestri, T., Bassi, A., Cattini, L., and Facchini, A. (1999) Flow cytometric analysis of intracellular chemokines in chondrocytes in vivo: constitutive expression and enhancement in osteoarthritis and rheumatoid arthritis, *FEBS Lett. 455*, 238–242.

115. Olson, T. S. and Ley, K. (2002) Chemokines and chemokine receptors in leukocyte trafficking, *Am. J. Physiol. Regul. Integr. Comp. Physiol. 283*, R7–R28.

116. Merz, D., Liu, R., Johnson, K., and Terkeltaub, R. (2003) IL-8/CXCL8 and growth-related oncogene alpha/CXCL1 induce chondrocyte hypertrophic differentiation, *J. Immunol. 171*, 4406–4415.

117. Roman-Blas, J. A. and Jimenez, S. A. (2006) NF-kappaB as a potential therapeutic target in osteoarthritis and rheumatoid arthritis, *Osteoarthritis Cartilage 14*, 839–848.

118. Pulai, J. I., Chen, H., Im, H. J., Kumar, S., Hanning, C., Hegde, P. S., and Loeser, R. F. (2005) NF-kappa B mediates the stimulation of cytokine and chemokine expression by human articular chondrocytes in response to fibronectin fragments, *J. Immunol. 174*, 5781–5788.

119. Forsyth, C. B., Cole, A., Murphy, G., Bienias, J. L., Im, H. J., and Loeser, R. F., Jr. (2005) Increased matrix metalloproteinase-13 production with aging by human articular chondrocytes in response to catabolic stimuli, *J. Gerontol. A Biol. Sci. Med. Sci. 60*, 1118–1124.

120. Loeser, R. F., Yammani, R. R., Carlson, C. S., Chen, H., Cole, A., Im, H. J., Bursch, L. S., and Yan, S. D. (2005) Articular chondrocytes express the receptor for advanced glycation end products: Potential role in osteoarthritis, *Arthritis Rheum. 52*, 2376–2385.

121. Johnson, K. A. and Terkeltaub, R. A. (2005) External GTP-bound transglutaminase 2 is a molecular switch for chondrocyte hypertrophic differentiation and calcification, *J. Biol. Chem. 280*, 15004–15012.

122. Lorand, L. and Graham, R. M. (2003) Transglutaminases: crosslinking enzymes with pleiotropic functions, *Nat. Rev. Mol. Cell Biol. 4*, 140–156.

123. Zemskov, E. A., Janiak, A., Hang, J., Waghray, A., and Belkin, A. M. (2006) The role of tissue transglutaminase in cell-matrix interactions, *Front. Biosci. 11*, 1057–1076.

124. Johnson, K. A., van Etten, D., Nanda, N., Graham, R. M., and Terkeltaub, R. A. (2003) Distinct transglutaminase 2-independent and transglutaminase 2-dependent pathways mediate articular chondrocyte hypertrophy, *J. Biol. Chem. 278*, 18824–18832.

125. Cecil, D. L. and Terkeltaub, R. (2008) Transamidation by transglutaminase 2 transforms S100A11 calgranulin into a procatabolic cytokine for chondrocytes, *J. Immunol. 180*, 8378–8385.

126. Orlandi, A., Oliva, F., Taurisano, G., Candi, E., Di Lascio, A., Melino, G., Spagnoli, L. G., and Tarantino, U. (2009) Transglutaminase-2 differently regulates cartilage destruction and osteophyte formation in a surgical model of osteoarthritis, *Amino Acids 36*, 755–763.

127. Bakker, A. C., van de Loo, F. A., van Beuningen, H. M., Sime, P., van Lent, P. L., Van Der Kraan, P. M., Richards, C. D., and Van Den Berg, W. B. (2001) Overexpression of active TGF-beta-1 in the murine knee joint: evidence for synovial-layer-dependent chondro-osteophyte formation, *Osteoarthritis Cartilage 9*, 128–136.

128. Gouze, J. N., Bianchi, A., Becuwe, P., Dauca, M., Netter, P., Magdalou, J., Terlain, B., and Bordji, K. (2002) Glucosamine modulates IL-1-induced activation of rat chondrocytes at a receptor level, and by inhibiting the NF-kappa B pathway, *FEBS Lett. 510*, 166–170.

129. Largo, R., Alvarez-Soria, M. A., Diez-Ortego, I., Calvo, E., Sanchez-Pernaute, O., Egido, J., and Herrero-Beaumont, G. (2003) Glucosamine inhibits IL-1beta-induced NFkappaB activation in human osteoarthritic chondrocytes, *Osteoarthritis Cartilage 11*, 290–298.

130. Karin, M. (2006) Nuclear factor-kappaB in cancer development and progression, *Nature 441*, 431–436.

131. Karin, M. (2008) The IkappaB kinase - a bridge between inflammation and cancer, *Cell Res. 18*, 334–342.

132. Nakanishi, C. and Toi, M. (2005) Nuclear factor-kappaB inhibitors as sensitizers to anticancer drugs, *Nat. Rev. Cancer 5*, 297–309.

133. Tergaonkar, V., Bottero, V., Ikawa, M., Li, Q., and Verma, I. M. (2003) IkappaB kinase-independent IkappaBalpha degradation pathway: functional NF-kappaB activity and implications for cancer therapy, *Mol. Cell Biol. 23*, 8070–8083.

134. Pacifico, F. and Leonardi, A. (2006) NF-kappaB in solid tumors, *Biochem. Pharmacol. 72*, 1142–1152.

135. Mehta, K. (1994) High levels of transglutaminase expression in doxorubicin-resistant human breast carcinoma cells, *Int. J. Cancer 58*, 400–406.

136. Han, J. A. and Park, S. C. (1999) Hydrogen peroxide mediates doxorubicin-induced transglutaminase 2 expression in PC-14 human lung cancer cell line, *Exp. Mol. Med. 31*, 83–88.

137. Antonyak, M. A., Miller, A. M., Jansen, J. M., Boehm, J. E., Balkman, C. E., Wakshlag, J. J., Page, R. L., and Cerione, R. A. (2004) Augmentation of tissue transglutaminase expression and activation by epidermal growth factor inhibit doxorubicin-induced apoptosis in human breast cancer cells, *J. Biol. Chem. 279*, 41461–41467.

138. Yuan, L., Choi, K., Khosla, C., Zheng, X., Higashikubo, R., Chicoine, M. R., and Rich, K. M. (2005) Tissue transglutaminase 2 inhibition promotes cell death and chemosensitivity in glioblastomas, *Mol. Cancer Ther. 4*, 1293–1302.

139. Yuan, L., Siegel, M., Choi, K., Khosla, C., Miller, C. R., Jackson, E. N., Piwnica-Worms, D., and Rich, K. M. (2007) Transglutaminase 2 inhibitor, KCC009, disrupts fibronectin assembly in the extracellular matrix and sensitizes orthotopic glioblastomas to chemotherapy, *Oncogene 26*, 2563–2573.

140. Hwang, J. Y., Mangala, L. S., Fok, J. Y., Lin, Y. G., Merritt, W. M., Spannuth, W. A., Nick, A. M., Fiterman, D. J., Vivas-Mejia, P. E., Deavers, M. T., Coleman, R. L., Lopez-Berestein, G., Mehta, K., and Sood, A. K. (2008) Clinical and biological significance of tissue transglutaminase in ovarian carcinoma, *Cancer Res. 68*, 5849–5858.

141. Kim, D. S., Park, S. S., Nam, B. H., Kim, I. H., and Kim, S. Y. (2006) Reversal of drug resistance in breast cancer cells by transglutaminase 2 inhibition and nuclear factor-kappaB inactivation, *Cancer Res. 66*, 10936–10943.

142. Kim, D. S., Park, K. S., and Kim, S. Y. (2009) Silencing of TGase 2 sensitizes breast cancer cells to apoptosis by regulation of survival factors, *Front. Biosci. 14*, 2514–2521.

143. Muma, N. A. (2007) Transglutaminase is linked to neurodegenerative diseases, *J. Neuropathol. Exp. Neurol. 66*, 258–263.

144. Gilad, G. M. and Varon, L. E. (1985) Transglutaminase activity in rat brain: characterization, distribution, and changes with age, *J. Neurochem. 45*, 1522–1526.

145. Appelt, D. M., Kopen, G. C., Boyne, L. J., and Balin, B. J. (1996) Localization of transglutaminase in hippocampal neurons: implications for Alzheimer's disease, *J. Histochem. Cytochem. 44*, 1421–1427.

146. Johnson, G. V., Cox, T. M., Lockhart, J. P., Zinnerman, M. D., Miller, M. L., and Powers, R. E. (1997) Transglutaminase activity is increased in Alzheimer's disease brain, *Brain Res. 751*, 323–329.

147. Lesort, M., Chun, W., Johnson, G. V., and Ferrante, R. J. (1999) Tissue transglutaminase is increased in Huntington's disease brain, *J. Neurochem. 73*, 2018–2027.

148. Kim, S. Y., Grant, P., Lee, J. H., Pant, H. C., and Steinert, P. M. (1999) Differential expression of multiple transglutaminases in human brain. Increased expression and cross-linking by transglutaminases 1 and 2 in Alzheimer's disease, *J. Biol. Chem. 274*, 30715–30721.

149. Junn, E., Ronchetti, R. D., Quezado, M. M., Kim, S. Y., and Mouradian, M. M. (2003) Tissue transglutaminase-induced aggregation of alpha-synuclein: Implications for Lewy body formation in Parkinson's disease and dementia with Lewy bodies, *Proc. Natl. Acad. Sci. U.S.A. 100*, 2047–2052.

150. Karpuj, M. V., Garren, H., Slunt, H., Price, D. L., Gusella, J., Becher, M. W., and Steinman, L. (1999) Transglutaminase aggregates huntingtin into nonamyloidogenic polymers, and its enzymatic activity increases in Huntington's disease brain nuclei, *Proc. Natl. Acad. Sci. U.S.A. 96*, 7388–7393.

151. Kim, S. Y., Marekov, L., Bubber, P., Browne, S. E., Stavrovskaya, I., Lee, J., Steinert, P. M., Blass, J. P., Beal, M. F., Gibson, G. E., and Cooper, A. J. (2005) Mitochondrial aconitase is a transglutaminase 2 substrate: transglutamination is a probable mechanism contributing to high-molecular-weight aggregates of aconitase and loss of aconitase activity in Huntington disease brain, *Neurochem. Res. 30*, 1245–1255.

152. Davis, D. G., Schmitt, F. A., Wekstein, D. R., and Markesbery, W. R. (1999) Alzheimer neuropathologic alterations in aged cognitively normal subjects, *J. Neuropathol. Exp. Neurol. 58*, 376–388.

153. Morris, J. C. (1999) Is Alzheimer's disease inevitable with age?: Lessons from clinicopathologic studies of healthy aging and very mild alzheimer's disease, *J. Clin. Invest. 104*, 1171–1173.

154. Gonzalez-Scarano, F. and Martin-Garcia, J. (2005) The neuropathogenesis of AIDS, *Nat. Rev. Immunol. 5*, 69–81.

155. Wajgt, A. and Gorny, M. (1983) CSF antibodies to myelin basic protein and to myelin-associated glycoprotein in multiple sclerosis. Evidence of the intrathecal production of antibodies, *Acta Neurol. Scand. 68*, 337–343.

156. Selkoe, D. J., Abraham, C., and Ihara, Y. (1982) Brain transglutaminase: in vitro crosslinking of human neurofilament proteins into insoluble polymers, *Proc. Natl. Acad. Sci. U.S.A. 79*, 6070–6074.

157. Rostasy, K., Monti, L., Yiannoutsos, C., Wu, J., Bell, J., Hedreen, J., and Navia, B. A. (2000) NFkappaB activation, TNF-alpha expression, and apoptosis in the AIDS-Dementia-Complex, *J. Neurovirol. 6*, 537–543.

158. Caccamo, D., Campisi, A., Curro, M., Aguennouz, M., Li Volti, G., Avola, R., and Ientile, R. (2005) Nuclear factor-kappab activation is associated with glutamate-evoked tissue transglutaminase up-regulation in primary astrocyte cultures, *J. Neurosci. Res. 82*, 858–865.

159. Stormer, L. (1952) Phylogeny and taxonomy of fossil horseshoe crab, *J. Paleontol. 26*, 630–639.

160. Tokunaga, F., Muta, T., Iwanaga, S., Ichinose, A., Davie, E. W., Kuma, K., and Miyata, T. (1993) Limulus hemocyte transglutaminase. cDNA cloning, amino acid sequence, and tissue localization, *J. Biol. Chem. 268*, 262-268.

161. Fuller, G. M. and Doolittle, R. F. (1971) Studies of invertebrate fibrinogen. I. Purification and characterization of fibrinogen from the spiny lobster, *Biochemistry 10*, 1305–1311.

162. Okada, Y., Nagase, H., and Harris, E. D., Jr. (1986) A metalloproteinase from human rheumatoid synovial fibroblasts that digests connective tissue matrix components. Purification and characterization, *J. Biol. Chem. 261*, 14245–14255.

163. Park, K. S., Kim, D. S., Ko, C., Lee, S. J., Oh, S. H., and Kim, S. Y. (2011) TNF-alpha mediated NF-kappaB activation is constantly extended by transglutaminase 2, *Front Biosci (Elite Ed). 3*, 341–354.

TRANSGLUTAMINASE 2 AT THE CROSSROADS BETWEEN CELL DEATH AND SURVIVAL

MAURO PIACENTINI
MANUELA D'ELETTO
LAURA FALASCA
MARIA GRAZIA FARRACE
CARLO RODOLFO

CONTENTS

I. TRANSGLUTAMINASES

Transglutaminases (TGases) are a peculiar family of enzymes that catalyse the post-translational modification of proteins either through protein

Advances in Enzymology and Related Areas of Molecular Biology, Volume 78.
Edited by Eric J. Toone.

cross-linking, via ε-(γ-glutamyl) lysine bonds, or through the incorporation of primary amines, at the level peptide-bound glutamine residues [1].

The cross-linked protein products become resistant to mechanical challenge and proteolytic degradation, their accumulation being found in a number of tissues, including skin and hair, in blood clotting, and wound healing [2].

In mammals, nine distinct TGases have been identified at the genomic level [3]; however, to date, only six have been characterized at the protein level. These isoforms are the products of different genes that display a wide structural homology and are members of the papain-like superfamily of cysteine proteases [4]. The characterized enzymes include (1) the plasmatic factor XIII, which is converted by thrombin-dependent proteolysis into the active TGase enzyme, the factor XIIIA, which is involved in the stabilization of fibrin clots and wound healing [5]; (2) the keratinocyte or type 1 transglutaminase (TG1), which exists in both membrane-bound and soluble form, requires proteolysis in order to be activated, and is also involved in the terminal differentiation of keratinocytes [6, 7]; (3) the ubiquitous "tissue" or type 2 transglutaminase (TG2) [8, 9]; (4) the epidermal/hair follicle or type 3 transglutaminase (TG3), which, like TG1, requires proteolysis and is involved in the terminal differentiation of keratinocytes [10]; (5) the prostatic secretory or type 4 transglutaminase (TG4), which is responsible for the formation of the copulatory plug and is essential for fertility in rodents [11]; and (6) the type 5 transglutaminase (TG5), which is involved in keratinocyte differentiation [12]. In tissues, the specific expression of the various isoenzymes is tightly regulated at the transcriptional level [13, 14].

All members of the family, with the exception of the nonenzymic erythrocyte band 4.2 protein, possess a catalytic triad of Cys-His-Asp or Cys-His-Asn. The first 3-D structural study on TGases, performed on the zymogenic A subunit of plasma factor XIII, revealed that each factor XIIIA subunit is composed of four domains: (1) *N*-terminal β-sandwich; (2) core domain, containing the catalytic triad; (3) *C*-terminal β-barrel 1 and (4) *C*-terminal β-barrel 2 (see Figure 1). These four domains' organization is highly conserved among all the TGase isoforms.

In addition to the well-characterized cross-linking enzymatic action, some members of the TGase family participate in a vast array of biological processes through actions not involving their transamidase catalytic activity [9]. In fact, TG2 [15, 16], TG4 [17], and TG5 [18] are able to bind and to

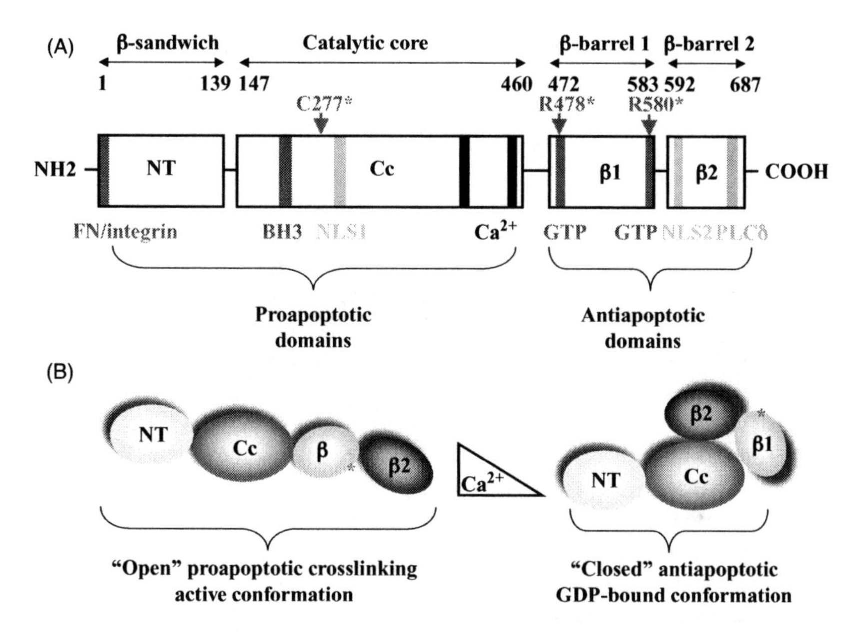

FIGURE 1. TG2 structure and domains. (A) Graphic representation of the four domains of TG2. The following regions are highlighted: FN/integrin binding motif; BH3 domain; C277, active site's cysteine; NLS1 and 2, nuclear localization signals; Ca^{2+} binding motif; GTP binding motif and relative binding impairing mutations (R478/580); and PLCδ binding motif. (B) It has been proposed that Ca^{2+} concentration may control the shift between the "open" proapoptotic (high [Ca^{2+}]) and the "closed" antiapoptotic (low [Ca^{2+}]) conformations of TG2. (See insert for color representation.)

hydrolyze GTP. The GTP binding causes a transition to a compact inactive conformation of the enzyme, leading to the inhibition of its transamidating activity (Figure 1) [19]. However, to date, only TG2 has been shown to utilize this function as a high-molecular-weight G protein (G_h) involved in signaling generated by certain G protein-coupled receptors [20]. TG2 has also been reported to act both as a serine/threonine kinase [21–23] and as a protein disulfide isomerase (PDI) [24, 25]. Finally, FXIIIA [26] and TG2 [27] also act as extra-cellular adaptor proteins to facilitate cell adhesion to the matrix.

Deregulation of these enzymes is associated with a number of human diseases, including autoimmune diseases, cancer, infectious diseases, neurodegeneration, progressive tissue fibrosis, and diseases related to the assembly of the stratum corneum of the epidermis of the skin [28].

A. TYPE 2 TRANSGLUTAMINASE

Tissue or type 2 transglutaminase (TG2) is the most ubiquitous member of the TGase family. The human gene is localized in the chromosome 20q12 and the mouse ortholog in the chromosome 2 [29, 30]. The human TG2 gene has been shown to be 32.5 kb in size and contains 13 exons and 12 introns [31, 32]. The full length TG2 protein consists of 687 amino acids, with a predicted molecular mass of about 78 kDa. This protein is present in all organs, due to its constitutive expression in endothelial cells, smooth muscle cells, and fibroblasts as well as in a number of organ-specific cell types [33, 34].

The structure of TG2 was resolved in a crystal complex with GDP [19]; similar to the other TGases, it showed the presence of four distinct domains (Figure 1): (1) an *N*-terminal β-sandwich, bearing fibronectin and integrin binding sites; (2) a catalytic core, bearing the catalytic triad for the acyl-transfer reaction and a BH3-like domain; and finally, the two *C*-terminal β-barrel domains-(3) *C*-terminal β-barrel 1, and (4) *C*-terminal β-barrel 2. A unique guanidine nucleotide-binding site, which has not been found in any TGase proteins, is located in a cleft between the catalytic core and the first β-barrel. The sequence coding for this domain is located in the exon 10 of the TG2 gene and shows very poor sequence homology with the same exon in the other TGases. The binding of GTP, but not that of Ca^{2+}, proves to be important for the stability of the enzyme's conformation, suggesting the molecular mechanism by which GTP inhibits TG2 activity. Accordingly, the GTP–TG2 adopts a conformation that decreases the accessibility of the protein matrix to the solvent, thus rendering the accessibility of the active site more difficult [35].

The 3-D organization of the Cys^{277}, His^{335}, and Asp^{358} in the catalytic triad is similar to that of thiol proteases, such as papain [36, 37], and confers high reactivity to Cys^{277}, which might form thioesters with peptidyl-glutamine moieties of the protein substrates. This high reactivity of Cys^{277} has been employed to develop a wide range of active-site-directed irreversible inhibitors of the enzyme. In the absence of Ca^{2+}, TG2 assumes the basic latent conformation and the reactivity of Cys^{277} is decreased either by hydrogen bonding, with the phenolic hydroxy group of Tyr^{516}, or by formation of a disulphide, with a neighboring cysteine residue, namely Cys^{336} [38]. It is worth noting that, the pro-apoptotic and pro-survival activities of

TG2 are confined in discrete domains along the primary sequence of the protein, suggesting a modular and independent phylogenetic evolution.

TG2 is located mostly in the cytosol, but is also found associated with the inner face of the plasma membrane [15, 16], with nuclear membrane [39], and in the mitochondria [40, 41]. In specific settings, the enzyme has been detected in the extracellular matrix (ECM), or at the cell surface, in association with proteins of the ECM [42, 43]. In particular, under pathological conditions, TG2 plays an important role in ECM stabilization, during wound healing, angiogenesis, fibrosis, and bone remodeling [44]. The enzyme exerts this extracellular stabilizing function by forming complexes with fibronectin and collagen [45], with β1/β3/β5 integrins [46], with the heparan sulfate chains of the heparan sulfate proteoglycan receptor, with syndecan 4 [47], or with the orphan G protein-coupled cell-adhesion receptor GPR56 [48]. On the other hand, until today, no *in vivo* evidences of the physiological role for the extracellular TG2 has been reported.

Inside the cell, TG2 is able to carry out three different enzymatic functions in addition to its transamidase activity. These functions are (1) the GTPase activity and its related intracellular G protein signaling, via coupling to the $\alpha_{1\beta}/\alpha_{1\delta}$-adrenergic receptors [20], the TPA thromboxane A2 receptor [49], and the oxytocin receptor [50]; (2) the serine/threonine kinase activity, with insulin-like growth factor binding protein (IGFBP)-3, p53, or histones as substrates [21–23]; (3) the PDI activity [24, 25], for which several mitochondrial substrates have been identified [40, 41, 51].

B. TG2 FUNCTIONS IN THE CELL

The various TG2 subcellular localizations as well as the large number of identified protein substrates clearly imply the involvement of the enzyme in multiple biological functions. In keeping with this assumption, studies carried out in different cellular populations or under physiological vs. pathological settings failed to provide a unique paradigm. In fact, while some cell types (e.g., endothelial and smooth muscle cells) express constitutive high level of TG2 protein [34, 52], other cell types showed an induction of the enzyme as a response to the activation of distinct signaling pathways related to the induction of cellular stress, defence, and/or differentiation. In fact, retinoic acid (RA), TGFβ, NF-κB, and AP-responsive elements have been identified in the regulatory region of the TG2 gene [53]. Nevertheless, the regulation of the various activities of the enzyme relies on multiple heterogeneous factors. The role played by TG2 as a G-protein has been clearly described [54, 55] in the transduction of the signal from the seven transmembrane-helix receptors to phospholipase C δ-1 (PLC), which becomes activated after binding of GTP to TG2 [56]. Conversely, high Ca^{2+} levels might induce the release of the bound GTP/GDP molecule, inhibiting signaling and promoting the transamidating activity. The interaction of TG2 with specific molecules (e.g., with sphingosylphosphocholine) might reduce the Ca^{2+} requirement for the transglutaminase activity [57]. TG2 activity may also be strongly influenced by nitric oxide: up to 15 of the 18 cysteine residues can be nitrosylated and denitrosylated, in a Ca^{2+}-dependent manner, leading to the inhibition and the activation of the enzyme, respectively [58].

The calcium-activated form of TG2 allows the enzyme to interact and modify major components of the cytoskeleton. Accordingly, after RA induction, TG2 can modify RhoA, a member of Rho GTPases widely involved in cytoskeletal modifications [59, 60]. These changes result in an increase of the binding of RhoA to ROCK-2 protein kinase and in the autophosophorylation of ROCK-2. Such changes lead to phosphorylation of vimentin, which causes stress fibers formation and increased cell adhesion. Moreover, TG2 can interact with beta-tubulin and with microtubule binding proteins, including tau, which can be eventually cross-linked by the enzyme [61, 62].

Cytoskeleton proteins do not represent the only target of TG2 enzymatic activity. An interesting aspect of action exerted by TG2 in the cell relies on its ability to localize at different cell compartments. It has been proposed that, under certain conditions, TG2 might translocate in the nucleus, through its own nuclear localization sequences (NLS) (Figures 1 and 2) and with the help of importin-α3 [63]. Once in the nucleus, TG2 might function either as a G-protein or as a transamidase activated by nuclear Ca^{2+}-signals, in order to cross-link histones, retinoblastoma (Rb), and SP1 proteins [62, 64–66]. These observations support the hypothesis that TG2 might have a direct role in chromatin post-translational modifications and/or gene expression regulation [40, 41].

TG2 protein expression and activity has been shown to be induced in cells undergoing apoptosis *in vivo* [67, 68]. Under physiological conditions, TG2 overexpression primes cells for suicide, while its inhibition, through several strategies, results in decreased cell death [40, 69]. It has been reported that the enzyme might sensitize cells towards apoptosis by interacting with mitochondria and by shifting them to a higher polarized state [40, 41]. This event triggers an alteration of the redox status that might provoke the activation of the TG2's PDI activity. At the same time, the massive increase of cytosolic Ca^{2+} concentration, observed during the later stages of apoptosis,

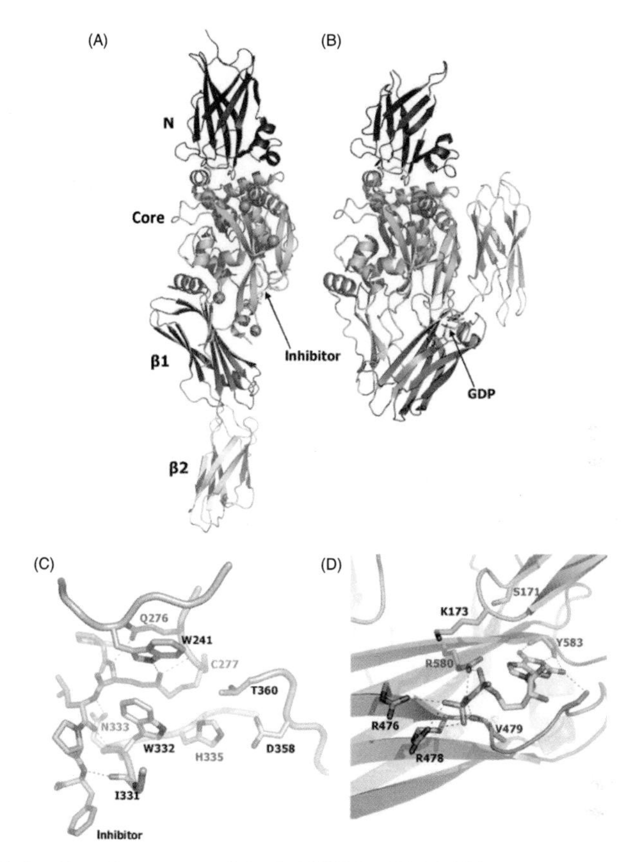

FIGURE 2. Crystal structures of Ac-P(DON)LPF-NH2 inhibitor-bound and GDP-bound TG2 represented as ribbons. The *N*-terminal sandwich domain (N) is shown in red, the catalytic core domain (core) in green, β-barrel 1 (β1) in magenta, and β-barrel 2 (β2) in cyan. (A) Open conformation of inhibitor-bound TG2 [20] showing the proposed positions of Ca^{2+} binding (orange spheres) according to Kiraly et al. [22]. (B) Closed conformation of GDP-bound TG2 [19] showing the degree of movement of β1 and β2 after nucleotide binding. (C) Structure of the active site of inhibitor-bound TG2 showing putative contacts between Ac-P(DON)LPF-NH2 active site residues. The hydrophobic tunnel for primary amine substrate access to C277 is formed by W241, W332, and T360, which disfavors entry of water. (D) The GTP binding site of TG2 showing bound GDP according to the crystal structure [19]. Putative contacts between GTP and TG2 are represented by dashed lines. The side chains of V479 and Y583 have been omitted for clarity. Residue S171 is involved in stabilizing H-bonds between the core domain and β-barrel 1 and also between neighboring residues in the loop that it resides in. The unfavorable proximity of K173 to R580 is thought to be important in control of TG2 conformation, being stabilized in the presence of guanine nucleoside phosphates.

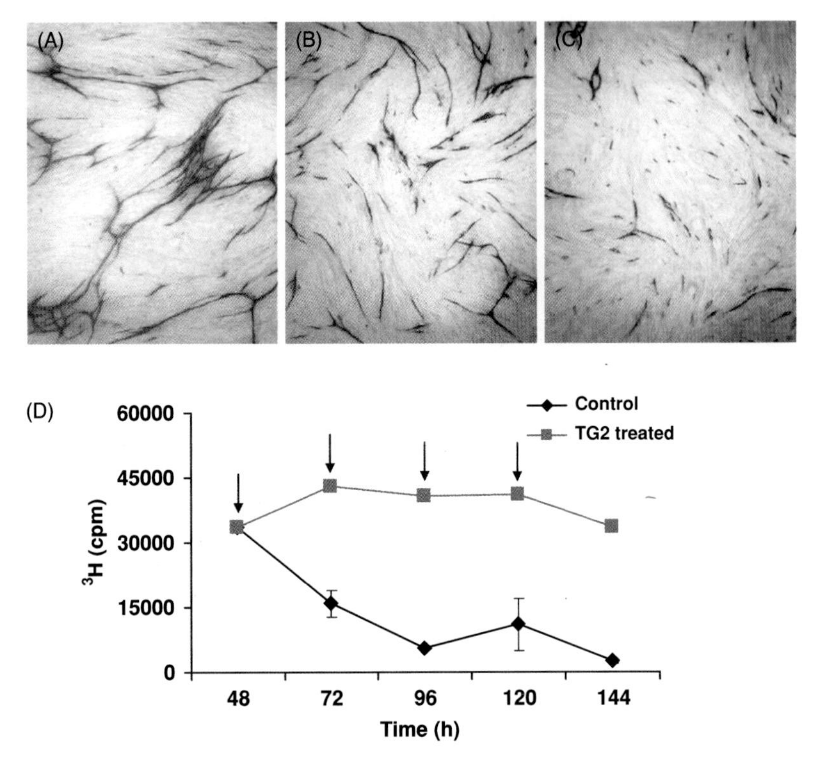

FIGURE 4. Inhibition of angiogenesis in a HUVEC cell coculture model by the addition of exogenous TG2. TG2 additions were made on days 2, 5, 7, and 10 of culture. Vessel formation was observed by staining for von Willebrand factor (vWF). (A) untreated, (b) TG2 treatment after day 5, (C) TG2 treatment after day 10. (D) Total ECM turnover following ^{3}H amino acid mixture pulse labeling. Arrows indicate addition of exogenous TG2 [127].

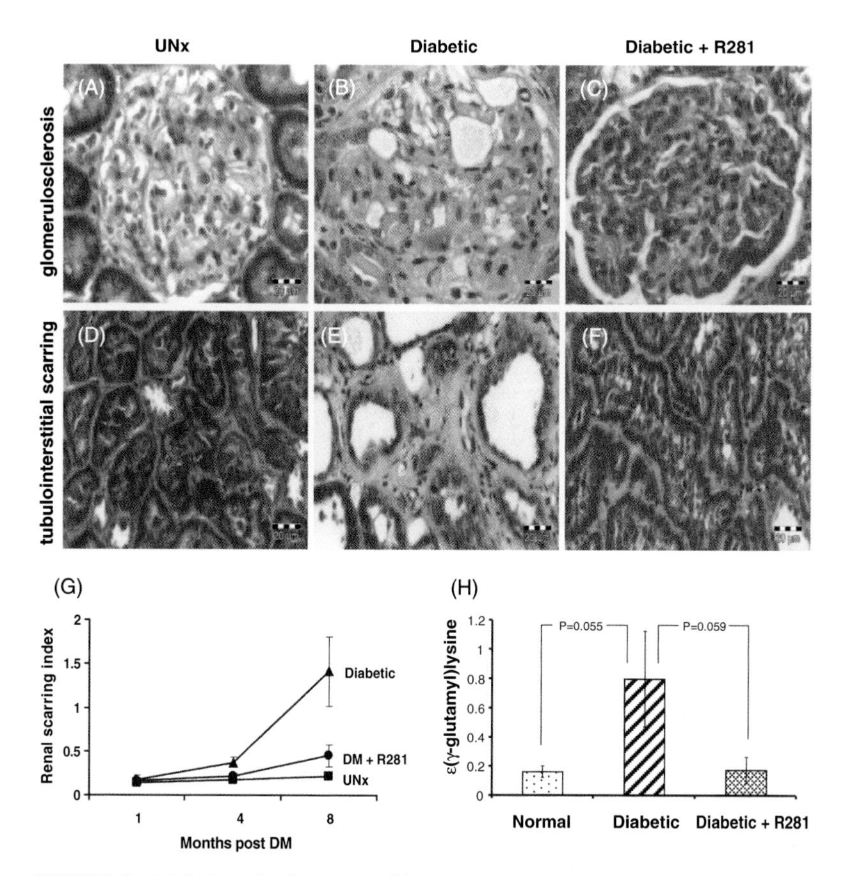

FIGURE 5. Inhibition of collagen deposition and ε(γ-glutamyl)-lysine cross-links into the ECM by TG inhibitors in a rat streptozotocin-induced model of diabetic kidney disease. Rats were subjected to unilateral nephrectomy followed by streptozotocin injection after 7 days, with blood glucose controlled between 10 and 25 mM using insulin implants (diabetic or DM). Control animals were subjected to unilateral nephrectomy, but did not receive streptozotocin or blood glucose control (UNx). TG2-inhibitor-treated rats (diabetic + R281 or DM + R281) had R281 [157] infused directly into the kidney for the duration of the experiment. (A–F) Masson's trichrome stained kidney sections showing degree of glomerulosclerosis (A–C) and tubulointerstitial scarring (D–F). Collagenous material is stained blue, with nuclei and cytoplasm red/pink. (G) Renal scarring index determined by analysis of Masson's trichrome stained sections. (H) ε(γ-glutamyl)-lysine cross-link levels in kidney sections. (Adapted from [161].)

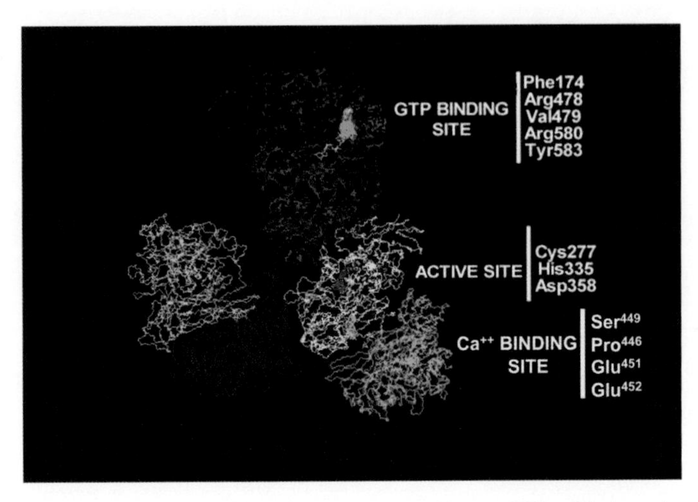

FIGURE 3. Representation of the tertiary structure of human TG2, with the amino acids responsible for the active site, the Ca^{++} binding site, and the GTP binding site.

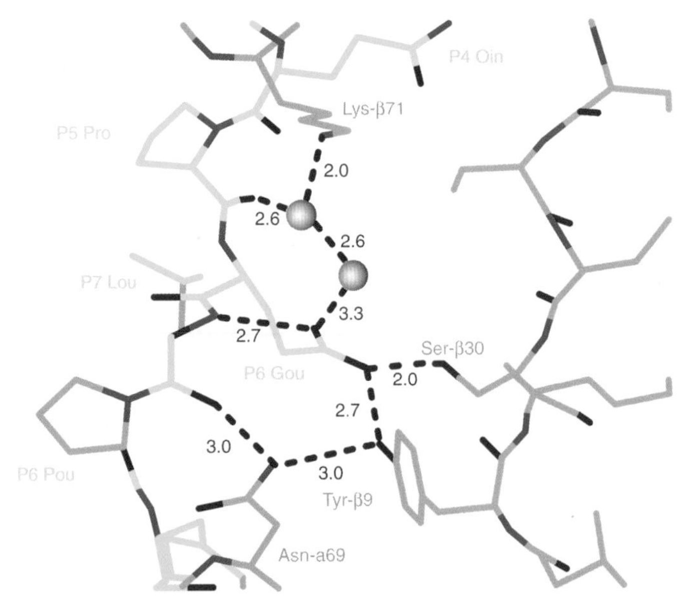

FIGURE 5. Picture of the structural binding between HLA DQ2 and deamidated gliadin epitope (see [4]).

TRANSGLUTAMINASE IN EPIDERMIS AND NEUROLOGICAL DISEASE OR WHAT MAKES A GOOD CROSS-LINKING SUBSTRATE

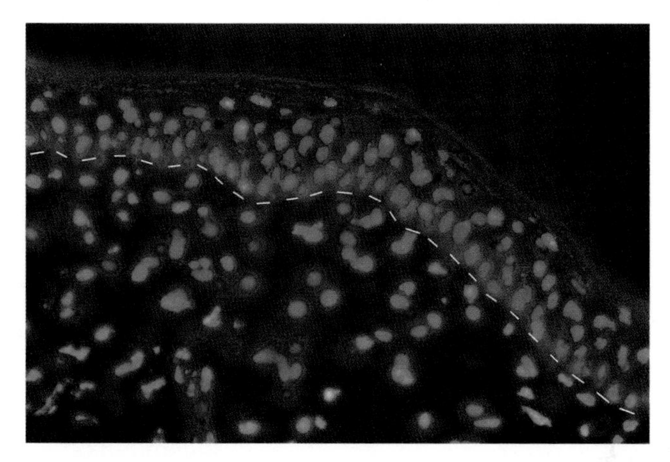

FIGURE 1. Detection of chicken involucrin with an antipeptide antibody. A section through *Gallus* skin was stained with an antiserum (red) prepared by Primm (Italy) and counterstained with DAPI (blue) for DNA. The cornified layer of the epidermis is strongly stained. The envelope precursor is concentrated peripherally.

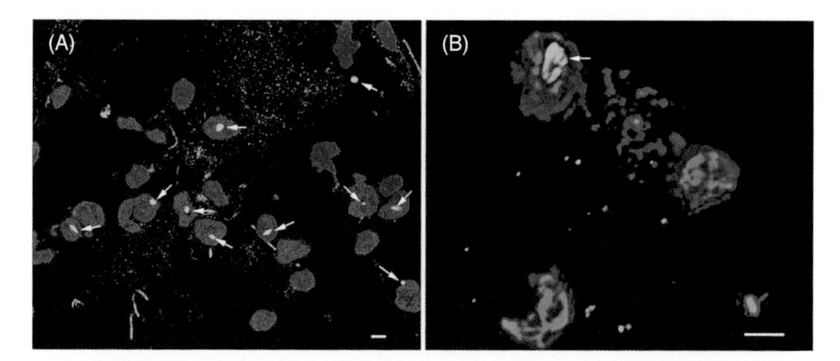

FIGURE 2. Neuronal inclusions in cerebral cortex of patient with juvenile Huntington's disease. Frozen sections were stained with an antibody directed against the N-terminus of huntingtin (green). (A) Nuclear and cytoplasmic inclusions (arrows) visualized with an LSM 510 confocal microsocope (Zeiss). DNA (red) was stained with TO-PRO-3. (B) Image of nuclear inclusion was obtained with a TE2000E microscope (Nikon). Spectral deconvolution of the image was performed using the ImageQuant software (GE Heathcare). DNA stained blue with DAPI. Bar: 5 μm.

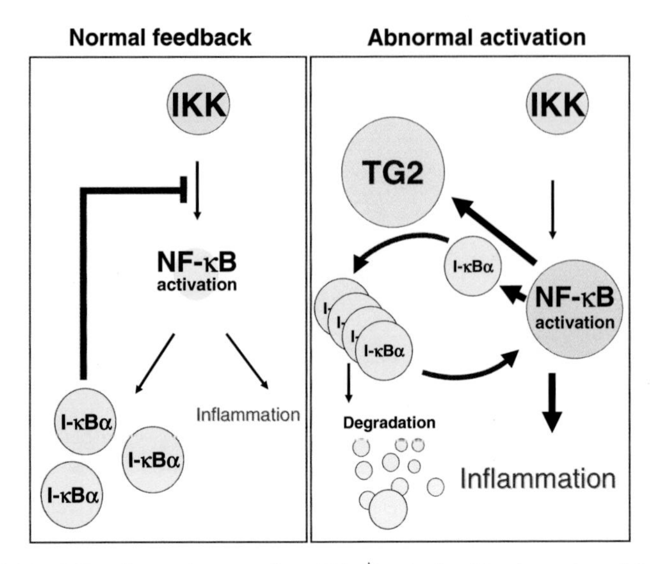

FIGURE 1. TGase 2 activation exacerbates NF-κB activation (Ouroboros theory). In normal stress, feedback mechanisms allow recovery from inflammation. However, under conditions of highly induced TGase 2 expression, inflammatory stress is increased by the depletion of I-κBα via TGase 2 polymerization. TGase 2 can be induced by various epigenetic mechanisms, including demethylation of the TGM2 promoter. TGase 2 expression can also be induced directly by NF-κB activation, because the TGase 2 promoter has an NF-κB binding motif. This is an intriguing finding because many stimuli in addition to LPS trigger NF-κB activation, including viral infection, UV, oxidative stress, and inflammatory cytokines. In the case of high TGase 2 expression, inflammation is maintained by a vicious cycle, which I have termed Ouroboros.

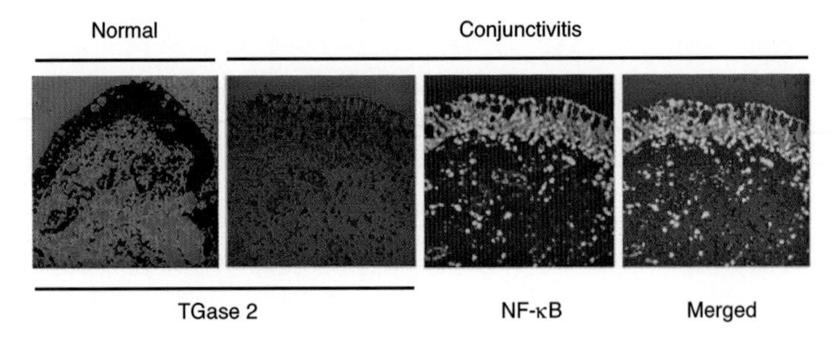

FIGURE 3. Colocalization of TGase 2 and NF-κB in conjunctivitis. An in vivo experimental model of allergic conjunctivitis was employed, as previously described [7]. Conjuctiva tissue was double-immunostained using FITC-conjugated anti-TGase 2 (red) and Alexafluor594-conjugated anti-NF-κB (green) antibodies. This model showed markedly increased colocalization of TGase 2 and NF-κB compared to that seen in normal conjuctiva.

TRANSGLUTAMINASE 2 AT THE CROSSROADS BETWEEN CELL DEATH AND SURVIVAL

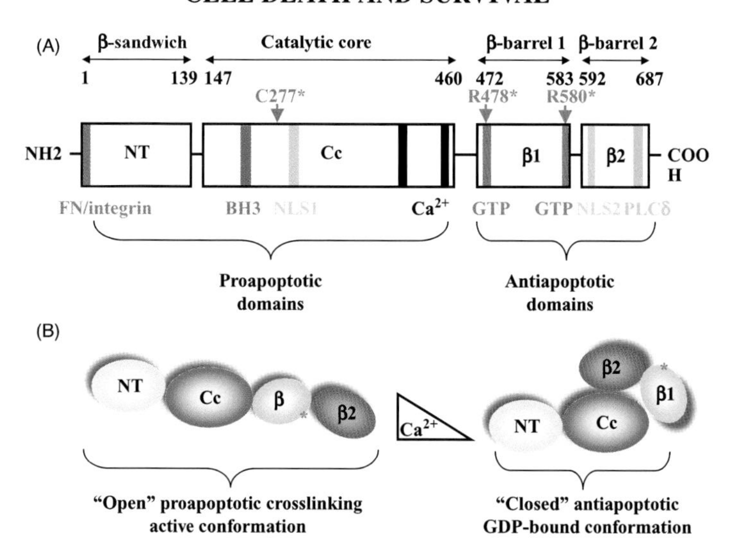

FIGURE 1. TG2 structure and domains. (A) Graphic representation of the four domains of TG2. The following regions are highlighted: FN/integrin binding motif; BH3 domain; C277, active site's cysteine; NLS1 and 2, nuclear localization signals; Ca^{2+} binding motif; GTP binding motif and relative binding impairing mutations (R478/580); and PLCδ binding motif. (B) It has been proposed that Ca^{2+} concentration may control the shift between the "open" proapoptotic (high [Ca^{2+}]) and the "closed" antiapoptotic (low [Ca^{2+}]) conformations of TG2.

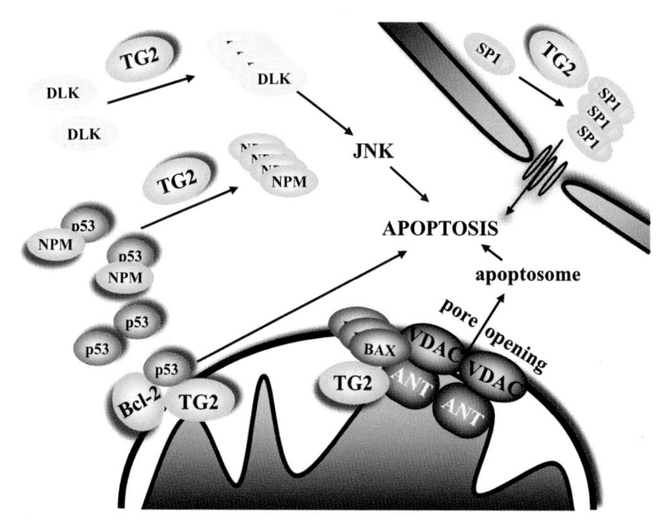

FIGURE 2. TG2 exerts its proapoptotic function by acting on different protein substrates in the cytosol, nucleus, and at mitochondrial level.

TRANSGLUTAMINASE 2 DYSFUNCTIONS IN THE DEVELOPMENT OF AUTOIMMUNE DISORDERS: CELIAC DISEASE AND TG2$^{-/-}$ MOUSE

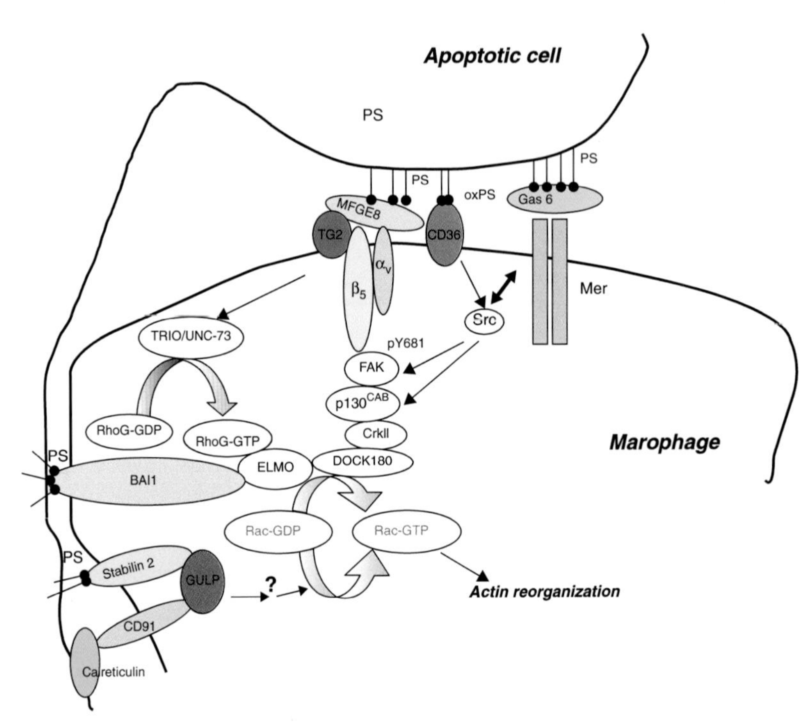

FIGURE 2. Signaling pathways regulating engulfment of apoptotic cells. GTP loading of the small GTPase Rac-1 plays a central role in initiation of actin reorganization leading to the apoptotic cell uptake. Rac-1 can be activated via the GULP pathway by unknown mechanisms or by the formation and activation of the bipartite guanine exchange factor DOCK180/ELMO. Activation of DOCK180/ELMO can be induced directly by the BAI1 receptor or via two main signaling pathways, which involve either GTP bound RhoG, or the CRKII/pCas complex. In addition, for full activation, the DOCK180/ELMO complex has to translocate to the membrane 3-phosphoinositide sites formed by the phosphoinositide-3-kinase activated during the phagocytosis process.

TRANSGLUTAMINASE-MEDIATED REMODELING OF THE HUMAN ERYTHROCYTE MEMBRANE SKELETON: RELEVANCE FOR ERYTHROCYTE DISEASES WITH SHORTENED CELL LIFESPAN

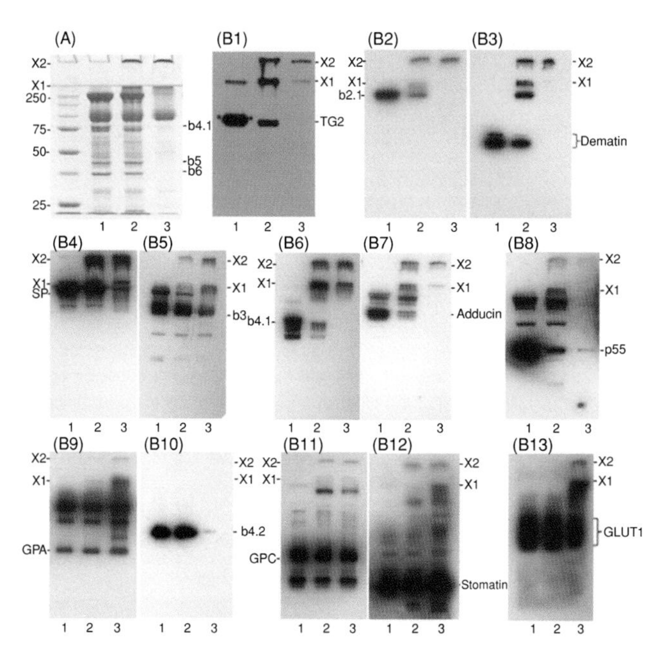

FIGURE 8. Cross-linked polymers, formed in human erythrocytes upon elevation of the concentration of internal Ca^{2+}, are recognized by several antibodies from our new antibody repertoire. The polymers on the SDS-PAGE protein profiles of membrane proteins at theSignaling pathways regulating engulfment of apoptotic cells. GTP loading of the small GTPase Rac-1 plays a central role in initiation of actin reorganization leading to the apoptotic cell uptake. Rac-1 can be activated via the GULP pathway by unknown mechanisms or by the formation and activation of the bipartite guanine exchange factor DOCK180/ELMO. Activation of DOCK180/ELMO can be induced directly by the BAI1 receptor or via two main signaling pathways, which involve either GTP bound RhoG, or the CRKII/pCas complex. In addition, for full activation, the DOCK180/ELMO complex has to translocate to the membrane 3-phosphoinositide sites formed by the phosphoinositide-3-kinase activated during the phagocytosis process. top of separating and stacking gels are marked as X1 and X2. Part (A) shows the Coomassie blue-stained gels of whole ghosts of control erythrocytes (lane 1), those with Ca^{2+} overload (lane 2), and of the alkali-stripped membranes of the latter (lane 3); molecular weight marker values are given in kDa; the position of actin is marked as band 5 or b5 and that of glyceraldehyde-3-phosphate dehydrogenase as band 6 or b6. PVDF transblots of the gels were probed with various dilutions of antibodies targeting transglutaminase 2 (TG2, B1; 1:20000), ankyrin (b2.1, B2; 1:20,000), dematin (B3; 1:5000), spectrins (SP, B4; 1:10,000), band 3 (b3, B5; 1:20,000), band 4.1 (b4.1, B6; 1:4000), adducin (B7; 1:3000), p55 (B8; 1:4000), glycophorin A (GPA, B9; 1:1000), band 4.2 (b4.2, B10; 1:200,000), glycophorin C (GPC, B11; 1:1000), stomatin (B12; 1:100), and glucose transporter 1 (GLUT1, B13; 1:20,000).

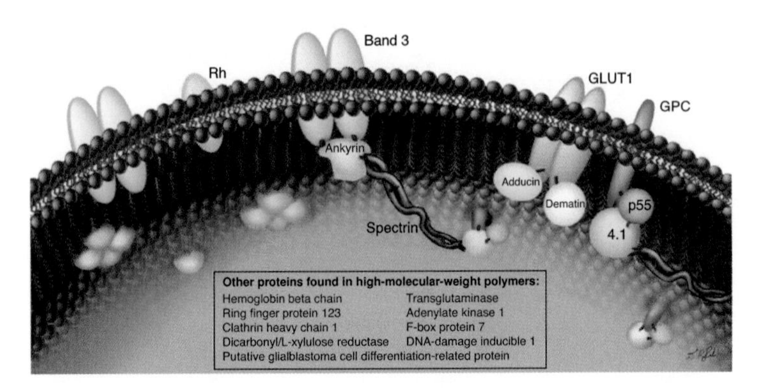

FIGURE 9. Illustration of a section of the inner surface of alkali-stripped ghosts from Ca^{2+}-loaded erythrocytes indicating scattered clusters of protein polymers, covalently linked to transmembrane proteins or to p55, which—through its palmitoyl moiety—is partially embedded in the lipid bi-layer. The N^{ε}(γ-glutamyl)lysine side chain bridges between constituent polypeptide chains of the polymer are shown by red lines. Without remodeling of the membrane skeleton (as in the left portion of the picture), only the cytoplasmic domains of proteins such as band 3, GLUT1, the Rhesus blood group CcEe antigen, and p55 are seen.

METHIONINE ADENOSYLTRANSFERASE (*S*-ADENOSYLMETHIONINE SYNTHETASE)

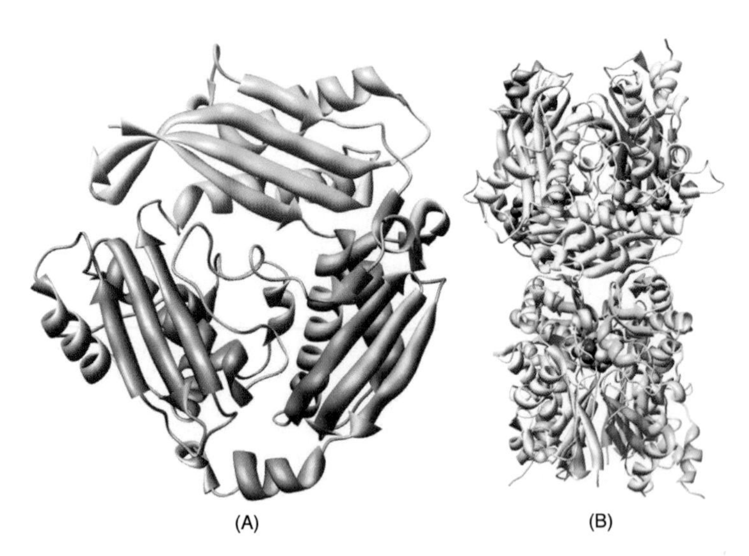

FIGURE 4. (A) Ribbon diagram of the monomer of the human MAT II illustrating the near three fold symmetry of the fold and the central loop (pdb code 2PO2); (B) the tetramer of cMAT (pdb code 1RL7) in complex formed with AMPPNP and methionine; half of the active sites had bound products AdoMet and PPNP. The active site is buried at a subunit interface.

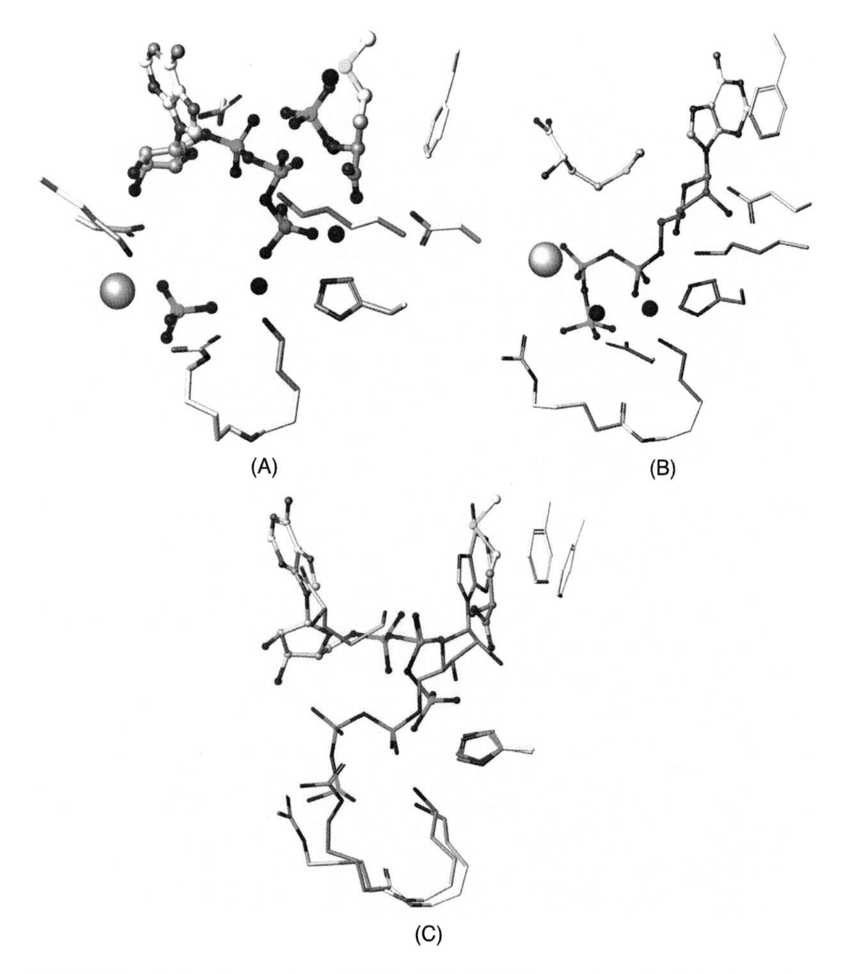

FIGURE 5. Active site structures of MATs: (A) MAT I B in complex with ATP and methionine; Mg^{2+}, K^{+}, and Pi are also in the active site (pdb code 1O9T); (B) *E. coli* MAT in complex with AMPPNP and methionine; the 2 $Mg^{2+}$and 1 K^{+} are not shown (pdb code 1RL7); and (C) overlay of selected regions of the active sites of cMAT and rMAT. The exchanged orientation of the substrates is shown when the protein portion of the structures is aligned, illustrated by the position of the active site histidine.

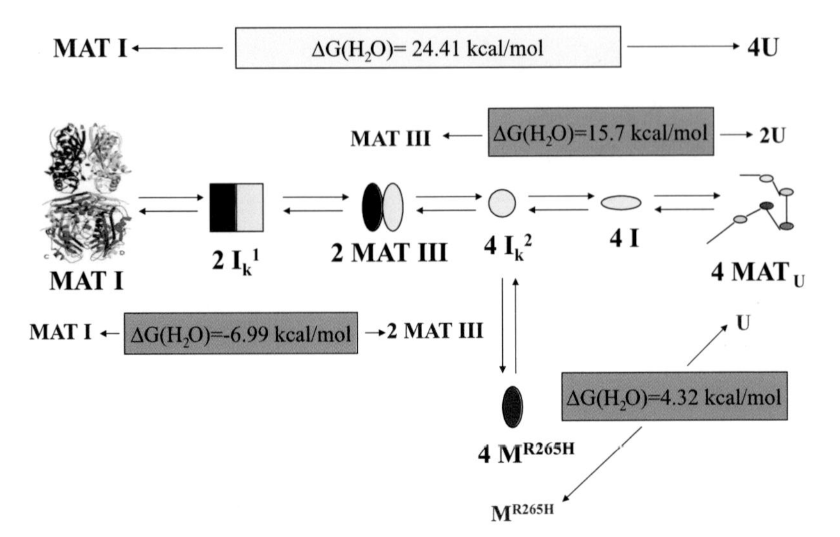

FIGURE 7. Folding pathways for MAT I/III. The figure shows a schematic representation of the data available for MAT I/III unfolding. Free energy changes calculated for each step are also included. The different intermediates are denoted I, $I_k{}^1$, and $I_k{}^2$, whereas M represents the monomeric mutant R265H and U the unfolded state.

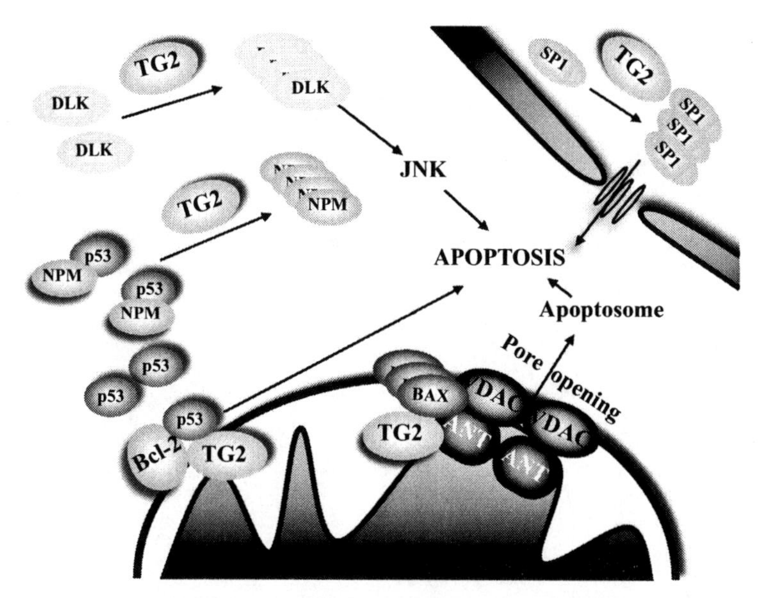

FIGURE 2. TG2 exerts its proapoptotic function by acting on different protein substrates in the cytosol, nucleus, and at mitochondrial level. (See insert for color representation.)

determines the switch of TG2 to its cross-linking configuration in all subcellular compartments. This transition results in an extensive polymerization of intracellular proteins, including actin, Sp1 and Rb, and the formation of detergent-insoluble structures (Table 1) [67, 69, 70]. This protein scaffold stabilizes the dying cell before its clearance by phagocytosis; thus preventing the release of harmful intracellular components and the activation of an inflammatory or autoimmune response [62]. Under pathological conditions, the death of cells expressing high amounts of TG2 may occur as a result of a "mummification" event caused by an extensive cross-linking of cytosolic proteins without signs of either apoptosis or necrosis [95].

II. CELL DEATH

The balance between cell proliferation and death allows multicellular organisms to model their shape during development and to maintain their own homeostasis in adulthood. Cell death might be accomplished in various ways: apoptosis, autophagy-associated cell death, necrosis, mitotic

TABLE 1
Proapoptotic Activity of TG2 in Cells and Tissues under Physiological and Pathological Conditions

Cells/Tissue	Stimulus	Reference
Embryonic and adult tissues		
Embryonic fibroblasts	STS	[40]
Periderm cells		[71]
Thymocytes	Various	[72, 73]
Limb development		[74, 75]
Erythrocytes	Ca^{2+}	[76]
Mammary gland	Hormones	[77]
Experimental settings		
Striatal neurons	Thapsigargin	[78]
Macrophages	RA	[79]
Smooth muscle	RA	[80]
Hepatocytes/liver	Ethanol	[70]
Cardiomyocytes	Ca^{2+}	[81, 82]
Frog liver	Hormones	[83]
Keratinocytes	Camptothecin	[84]
Photoreceptors	Light	[85]
Hepatocytes	Lead Nitrate	[86, 87]
Pathological settings		
Lymph nodes	HIV	[88, 89]
Encephalitis		
Hippocampus	Ischemia	[90, 91]
Brain	Huntington	[92]
Myoenteric neurons	Colitis	[93]
Dopaminergic neurons	Parkinson	[94]

catastrophe, anoikis, excitotoxicity, Wallerian degeneration, and cornification of the skin [96]. The end point of each of these processes consists in the removal of the dying/dead cells from the living tissues through the action of professional or nonprofessional phagocytes. This final process is of utmost importance both for development and homeostasis of organisms as well as for preventing the onset of autoimmune pathologies [97, 98].

During the last 20 years, most of the morphological and biochemical aspects of apoptosis, necrosis, and autophagy-associated cell death have been successfully characterized. Cell death by apoptosis is characterized by well-defined morphological changes, such as plasma membrane blebbing, chromatin condensation, nuclear fragmentation, and the final formation of the apoptotic bodies [99]. These drastic morphological changes are associated with several biochemical ones, such as phosphatidylserine exposure on the outer leaflet of the plasma membrane, changes in the mitochondrial membrane permeability, and the release of the intermembrane mitochondrial proteins [100]. All these changes lead to the activation of effector caspases and caspases-activated DNase, which results in DNA cleavage [101]. These events lead the cell to be dismantled and eliminated selectively, without affecting tissue's integrity.

Autophagy, originally described by Christian De Duve in 1966 [102, 103], has been known to be a survival mechanism activated by cells subjected to nutrients or growth factors deprivation. Autophagy allows cells to self-digest their own macromolecules and organelles, through the activity of their lysosomes, in order to recycle metabolites and provide a source of energy. Nevertheless, autophagy is a tightly regulated process and the characterization of the molecular players involved in its control and execution revealed the existence of a complex cross talk between autophagy and apoptosis [104]. In fact, under prolonged stress conditions, such as in neurodegenerative disease or cancer, autophagy might play a complex role as a death process in itself or show features of apoptotic or necrotic cell death [105].

Necrosis is a process characterized by the swelling of the intracellular organelles and the rupture of the plasma membrane. The release of proteolytic enzymes from lysosomes, together with the activation of intracellular enzymes, leads to cell demolition. This event is associated with the release of the intracellular material and, in most cases, with the onset of inflammation processes [106]. Necrosis is usually considered a non-programmed form of death, resulting from metabolic failure associated with a rapid depletion of ATP, such as after ischemia and stroke. The alteration of the permeability of cellular membranes results in an increased intracellular concentration of calcium ions, which activate cellular proteases, such as calpains and cathepsins. Intriguingly, if the calcium efflux crosses the plasma membrane, it might trigger necrosis, while if the calcium comes from the endoplasmic reticulum (ER), it might trigger apoptosis [107, 108]. ATP also plays a different role in the two processes, as energy is necessary for the activation of various apoptosis effectors (i.e., the apoptosome), while ATP depletion shifts cells from apoptosis to necrosis. Accumulating evidence indicate that necrosis is also more ordered, if not controlled, than was originally thought [107].

These three major and most characterized types of cell death are not mutually exclusive. In fact, the type and intensity of the death signal, the ATP concentration, the cell type as well as many other factors are all able to determine which type of death occurs and, eventually, the shift from necrosis to apoptosis and vice versa [109]. In addition, the blockade of one of these cell death pathways rather than preventing the destruction of the cell may actually activate an alternative pathway [110–112]. This picture is supported by the growing numbers of evidence involving protein factors, such as p53 and Bcl-2 family members, which are able not only to modulate the onset of cell death but also to determine the cell's choice between the various types of death.

To date, more than 500 publications have shown that TG2 is involved both in apoptosis and autophagy. The detection of the enzyme upregulation is used as a marker of apoptosis for *in vivo* studies [113]. Although the molecular mechanism(s) of its action is still lacking, it is clear that TG2 can play both pro- and anti-death functions, depending on the cell type, the subcellular localization, and the type of the death stimulus. In general, under physiological conditions, TG2 plays a proapoptotic function (see Table 1), while, in highly transformed cells, it can also switch its activity toward cell protective functions. It is not clear how this is possible and which of the various TG2 activities and/or protein partners are involved in this prosurvival action. Indeed, in some tumors, TG2 expression has been shown not only to protect cells from death, but also to carry out a proapoptotic function (see Table 2). This is probably due to the different cell clones derived from the original tumor as well as to their adaptation to *in vitro* culture conditions or to the apoptotic stimulus used (i.e., agents, concentrations, time frame, etc.). An explanation for these controversial results could be found in the recently described role played by TG2 in autophagy (see the "Autophagy" section). In fact, it is well-established that autophagy can play an important survival role in cancer cells, its downregulation constituting a new frontier for cancer chemotherapy.

A. TG2 AS A PROAPOPTOTIC FACTOR

Apoptosis is a genetically programmed and controlled form of cell death, accomplished through the activation of specific intracellular pathways in response to various death stimuli. Two distinct but convergent pathways have been extensively characterized: (1) the extrinsic and (2) the intrinsic pathway [106]. The extrinsic or "death-receptor" pathway receives and decoys signals coming from outside the cell, through the activation of specific plasma membrane proteins belonging to the tumor necrosis factor (TNF) receptor family, the "death receptors." The binding of the TNF ligand to its

TABLE 2
Proapoptotic and Prosurvival Functions of TG2 in Cancer

Proapoptotic			Antiapoptotic		
Cells/Tissue	Stimulus	Reference	Cells/Tissue	Stimulus	Reference
Breast cancer (MDA-MB-231)	Calphostin C	[114]	Ovarian cancer	various	[115, 116]
Epithelial tumors	Theophyllin	[117]	Caki	TRAIL	[118]
Pancreatic cancer	A23187	[119]	Ovarian cancer	Cisplatin	[120]
Myeloma	RA	[121, 122]	Meningioma	Radiation	[123]
U937	RA	[69]	Glioblastoma	BCNU	[124]
Lung cancer	IFN alpha	[125]	Pancreatic cancer (PDAC)	RA	[126, 127]
Neuroblastoma	RA	[128, 129]	Breast cancer	Doxorubicin	[130–132]
Carcinoma cells (HeLa)	RA and DFMO	[129]	Nonsmall cell lung cancer	Cisplatin	[133]
Pancreatic adenocarcinoma	RA	[134]	Neuroblastoma	TNF alpha	[135]

receptor triggers a conformational change in the receptor itself and induces the intracellular recruitment of modulator proteins, such as FADD and c-FLIP, which assemble to form the DISC complex at the cytoplasmic side of the plasma membrane. The DISC complex is able to recruit and activate caspase 8, which ultimately activates the executioner caspase 3 [136].

The intrinsic or mitochondrial pathway receives and decoys signals coming from inside the cell, such as increased intracellular reactive oxygen species (ROS), DNA damage, the unfolded protein response, and growth factors deprivation. All these initiators lead to increased mitochondrial permeability and promote the release of proteins from the mitochondrial intermembrane space into the cytosol [100, 137]. The release of proapoptotic factors (e.g., cytochrome *c*) and of antagonists of cytosolic antiapoptotic proteins (i.e., SMAC/DIABLO) allows the activation of the APAF-1 complex, which recruits and activates caspase 9. The two pathways converge, since the activated caspase 8 and 9 in turn activate the executioner caspase 3, 6, and 7, which dismantle the cell by cleaving various protein substrates and activate DNases [106]. The two pathways are not only convergent at the final steps, but are also cross talking at mitochondrial level. It has been shown that, death-receptors may activate the release of cytochrome *c* via the caspase 8-dependent processing of the proapoptotic BH3-only protein Bid [138].

At once very delicate and highly coordinated, the regulation of these pathways involves cellular organelles, such as mitochondria and the ER, as well as specific protein factors, such as the members of the Bcl-2 family. The proteins belonging to this family are characterized by the presence of shared domains, the Bcl-2-Homology (BH) domains, which allow them to interact both with each other and with other proteins, in order to modulate the permeabilization of the outer mitochondrial membrane [139, 140]. These proteins might then behave as pro- or anti-apoptotic factors, by acting on mitochondrial and ER membranes [141]. During the last 5 years, the characterization of a large number of proteins sharing homology with the BH3 domain of Bcl-2 has greatly increased the number of the member of the BH3-only protein family. All these newly discovered proteins can act as inducers or sensitizers of apoptosis, even if normally they perform different intracellular activities [142–146].

The initial observation that both TG2's levels and activity were induced during apoptosis, both *in vitro* and *in vivo* (see Table 1 and [52, 147]), suggested a strong involvement of TG2 in cell death. An early hypothesis postulated that the formation of highly stable cross-links between glutamine and lysine residues of cellular proteins stabilized the dying cell and prevented the release of the cytosolic components [67, 148]. The observed induction of TG2 gene expression during apoptosis onset *in vivo* [53, 83, 149] was coupled with *in vitro* observations of increased enzymatic sensitivity [41, 128, 149] as well as protection [69] against apoptosis upon induction or downregulation of TG2 expression, respectively. Such evidence, suggesting the enzyme acting as a proapoptotic factor, was further supported by the discovery that TG2 might localize on mitochondria where it could act both as a BH3-only protein [51] or as a PDI [25, 40]. The BH3-only proteins belong to the Bcl-2 family and might exert their proapoptotic activity by activating Bax and Bak in order to induce the mitochondrial membrane permeabilization. TG2 possesses a domain (Figure 1), which shares more than 70% of homology with the consensus for the BH3 domain; it is through this domain that the enzyme is able to interact with Bax and Bak, which might be favored in their mitochondrial localization and proapoptotic activity (Figure 2). Activation of the TG2's cross-linking activity, upon apoptosis induction, results in the TG2-dependent polymerization of these proapoptotic factors on the mitochondrial outer membrane, as well as to a faster loss of the mitochondrial membrane potential and to the execution of the apoptotic program [51]. It has also been shown that the interaction of Bax with Adenine Nucleotide Translocase 1 (ANT1), which represents an essential step in the opening of the mitochondrial permeability transition pore, is largely reduced in MEFs lacking TG2, thus resulting in the impairment of apoptosis [40]. Another aspect of the complex interaction between TG2 and mitochondria that might lead to a proapoptotic outcome, in drug-resistant cancer cells, is related to the localization of nucleophosmin 1 (NPM1) in the cytosol. NPM1 gene mutations causing aberrant cytosolic expression of nucleophosmin are the most frequent genetic alteration observed in acute myeloid leukemia (AML), such an alteration being found in about 30% cases. It has recently been shown that NPM1 expression enhances the levels of p53 in the nucleus but reduces p53 levels in mitochondria. It is well known that the presence of p53 in mitochondria plays an important role in stress-induced apoptosis, suggesting that NPM1 may protect cells from apoptosis, by reducing the mitochondrial level of p53 [150]. Interestingly, it has been reported that NPM1 acts a TG2 substrate [133], suggesting that TG2 can inhibit NPM1 accumulation in the cytoplasm, through its polymerization, and thus protecting cells from apoptosis, by reducing the mitochondrial level of p53 (Figure 2).

Another interesting finding is that TG2 might also localize in the mitochondrial intermembrane space, where, through its PDI activity, it may modify and stabilize the assembly and the activity of some members of the respiratory chain complexes as well as of ANT1, in the inner mitochondrial membrane [25, 40]. In fact, the ablation of the TG2 gene in mice leads to an impairment of the respiratory complex chain coupled with a decrease in the ATP production and an effect on the cell's sensitivity towards death induction, dependent on the cell type and on the kind of death stimuli [151, 152].

Further important evidence, highlighting the proapoptotic function of TG2 and its effect on mitochondria, has been reported in pancreatic ductal adenocarcinoma (PDAC), a lethal malignant disease with poor long-term survival rates. In these highly transformed cells, the activation of endogenous TG2, by the calcium ionophore A23187, results in rapid and spontaneous apoptosis, which is associated with the release of the apoptosis-inducing factor (AIF). The translocation of AIF from mitochondria to the nucleus leads to the execution of a caspase-independent form apoptosis [119].

The action exerted by TG2 at mitochondrial level highlights a more general contribution of this multifunctional enzyme both to the maintenance of mitochondrial physiology and to the modifications taking place during apoptosis onset and execution [147, 153]. This aspect is of potential relevance in all those pathologies, such as neurodegenerative diseases and stroke, in which the mitochondrial functionality plays a crucial role in the cell's decision between survival and death. In addition, TG2 overexpression leads to the accumulation of ROS associated with a large depletion of GSH [41]. In keeping with these findings, GST P1-1 acts as a very efficient TG2 substrate both in cells and *in vitro*, and its TG2-dependent polymerization causes a functional inactivation of the enzyme [62]. GSH depletion occurs during the early phases of apoptosis and the functional inactivation of GST P1-1, by TG2-catalyzed oligomerization, indicates a further potential proapoptotic role for TG2 (Figure 2).

TG2 is largely localized in the cytosol at the level of the plasma membrane, the cytoskeleton, the ER, and the mitochondria, although, depending on cell type, a small amount of the enzyme is also present in the nucleus [154]. It has been shown that both the intracellular localization and the activation of its transamidating activity are important factors in the modulation of TG2's effects on apoptosis induction (Figures 2 and 3). Transfection of cells with cDNAs coding for wild type or mutant TG2, which lack the transamidating activity and were targeted to different intracellular compartments, confirmed the proapoptotic nature of the cytosolic form of TG2. Nevertheless, the nuclear localization of cross-linking-inactive TG2 reduced apoptosis, thus indicating how the intracellular localization influences its effect on cell death. Further evidence derives from the observation that the

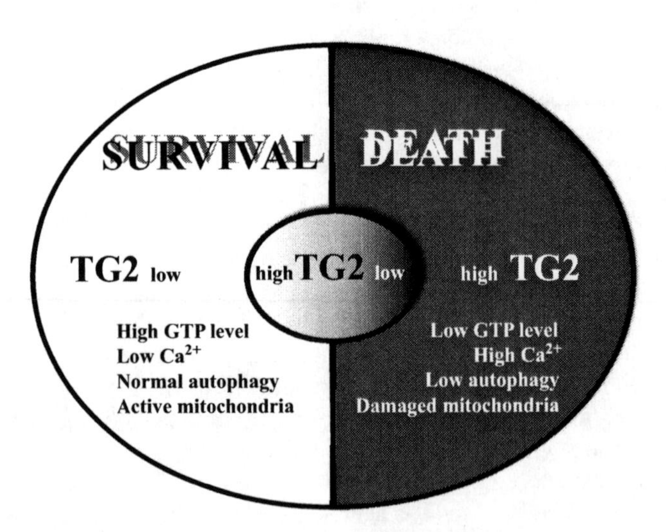

FIGURE 3. The pro-survival and pro-death features of TG2 rely on its localization in the cytosol or in the nucleus.

DAP-like kinase (DLK) undergoes TG2-dependent oligomerization in cells treated with calphostin C [114, 155, 156]. DLK is a nuclear serine/threonine-specific kinase, belonging to the subgroup of the mixed-lineage kinases (MLKs). These proteins act as key regulators of the stress-activated c-Jun *N*-terminal kinase (JNK) mitogen-activated protein kinase (MAPK) signaling pathway. The induction of apoptosis by DLK requires its relocation to the cytoplasm and its association with the actin cytoskeleton, which is achieved through the interaction with the proapoptotic protein Par-4 [157]. DLK overexpression in neural cells and in sympathetic neurons induces apoptosis via the mitochondrial pathway. Conversely, overexpression of a dominant-negative form of DLK in the same cells prevents apoptosis, thus indicating that DLK is involved in the control of cell death [158]. Interestingly, the TG2-dependent oligomerization of DLK occurs early in the apoptotic response and significantly enhances its kinase activity and consequently its ability to activate the JNK pathway [114]. Moreover, functional studies demonstrate that TG2-mediated oligomerization of wild-type DLK sensitizes cells to calphostin C-induced apoptosis, while cross-linking of a kinase-inactive variant of DLK does not [114]. These findings suggest that TG2 proapoptotic activity is at least partially mediated by the oligomerization and activation of the proapoptotic kinase DLK, which in turn will activate the proapoptotic JNK pathway (Figure 2). Recently, it has been shown that TG2 mediates alcohol-induced hepatocytes apoptosis *in vivo* by cross-linking SP1 [70]. This TG2-dependent modification leads to SP1 inactivation and results in a reduction of c-Met expression, which is required for hepatocytes cell survival.

B. TG2 AS AN ANTIAPOPTOTIC FACTOR

TG2 acts as a proapoptotic factor under physiological conditions, during both embryonic and adult life. Conversely, in transformed cells, the enzyme can also play a dual role, acting as an antiapoptotic factor. This alternative behavior has often been described in the same tumor type, such as neuroblastoma, pancreatic, and ovarian cancer. Such observations suggest that the switch between the two opposite actions exerted by the enzyme might be dictated by environmental as well as genetic background of the cell line studied. Nevertheless, during recent years, various papers have reported the ability of TG2 to perform as a G protein [20], as a kinase [21], and as a modulator of the cell/ECM adhesion processes [159]. The prevalence of a specific activity coupled with different localization, both inside and outside the cell, very likely account for the switch between pro- and anti-apoptotic functions of TG2 [153, 154, 120, 160, 161]. This differential regulation of cell death by TG2 is highly relevant in cancer (see Table 2), in which TG2's antiapoptotic activity might lead to the survival of cells that have to be eliminated by the organism. Some clues about the possible mechanisms at the basis of this prosurvival function of TG2 have been recently published [161–165]. We first demonstrated that TG2 post-translationally modifies the Rb protein, an important suppressor of apoptosis [66]. Recently, it has been proposed that TG2/Rb interaction increases significantly, concomitant with an attenuation of apoptosis. The TG2/Rb interaction is emerging as an important aspect of the antiapoptotic effects of TG2. In fact, in cells undergoing apoptosis, Rb is degraded by the action of caspases, this degradation proving to be blocked when cells have been pretreated with RA, an important transcriptional inducer of TG2 [166]. Experiments performed with $Rb^{-/-}$ fibroblasts demonstrated that Rb is required for TG2 to exhibit antiapoptotic activity in response to RA treatment. These results imply that the ability of TG2 to modify Rb represents a key step for TG2's survival activity. It has also been hypothesized that TG2 might protect tumor cells against apoptosis by translocating into the nucleus. However, controversy remains as to whether the transamidating activity of nuclear TG2 is required to bind and protect Rb from the degradation occurring during apoptosis.

It has been shown that the exposure of TG2-expressing cells to the phosphoinositide 3-kinase (PI3K) inhibitor LY294002 reduces the ability of the enzyme to bind GTP. This observation suggests that PI3K regulates the shift between the transamidating and the GTP binding activity of TG2, thus inhibiting TG2's cross-linking activity (Figures 1 and 3) [167]. These findings imply that PI3K activity, a well-established cell survival factor, is required for the modulation of the GTP binding activity of TG2. Accordingly, TG2 expression and its GTP binding activity prove to be very high in a number of transformed cell lines. Interestingly, the switch between the prosurvival TG2/GTPase and the proapoptotic transamidating activity can be influenced by the Ras-ERK pathway [168]. These observations suggest that the TG2 pro- and anti-apoptotic conflicting functions may be regulated by conformational changes of the protein. In fact, the binding of GTP can convert the enzyme from an "open" cell death-promoting to a "closed" protein conformation able to provide protection against apoptotic stimuli (Figure 1). This hypothesis has been supported by the observation that the expression of full-length TG2 in tumor cells confers protection against cell death, while the expression of a shorter version of TG2, truncated at the 3′ end and unable to bind GTP, turns out to be cytotoxic [169]. The proapoptotic activity of the short form of TG2 does not rely on its transamidating activity, because the mutation of the cysteine 277 residue, essential for catalyzing this reaction, does not compromise the ability of this short form of TG2 to induce cell death. Notably, a shorter TG2 transcript, encoding for a truncated form of TG2 (TGase-S), which shows strong proapoptotic activity, has been identified from the brains of Alzheimer's patients. This TGase-S exhibits no detectable GTP-binding capability, further suggesting that TG2's ability to induce cell death is due to its inability to bind GTP. These results are particularly relevant considering TG2's well-known involvement in neurodegenerative diseases (see below).

III. AUTOPHAGY

Autophagy is the cellular metabolic pathway implicated in the recycling of portions of cytosol as well as in the removal of superfluous or damaged organelles. In addition to proteins, this transport route is uniquely able to catabolize other cellular constituents such as lipids, carbohydrates, and nucleic acids. This essential biological process occurs at a basal level in most tissues and contributes to the routine turnover of cytoplasmic components of the cell [170, 171]. However, it can also be massively induced by a change in the environmental conditions or by cytokines and other signaling molecules, to help the cell to adapt and/or cope with various physiological and pathological situations [172]. Autophagy is very important for cellular remodeling and development and is also involved in preventing ageing and controlling cell growth [173]. Moreover, it plays an important role in several human diseases, such as cancer, neurodegeneration (Huntington's diseases (HD), Parkinson's diseases (PD), and Alzheimer's diseases (AD)), and muscular disorders [105, 174].

Autophagy is also used by cells to defend themselves from invasion by pathogenic bacteria such as *Mycobacterium tuberculosis*, viruses such as the *Herpes simplex* and the *Tobacco mosaic* viruses, and intracellular parasites like *Toxoplasma gondii* [105, 175, 176]. Finally, autophagy may play a controversial role as a cell death mechanism (i.e., type II programmed cell death) and in some cases appears to be regulated in conjunction with apoptosis [177, 178].

To date, three forms of autophagy have been defined, according to how lysosomes receive the material to be degraded [106, 179]. In macroautophagy, a double membrane structure, called autophagosome, envelops the cargo and fuses with the lysosome (Figure 4). In microautophagy, the lysosome directly engulfs the material by means of an invagination of the organelle's membrane. In chaperone-mediated autophagy, heat shock cognate proteins deliver the substrates to the lysosome. The most characterized form of autophagy is macro-autophagy in which cells are characterized by the formation of double membranes autophagic vacuoles, containing cytoplasmic material (cytosol and/or organelles), which are then delivered to fusion with lysosome, in order to achieve bulk degradation [180]. This process might occur in a generalized fashion; alternatively, it might target specific organelles, such as mitochondria and ER, thereby eliminating supernumerary, outlived, or damaged structures in the stressed cell [181]. Macro-autophagy's initial steps include the formation (vesicle nucleation) and the expansion (vesicle elongation) of an isolation membrane, which is also called phagophore. The edges of the phagophore then fuse (vesicle completion) in order to form the autophagosome, a double-membraned vesicle that sequesters the cytoplasmic material. These steps are followed by the fusion of the autophagosome with a lysosome, in order to form an autolysosome, where the captured material, together with the inner membrane, is degraded (Figure 4).

Nutritional status, hormonal factors, and other cues such as temperature, oxygen concentrations, and cell density are important factors involved in the control of autophagy onset. At the molecular level, one of the key regulators of the autophagy process is the target of rapamycin, mTOR kinase, this is the major inhibitory signal that shuts off autophagy in the presence of growth factors and abundant nutrients [182].

The class I PI3K/Akt signaling molecules link receptor tyrosine kinases to TOR activation and thereby repress autophagy in response to insulin-like and other growth factor signals [183]. Some of the other regulatory molecules that control autophagy include 5′-AMP-activated protein kinase (AMPK), which responds to low energy state; the eukaryotic initiation factor 2α (eIF2α), which responds to nutrient starvation, double-stranded RNA, and ER stress; the BH3-only proteins, which interfere with the Bcl-2/Bcl-X_L inhibition of the Beclin 1/class III PI3K complex; the tumor suppressor protein p53; the death-associated protein kinases; the ER membrane-associated protein Ire-1; the stress-activated kinase, c-Jun-*N*-terminal kinase; the inositol–trisphosphate receptor; GTPases; Erk1/2; ceramide; and calcium [184–187].

The characterization of autophagy at molecular level, extensively carried out in yeast, revealed the existence of more than 20 genes, known as the ATG genes, downstream of mTOR kinase. Those genes encoded evolutionarily

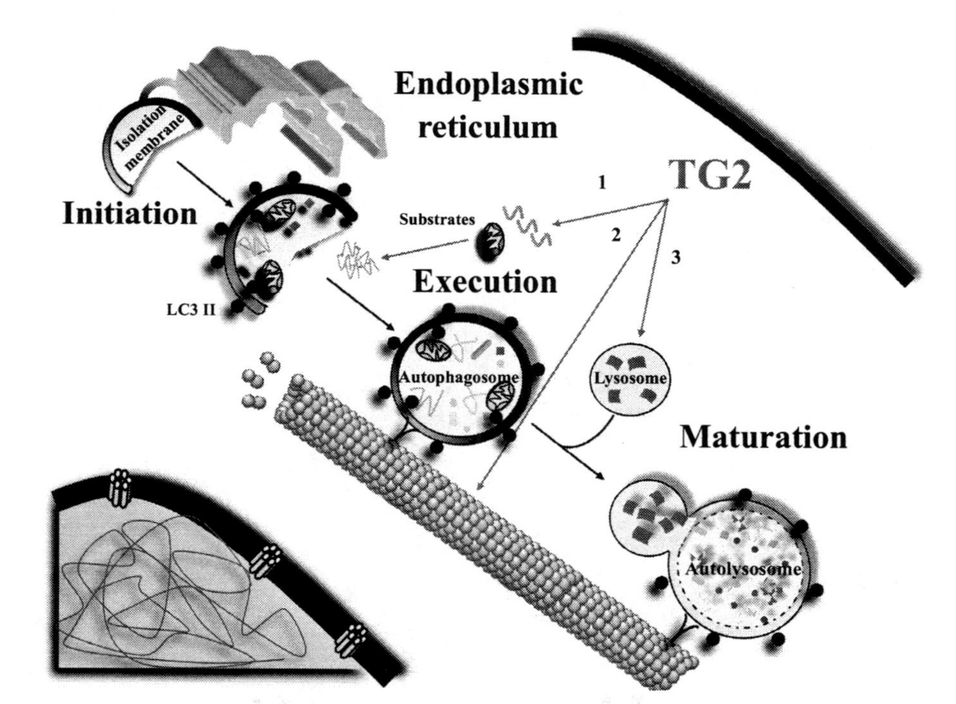

FIGURE 4. Putative role for TG2 in the autophagy pathway. It has been proposed that TG2 may play a role in autophagy by acting at different levels. (1) TG2 may be involved in the post-translational modifications of putative autophagy substrates, such as aggregated and mitochondrial proteins; (2) TG2 may modify the cell's cytoskeleton by acting on actin and tubulin; and (3) TG2 is involved in the maturation of autophagosomes, its absence resulting in an accumulation of lysosomes.

conserved proteins essential for the execution of the autophagy process [171]. These proteins include a protein serine/threonine kinase complex, which responds to upstream signals such as mTOR kinase (Atg1, Atg13, Atg17); a lipid kinase signaling complex, which mediates vesicle nucleation (Atg6, Atg14, Vps34 and Vps15); two ubiquitin-like conjugation pathways, which mediate vesicle expansion (Atg8 and Atg12); a recycling pathway, which mediates the disassembly of Atg proteins from mature autophagosomes (Atg2, Atg9, Atg18); and vacuolar permeases, which permit the efflux of amino acids from the degradative compartment (Atg22).

In mammals, proper fusion with autophagosomes requires proteins that act more generally in lysosomal functions, such as the lysosomal transmembrane proteins, LAMP-2, and CLN3. Accordingly, proper degradation of the autophagosomal contents requires the activity of lysosomal cysteine proteases and cathepsins B, D, and L.

The detection and manipulation of autophagy pathways has been greatly enhanced by the identification of, first, the signals able to induce autophagy and, second, the genes involved in carrying out this process. One of the key events occurring in autophagy is the phosphatidyl-ethanolamine (PE) conjugation of yeast Atg8 or mammalian LC3 protein. The result of this reaction is a nonsoluble form of Atg8 (Atg8-PE) or LC3 (LC3 II) that stably associates with the membrane of the autophagosome. This event allowed researchers to detect the onset of autophagy either by assessing the generation of Atg8-PE or LC3 II, on western blot, or by observing, with fluorescence or confocal microscopy, the localization pattern of fluorescently tagged Atg8 or LC3 [188]. These approaches must be coupled with ancillary measures to discriminate between two physiologically distinct scenarios: (1) the increased autophagic flux, without impairment in autophagic turnover, versus (2) the impaired clearance of autophagosomes, which results in a functional defect in autophagic catabolism.

On the other hand, autophagy can be pharmacologically induced by targeting the following elements: (1) negative regulators of the process, such as mTOR, with rapamycin [187]; (2) proteins involved in the autophagosome formation, such as the class III PI3K, with 3-methyladenine; (3) the fusion of autophagosomes with lysosomes, by inhibiting the lysosomal proton pump with bafilomycin A1 or by decreasing the lysosomal protease activities with NH_4Cl that neutralize lysosomal pH [189].

Very little is known about the relationship existing between TG2 and the execution/regulation of the autophagy process. Previous studies, carried out in a mice model for Huntington's disease (HD), suggested a possible involvement of TG2 in autophagy associated with the neurodegenerative processes observed in this disease [92]. In fact, the analysis of brains belonging to patients and animals suffering from CAG-repeat-based genetic diseases, such as HD, revealed clear neurodegenerative features unassociated with any hallmark of classical apoptosis [190, 191]. Degenerating neurons show lysosome-associated responses, including induction of autophagic granules and electron-dense lysosomes, thus indicating that mutated huntingtin can induce autophagy. Therefore, autophagy could be the major route for the clearance of intracellular misfolded protein aggregates that underlie these neurodegenerative diseases. Analysis of the $TG2^{-/-}$/HD transgenic mice showed that HD onset is associated with a large reduction in nonapoptotic cell death and with an increased number of nuclear protein inclusions, suggesting an impairment in their clearance by means of autophagy [92]. In keeping with this hypothesis, ablation of TG2 protein has recently been shown to result in an evident accumulation of LC3 II on pre-autophagic vesicles, pointing to a marked induction of autophagy [192]. By contrast, the formation of the acidic vesicular organelles was very limited, as detected by the lack of acidification of the autophago-lysosomes observed in $TG2^{-/-}$ MEFs, after acridine orange staining. These findings suggest an impairment in the autophagosomes maturation process. This conclusion is supported by the fact that treatment of TG2 proficient cells with NH_4Cl, in order to inhibit lysosomal activity, results in a marked accumulation of LC3 II and damaged mitochondria, as also observed in $TG2^{-/-}$ cells. It is well known that autophagy plays a crucial role in the turnover of cellular organelles; in particular, it has been proposed that autophagy selectively degrades dysfunctional mitochondria [193]. Taken together, these data indicate a role for the TG2-mediated post-translational modifications of proteins involved in the fusion between autophagosomes and lysosomes. In fact, a drastic reduction in the colocalization between autophagosomes and lysosomes has been detected in $TG2^{-/-}$ cells, even in presence of the inhibitor of intralysosomal protein degradation. It is worth noting that the inhibition of the transamidating activity of TG2, by means of the specific inhibitor R283 [194], results in a marked decrease of the number of acidic vesicular organelles, possibly highlighting the TG2's cross-linking activity as an important player in the autophagy pathway. In line with this finding is the use of a well-known competitive inhibitor of transglutaminase, the autofluorescent compound mono-dansyl-cadaverine (MDC), for *in vivo* labeling of autophagic vacuoles. MDC accumulates as a selective marker for autophagic vacuoles under *in vivo* conditions and is not present in the early and late endosome. Interestingly in this regard, the presence of a latent transglutaminase activity in rat liver lysosomes has been reported [195].

TABLE 3
TG2 Protein Substrates Involved in Autophagy (For reference go to http://genomics.dote.hu/wiki/ [198].)

Regulatory Proteins	Autophagy Substrates
TGF-beta	Huntingtin
Bax	Alpha synuclein
Rho A	Ataxin 1
Bip	Hsp 70
Sinapsin 1	Hsp 90
Rock	Valosin-containing protein
	Parkin
	Ubiquitin
	Phosphorylase kinase
	Fructose 1,6-bisphosphatase
	Phosphoglycerate dehydrogenase
	GAPDH

The involvement of TG2 in autophagy has also been shown in highly metastatic pancreatic carcinoma cells. In this cells, the inhibition of PKC-δ, by rottlerin, or the knockdown of TG2 protein, by siRNA, leads to the accumulation of autophagic vacuoles in the cytoplasm and a marked induction of LC3 II [196].

Although the molecular mechanism(s) by which TG2 can regulate autophagy is not yet known, one attractive hypothesis is that this effect could be exerted at the level of cytoskeleton (Figure 4). Indeed, TG2-dependent post-translational modification of the cytoskeleton's major components, such as tubulin, actin, and vimentin, has been shown to be able to influence its regulation [8]. A functional cytoskeleton is of fundamental importance for the proper intra-cellular traffic of the autophagic vesicles and for their fusion with lysosomes. Accordingly, it has been shown that, in mammalian cells, the disruption of the microtubule network results in a delay, rather than a complete block, in the autophagy process [197].

Finally, it is relevant to note that several TG2 candidate protein substrates (Table 3) have been shown to be enriched on the autophagosome membrane; this provides further evidence that TG2-catalyzed post-translational modifications may have a role in the maturation of the autophagosomes [192].

IV. TG2 IN DISEASES

Considering the multifunctionality and the unique cellular biochemistry of TG2, it comes as a surprise that homozygous deletion of TG2 does not result in an embryonic lethal phenotype [199, 200]. The homozygous null animals are viable, of normal size and weight, and born with Mendelian frequency. The most probable explanation for the lack of severe phenotypes is that other TGs in mammalian tissues can compensate for the loss of TG2. However, such compensation is partial, since the other mammalian TGs do not have all the enzymatic activities carried out by TG2.

In line with this, several alterations have been detected in $TG2^{-/-}$ mice, especially under stressful and pathological conditions. In fact, decreased adherence of primary fibroblasts [200] and impaired wound healing, related to altered cytoskeleton dynamics of fibroblasts, have been observed in these mice, consistent with the suggested extra- and intra-cellular functions of TG2. Moreover, when cell death is induced in $TG2^{-/-}$ mice, the clearance of apoptotic cells by phagocytosis is defective in the thymus and liver and both inflammatory and autoimmune reactions develop [201].

TG2-deficient mice also show glucose intolerance and hyperglycemia, because of a reduced insulin secretion, a phenomenon similar to a subtype of diabetes called MODY (maturity-onset diabetes of the young) [202]. Furthermore, the ablation of TG2 in mice causes an increased vulnerability of cardiomyocytes to ischemia/reperfusion injury [152]. This effect is associated with a decreased capacity of $TG2^{-/-}$ mice to synthesize ATP, due to a reduced activity of mitochondrial complex I [151].

All the major human diseases are associated with dysregulation of pathways controlling cell death and survival. As previously described, apoptosis and autophagy constitute the two processes through which superfluous, damaged, or aged cells and/or organelles are eliminated. The functional relationship between these two processes is highly complex. In fact, autophagy constitutes a stress adaptation that in general avoids cell death, whereas in different settings, it might constitute an alternative cell-death pathway [104]. Because of TG2's involvement in these two basic cellular events, the enzyme has been clearly implicated in the pathogenesis of a number of pathological conditions, such as celiac sprue [203], neurodegenerative disorders [204], diabetes [202], liver cirrhosis and fibrosis [205, 206], renal scarring [207], and certain types of cancer [208].

A. INFLAMMATORY DISEASES

Many reports have shown that the expression of TG2 is increased in inflammatory diseases and/or autoimmune diseases [209]. Interestingly, both apoptosis and autophagy have been shown to play an essential role in either the pathogenesis or the response to these diseases. Results obtained in recent years have shown a participation of TG2 in these self-eating and self-killing pathways, thus making TG2 a potential therapeutic target.

A characteristic feature of chronic inflammatory diseases is the persistent presence of apoptotic cells resulting from impaired phagocytosis [210, 211]. It has been shown that the lack of TG2 function is associated with a defective clearance of apoptotic cells. In fact, $TG2^{-/-}$ mice showed a significant perturbation in the phagocytosis of apoptotic cells. The long-term consequence of the absence of the enzyme results in the development of splenomegaly, autoantibodies, and glomerulonephritis [201]. Under physiological conditions, macrophages play a pivotal role in both recognition and clearance of apoptotic cells. An essential aspect of this process is the absence of an inflammatory response [212]. The lack of TG2 results in an impaired capacity of macrophages to engulf apoptotic cells and also affects the release of proinflammatory cytokine from macrophages, leading to an abnormal inflammatory response. Accordingly, both TGF-β and IL-12 regulations were significantly altered in the $TG2^{-/-}$ mice [213]. These results help to explain the autoimmune phenotype developed by these mice and suggest that TG2 is a key regulatory element of the anti-inflammatory features of apoptosis. Indeed, it has been reported that the production of inflammatory cytokines by macrophages requires the activation of the p38 MAPK pathway, in order to promote the stability and availability of inflammatory cytokine transcripts for translation [214, 215], whereas selective inhibitors of p38 MAPK prevent proinflammatory cytokine release [216]. In keeping with these findings, it is interesting to note that TG2 activity is required for the activation of p38 MAPKs. In fact, it has been shown that TG2 mediates the activation of RhoA and MAP kinase pathways during RA-induced neuronal differentiation of SH-SY5Y cells [60]. Consistent with these evidences, in $TG2^{-/-}$ mice, the expression of both IL-12 and sTNF-RI is significantly increased. Interestingly, the dysregulation of IL-12 has been described in a wide range of autoimmune-prone mouse models such as the TG2 ones [217–219].

The recently discovered function of autophagy in the ATP-dependent generation of engulfment signals and heterophagic removal of apoptotic corpses [220], highlights the potential role for autophagy in the prevention of inflammation and autoimmunity [174].

The inability to regulate inflammation is of great importance in the pathogenesis of sepsis, during which dysregulated inflammatory processes might induce organ's impairment and death. The progress from the disease to septic shock and death depends on a complex interaction of the inflammatory cascades with several intracellular signaling pathways [221, 222]. Inflammation and apoptosis appear to be closely linked in the immuno-pathogenesis of sepsis. Previously, it was generally thought that sepsis represented an unbridled immune response with excessive cytokine production. Now, increasing evidence suggest that extensive apoptotic death, leading to a depletion of the cells of the immune system, weakens patient's ability to eradicate infections. Recent findings show that apoptosis of the parenchymal cells in the lung, gut, and liver contributes substantially to the failure of these organs in response to sepsis. Hopefully, therapeutic approaches, involving the inhibition of the apoptotic processes, will prove capable of restoring necessary immune functions and improving the chances to survive the septic insult [223, 224]. TG2 expression has been reported to be induced by LPS in several tissues and organs [225–227]. TG2's possible contribution to the pathological inflammatory dysregulation occurring in septic shock has recently been investigated, demonstrating that TG2, acting at multiple levels, could be an important factor in the mechanism through which sepsis develops [228]. Hence, $TG2^{-/-}$ mice displayed enhanced survival to LPS challenge, the absence of TG2 being associated with profound reduction of the inflammatory response and attenuated organ damage. Although proinflammatory mediator's production is elicited in $TG2^{-/-}$ mice by LPS treatment, these mice have the capacity to restore the initial equilibrium. Activation of the nuclear transcription factor NF-kB plays a key role in the inflammatory process by inducing the transcription of proinflammatory mediators [229]. This activation is the result of phosphorylation and subsequent degradation of the inhibitory factor I-kBα, induced by the I-kB kinase complex [230]. In wild-type mice, TG2 expression is increased during endotoxemia and, being the enzyme directly involved in the mechanisms of NF-kB activation through an I-kB kinase independent pathway, which is mediated by the polymerization of I-kBα [231], it may cause a continuous activation cycle in the inflammatory process, thus contributing to the development of sepsis pathogenesis. The increased survival of $TG2^{-/-}$ mice was also reflected in a drastic reduction of organ injury, which is characterized by a limited infiltration of neutrophils, in the kidney and the peritoneum, and by a better homeostasis of the proinflammatory mediators and of mitochondrial functionality [228]. In keeping with this assumption, it has been shown that the TG2 inhibition switches off inflammation both *in vitro* and *in vivo*, in the homozygous F508del-CFTR mice model for cystic fibrosis (CF). CF is a monogenic disease caused by mutations in the CF transmembrane conductance regulator gene and characterized by chronic bacterial infections in the lungs and by inflammation. TG2 has been shown to be constitutively upregulated in CF and to play a key role in the onset of chronic inflammation [232, 233]. It has been proposed that the persistent TG2 activation observed relies on the action of the TG2-dependent SUMOylation of the protein inhibitor of activated STAT y and thus to inhibition of TG2 ubiquitination and proteasome degradation. This prevents peroxisome proliferator-activated receptor (PPAR) gamma and IkBα SUMOylation, triggering NF-kB activation and an uncontrolled inflammatory response [234]. These data suggested that TG2 might function as a link between oxidative stress and inflammation, by driving the decision as to whether a protein should undergo SUMO-mediated regulation or degradation. Targeting TG2-SUMO interactions may represent a new option in controling disease evolution in patients with CF and in other chronic inflammatory diseases. Indeed, the variety of TG2's protein substrates and of the biological functions it exerts inside the cell suggests the enzyme's involvement at different levels of the inflammation cascade (i.e., NF-kB activation, cytokine homeostasis, mitochondrial function). In this picture, the development of specific TG2 inhibitors could represent a novel approach in the treatment of the inflammatory process with important clinical implications.

B. INTRACELLULAR PROTEIN INCLUSIONS-RELATED DISORDERS

A number of severe human diseases are associated with the presence of non-correctly folded proteins that exhibit decreased solubility under physiological conditions. These diseases result either from mutations, able to change the amino acid sequence of a protein, or from misfolding of wild-type proteins that cause the formation of nuclear and/or cytoplasmic aggregates of stainable substances known as inclusion bodies.

1. Neurodegenerative Diseases

Accumulation of misfolded proteins in proteinaceous inclusions is a prominent pathological feature common to many age-related neurodegenerative diseases, including PD, AD, HD, and amyotrophic lateral sclerosis (ALS) [235].

AD, the most common age-related neurodegenerative disorder, is associated with the selective damage of brain regions and neural circuits including, but not exclusively, neurons in the neo-cortex, hippocampus, and amygdala. Dysfunction and loss of neurons in these neural circuits result in impaired memory, thinking, and behavior. Two major hallmarks of AD pathology are extracellular neuritic senile plaques and intraneuronal neurofibrillary tangles. Senile plaques and neurofibrillary tangles are formed by abnormally polymerized proteins in the brain, and both these lesions are extremely insoluble structures. The purification and analysis of senile plaques demonstrated that they are comprised of fibrils of the amyloid β-protein (Aβ). Aβ is a 39–42 amino acid peptide derived from the proteolysis of a larger transmembrane glycoprotein, the amyloid precursor protein [236–238]. Neurofibrillary tangles are composed primarily of paired helical filaments (PHFs) [239, 240]; a major component of the PHFs is the hyperphosphorylated form of the microtubule-associated protein tau [241–243]. It has been hypothesized that TG2 may be involved in the pathogenesis of AD by facilitating the formation of one or both of these insoluble lesions. Biochemical and immuno-cytochemical assays have revealed TG2 expression in neurons of both normal and AD-affected elderly individuals [244, 245]. Miller and Anderton investigated the possible role of TG2 in the cross-linking of neurofilaments and extended the previous findings by demonstrating that all three neurofilaments polypeptides are TG2 substrates and can be cross-linked into insoluble, but nonfilamentous aggregates. Neurofibrillary tangles' major component is an abnormally phosphorylated form of the microtubule-associated protein tau but not neurofilament; therefore, it was hypothesized that the pathological aggregation of tau into insoluble neurofibrillary tangles may be enzymatically facilitated by TG2. Accordingly, several studies have demonstrated that tau is readily cross-linked by TG2 [244, 246, 247].

HD is a dominantly inherited disorder characterized by a progressive degeneration of motor neuron coupled to an impairment of coordination and with variable mental syndromes [191]. The disease's molecular basis relies on the expansion of the CAG tri-nucleotide repeat in the gene encoding for huntingtin (htt). This expansion leads to the presence of a stretch of polyglutamine in the expressed protein. In individuals not affected by HD, the number of CAG repeats varies from 6 to 35, whereas lengths of 40 and over invariably cause the onset of the pathology [248]; the longer is the repeat the more severe are the patient's clinical symptoms [249].

Several studies have proposed that HD might be caused in part by abnormal protein–protein interactions related to the elongated poly-glutamine stretch of huntingtin. The presence of both neuronal intranuclear and cytosolic inclusions, composed of mutant huntingtin, has been detected in brains of both HD patients and HD animal models [250, 251], although it is not clear whether such inclusions are harmful or beneficial gent [252]. Nevertheless, they are likely to play a role in the onset and progression of the disease and a role for TG2 in their formation, growth and/or stabilization has been proposed [253]. It has also been hypothesized that one of mechanisms for htt aggregation is based on the action of TG2 on the expanded poly-glutamine repeats, which might act as excellent glutaminyl-donor substrates for the TG2-catalyzed cross-linking reaction [254]. On the other hand, the analysis of a mice model derived by crossing HD R6/1 transgenic mice with $TG2^{-/-}$ ones highlighted an increase in the formation of htt aggregates. The HD R6/1/$TG2^{-/-}$ mice showed a reduction of neuronal cell death coupled with a significant improvement in both locomotory performance and survival, as compared with the HD R6/1/$TG2^{+/+}$ [92]. This suggests that the involvement of TG2 in the loss of neurons in HD is not related to the formation of htt aggregates.

Although TG2's role in the aggregate formation is yet to be fully established, it is clear that TG2 could contribute to the pathogenesis of HD through other mechanisms than the formation of aggregates. The interplay between TG2 and some mitochondrial functions could be one of the mechanisms involved in the pathogenesis of HD, although supporting evidence for this hypothesis is still lacking. It remains the case that an impairment in the mitochondrial function, which has been proposed as one of the pathological mechanisms of HD, resulted in a significant increase of TGase activity in situ [255]. In addition, it has been shown that TG2 might act as a "sensitizer" towards apoptotic stimuli by modulating mitochondrial function [41]. This evidence is also supported by the fact that the $TG2^{-/-}$ mice showed a defect in the activity of the mitochondrial respiratory complex I, this defect being partially compensated by an increase in the activity of complex II [151].

Recent studies have proved that autophagy plays a protective role towards the progression of HD, being the fundamental mechanism for removing protein aggregates [256]. The hypothesis that autophagy might contribute to the pathogenesis of neurodegenerative diseases was initially supported by reports demonstrating that autophagosomes accumulate in the brains of these patients [187, 257]. In addition, studies on mice with cerebellar degeneration, due to mutations in the glutamate receptor, suggested that autophagy might be a mechanism of nonapoptotic cell death [258]. In contrast, more recent studies provide compelling evidence that, at least in model organisms, autophagy might be a protection mechanism against various neurodegenerative diseases and that the accumulation of autophagosomes represents the activation of autophagy as a beneficial physiological response or, as in the case of AD, the consequence of a defect in the maturation of autophagosomes. In this context, autophagy could be responsible not only for the clearance of spontaneously generated misfolded proteins and for routine protein turnover but it may also play an important role in the clearance of aggregate-prone mutant proteins often associated with neurodegenerative diseases. Normal protein turnover via proteasome requires the substrates to be unfolded in order to pass through the narrow pore of the proteasomal barrel. The bulk aggregates formed by poly-glutamine (poly-Q) expansion, as in HD; mutant α-synuclein, as in familial PD; and different forms of tau, including mutations causing fronto-temporal dementia [257] result to be poor substrates for this degradation pathway and might be better targets for autophagic degradation.

The mechanism by which these proteins exert their cellular toxicity is still controversial, but it is generally believed that they are particularly toxic in oligomeric complexes and that higherorder protein aggregates may be formed as a last attempt to prevent toxicity, in the absence of a properly functioning quality-control system [259].

Autophagy has fundamental functions in cellular homeostasis and its modulation has been proposed as a therapeutic strategy for neurodegenerative diseases associated with aggregate formation, such as HD [187]. Considering that $TG2^{-/-}$ mice showed an inhibition of the maturation of autophagolysosomes, the increased number of NII observed in the $TG2^{-/-}$/HD mice [92] is very likely explained by the inhibitory effect on autophagy progression caused by TG2 absence. Although the molecular mechanism(s) by which TG2 can regulate autophagy is not yet known, the most plausible hypothesis suggests a possible effect at the level of cytoskeleton. Since TG2 can post-translationally modify the major component of cytoskeleton such as tubulin, actin, and vimentin and so deeply influence its regulation [8, 59, 209], future studies should verify this hypothesis as well as identifying other possible biochemical pathways by which the enzyme cross-linking activity regulates autophagy.

2. *Liver Steatohepatitis*

Mallory body (MB) inclusions are a characteristic feature of several liver disorders and share similarities with cytoplasmic inclusions observed in neural diseases and myopathies. MBs are composed mainly of keratins 8 and

18 (K8/K18) and contain glutamine-lysine cross-links catalyzed by TG2. In a mouse model of MBs, the inhibition of TG2, achieved by the use of the specific inhibitor KCC009, causes a reduction of the 3,5-diethoxycarbonyl-1,4-dihydrocollidine-induced liver enlargement, without affecting MB formation or the extent of the liver injury. The observed hepatomegaly is due to increased hepatocytes cell size rather than due to their proliferation [260]. Hence, once again, TG2 inhibition does not affect the MB formation, similar to what has been reported for the NII in the HD/TG2$^{-/-}$ mice. However, in clear contrast with these results, the inhibition of MB formation has been observed in an experimental model based on the TG2$^{-/-}$ mice. Further studies should clarify this controversial aspect of the problem.

It is important to note that a common feature of several liver disorders (including alcoholic and non-alcoholic steatohepatitis) is the accumulation of intra-cellular protein aggregates, which are known to be cleared by autophagy. It has been observed that rapamycin treatment decreased the number of inclusions significantly. Thus, TG2 inhibition associated with autophagy provides a potential cellular approach for the resorption of cytoplasmic inclusions.

Recently, it has been shown that autophagy is involved in the normal physiology of pancreatic β cells and thus in the maintenance of the glucose homeostasis. Accordingly, Atg7 mutant mice develop impaired glucose tolerance and decreased serum insulin level [261]. Morphological analysis of autophagy-deficient β cells in these mice revealed the accumulation of ubiquitinated protein aggregates that colocalize with p62. These features associate with mitochondrial swelling and vacuolar changes [262]. Given that TG2$^{-/-}$ mice develop a type II diabetes syndrome [202], it would be worth to investigate the involvement of TG2 and autophagy induction in the onset of this disease.

3. Cardiac Diseases

Constitutive autophagy plays an important role in the maintenance of cellular homeostasis in the heart, whereas unrestrained autophagic activity accentuates the maladaptive cardiac remodeling in response to stress (e.g., hypertension) and may contribute to heart failure [263]. Recent reports demonstrate that multiple forms of cardiovascular stress, including pressure overload, chronic ischemia, and infarction-reperfusion injury, lead to an increase of autophagy in cardiomyocytes [264–266]. In this regard, an accumulation of autophagosomes has been detected in biopsies of cardiac tissues belonging to patients suffering from these cardiac diseases, as well as in rodent models and isolated stressed cardiomyocytes [267, 268]. Since cardiac myocytes are terminally differentiated cells, the role of autophagy is essential for maintaining the homeostasis of the myocardium. Autophagy supplies nutrients for the synthesis of essential proteins during starvation and thus helps to extend cell survival. Under pathological settings, protein aggregation occurs in the heart in response to hemodynamic stress, putting pressure-overload heart disease in the category of proteinopathies [269]. Protein aggregation is the trigger of cardiomyocyte autophagy, which functions to attenuate aggregate/aggresome formation in the heart [270].

It has been recently shown that angiotensin II increases autophagosome formation, via the angiotensin II type 1 (AT1) receptor, and that, in neonatal cardiomyocytes, this response is constitutively antagonized by the coexpression of the angiotensin II type 2 (AT2) receptor [271, 272]. Interestingly, TG2 cross-links the carboxyl-terminal tail of the AT1 receptor and the so-formed receptor dimers display enhanced signaling as well as desensitization both *in vitro* and *in vivo* [273].

The heart consumes more energy per gram of tissue than any other organ in the body; thus the most common cardiac disorders (e.g., cardiac ischemia and heart failure) are characterized by a reduction in the availability of energy substrates, a factor that contributes to transient or sustained impairment of cardiac function. Furthermore, when cardiac stresses are sustained for long periods of time, myocytes remodel their cellular architecture (e.g., undergo elongation and hypertrophy) in order to adapt to stress. It has been hypothesized that, when the heart is under stress, its need for more energy and for cellular remodeling could be met, at least partly, by via activation of the autophagy pathway. In keeping with this hypothesis, cardiac-specific deficiency of Atg5, early in cardiogenesis, results in no phenotypic abnormality under basal conditions; instead, after treatment with pressure overload or α-adrenergic stress, severe cardiac dysfunction becomes evident. Interestingly, the stimulation of the $\alpha(1)$-adrenoreceptor in cardiac myocytes prompts the appearance of a hyper-trophic phenotype [269].

Ablation of TG2 in mice has been shown to cause an increased vulnerability of cardiomyocytes to ischemia/reperfusion injury. This effect is associated with a decreased capacity of $TG2^{-/-}$ mice to synthesize ATP, due to both a reduced activity of mitochondrial complex I and to defects in the functions of the ATP/ADP transporter [40, 151, 152, 153]. Taking into account the role played by TG2 in the autophagosome formation, it is tempting to hypothesize that the vulnerability displayed by the cardiac muscle of the $TG2^{-/-}$ mice could be due not only to their congenital mitochondrial defects, but also to an impaired autophagic response to cellular stress [274].

V. CONCLUSIONS

Despite the extensive research carried out to date, the comprehension of the physiological role played by TG2 remains largely unknown as does the significance of the post-translational modifications of its protein substrates. This review offers further compelling evidence that under physiological and pathological circumstances, the enzyme is mainly involved in the regulation of cell death. Accordingly, it seems clear that under physiological settings the up-regulation and the activation of TG2 are associated with the induction of apoptosis. On the other hand, in pathological conditions, such as in cancer cells, the enzyme may act both as a pro-survival and a pro-death factor, depending upon the cell type and/or its subcellular localization (Figure 3). In addition, a deregulation of TG2 has been proposed as playing a major role in various diseases, including cardiomyopathy, hepatitis, diabetes, and neurodegenerative disorders. A common feature of all these pathologies relies on the accumulation of misfolded proteins in proteinaceous inclusions. It has become increasingly evident that, in all these disease, the affected cell activates autophagy in order to eliminate these large protein aggregates. It is worth noting that several TG2 candidate protein substrates have been shown to be enriched on the membrane of the autophagosomes (Table 3). These findings, taken together with the blockade of the autophagosome maturation observed in the $TG2^{-/-}$ mice, support TG2's involvement in the final process of the autolysosomes maturation in multiple ways. Future studies should address the different controversial points in order to establish whether TG2 could indeed be used as a target for treatment of the abovementioned degenerative diseases.

ACKNOWLEDGMENTS

This work was partly funded by grants from European Community "APO-SYS" and "TRAKS;" Ricerca corrente e finalizzata del "Ministero della Salute." The financial support from Italian Telethon Foundation, A.I.R.C. (grants to M.P.), Compagnia di San Paolo and CHDI Foundation Inc. (USA) is gratefully acknowledged.

REFERENCES

1. Folk, J. E. and Finlayson, J. S. (1977) The epsilon-(gamma-glutamyl)lysine crosslink and the catalytic role of transglutaminases, *Adv. Protein Chem. 31*, 1.
2. Lorand, L. and Conrad, S. M. (1984) Transglutaminases, *Mol. Cell. Biochem. 58*, 9.
3. Grenard, P., Bates, M. K., and Aeschlimann, D. (2001) Evolution of transglutaminase genes: identification of a transglutaminase gene cluster on human chromosome 15q15. Structure of the gene encoding transglutaminase X and a novel gene family member, transglutaminase Z, *J. Biol. Chem. 276*, 33066.
4. Makarova, K. S., Aravind, L., and Koonin, E. V. (1999) A superfamily of archaeal, bacterial, and eukaryotic proteins homologous to animal transglutaminases, *Protein Sci. 8*, 1714.
5. Mosher, D. F., Schad, P. E., and Kleinman, H. K. (1979) Inhibition of blood coagulation factor XIIIa-mediated cross-linking between fibronectin and collagen by polyamines, *J. Supramol. Struct. 11*, 227.
6. Kim, S. Y., Kim, I. G., Chung, S. I., and Steinert, P. M. (1994) The structure of the transglutaminase 1 enzyme. Deletion cloning reveals domains that regulate its specific activity and substrate specificity, *J. Biol. Chem. 269*, 27979.
7. Rice, R. H. and Green, H. (1978) Relation of protein synthesis and transglutaminase activity to formation of the cross-linked envelope during terminal differentiation of the cultured human epidermal keratinocyte, *J. Cell Biol. 76*, 705.
8. Fesus, L. and Piacentini, M. (2002) Transglutaminase 2: an enigmatic enzyme with diverse functions, *Trends Biochem. Sci. 27*, 534.
9. Lorand, L. and Graham, R. M. (2003) Transglutaminases: crosslinking enzymes with pleiotropic functions, *Nat. Rev. Mol. Cell Biol. 4*, 140.
10. Lee, J. H., Jang, S. I., Yang, J. M., Markova, N. G., and Steinert, P. M. (1996) The proximal promoter of the human transglutaminase 3 gene. Stratified squamous epithelial-specific expression in cultured cells is mediated by binding of Sp1 and ets transcription factors to a proximal promoter element, *J. Biol. Chem. 271*, 4561.
11. Dubbink, H. J., de Waal, L., van Haperen, R., Verkaik, N. S., Trapman, J., and Romijn, J. C. (1998) The human prostate-specific transglutaminase gene (TGM4): genomic organization, tissue-specific expression, and promoter characterization, *Genomics 51*, 434.
12. Candi, E., Oddi, S., Terrinoni, A., Paradisi, A., Ranalli, M., Finazzi-Agro, A., and Melino, G. (2001) Transglutaminase 5 cross-links loricrin, involucrin, and small proline-rich proteins in vitro, *J. Biol. Chem. 276*, 35014.
13. Lee, C. J., Do, B. R., Lee, J. M., Song, K. W., Kang, J. S., and Park, M. H. (2003) Differential expression of tissue transglutaminase protein in mouse ovarian follicle, *In Vivo 17*, 235.
14. Polakowska, R. R., Graf, B. A., Falciano, V., and LaCelle, P. (1999) Transcription regulatory elements of the first intron control human transglutaminase type I gene expression in epidermal keratinocytes, *J. Cell Biochem. 73*, 355.
15. Im, M. J. and Graham, R. M. (1990) A novel guanine nucleotide-binding protein coupled to the alpha 1-adrenergic receptor. I. Identification by photolabeling or membrane and ternary complex preparation, *J. Biol. Chem. 265*, 18944.

16. Im, M. J., Riek, R. P., and Graham, R. M. (1990) A novel guanine nucleotide-binding protein coupled to the alpha 1-adrenergic receptor. II. Purification, characterization, and reconstitution, *J. Biol. Chem. 265*, 18952.
17. Spina, A. M., Esposito, C., Pagano, M., Chiosi, E., Mariniello, L., Cozzolino, A., Porta, R., and Illiano, G. (1999) GTPase and transglutaminase are associated in the secretion of the rat anterior prostate, *Biochem. Biophys. Res. Commun. 260*, 351.
18. Candi, E., Paradisi, A., Terrinoni, A., Pietroni, V., Oddi, S., Cadot, B., Jogini, V., Meiyappan, M., Clardy, J., Finazzi-Agro, A., and Melino, G. (2004) Transglutaminase 5 is regulated by guanine-adenine nucleotides, *Biochem. J. 381*, 313.
19. Liu, S., Cerione, R. A., and Clardy, J. (2002) Structural basis for the guanine nucleotide-binding activity of tissue transglutaminase and its regulation of transamidation activity, *Proc. Natl. Acad. Sci. U.S.A. 99*, 2743.
20. Nakaoka, H., Perez, D. M., Baek, K. J., Das, T., Husain, A., Misono, K., Im, M. J., and Graham, R. M. (1994) Gh: a GTP-binding protein with transglutaminase activity and receptor signaling function, *Science 264*, 1593.
21. Mishra, S. and Murphy, L. J. (2004) Tissue transglutaminase has intrinsic kinase activity: identification of transglutaminase 2 as an insulin-like growth factor-binding protein-3 kinase, *J. Biol. Chem. 279*, 23863.
22. Mishra, S. and Murphy, L. J. (2006) The p53 oncoprotein is a substrate for tissue transglutaminase kinase activity, *Biochem. Biophys. Res. Commun. 339*, 726.
23. Mishra, S., Saleh, A., Espino, P. S., Davie, J. R., and Murphy, L. J. (2006) Phosphorylation of histones by tissue transglutaminase, *J. Biol. Chem. 281*, 5532.
24. Hasegawa, G., Suwa, M., Ichikawa, Y., Ohtsuka, T., Kumagai, S., Kikuchi, M., Sato, Y., and Saito, Y. (2003) A novel function of tissue-type transglutaminase: protein disulphide isomerase, *Biochem. J. 373*, 793.
25. Mastroberardino, P. G., Farrace, M. G., Viti, I., Pavone, F., Fimia, G. M., Melino, G., Rodolfo, C., and Piacentini, M. (2006) "Tissue" transglutaminase contributes to the formation of disulphide bridges in proteins of mitochondrial respiratory complexes, *Biochim. Biophys. Acta 1757*, 1357.
26. Ueki, S., Takagi, J., and Saito, Y. (1996) Dual functions of transglutaminase in novel cell adhesion, *J. Cell Sci. 109(Pt 11)*, 2727.
27. Lorand, L. (1988) Transglutaminase-mediated cross-linking of proteins and cell ageing: the erythrocyte and lens models, *Adv. Exp. Med. Biol. 231*, 79.
28. Kim, S. Y., Jeitner, T. M., and Steinert, P. M. (2002) Transglutaminases in disease, *Neurochem. Int. 40*, 85.
29. D'Amato, M., Iannicola, C., Monteriu, G., and Piacentini, M. (1999), Mapping and sequencing of the murine 'tissue' transglutaminase (Tgm2) gene: absence of mutations in MRLlpr/lpr mice, *Cell Death Differ. 6*, 216.
30. Gentile, V., Davies, P. J., and Baldini, A. (1994), The human tissue transglutaminase gene maps on chromosome 20q12 by in situ fluorescence hybridization, *Genomics 20*, 295.
31. Fraij, B. M. and Gonzales, R. A. (1997), Organization and structure of the human tissue transglutaminase gene, *Biochim. Biophys. Acta 1354*, 65.

32. Gentile, V., Saydak, M., Chiocca, E. A., Akande, O., Birckbichler, P. J., Lee, K. N., Stein, J. P., and Davies, P. J. (1991) Isolation and characterization of cDNA clones to mouse macrophage and human endothelial cell tissue transglutaminases, *J. Biol. Chem. 266*, 478.

33. Fesus, L. and Arato, G. (1986), Quantitation of tissue transglutaminase by a sandwich ELISA system, *J. Immunol. Methods 94*, 131.

34. Thomazy, V. and Fesus, L. (1989) Differential expression of tissue transglutaminase in human cells. An immunohistochemical study, *Cell Tissue Res. 255*, 215.

35. Di Venere, A. , Rossi, A., De Matteis, F., Rosato, N., Agro, A. F., and Mei, G. (2000) Opposite effects of Ca(2+) and GTP binding on tissue transglutaminase tertiary structure, *J. Biol. Chem. 275*, 3915.

36. Pedersen, L. C., Yee, V. C., Bishop, P. D., Le Trong, I., Teller, D. C., and Stenkamp, R. E. (1994) Transglutaminase factor XIII uses proteinase-like catalytic triad to crosslink macromolecules, *Protein Sci. 3*, 1131.

37. Yee, V. C., Pedersen, L. C., Le Trong, I., Bishop, P. D., Stenkamp, R. E., and Teller, D. C. (1994) Three-dimensional structure of a transglutaminase: human blood coagulation factor XIII, *Proc. Natl. Acad. Sci. U.S.A. 91*, 7296.

38. Noguchi, K., Ishikawa, K., Yokoyama, K., Ohtsuka, T., Nio, N., and Suzuki, E. (2001) Crystal structure of red sea bream transglutaminase, *J. Biol. Chem. 276*, 12055.

39. Singh, U. S., Erickson, J. W., and Cerione, R. A. (1995) Identification and biochemical characterization of an 80 kilodalton GTP-binding/transglutaminase from rabbit liver nuclei, *Biochemistry 34*, 15863.

40. Malorni, W., Farrace, M. G., Matarrese, P., Tinari, A., Ciarlo, L., Mousavi-Shafaei, P., D'Eletto, M., Di Giacomo, G., Melino, G., Palmieri, L., Rodolfo, C., and Piacentini, M. (2009) The adenine nucleotide translocator 1 acts as a type 2 transglutaminase substrate: implications for mitochondrial-dependent apoptosis, *Cell Death Differ. 16*, 1480.

41. Piacentini, M., Farrace, M. G., Piredda, L., Matarrese, P., Ciccosanti, F., Falasca, L., Rodolfo, C., Giammarioli, A. M., Verderio, E., Griffin, M., and Malorni, W. (2002), Transglutaminase overexpression sensitizes neuronal cell lines to apoptosis by increasing mitochondrial membrane potential and cellular oxidative stress, *J. Neurochem. 81*, 1061.

42. Gaudry, C. A., Verderio, E., Aeschlimann, D., Cox, A., Smith, C., and Griffin, M. (1999) Cell surface localization of tissue transglutaminase is dependent on a fibronectin-binding site in its N-terminal beta-sandwich domain, *J. Biol. Chem. 274*, 30707.

43. Gaudry, C. A., Verderio, E., Jones, R. A., Smith, C., and Griffin, M. (1999) Tissue transglutaminase is an important player at the surface of human endothelial cells: evidence for its externalization and its colocalization with the beta(1) integrin, *Exp. Cell Res. 252*, 104.

44. Stevens, H. Y., Reeve, J., and Noble, B. S. (2000) Bcl-2, tissue transglutaminase and p53 protein expression in the apoptotic cascade in ribs of premature infants, *J. Anat. 196(Pt 2)*, 181.

45. Radek, J. T., Jeong, J. M., Murthy, S. N., Ingham, K. C., and Lorand, L. (1993) Affinity of human erythrocyte transglutaminase for a 42-kDa gelatin-binding fragment of human plasma fibronectin, *Proc. Natl. Acad. Sci. U.S.A. 90*, 3152.

46. Akimov, S. S., Krylov, D., Fleischman, L. F., and Belkin, A. M. (2000) Tissue transglutaminase is an integrin-binding adhesion coreceptor for fibronectin, *J. Cell Biol. 148*, 825.

47. Telci, D., Wang, Z., Li, X., Verderio, E. A., Humphries, M. J., Baccarini, M., Basaga, H., and Griffin, M. (2008) Fibronectin-tissue transglutaminase matrix rescues RGD-impaired cell adhesion through syndecan-4 and beta1 integrin co-signaling, *J. Biol. Chem. 283*, 20937.

48. Xu, L., Begum, S., Hearn, J. D., and Hynes, R. O. (2006) GPR56, an atypical G protein-coupled receptor, binds tissue transglutaminase, TG2, and inhibits melanoma tumor growth and metastasis, *Proc. Natl. Acad. Sci. U.S.A. 103*, 9023.

49. Vezza, R., Habib, A., and FitzGerald, G. A. (1999) Differential signaling by the thromboxane receptor isoforms via the novel GTP-binding protein, Gh, *J. Biol. Chem. 274*, 12774.

50. Park, E. S., Won, J. H., Han, K. J., Suh, P. G., Ryu, S. H., Lee, H. S., Yun, H. Y., Kwon, N. S., and Baek, K. J. (1998) Phospholipase C-delta1 and oxytocin receptor signalling: evidence of its role as an effector, *Biochem. J. 331(Pt 1)*, 283.

51. Rodolfo, C., Mormone, E., Matarrese, P., Ciccosanti, F., Farrace, M. G., Garofano, E., Piredda, L., Fimia, G. M., Malorni, W., and Piacentini, M. (2004) Tissue transglutaminase is a multifunctional BH3-only protein, *J. Biol. Chem. 279*, 54783.

52. Piacentini, M., Rodolfo, C., Farrace, M. G., and Autuori, F. (2000) "Tissue" transglutaminase in animal development, *Int. J. Dev. Biol. 44*, 655.

53. Szegezdi, E., Szondy, Z., Nagy, L., Nemes, Z., Friis, R. R., Davies, P. J., and Fesus, L. (2000) Apoptosis-linked in vivo regulation of the tissue transglutaminase gene promoter, *Cell Death Differ. 7*, 1225.

54. Iismaa, S. E., Chung, L., Wu, M. J., Teller, D. C., Yee, V. C., and Graham, R. M. (1997) The core domain of the tissue transglutaminase Gh hydrolyzes GTP and ATP, *Biochemistry 36*, 11655.

55. Iismaa, S. E., Wu, M. J., Nanda, N., Church, W. B., and Graham, R. M. (2000) GTP binding and signaling by Gh/transglutaminase II involves distinct residues in a unique GTP-binding pocket, *J. Biol. Chem. 275*, 18259.

56. Murthy, S. N., Lomasney, J. W., Mak, E. C., and Lorand, L. (1999) Interactions of G(h)/transglutaminase with phospholipase Cdelta1 and with GTP, *Proc. Natl. Acad. Sci. U.S.A. 96*, 11815.

57. Lai, T. S., Bielawska, A., Peoples, K. A., Hannun, Y. A., and Greenberg, C. S. (1997) Sphingosylphosphocholine reduces the calcium ion requirement for activating tissue transglutaminase, *J. Biol. Chem. 272*, 16295.

58. Lai, T. S., Hausladen, A., Slaughter, T. F., Eu, J. P., Stamler, J. S., and Greenberg, C. S. (2001) Calcium regulates S-nitrosylation, denitrosylation, and activity of tissue transglutaminase, *Biochemistry 40*, 4904.

59. Singh, U. S., Kunar, M. T., Kao, Y. L., and Baker, K. M. (2001), Role of transglutaminase II in retinoic acid-induced activation of RhoA-associated kinase-2, *Embo. J. 20*, 2413.

60. Singh, U. S., Pan, J., Kao, Y. L., Joshi, S., Young, K. L., and Baker, K. M. (2003), Tissue transglutaminase mediates activation of RhoA and MAP kinase pathways during

retinoic acid-induced neuronal differentiation of SH-SY5Y cells, *J. Biol. Chem. 278*, 391.

61. Murthy, S. N., Wilson, J. H., Lukas, T. J., Kuret, J., and Lorand, L. (1998) Cross-linking sites of the human tau protein, probed by reactions with human transglutaminase, *J. Neurochem. 71*, 2607.
62. Piredda, L., Farrace, M. G., Lo Bello, M., Malorni, W., Melino, G., Petruzzelli, R., and Piacentini, M. (1999) Identification of 'tissue' transglutaminase binding proteins in neural cells committed to apoptosis, *Faseb. J. 13*, 355.
63. Peng, X., Zhang, Y., Zhang, H., Graner, S., Williams, J. F., Levitt, M. L., and Lokshin, A. (1999) Interaction of tissue transglutaminase with nuclear transport protein importin-alpha3, *FEBS Lett. 446*, 35.
64. Ballestar, E., Abad, C., and Franco, L. (1996) Core histones are glutaminyl substrates for tissue transglutaminase, *J. Biol. Chem. 271*, 18817.
65. Ballestar, E., Boix-Chornet, M., and Franco, L. (2001) Conformational changes in the nucleosome followed by the selective accessibility of histone glutamines in the transglutaminase reaction: effects of ionic strength, *Biochemistry 40*, 1922.
66. Oliverio, S., Amendola, A., Di Sano, F., Farrace, M. G., Fesus, L., Nemes, Z., Piredda, L., Spinedi, A., and Piacentini, M. (1997) Tissue transglutaminase-dependent post-translational modification of the retinoblastoma gene product in promonocytic cells undergoing apoptosis, *Mol. Cell Biol. 17*, 6040.
67. Fesus, L., Thomazy, V., Autuori, F., Ceru, M. P., Tarcsa, E., and Piacentini, M. (1989) Apoptotic hepatocytes become insoluble in detergents and chaotropic agents as a result of transglutaminase action, *FEBS Lett. 245*, 150.
68. Fesus, L., Thomazy, V., and Falus, A. (1987) Induction and activation of tissue transglutaminase during programmed cell death, *FEBS Lett. 224*, 104.
69. Oliverio, S., Amendola, A., Rodolfo, C., Spinedi, A., and Piacentini, M. (1999) Inhibition of "tissue" transglutaminase increases cell survival by preventing apoptosis, *J. Biol. Chem. 274*, 34123.
70. Tatsukawa, H., Fukaya, Y., Frampton, G., Martinez-Fuentes, A., Suzuki, K., Kuo, T. F., Nagatsuma, K., Shimokado, K., Okuno, M., Wu, J., Iismaa, S., Matsuura, T., Tsukamoto, H., Zern, M. A., Graham, R. M., and Kojima, S. (2009) Role of transglutaminase 2 in liver injury via cross-linking and silencing of transcription factor Sp1, *Gastroenterology* 136, 1783.
71. Polakowska, R. R., Piacentini, M., Bartlett, R., Goldsmith, L. A., and Haake, A. R. (1994) Apoptosis in human skin development: morphogenesis, periderm, and stem cells, *Dev. Dyn. 199*, 176.
72. Szondy, Z., Molnar, P., Nemes, Z., Boyiadzis, M., Kedei, N., Toth, R., and Fesus, L. (1997) Differential expression of tissue transglutaminase during in vivo apoptosis of thymocytes induced via distinct signalling pathways, *FEBS Lett. 404*, 307.
73. Szondy, Z., Reichert, U., Bernardon, J. M., Michel, S., Toth, R., Ancian, P., Ajzner, E., and Fesus, L. (1997) Induction of apoptosis by retinoids and retinoic acid receptor gamma-selective compounds in mouse thymocytes through a novel apoptosis pathway, *Mol. Pharmacol. 51*, 972.

74. Nagy, L., Thomazy, V. A., and Davies, P. J. (1998) A transgenic mouse model for the study of apoptosis during limb development, *Cell Death Differ.* *5*, 126.

75. Thomazy, V. A. and Davies, P. J. (1999) Expression of tissue transglutaminase in the developing chicken limb is associated both with apoptosis and endochondral ossification, *Cell Death Differ.* *6*, 146.

76. Sarang, Z., Madi, A., Koy, C., Varga, S., Glocker, M. O., Ucker, D. S., Kuchay, S., Chishti, A. H., Melino, G., Fesus, L., and Szondy, Z. (2007) Tissue transglutaminase (TG2) facilitates phosphatidylserine exposure and calpain activity in calcium-induced death of erythrocytes, *Cell Death Differ.* *14*, 1842.

77. Nemes, Z., Jr., Friis, R. R., Aeschlimann, D., Saurer, S., Paulsson, M., and Fesus, L. (1996) Expression and activation of tissue transglutaminase in apoptotic cells of involuting rodent mammary tissue, *Eur. J. Cell Biol.* *70*, 125.

78. Ruan, Q., Quintanilla, R. A., and Johnson, G. V. (2007) Type 2 transglutaminase differentially modulates striatal cell death in the presence of wild type or mutant huntingtin, *J. Neurochem.* *102*, 25.

79. Rebe, C., Raveneau, M., Chevriaux, A., Lakomy, D., Sberna, A. L., Costa, A., Bessede, G., Athias, A., Steinmetz, E., Lobaccaro, J. M., Alves, G., Menicacci, A., Vachenc, S., Solary, E., Gambert, P., and Masson, D. (2009) Induction of transglutaminase 2 by a liver X receptor/retinoic acid receptor alpha pathway increases the clearance of apoptotic cells by human macrophages, *Circ. Res.* *105*, 393.

80. Ou, H., Haendeler, J., Aebly, M. R., Kelly, L. A., Cholewa, B. C., Koike, G., Kwitek-Black, A., Jacob, H. J., Berk, B. C., and Miano, J. M. (2000) Retinoic acid-induced tissue transglutaminase and apoptosis in vascular smooth muscle cells, *Circ. Res.* *87*, 881.

81. Gorza, L., Menabo, R., Di Lisa, F., and Vitadello, M. (1997) Troponin T cross-linking in human apoptotic cardiomyocytes, *Am. J. Pathol.* *150*, 2087.

82. Gorza, L., Menabo, R., Vitadello, M., Bergamini, C. M., and Di Lisa, F. (1996) Cardiomyocyte troponin T immunoreactivity is modified by cross-linking resulting from intracellular calcium overload, *Circulation* *93*, 1896.

83. Assisi, L., Autuori, F., Botte, V., Farrace, M. G., and Piacentini, M. (1999) Hormonal control of "tissue" transglutaminase induction during programmed cell death in frog liver, *Exp. Cell Res.* *247*, 339.

84. Lin, X. R., Wilkinson, D. I., and Farber, E. M. (1998) Camptothecin induces differentiation, tissue transglutaminase and apoptosis in cultured keratinocytes, *Exp. Dermatol.* *7*, 179.

85. Zhang, S. R., Li, S. H., Abler, A., Fu, J., Tso, M. O., and Lam, T. T. (1996) Tissue transglutaminase in apoptosis of photoreceptor cells in rat retina, *Invest. Ophthalmol. Vis. Sci.* *37*, 1793.

86. Dini, L., Falasca, L., Lentini, A., Mattioli, P., Piacentini, M., Piredda, L., and Autuori, F. (1993) Galactose-specific receptor modulation related to the onset of apoptosis in rat liver, *Eur. J. Cell Biol.* *61*, 329.

87. Fesus, L., Tarcsa, E., Kedei, N., Autuori, F., and Piacentini, M. (1991) Degradation of cells dying by apoptosis leads to accumulation of epsilon(gamma-glutamyl)lysine isodipeptide in culture fluid and blood, *FEBS Lett.* *284*, 109.

88. Amendola, A., Gougeon, M. L., Poccia, F., Bondurand, A., Fesus, L., and Piacentini, M. (1996) Induction of "tissue" transglutaminase in HIV pathogenesis: evidence for high rate of apoptosis of CD4+ T lymphocytes and accessory cells in lymphoid tissues, *Proc. Natl. Acad. Sci. U.S.A. 93*, 11057.

89. Nardacci, R., Antinori, A., Larocca, L. M., Arena, V., Amendola, A., Perfettini, J. L., Kroemer, G., and Piacentini, M. (2005) Characterization of cell death pathways in human immunodeficiency virus-associated encephalitis, *Am. J. Pathol. 167*, 695.

90. Fujita, K., Kato, T., Shibayama, K., Imada, H., Yamauchi, M., Yoshimoto, N., Miyachi, E., and Nagata, Y. (2006) Protective effect against 17beta-estradiol on neuronal apoptosis in hippocampus tissue following transient ischemia/recirculation in mongolian gerbils via down-regulation of tissue transglutaminase activity, *Neurochem. Res. 31*, 1059.

91. Hwang, I. K., Yoo, K. Y., Yi, S. S., Kim, I. Y., Hwang, H. S., Lee, K. Y., Choi, S. M., Lee, I. S., Yoon, Y. S., Kim, S. Y., Won, M. H., and Seong, J. K. (2009) Expression of tissue-type transglutaminase (tTG) and the effect of tTG inhibitor on the hippocampal CA1 region after transient ischemia in gerbils, *Brain Res. 1263*, 134.

92. Mastroberardino, P. G., Iannicola, C., Nardacci, R., Bernassola, F., De Laurenzi, V., Melino, G., Moreno, S., Pavone, F., Oliverio, S., Fesus, L., and Piacentini, M. (2002) 'Tissue' transglutaminase ablation reduces neuronal death and prolongs survival in a mouse model of Huntington's disease, *Cell Death Differ. 9*, 873.

93. Sarnelli, G., De Giorgio, R., Gentile, F., Cali, G., Grandone, I., Rocco, A., Cosenza, V., Cuomo, R., and D'Argenio, G. (2009) Myenteric neuronal loss in rats with experimental colitis: role of tissue transglutaminase-induced apoptosis, *Dig. Liver Dis. 41*, 185.

94. Nemes, Z., Petrovski, G., Aerts, M., Sergeant, K., Devreese, B., and Fesus, L. (2009) Transglutaminase-mediated intramolecular cross-linking of membrane-bound alpha-synuclein promotes amyloid formation in Lewy bodies, *J. Biol. Chem. 284*, 27252.

95. Griffin, M. and Verderio, E. (2000) Tissue transglutaminase in cell death, *Symp. Soc. Exp. Biol. 52*, 223.

96. Kroemer, G., El-Deiry, W. S., Golstein, P., Peter, M. E., Vaux, D., Vandenabeele, P., Zhivotovsky, B., Blagosklonny, M. V., Malorni, W., Knight, R. A., Piacentini, M., Nagata, S., and Melino, G. (2005) Classification of cell death: recommendations of the Nomenclature Committee on Cell Death, *Cell Death Differ. 12(Suppl 2)*, 1463.

97. Krysko, D. V., D'Herde, K., and Vandenabeele, P. (2006) Clearance of apoptotic and necrotic cells and its immunological consequences, *Apoptosis 11*, 1709.

98. Piacentini, M. (1999) Apoptosis and autoimmunity: two sides to the coin, *Cell Death Differ. 6*, 1.

99. Kerr, J. F., Wyllie, A. H., and Currie, A. R. (1972) Apoptosis: a basic biological phenomenon with wide-ranging implications in tissue kinetics, *Br. J. Cancer 26*, 239.

100. Kroemer, G. and Reed, J. C. (2000) Mitochondrial control of cell death, *Nat. Med. 6*, 513.

101. Enari, M., Sakahira, H., Yokoyama, H., Okawa, K., Iwamatsu, A., and Nagata, S. (1998) A caspase-activated DNase that degrades DNA during apoptosis, and its inhibitor ICAD, *Nature 391*, 43.

102. De Duve, C. (1966) The significance of lysosomes in pathology and medicine, *Proc. Inst. Med. Chic. 26*, 73.

103. De Duve, C. and Wattiaux, R. (1966) Functions of lysosomes, *Annu. Rev. Physiol. 28*, 435.

104. Maiuri, M. C., Zalckvar, E., Kimchi, A., and Kroemer, G. (2007) Self-eating and self-killing: crosstalk between autophagy and apoptosis, *Nat. Rev. Mol. Cell Biol. 8*, 741.

105. Huang, J. and Klionsky, D. J. (2007) Autophagy and human disease, *Cell Cycle 6*, 1837.

106. Hotchkiss, R. S., Strasser, A., McDunn, J. E., and Swanson, P. E. (2009) Cell death, *N. Engl. J. Med. 361*, 1570.

107. Vanlangenakker, N., Vanden Berghe, T., Krysko, D. V., Festjens, N., and Vandenabeele, P. (2008) Molecular mechanisms and pathophysiology of necrotic cell death, *Curr. Mol. Med. 8*, 207.

108. Zong, W. X. and Thompson, C. B. (2006) Necrotic death as a cell fate, *Genes Dev. 20*, 1.

109. Galluzzi, L., Maiuri, M. C., Vitale, I., Zischka, H., Castedo, M., Zitvogel, L., and Kroemer, G. (2007) Cell death modalities: classification and pathophysiological implications, *Cell Death Differ. 14*, 1237.

110. Boya, P., Gonzalez-Polo, R. A., Casares, N., Perfettini, J. L., Dessen, P., Larochette, N., Metivier, D., Meley, D., Souquere, S., Yoshimori, T., Pierron, G., Codogno, P., and Kroemer, G. (2005) Inhibition of macroautophagy triggers apoptosis, *Mol. Cell Biol. 25*, 1025.

111. Gonzalez-Polo, R. A., Boya, P., Pauleau, A. L., Jalil, A., Larochette, N., Souquere, S., Eskelinen, E. L., Pierron, G., Saftig, P., and Kroemer, G. (2005) The apoptosis/autophagy paradox: autophagic vacuolization before apoptotic death, *J. Cell Sci. 118*, 3091.

112. Vandenabeele, P., Vanden Berghe, T., and Festjens, N. (2006) Caspase inhibitors promote alternative cell death pathways, *Sci. STKE 2006*, pe44.

113. Grabarek, J., Ardelt, B., Kunicki, J., and Darzynkiewicz, Z. (2002) Detection of in situ activation of transglutaminase during apoptosis: correlation with the cell cycle phase by multiparameter flow and laser scanning cytometry, *Cytometry 49*, 83.

114. Robitaille, K., Daviau, A., Tucholski, J., Johnson, G. V., Rancourt, C., and Blouin, R. (2004) Tissue transglutaminase triggers oligomerization and activation of dual leucine zipper-bearing kinase in calphostin C-treated cells to facilitate apoptosis, *Cell Death Differ. 11*, 542.

115. Hwang, J. Y., Mangala, L. S., Fok, J. Y., Lin, Y. G., Merritt, W. M., Spannuth, W. A., Nick, A. M., Fiterman, D. J., Vivas-Mejia, P. E., Deavers, M. T., Coleman, R. L., Lopez-Berestein, G., Mehta, K., and Sood, A. K. (2008) Clinical and biological significance of tissue transglutaminase in ovarian carcinoma, *Cancer Res. 68*, 5849.

116. Shao, M., Cao, L., Shen, C., Satpathy, M., Chelladurai, B., Bigsby, R. M., Nakshatri, H., and Matei, D. (2009) Epithelial-to-mesenchymal transition and ovarian tumor progression induced by tissue transglutaminase, *Cancer Res. 69*, 9192.

117. Caraglia, M., Marra, M., Giuberti, G., D'Alessandro, A. M., Beninati, S., Lentini, A., Pepe, S., Boccellino, M., and Abbruzzese, A. (2002) Theophylline-induced apoptosis is paralleled by protein kinase A-dependent tissue transglutaminase activation in cancer cells, *J. Biochem. (Tokyo) 132*, 45.

118. Jang, J. H., Park, J. S., Lee, T. J., and Kwon, T. K. (2009) Transglutaminase 2 expression levels regulate sensitivity to cystamine plus TRAIL-mediated apoptosis, *Cancer Lett. 287*, 224.

119. Fok, J. Y. and Mehta, K. (2007) Tissue transglutaminase induces the release of apoptosis inducing factor and results in apoptotic death of pancreatic cancer cells, *Apoptosis 12*, 1455.

120. Cao, L., Petrusca, D. N., Satpathy, M., Nakshatri, H., Petrache, I., and Matei, D. (2008) Tissue transglutaminase protects epithelial ovarian cancer cells from cisplatin-induced apoptosis by promoting cell survival signaling, *Carcinogenesis 29*, 1893.

121. Joseph, B., Lefebvre, O., Mereau-Richard, C., Danze, P. M., Belin-Plancot, M. T., and Formstecher, P. (1998) Evidence for the involvement of both retinoic acid receptor- and retinoic X receptor-dependent signaling pathways in the induction of tissue transglutaminase and apoptosis in the human myeloma cell line RPMI 8226, *Blood 91*, 2423.

122. Lefebvre, O., Wouters, D., Mereau-Richard, C., Facon, T., Zandecki, M., Formstecher, P., and Belin, M. T. (1999) Induction of apoptosis by all-trans retinoic acid in the human myeloma cell line RPMI 8226 and negative regulation of some of its typical morphological features by dexamethasone, *Cell Death Differ. 6*, 433.

123. Yuan, L., Behdad, A., Siegel, M., Khosla, C., Higashikubo, R., and Rich, K. M. (2008) Tissue transgluaminase 2 expression in meningiomas, *J. Neurooncol. 90*, 125.

124. Yuan, L., Choi, K., Khosla, C., Zheng, X., Higashikubo, R., Chicoine, M. R., and Rich, K. M. (2005) Tissue transglutaminase 2 inhibition promotes cell death and chemosensitivity in glioblastomas, *Mol. Cancer Ther. 4*, 1293.

125. Esposito, C., Marra, M., Giuberti, G., D'Alessandro, A. M., Porta, R., Cozzolino, A., Caraglia, M., and Abbruzzese, A. (2003) Ubiquitination of tissue transglutaminase is modulated by interferon alpha in human lung cancer cells, *Biochem. J. 370*, 205.

126. Verma, A., Guha, S., Diagaradjane, P., Kunnumakkara, A. B., Sanguino, A. M., Lopez-Berestein, G., Sood, A. K., Aggarwal, B. B., Krishnan, S., Gelovani, J. G., and Mehta, K. (2008) Therapeutic significance of elevated tissue transglutaminase expression in pancreatic cancer, *Clin. Cancer Res. 14*, 2476.

127. Verma, A., Guha, S., Wang, H., Fok, J. Y., Koul, D., Abbruzzese, J., and Mehta, K. (2008) Tissue transglutaminase regulates focal adhesion kinase/AKT activation by modulating PTEN expression in pancreatic cancer cells, *Clin. Cancer Res. 14*, 1997.

128. Melino, G., Annicchiarico-Petruzzelli, M., Piredda, L., Candi, E., Gentile, V., Davies, P. J., and Piacentini, M. (1994) Tissue transglutaminase and apoptosis: sense and antisense transfection studies with human neuroblastoma cells, *Mol. Cell Biol. 14*, 6584.

129. Piacentini, M., Fesus, L., Farrace, M. G., Ghibelli, L., Piredda, L., and Melino, G. (1991) The expression of "tissue" transglutaminase in two human cancer cell lines is related with the programmed cell death (apoptosis), *Eur. J. Cell Biol. 54*, 246.

130. Antonyak, M. A., Miller, A. M., Jansen, J. M., Boehm, J. E., Balkman, C. E., Wakshlag, J. J., Page, R. L., and Cerione, R. A. (2004) Augmentation of tissue transglutaminase expression and activation by epidermal growth factor inhibit doxorubicin-induced apoptosis in human breast cancer cells, *J. Biol. Chem. 279*, 41461.

131. Datta, S., Antonyak, M. A., and Cerione, R. A. (2006) Importance of Ca(2+)-dependent transamidation activity in the protection afforded by tissue transglutaminase against doxorubicin-induced apoptosis, *Biochemistry 45*, 13163.

132. Kim, D. S., Park, K. S., and Kim, S. Y. (2009) Silencing of TGase 2 sensitizes breast cancer cells to apoptosis by regulation of survival factors, *Front. Biosci. 14*, 2514.

133. Park, K. S., Han, B. G., Lee, K. H., Kim, D. S., Kim, J. M., Jeon, H., Kim, H. S., Suh, S. W., Lee, E. H., Kim, S. Y., and Lee, B. I. (2009) Depletion of nucleophosmin via transglutaminase 2 cross-linking increases drug resistance in cancer cells, *Cancer Lett. 274*, 201.

134. El-Metwally, T. H., Hussein, M. R., Pour, P. M., Kuszynski, C. A., and Adrian, T. E. (2005) Natural retinoids inhibit proliferation and induce apoptosis in pancreatic cancer cells previously reported to be retinoid resistant, *Cancer Biol. Ther. 4*, 474.

135. Kweon, S. M., Lee, Z. W., Yi, S. J., Kim, Y. M., Han, J. A., Paik, S. G., and Ha, S. S. (2004) Protective role of tissue transglutaminase in the cell death induced by TNF-alpha in SH-SY5Y neuroblastoma cells, *J. Biochem. Mol. Biol. 37*, 185.

136. Yu, J. W. and Shi, Y. (2008) FLIP and the death effector domain family, *Oncogene 27*, 6216.

137. Green, D. R. (2005) Apoptotic pathways: ten minutes to dead, *Cell 121*, 671.

138. Scorrano, L. (2008) Caspase-8 goes cardiolipin: a new platform to provide mitochondria with microdomains of apoptotic signals? *J. Cell Biol. 183*, 579.

139. Chipuk, J. E. and Green, D. R. (2008) How do BCL-2 proteins induce mitochondrial outer membrane permeabilization?, *Trends Cell Biol. 18*, 157.

140. Scorrano, L. and Korsmeyer, S. J. (2003) Mechanisms of cytochrome c release by proapoptotic BCL-2 family members, *Biochem. Biophys. Res. Commun. 304*, 437.

141. Scorrano, L., Oakes, S. A., Opferman, J. T., Cheng, E. H., Sorcinelli, M. D., Pozzan, T., and Korsmeyer, S. J. (2003) BAX and BAK regulation of endoplasmic reticulum Ca2+: a control point for apoptosis, *Science 300*, 135.

142. Chipuk, J. E., Fisher, J. C., Dillon, C. P., Kriwacki, R. W., Kuwana, T., and Green, D. R. (2008) Mechanism of apoptosis induction by inhibition of the anti-apoptotic BCL-2 proteins, *Proc. Natl. Acad. Sci. U.S.A. 105*, 20327.

143. Danial, N. N. (2007) BCL-2 family proteins: critical checkpoints of apoptotic cell death, *Clin Cancer Res 13*, 7254.

144. Lomonosova, E. and Chinnadurai, G. (2008) BH3-only proteins in apoptosis and beyond: an overview, *Oncogene 27(Suppl 1)*, S2.

145. Susnow, N., Zeng, L., Margineantu, D., and Hockenbery, D. M. (2009) Bcl-2 family proteins as regulators of oxidative stress, *Semin. Cancer Biol. 19*, 42.

146. Youle, R. J. and Strasser, A. (2008) The BCL-2 protein family: opposing activities that mediate cell death, *Nat. Rev. Mol. Cell Biol. 9*, 47.

147. Piacentini, M., Amendola, A., Ciccosanti, F., Falasca, L., Farrace, M. G., Mastroberardino, P. G., Nardacci, R., Oliverio, S., Piredda, L., Rodolfo, C., and Autuori, F. (2005) Type 2 transglutaminase and cell death, *Prog. Exp. Tumor Res. 38*, 58.

148. Piacentini, M., Autuori, F., Dini, L., Farrace, M. G., Ghibelli, L., Piredda, L., and Fesus, L. (1991) "Tissue" transglutaminase is specifically expressed in neonatal rat liver cells undergoing apoptosis upon epidermal growth factor-stimulation, *Cell Tissue Res. 263*, 227.

149. Piacentini, M., Ceru, M. P., Dini, L., Di Rao, M., Piredda, L., Thomazy, V., Davies, P. J., and Fesus, L. (1992) In vivo and in vitro induction of 'tissue' transglutaminase in rat hepatocytes by retinoic acid, *Biochim. Biophys. Acta 1135*, 171.

150. Dhar, S. K. and St Clair, D. K. (2009) Nucleophosmin blocks mitochondrial localization of p53 and apoptosis, *J. Biol. Chem. 284*, 16409.

151. Battaglia, G., Farrace, M. G., Mastroberardino, P. G., Viti, I., Fimia, G. M., Van Beeumen, J., Devreese, B., Melino, G., Molinaro, G., Busceti, C. L., Biagioni, F., Nicoletti, F., and Piacentini, M. (2007) Transglutaminase 2 ablation leads to defective function of mitochondrial respiratory complex I affecting neuronal vulnerability in experimental models of extrapyramidal disorders, *J. Neurochem. 100*, 36.

152. Szondy, Z., Mastroberardino, P. G., Varadi, J., Farrace, M. G., Nagy, N., Bak, I., Viti, I., Wieckowski, M. R., Melino, G., Rizzuto, R., Tosaki, A., Fesus, L., and Piacentini, M. (2006) Tissue transglutaminase (TG2) protects cardiomyocytes against ischemia/reperfusion injury by regulating ATP synthesis, *Cell Death Differ. 13*, 1827.

153. Malorni, W., Farrace, M. G., Rodolfo, C., and Piacentini, M. (2008) Type 2 transglutaminase in neurodegenerative diseases: the mitochondrial connection, *Curr. Pharm. Des. 14*, 278.

154. Lesort, M., Attanavanich, K., Zhang, J., and Johnson, G. V. (1998) Distinct nuclear localization and activity of tissue transglutaminase, *J. Biol. Chem. 273*, 11991.

155. Robitaille, H., Proulx, R., Robitaille, K., Blouin, R., and Germain, L. (2005) The mitogen-activated protein kinase kinase kinase dual leucine zipper-bearing kinase (DLK) acts as a key regulator of keratinocyte terminal differentiation, *J. Biol. Chem. 280*, 12732.

156. Robitaille, K., Daviau, A., Lachance, G., Couture, J. P., and Blouin, R. (2008) Calphostin C-induced apoptosis is mediated by a tissue transglutaminase-dependent mechanism involving the DLK/JNK signaling pathway, *Cell Death Differ. 15*, 1522.

157. Boosen, M., Vetterkind, S., Kubicek, J., Scheidtmann, K. H., Illenberger, S., and Preuss, U. (2009) Par-4 is an essential downstream target of DAP-like kinase (Dlk) in Dlk/Par-4-mediated apoptosis, *Mol. Biol. Cell 20*, 4010.

158. Kogel, D., Reimertz, C., Mech, P., Poppe, M., Fruhwald, M. C., Engemann, H., Scheidtmann, K. H., and Prehn, J. H. (2001), Dlk/ZIP kinase-induced apoptosis in human medulloblastoma cells: requirement of the mitochondrial apoptosis pathway, *Br. J. Cancer 85*, 1801.

159. Verderio, E. A., Telci, D., Okoye, A., Melino, G., and Griffin, M. (2003) A novel RGD-independent cell adhesion pathway mediated by fibronectin-bound tissue transglutaminase rescues cells from anoikis, *J. Biol. Chem. 278*, 42604.

160. Collighan, R. J. and Griffin, M. (2009) Transglutaminase 2 cross-linking of matrix proteins: biological significance and medical applications, *Amino Acids 36*, 659.

161. Gundemir, S. and Johnson, G. V. (2009) Intracellular localization and conformational state of transglutaminase 2: implications for cell death, *PLoS One 4*, e6123.

162. Antonyak, M. A., Singh, U. S., Lee, D. A., Boehm, J. E., Combs, C., Zgola, M. M., Page, R. L., and Cerione, R. A. (2001) Effects of tissue transglutaminase on retinoic acid-induced cellular differentiation and protection against apoptosis, *J. Biol. Chem. 276*, 33582.

163. Beck, K. E., De Girolamo, L. A., Griffin, M., and Billett, E. E. (2006) The role of tissue transglutaminase in 1-methyl-4-phenylpyridinium (MPP+)-induced toxicity in differentiated human SH-SY5Y neuroblastoma cells, *Neurosci. Lett. 405*, 46.

164. Verma, A. and Mehta, K. (2007) Transglutaminase-mediated activation of nuclear transcription factor-kappaB in cancer cells: a new therapeutic opportunity, *Curr. Cancer Drug Targets 7*, 559.

165. Verma, A. and Mehta, K. (2007) Tissue transglutaminase-mediated chemoresistance in cancer cells, *Drug Resist. Update 10*, 144.

166. Boehm, J. E., Singh, U., Combs, C., Antonyak, M. A., and Cerione, R. A. (2002) Tissue transglutaminase protects against apoptosis by modifying the tumor suppressor protein p110 Rb, *J. Biol. Chem. 277*, 20127.

167. Antonyak, M. A., Boehm, J. E., and Cerione, R. A. (2002) Phosphoinositide 3-kinase activity is required for retinoic acid-induced expression and activation of the tissue transglutaminase, *J. Biol. Chem. 277*, 14712.

168. Antonyak, M. A., McNeill, C. J., Wakshlag, J. J., Boehm, J. E., and Cerione, R. A. (2003) Activation of the Ras-ERK pathway inhibits retinoic acid-induced stimulation of tissue transglutaminase expression in NIH3T3 cells, *J. Biol. Chem. 278*, 15859.

169. Antonyak, M. A., Jansen, J. M., Miller, A. M., Ly, T. K., Endo, M., and Cerione, R. A. (2006) Two isoforms of tissue transglutaminase mediate opposing cellular fates, *Proc. Natl. Acad. Sci. U.S.A. 103*, 18609.

170. Mizushima, N. and Klionsky, D. J. (2007) Protein turnover via autophagy: implications for metabolism, *Annu. Rev. Nutr. 27*, 19.

171. Mizushima, N. (2007) Autophagy: process and function, *Genes Dev. 21*, 2861.

172. Kundu, M. and Thompson, C. B. (2008) Autophagy: basic principles and relevance to disease, *Annu. Rev. Pathol. 3*, 427.

173. Levine, B. and Klionsky, D. J. (2004) Development by self-digestion: molecular mechanisms and biological functions of autophagy, *Dev. Cell 6*, 463.

174. Levine, B. and Kroemer, G. (2008) Autophagy in the pathogenesis of disease, *Cell 132*, 27.

175. Amano, A., Nakagawa, I., and Yoshimori, T. (2006) Autophagy in innate immunity against intracellular bacteria, *J. Biochem. 140*, 161.

176. Levine, B. and Deretic, V. (2007) Unveiling the roles of autophagy in innate and adaptive immunity, *Nat. Rev. Immunol. 7*, 767.

177. Gorski, S. M., Chittaranjan, S., Pleasance, E. D., Freeman, J. D., Anderson, C. L., Varhol, R. J., Coughlin, S. M., Zuyderduyn, S. D., Jones, S. J., and Marra, M. A. (2003) A SAGE approach to discovery of genes involved in autophagic cell death, *Curr. Biol. 13*, 358.

178. Hou, Y. C., Hannigan, A. M., and Gorski, S. M. (2009) An executioner caspase regulates autophagy, *Autophagy 5*, 530.

179. Klionsky, D. J. (2007) Autophagy: from phenomenology to molecular understanding in less than a decade, *Nat. Rev. Mol. Cell Biol. 8*, 931.

180. Shintani, T. and Klionsky, D. J. (2004) Cargo proteins facilitate the formation of transport vesicles in the cytoplasm to vacuole targeting pathway, *J. Biol. Chem. 279*, 29889.

181. Baehrecke, E. H. (2005) Autophagy: dual roles in life and death? *Nat. Rev. Mol. Cell Biol. 6*, 505.

182. Lum, J. J., DeBerardinis, R. J., and Thompson, C. B. (2005) Autophagy in metazoans: cell survival in the land of plenty, *Nat. Rev. Mol. Cell Biol. 6*, 439.

183. Lum, J. J., Bauer, D. E., Kong, M., Harris, M. H., Li, C., Lindsten, T., and Thompson, C. B. (2005) Growth factor regulation of autophagy and cell survival in the absence of apoptosis, *Cell 120*, 237.

184. Levine, B. and Abrams, J. (2008) p53: The Janus of autophagy? *Nat. Cell Biol. 10*, 637.

185. Meijer, A. J. and Codogno, P. (2006) Signalling and autophagy regulation in health, aging and disease, *Mol. Aspects Med. 27*, 411.

186. Meley, D., Bauvy, C., Houben-Weerts, J. H., Dubbelhuis, P. F., Helmond, M. T., Codogno, P., and Meijer, A. J. (2006) AMP-activated protein kinase and the regulation of autophagic proteolysis, *J. Biol. Chem. 281*, 34870.

187. Rubinsztein, D. C., Gestwicki, J. E., Murphy, L. O., and Klionsky, D. J. (2007) Potential therapeutic applications of autophagy, *Nat. Rev. Drug Discov. 6*, 304.

188. Mizushima, N. and Yoshimori, T. (2007) How to interpret LC3 immunoblotting, *Autophagy 3*, 542.

189. Qiao, L. and Zhang, J. (2009) Inhibition of lysosomal functions reduces proteasomal activity, *Neurosci. Lett. 456*, 15.

190. Turmaine, M., Raza, A., Mahal, A., Mangiarini, L., Bates, G. P., and Davies, S. W. (2000) Nonapoptotic neurodegeneration in a transgenic mouse model of Huntington's disease, *Proc. Natl. Acad. Sci. U.S.A. 97*, 8093.

191. Vonsattel, J. P. and DiFiglia, M. (1998) Huntington disease, *J. Neuropathol. Exp. Neurol. 57*, 369.

192. D'Eletto, M., Farrace, M. G., Falasca, L., Reali, V., Oliverio, S., Melino, G., Griffin, M., Fimia, G. M., and Piacentini, M. (2009) Transglutaminase 2 is involved in autophagosome maturation, *Autophagy 5*, 1145.

193. Kim, I., Rodriguez-Enriquez, S., and Lemasters, J. J. (2007) Selective degradation of mitochondria by mitophagy, *Arch. Biochem. Biophys. 462*, 245.

194. Skill, N. J., Johnson, T. S., Coutts, I. G., Saint, R. E., Fisher, M., Huang, L., El Nahas, A. M., Collighan, R. J., and Griffin, M. (2004) Inhibition of transglutaminase activity reduces extracellular matrix accumulation induced by high glucose levels in proximal tubular epithelial cells, *J. Biol. Chem. 279*, 47754.

195. Juprelle-Soret, M., Wattiaux-De Coninck, S., and Wattiaux, R. (1984) Presence of a transglutaminase activity in rat liver lysosomes, *Eur. J. Cell Biol. 34*, 271.

196. Akar, U., Ozpolat, B., Mehta, K., Fok, J., Kondo, Y., and Lopez-Berestein, G. (2007) Tissue transglutaminase inhibits autophagy in pancreatic cancer cells, *Mol. Cancer Res. 5*, 241.

197. Monastyrska, I., Rieter, E., Klionsky, D. J., and Reggiori, F. (2009) Multiple roles of the cytoskeleton in autophagy, *Biol. Rev. Camb. Philos. Soc. 84*, 431.

198. Csosz, E., Mesko, B., and Fesus, L. (2009) Transdab wiki: the interactive transglutaminase substrate database on web 2.0 surface, *Amino Acids 36*, 615.

199. De Laurenzi, V. and Melino, G. (2001) Gene disruption of tissue transglutaminase, *Mol. Cell Biol. 21*, 148.

200. Nanda, N., Iismaa, S. E., Owens, W. A., Husain, A., Mackay, F., and Graham, R. M. (2001) Targeted inactivation of Gh/tissue transglutaminase II, *J. Biol. Chem. 276*, 20673.

201. Szondy, Z., Sarang, Z., Molnar, P., Nemeth, T., Piacentini, M., Mastroberardino, P. G., Falasca, L., Aeschlimann, D., Kovacs, J., Kiss, I., Szegezdi, E., Lakos, G., Rajnavolgyi, E., Birckbichler, P. J., Melino, G., and Fesus, L. (2003) Transglutaminase 2-/- mice reveal a phagocytosis-associated crosstalk between macrophages and apoptotic cells, *Proc. Natl. Acad. Sci. U.S.A. 100*, 7812.

202. Bernassola, F., Federici, M., Corazzari, M., Terrinoni, A., Hribal, M. L., De Laurenzi, V., Ranalli, M., Massa, O., Sesti, G., McLean, W. H., Citro, G., Barbetti, F., and Melino, G. (2002) Role of transglutaminase 2 in glucose tolerance: knockout mice studies and a putative mutation in a MODY patient, *Faseb J. 16*, 1371.

203. Molberg, O., McAdam, S. N., and Sollid, L. M. (2000) Role of tissue transglutaminase in celiac disease, *J. Pediatr. Gastroenterol. Nutr. 30*, 232.

204. Hoffner, G. and Djian, P. (2005) Transglutaminase and diseases of the central nervous system, *Front. Biosci. 10*, 3078.

205. Issa, R., Zhou, X., Constandinou, C. M., Fallowfield, J., Millward-Sadler, H., Gaca, M. D., Sands, E., Suliman, I., Trim, N., Knorr, A., Arthur, M. J., Benyon, R. C., and Iredale, J. P. (2004) Spontaneous recovery from micronodular cirrhosis: evidence for incomplete resolution associated with matrix cross-linking, *Gastroenterology 126*, 1795.

206. Mirza, A., Liu, S. L., Frizell, E., Zhu, J., Maddukuri, S., Martinez, J., Davies, P., Schwarting, R., Norton, P., and Zern, M. A. (1997) A role for tissue transglutaminase in hepatic injury and fibrogenesis, and its regulation by NF-kappaB, *Am. J. Physiol. 272*, G281.

207. Johnson, T. S., El-Koraie, A. F., Skill, N. J., Baddour, N. M., El Nahas, A. M., Njloma, M., Adam, A. G., and Griffin, M. (2003) Tissue transglutaminase and the progression of human renal scarring, *J. Am. Soc. Nephrol. 14*, 2052.

208. Mangala, L. S., Fok, J. Y., Zorrilla-Calancha, I. R., Verma, A., and Mehta, K. (2007) Tissue transglutaminase expression promotes cell attachment, invasion and survival in breast cancer cells, *Oncogene 26*, 2459.

209. Facchiano, F., Facchiano, A., and Facchiano, A. M. (2006) The role of transglutaminase-2 and its substrates in human diseases, *Front. Biosci. 11*, 1758.

210. Eguchi, K. (2001) Apoptosis in autoimmune diseases, *Intern. Med. 40*, 275.

211. Navratil, J. S., Sabatine, J. M., and Ahearn, J. M. (2004) Apoptosis and immune responses to self, *Rheum. Dis. Clin. North Am. 30*, 193.

212. Savill, J., Dransfield, I., Gregory, C., and Haslett, C. (2002) A blast from the past: clearance of apoptotic cells regulates immune responses, *Nat. Rev. Immunol. 2*, 965.

213. Falasca, L., Iadevaia, V., Ciccosanti, F., Melino, G., Serafino, A., and Piacentini, M. (2005) Transglutaminase type II is a key element in the regulation of the anti-inflammatory response elicited by apoptotic cell engulfment, *J. Immunol. 174*, 7330.

214. Neininger, A., Kontoyiannis, D., Kotlyarov, A., Winzen, R., Eckert, R., Volk, H. D., Holtmann, H., Kollias, G., and Gaestel, M. (2002) MK2 targets AU-rich elements and regulates biosynthesis of tumor necrosis factor and interleukin-6 independently at different post-transcriptional levels, *J. Biol. Chem. 277*, 3065.

215. Suzuki, K., Hino, M., Kutsuna, H., Hato, F., Sakamoto, C., Takahashi, T., Tatsumi, N., and Kitagawa, S. (2001) Selective activation of p38 mitogen-activated protein kinase cascade in human neutrophils stimulated by IL-1beta, *J. Immunol. 167*, 5940.

216. Lee, J. C. and Young, P. R. (1996) Role of CSB/p38/RK stress response kinase in LPS and cytokine signaling mechanisms, *J. Leukoc. Biol. 59*, 152.

217. Alleva, D. G., Johnson, E. B., Wilson, J., Beller, D. I., and Conlon, P. J. (2001) SJL and NOD macrophages are uniquely characterized by genetically programmed, elevated expression of the IL-12(p40) gene, suggesting a conserved pathway for the induction of organ-specific autoimmunity, *J. Leukoc. Biol. 69*, 440.

218. Alleva, D. G., Pavlovich, R. P., Grant, C., Kaser, S. B., and Beller, D. I. (2000) Aberrant macrophage cytokine production is a conserved feature among autoimmune-prone mouse strains: elevated interleukin (IL)-12 and an imbalance in tumor necrosis factor-alpha and IL-10 define a unique cytokine profile in macrophages from young nonobese diabetic mice, *Diabetes 49*, 1106.

219. Liu, J. and Beller, D. (2002) Aberrant production of IL-12 by macrophages from several autoimmune-prone mouse strains is characterized by intrinsic and unique patterns of NF-kappa B expression and binding to the IL-12 p40 promoter, *J. Immunol. 169*, 581.

220. Qu, X., Zou, Z., Sun, Q., Luby-Phelps, K., Cheng, P., Hogan, R. N., Gilpin, C., and Levine, B. (2007) Autophagy gene-dependent clearance of apoptotic cells during embryonic development, *Cell 128*, 931.

221. Lolis, E. and Bucala, R. (2003) Therapeutic approaches to innate immunity: severe sepsis and septic shock, *Nat. Rev. Drug Discov. 2*, 635.

222. Perl, M., Chung, C. S., Garber, M., Huang, X., and Ayala, A. (2006) Contribution of anti-inflammatory/immune suppressive processes to the pathology of sepsis, *Front. Biosci. 11*, 272.

223. Bantel, H. and Schulze-Osthoff, K. (2009) Cell death in sepsis: a matter of how, when, and where, *Crit. Care 13*, 173.

224. Perl, M., Chung, C. S., Swan, R., and Ayala, A. (2007) Role of Programmed Cell Death in the Immunopathogenesis of Sepsis, *Drug Discov. Today Dis. Mech. 4*, 223.

225. Bowness, J. M. and Tarr, A. H. (1997) Increase in transglutaminase and its extracellular products in response to an inflammatory stimulus by lipopolysaccharide, *Mol. Cell Biochem. 169*, 157.

226. Leu, R. W., Herriott, M. J., Moore, P. E., Orr, G. R., and Birckbichler, P. J. (1982) Enhanced transglutaminase activity associated with macrophage activation. Possible role in Fc-mediated phagocytosis, *Exp. Cell Res. 141*, 191.

227. Park, K. C., Chung, K. C., Kim, Y. S., Lee, J., Joh, T. H., and Kim, S. Y. (2004) Transglutaminase 2 induces nitric oxide synthesis in BV-2 microglia, *Biochem. Biophys. Res. Commun. 323*, 1055.

228. Falasca, L., Farrace, M. G., Rinaldi, A., Tuosto, L., Melino, G., and Piacentini, M. (2008) Transglutaminase type II is involved in the pathogenesis of endotoxic shock, *J. Immunol. 180*, 2616.

229. Liu, S. F. and Malik, A. B. (2006) NF-kappa B activation as a pathological mechanism of septic shock and inflammation, *Am. J. Physiol. Lung Cell Mol. Physiol. 290*, L622.

230. Baeuerle, P. A. and Baltimore, D. (1988) I kappa B: a specific inhibitor of the NF-kappa B transcription factor, *Science 242*, 540.

231. Lee, J., Kim, Y. S., Choi, D. H., Bang, M. S., Han, T. R., Joh, T. H., and Kim, S. Y. (2004) Transglutaminase 2 induces nuclear factor-kappaB activation via a novel pathway in BV-2 microglia, *J. Biol. Chem. 279*, 53725.

232. Hasosah, M., Davidson, G., and Jacobson, K. (2009) Persistent elevated tissue-transglutaminase in cystic fibrosis, *J. Paediatr. Child Health 45*, 172.

233. Maiuri, L., Luciani, A., Giardino, I., Raia, V., Villella, V. R., D'Apolito, M., Pettoello-Mantovani, M., Guido, S., Ciacci, C., Cimmino, M., Cexus, O. N., Londei, M., and Quaratino, S. (2008) Tissue transglutaminase activation modulates inflammation in cystic fibrosis via PPARgamma down-regulation, *J. Immunol. 180*, 7697.

234. Luciani, A., Villella, V. R., Vasaturo, A., Giardino, I., Raia, V., Pettoello-Mantovani, M., D'Apolito, M., Guido, S., Leal, T., Quaratino, S., and Maiuri, L. (2009) SUMOylation of tissue transglutaminase as link between oxidative stress and inflammation, *J. Immunol. 183*, 2775.

235. Johnson, W. G. (2000) Late-onset neurodegenerative diseases–the role of protein insolubility, *J. Anat. 196(Pt 4)*, 609.

236. Kang, J., Lemaire, H. G., Unterbeck, A., Salbaum, J. M., Masters, C. L., Grzeschik, K. H., Multhaup, G., Beyreuther, K., and Muller-Hill, B. (1987) The precursor of Alzheimer's disease amyloid A4 protein resembles a cell-surface receptor, *Nature 325*, 733.

237. Masters, C. L., Multhaup, G., Simms, G., Pottgiesser, J., Martins, R. N., and Beyreuther, K. (1985) Neuronal origin of a cerebral amyloid: neurofibrillary tangles of Alzheimer's disease contain the same protein as the amyloid of plaque cores and blood vessels, *Embo J. 4*, 2757.

238. Masters, C. L., Simms, G., Weinman, N. A., Multhaup, G., McDonald, B. L., and Beyreuther, K. (1985) Amyloid plaque core protein in Alzheimer disease and Down syndrome, *Proc. Natl. Acad. Sci. U.S.A. 82*, 4245.

239. Delacourte, A. and Defossez, A. (1986) Alzheimer's disease: Tau proteins, the promoting factors of microtubule assembly, are major components of paired helical filaments, *J. Neurol. Sci. 76*, 173.

240. Kosik, K. S., Joachim, C. L., and Selkoe, D. J. (1986) Microtubule-associated protein tau (tau) is a major antigenic component of paired helical filaments in Alzheimer disease, *Proc. Natl. Acad. Sci. U.S.A. 83*, 4044.

241. Biernat, J., et al. (1992) The switch of tau protein to an Alzheimer-like state includes the phosphorylation of two serine-proline motifs upstream of the microtubule binding region, *Embo J. 11*, 1593.

242. Crowther, R. A., Olesen, O. F., Jakes, R., and Goedert, M. (1992) The microtubule binding repeats of tau protein assemble into filaments like those found in Alzheimer's disease, *FEBS Lett. 309*, 199.

243. Goedert, M., Spillantini, M. G., Cairns, N. J., and Crowther, R. A. (1992) Tau proteins of Alzheimer paired helical filaments: abnormal phosphorylation of all six brain isoforms, *Neuron 8*, 159.

244. Appelt, D. M., Kopen, G. C., Boyne, L. J., and Balin, B. J. (1996) Localization of transglutaminase in hippocampal neurons: implications for Alzheimer's disease, *J. Histochem. Cytochem. 44*, 1421.

245. Miller, C. C. and Anderton, B. H. (1986) Transglutaminase and the neuronal cytoskeleton in Alzheimer's disease, *J. Neurochem. 46*, 1912.

246. Appelt, D. M. and Balin, B. J. (1997) The association of tissue transglutaminase with human recombinant tau results in the formation of insoluble filamentous structures, *Brain Res. 745*, 21.

247. Miller, M. L. and Johnson, G. V. (1995) Transglutaminase cross-linking of the tau protein, *J. Neurochem. 65*, 1760.

248. Gusella, J. F. and MacDonald, M. E. (1993) Hunting for Huntington's disease, *Mol. Genet. Med. 3*, 139.

249. Andrew, S. E., et al. (1993) The relationship between trinucleotide (CAG) repeat length and clinical features of Huntington's disease, *Nat. Genet. 4*, 398.

250. Davies, S. W., Turmaine, M., Cozens, B. A., DiFiglia, M., Sharp, A. H., Ross, C. A., Scherzinger, E., Wanker, E. E., Mangiarini, L., and Bates, G. P. (1997) Formation of neuronal intranuclear inclusions underlies the neurological dysfunction in mice transgenic for the HD mutation, *Cell 90*, 537.

251. DiFiglia, M., Sapp, E., Chase, K. O., Davies, S. W., Bates, G. P., Vonsattel, J. P., and Aronin, N. (1997) Aggregation of huntingtin in neuronal intranuclear inclusions and dystrophic neurites in brain, *Science 277*, 1990.

252. Saudou, F., Finkbeiner, S., Devys, D., and Greenberg, M. E. (1998) Huntingtin acts in the nucleus to induce apoptosis but death does not correlate with the formation of intranuclear inclusions, *Cell 95*, 55.

253. Gentile, V., Sepe, C., Calvani, M., Melone, M. A., Cotrufo, R., Cooper, A. J., Blass, J. P., and Peluso, G. (1998) Tissue transglutaminase-catalyzed formation of high-molecular-weight aggregates in vitro is favored with long polyglutamine domains: a possible mechanism contributing to CAG-triplet diseases, *Arch. Biochem. Biophys. 352*, 314.

254. Lesort, M., Tucholski, J., Miller, M. L., and Johnson, G. V. (2000) Tissue transglutaminase: a possible role in neurodegenerative diseases, *Prog. Neurobiol. 61*, 439.

255. Lesort, M., Tucholski, J., Zhang, J., and Johnson, G. V. (2000) Impaired mitochondrial function results in increased tissue transglutaminase activity in situ, *J. Neurochem. 75*, 1951.

256. Winslow, A. R. and Rubinsztein, D. C. (2008) Autophagy in neurodegeneration and development, *Biochim. Biophys. Acta 1782*, 723.

257. Williams, A., Jahreiss, L., Sarkar, S., Saiki, S., Menzies, F. M., Ravikumar, B., and Rubinsztein, D. C. (2006) Aggregate-prone proteins are cleared from the cytosol by autophagy: therapeutic implications, *Curr. Top. Dev. Biol. 76*, 89.

258. Yue, Z., Horton, A., Bravin, M., DeJager, P. L., Selimi, F., and Heintz, N. (2002) A novel protein complex linking the delta 2 glutamate receptor and autophagy: implications for neurodegeneration in lurcher mice, *Neuron 35*, 921.

259. Martinez-Vicente, M. and Cuervo, A. M. (2007) Autophagy and neurodegeneration: when the cleaning crew goes on strike, *Lancet Neurol 6*, 352.

260. Strnad, P., Harada, M., Siegel, M., Terkeltaub, R. A., Graham, R. M., Khosla, C., and Omary, M. B. (2007) Transglutaminase 2 regulates mallory body inclusion formation and injury-associated liver enlargement, *Gastroenterology 132*, 1515.

261. Fujitani, Y., Kawamori, R., and Watada, H. (2009) The role of autophagy in pancreatic beta-cell and diabetes, *Autophagy 5*, 280.

262. Jung, H. S., Chung, K. W., Won Kim, J., Kim, J., Komatsu, M., Tanaka, K., Nguyen, Y. H., Kang, T. M., Yoon, K. H., Kim, J. W., Jeong, Y. T., Han, M. S., Lee, M. K., Kim, K. W., Shin, J., and Lee, M. S. (2008) Loss of autophagy diminishes pancreatic beta cell mass and function with resultant hyperglycemia, *Cell Metab. 8*, 318.

263. Gottlieb, R. A. and Mentzer, R. M. (2010) Autophagy during cardiac stress: joys and frustrations of autophagy, *Annu. Rev. Physiol. 72*, 45.

264. Gustafsson, A. B. and Gottlieb, R. A. (2009) Autophagy in ischemic heart disease, *Circ. Res. 104*, 150.

265. Rothermel, B. A. and Hill, J. A. (2008) Autophagy in load-induced heart disease, *Circ. Res. 103*, 1363.

266. Wang, Z. V., Rothermel, B. A., and Hill, J. A. (2010), Autophagy in hypertensive heart disease, *J. Biol. Chem. 285*, 8509.

267. Terman, A. and Brunk, U. T. (2005) The aging myocardium: roles of mitochondrial damage and lysosomal degradation, *Heart Lung Circ. 14*, 107.

268. Terman, A. and Brunk, U. T. (2005) Autophagy in cardiac myocyte homeostasis, aging, and pathology, *Cardiovasc. Res. 68*, 355.

269. Nakai, A., Yamaguchi, O., Takeda, T., Higuchi, Y., Hikoso, S., Taniike, M., Omiya, S., Mizote, I., Matsumura, Y., Asahi, M., Nishida, K., Hori, M., Mizushima, N., and Otsu, K. (2007) The role of autophagy in cardiomyocytes in the basal state and in response to hemodynamic stress, *Nat. Med. 13*, 619.

270. Tannous, P., Zhu, H., Nemchenko, A., Berry, J. M., Johnstone, J. L., Shelton, J. M., Miller, F. J., Jr., Rothermel, B. A., and Hill, J. A. (2008) Intracellular protein aggregation is a proximal trigger of cardiomyocyte autophagy, *Circulation 117*, 3070.

271. Porrello, E. R., D'Amore, A., Curl, C. L., Allen, A. M., Harrap, S. B., Thomas, W. G., and Delbridge, L. M. (2009) Angiotensin II type 2 receptor antagonizes angiotensin II type 1 receptor-mediated cardiomyocyte autophagy, *Hypertension 53*, 1032.

272. Porrello, E. R. and Delbridge, L. M. (2009) Cardiomyocyte autophagy is regulated by angiotensin II type 1 and type 2 receptors, *Autophagy 5*, 1215.

273. AbdAlla, S., Lother, H., Langer, A., el Faramawy, Y., and Quitterer, U. (2004) Factor XIIIA transglutaminase crosslinks AT1 receptor dimers of monocytes at the onset of atherosclerosis, *Cell 119*, 343.

274. Sane, D. C., Kontos, J. L., and Greenberg, C. S. (2007) Roles of transglutaminases in cardiac and vascular diseases, *Front. Biosci. 12*, 2530.

TISSUE TRANSGLUTAMINASE AND ITS ROLE IN HUMAN CANCER PROGRESSION

BO LI
RICHARD A. CERIONE
MARC ANTONYAK

CONTENTS

I. INTRODUCTION

Tissue transglutaminase, or TGase-2, is a Ca^{2+}-dependent acyltransferase that catalyzes the formation of new amide bonds between the γ-carboxamide of glutamine and the ε-amino group of lysine or another primary amine (i.e., transamidation) [1, 2]. This protein cross-linking capability has been

Advances in Enzymology and Related Areas of Molecular Biology, Volume 78.
Edited by Eric J. Toone.

implicated in a wide range of cellular processes (many of which are described in this volume) including neuronal growth and regeneration, bone development, wound healing, angiogenesis, cellular differentiation, and apoptosis [3–20]. In this review, we will consider the roles played by TGase-2 in the growth, survival, and migration of cancer cells, and how the loss of the normal regulation of TGase-2 may have significant consequences for the development of the malignant state.

We will begin by considering the structural features of TGase-2. What makes this protein particularly interesting from a structure–function perspective is the fact that it is not only capable of acyl transferase activity, but that it also exhibits GTP-binding and GTP-hydrolytic activity and that there is a reciprocal relationship between its "G-protein-like functions" and its ability to catalyze protein cross-linking reactions. Indeed, the three-dimensional structures for guanine nucleotide-bound TGase-2, and for TGase-2 that has its transamidation active site accessible and capable of binding substrates, are dramatically different [21, 22]. This likely explains some of the marked differences in the cellular effects caused by GTP-binding-defective TGase-2 mutants versus mutants that are defective for transamidation activity [4, 23–28].

With this structure–function perspective as a backdrop, we will then describe some of the initial lines of study that have implicated TGase-2 in cell survival and then show how these findings have led to a growing appreciation that TGase-2 has interesting and important roles in cancer cell biology. In particular, previous work from our laboratory suggested that TGase-2 helps ensure that cells remain viable during their differentiation [19, 20]. This is a potentially important function because factors that normally cause cell-cycle arrest and differentiation can also trigger apoptotic programs and so it is necessary that these factors induce the expression and activation of additional gene products that can provide protection against apoptotic activities. TGase-2 appears to be one such protein, so that when cells receive signals triggering growth-arrest, the upregulation of TGase-2 expression occurs in order to help to tip the balance toward differentiation rather than cell death. From these initial results, we then went on to discover that the ability of TGase-2 to confer a survival response in cells extends beyond those conditions leading to cellular differentiation and in fact contributes to the ability of mitogenic stimuli such as EGF to provide a survival advantage to human breast cancer cells [29–31]. Moreover, we went on to find that TGase-2 also plays roles in stimulating the ability of human breast cancer cells to grow in the absence of a substratum (i.e., anchorage-independent growth) [32] and that it plays an important role in the EGF-stimulated migration and invasive activity of cancer cells [33].

These various studies will be elaborated upon further in the sections below. However, what we find to be very intriguing is that the different roles played by TGase-2 hold some potentially important implications for cancer and the development of potential new strategies for intervention. In particular, the overexpression of TGase-2 might act as a "double-edged sword" by making cancer cells resistant to the stress that accompanies their malignant transformation and to the apoptotic actions of various chemotherapeutic agents, as well as contributing in a strong way to their invasive behavior and metastatic potential. All of this makes it very interesting to consider that TGase-2 might serve as an extremely valuable therapeutic target, with the idea being that inhibitors of TGase-2 will make cancer cells more vulnerable to the actions of apoptotic agents, as well as reduce their metastatic capability.

II. STRUCTURAL OVERVIEW

A. CATALYTIC MECHANISM OF TGase-2

TGase-2 catalyzes the transamidation reaction between two substrates, a glutamine-containing protein substrate 1 and a lysine-containing protein substrate 2 or a polyamine, through the cooperation of a protease-like catalytic triad Cys277, His335, and Asp358. Similar to the cysteine protease family, the catalytic process comprises two steps: (1) the transient acylation of Cys277 by its nucleophilic attack on the glutamine side chain of substrate 1 to form an acyl-thiolate intermediate, and (2) the subsequent deacylation of Cys277 by the nucleophilic attack from the amino group of substrate 2, leading to the release of cross-linked products from the catalytic site. Specifically, as the first step, His335 attacks Cys277 to create a thiolate–imidazolium ion pair, with the help of Asp358 to counterbalance the positive change on His335 through its carboxylate group formed under conditions of physiological pH. The glutamine side chain of substrate 1 approaches the vicinity of Cys277 and is attacked by the thiolate ion, resulting in the formation of an acyl-thiolate intermediate, accompanied by the release of an ammonia molecule. Because of the high-energy consumption required to produce the tetrahedral adduct, the formation of an acyl-thiolate intermediate is the rate-limiting step of the transamidation reaction. The amino group of the second substrate then generates the amide ion by its interaction with the nearby His335 and Asp358 through the same mechanism as in the first step and then participates in a nucleophilic attack on the acyl-thiolate intermediate. The formation of the tetrahedral adduct in this step involves a low-energy barrier; therefore, the specificity of the attacking amino group is not very high. It either comes from a protein-bound lysine residue or a small polyamine compound. Finally, the breakdown of the tetrahedral adduct leads to the release of the cross-linked product from the enzyme, while restoring the catalytic triad and completing the transamidation reaction. The detailed mechanism is illustrated in Figure 1.

B. STRUCTURAL CONSIDERATIONS OF TGase-2: A COMPARISON BETWEEN ITS ACTIVE AND INACTIVE CONFORMATION

Although the catalytic mechanism of TGase-2 was elucidated some years ago, how the noncore residues cooperate to facilitate and regulate the transamidation activity has been less clear. Efforts to address this issue led to the sequence alignment between TGase-2 and human factor XIIIA, a hetero-tetrameric transglutaminase family member implicated in blood coagulation. The X-ray crystal structure of the human recombinant, dimeric human factor XIIIA was determined in 1994 [34], and sequence homology between human factor XIIIA and TGase-2 predicted that TGase-2 probably folded into four sequential domains: an *N*-terminal β-sandwich domain, the central catalytic core domain, and two β-barrel domains at the *C*-terminal end of the protein [35]. This result was confirmed by the first TGase-2 structure that was solved in 2002 to 2.8 Å resolution, which showed TGase-2 folded in a "closed" or inactive conformation [21]. In this structure, the carboxylic group of Asp358 formed a hydrogen bond with the imidazole ring of His335, and the hydroxyl group of Cys277 was positioned close to His335, waiting to generate the thiolate–imidazolium ion pair through the resonant cooperation of the catalytic triad (Figure 2). The reason that this structure was thought to be catalytically inactive was because of the inaccessibility of its catalytic core to the outer environment. The proposed substrate entrance was blocked by two loops within the first β-barrel domain at the *C*-terminal end of TGase-2. Most importantly, Tyr516, which is located within this loop region, formed a tight hydrogen bond with the hydroxyl group of Cys277, thus negating the catalytic activity (Figure 2). As a result, a plausible hypothesis for TGase-2 activation involves the withdrawal of the two loops

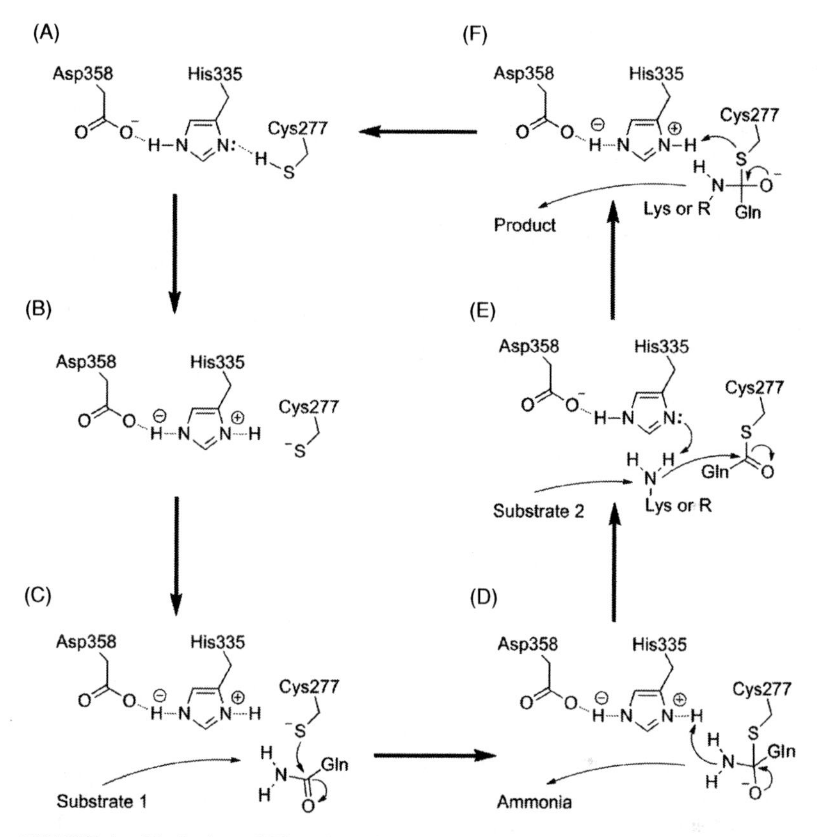

FIGURE 1. Mechanism of TGase-2-catalyzed transamidation reaction. First, a thiolate anion is formed at the side chain of Cys277, which then attacks the glutamine group of substrate 1. Release of ammonia occurs after the proton donation from His335, which creates an acyl-thiolate intermediate. Next, the lysine side chain of substrate 2 approaches the catalytic site and undergoes a nucleophilic attack upon the intermediate, leading to the formation of a tetrahedral adduct and the subsequent release of the cross-linked product. As a result, substrates 1 and 2 are covalently linked to each other through a newly formed peptide bond.

from the catalytic core, and the concomitant disruption of the hydrogen bond between Cys277 and Tyr516. This hypothesis was confirmed by the more recently determined TGase-2 structure in its "open" state, where the active conformation of TGase-2 was, in effect, trapped by incubating this enzyme with an inhibitor that mimicked its gluten peptide substrate [22].

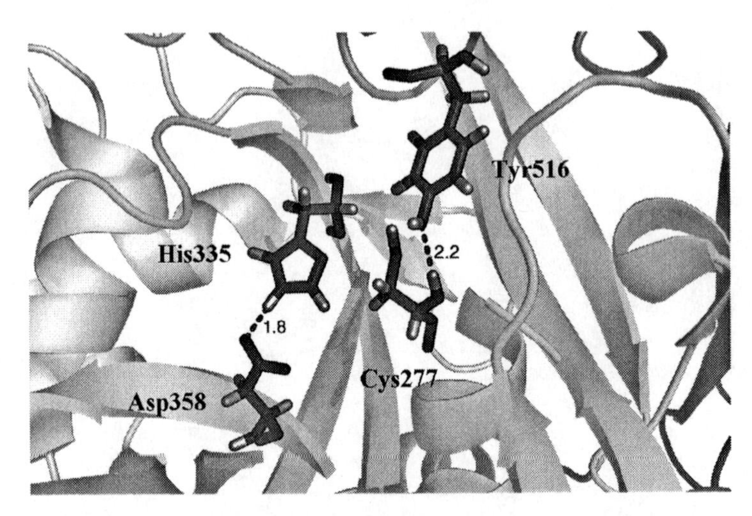

FIGURE 2. The catalytic site of TGase-2 in the "closed," inactive state. Shown is a close-up view of the catalytic triad Cys277, His335, Asp358 of TGase-2, as well as the regulatory residue Tyr516 within the inactive TGase-2 structure (PDB ID: 1KV3). The four indicated residues are displayed in stick mode and the hydrogen bonds are shown as dashed lines. All the surrounding residues are displayed in opaque ribbon mode.

Surprisingly, the inhibitor stabilized TGase-2 in a conformation that significantly differed from its "closed" state (Figure 3). The *N*-terminal superimposition of the "open" and "closed" TGase-2 structures revealed that the loops containing Tyr516 move away from the catalytic core region when the enzyme is in an active conformation, but with an unexpected swing angle of 120 degrees (Figure 4). How TGase-2 achieves these dramatic structure changes is not yet fully understood, but it is believed that certain allosteric regulatory factors play important roles in this process, which will be discussed in the following sections.

Another significant finding associated with the "open" structure of TGase-2 was the visualization of the "tryptophan tunnel." On the basis of the comparison of the active sites between TGase-2 and the structurally related cysteine proteases, Trp241 of TGase-2 was proposed to be a critical residue for catalyzing the transamidation reaction through its ability to stabilize the first transition-state intermediate [36, 37]. Consistent with this idea, Cys277 was found to be located within a hydrophobic tunnel bridged by Trp241 and

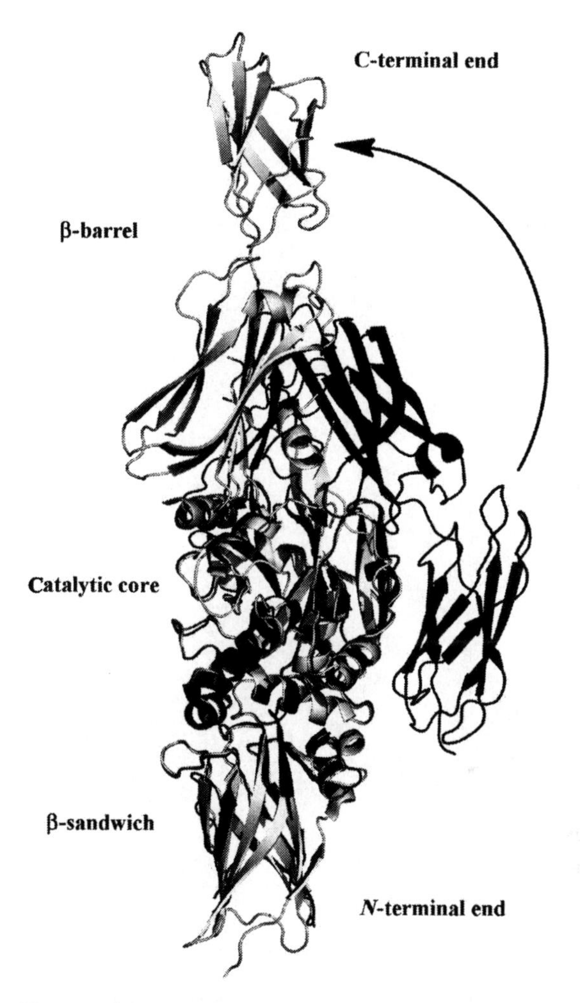

FIGURE 3. The structural superimposing and comparison of inactive TGase-2 to active TGase-2. The overall structure of inactive TGase-2 (PDB ID: 1KV3) is shown in dark gray color and the overall structure of active TGase-2 (PDB ID: 2Q3Z) is shown in light gray color. Both of them have a β-sandwich domain, a catalytic core domain and two β-barrel domains arranged from the *N*-terminal end to the *C*-terminal end. The *N*-terminal β-sandwich domains of these two structures are superimposed, highlighting the conformational change at the *C*-terminal end. The arrow indicates the 120° swing angle of the *C*-terminal β-barrel domains when TGase-2 changes from its inactive conformation to its active conformation.

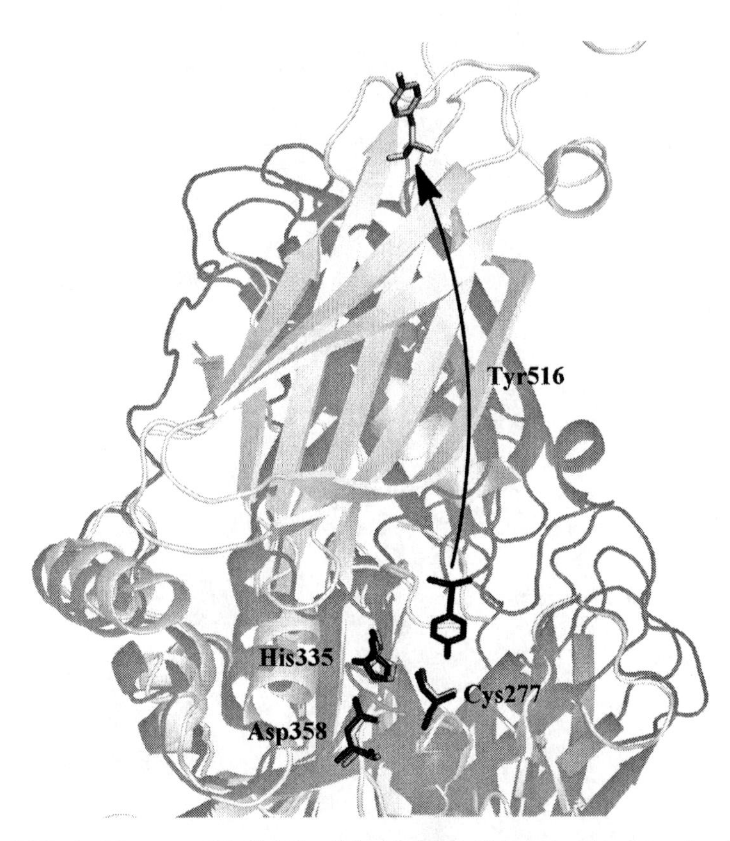

FIGURE 4. Movement of Tyr516 when TGase-2 changes from its inactive state to its active state. The catalytic triad Cys277, His335, and Asp358 of TGase-2, as well as the regulatory residue Tyr516 are displayed in stick mode, and all the other residues are drawn in opaque ribbon mode. The inactive conformation of TGase-2 is in dark gray, while the active conformation of TGase-2 is in light gray. These two structures are superimposed through the alignment of respective *N*-terminal β-sandwich domain as illustrated in Figure 3. The arrow indicates the movement of Tyr516 when TGase-2 changes from its inactive state to its active state.

Trp332, where the glutamine-containing substrate was proposed to enter from one side, while the lysine-containing substrate entered from the opposite side (Figure 5). Substitution of Trp241 with Ala or Gly caused a more than 300-fold decrease in the catalytic rate constant [37], demonstrating the critical role of this tryptophan tunnel for transamidation activity.

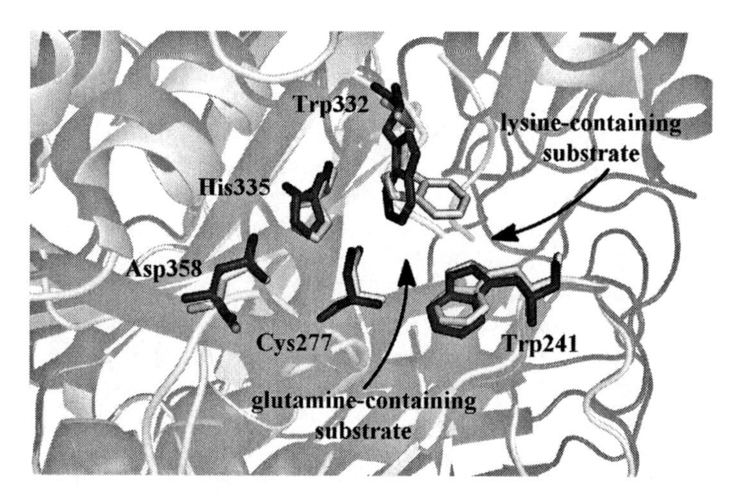

FIGURE 5. The "tryptophan tunnel" Trp241 and Trp332 bridges the access of the two substrates to the catalytic site of TGase-2. Trp241 and Trp332, as well as the catalytic triad Cys277, His335, and Asp358 of TGase-2, are displayed in stick mode, and all the other residues are drawn in opaque ribbon mode. Inactive TGase-2 is presented in dark gray, while active TGase-2 is presented in light gray, which are superimposed through the alignment of their respective *N*-terminal β-sandwich domains. The two arrows within the figure point out the approaching track of the glutamine-containing substrate and the lysine-containing substrate to TGase-2's catalytic site, respectively.

C. FUNCTION OF TGase-2 AS A G-PROTEIN

In 1992, Graham and his coworkers made the unexpected finding that rat TGase-2 was involved in receptor-coupled signal transduction by functioning as a classical GTP-binding protein (G-protein), specifically through its ability to mediate agonist-induced signaling events from the $\alpha_{1B/D}$-adrenergic receptor to its downstream effector phospholipase Cδ1 (PLCδ1) [38]. Subsequently, TGase-2 was shown to mediate the activation of Ca^{2+}-activated K^+ channels in vascular smooth muscle cells [39]. What made these findings particularly surprising was that the primary sequence of TGase-2 did not contain any of the conventional guanine nucleotide-interacting motifs featured in the small or large G-proteins. However, the intrinsic ability of TGase-2 to bind GDP/GTP and hydrolyze GTP to GDP argues that it contains a unique guanine nucleotide-interacting pocket that lacks obvious sequence similarity with those of the classical G-proteins.

This was confirmed when the three dimensional structure for the "closed" conformation of TGase-2 was determined, where one molecule of GDP was found to bind and cocrystalize with the inactive form of TGase-2. In large G-proteins, the Gα subunit often contains a helical domain adjacent to its guanine nucleotide-binding region to ensure the high affinity binding of guanine nucleotides, whereas small G-proteins lack the helical domain and instead use a bound Mg^{2+} ion to strengthen nucleotide-binding [40, 41]. However, neither of these classical features was present in the TGase-2 structure. Instead, surrounding the bound GDP molecule was a sequence of positively charged residues, which were assumed to contribute to nucleotide binding (Figure 6). These residues were located within the first β-strand and the following loop region of the first *C*-terminal β-barrel domain. Perhaps the most important of these residues was Arg580, which formed three hydrogen bonds with the guanine ring, the α-phosphate, and the β-phosphate of the GDP molecule, respectively. Substitutions at the corresponding position in rat TGase-2 (Arg579) resulted in as much as an 100-fold reduction in GTP-binding affinity, as determined by isothermal titration calorimetry

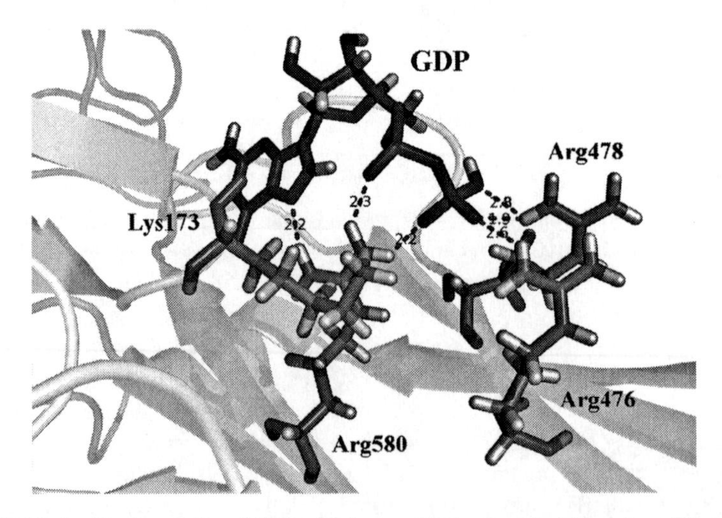

FIGURE 6. The interaction of GDP with four positively charged residues from TGase-2. Lys173, Arg476, Arg478, and Arg580 from inactive TGase-2 and the bound-GDP are shown in stick mode. The remaining residues are shown in opaque ribbon mode. The four positively charged residues are displayed in light gray, while GDP is drawn in dark gray. The hydrogen bonds are shown as dashed lines.

[42]. In addition to Arg580, two other residues, Arg476 and Arg478, also helped to properly position the GDP molecule in the pocket, through the formation of two hydrogen bonds between Arg478 and the β-phosphate group, and a single hydrogen bond contributed by Arg476 to the β-phosphate. The Arg476Ala and Arg478Ala substitutions attenuated the photoaffinity labeling of TGase-2 with radioactive GTP, thus indicating that these residues contributed to the binding of guanine nucleotides [42]. In addition to these three arginine residues that made direct contact with GDP, Lys173 was another positively charged residue that drew significant attention. Changing Lys173 to asparagine did not alter the ability of TGase-2 to bind guanine nucleotides, but it greatly impaired its GTP-hydrolytic capability [43]. A plausible mechanism underlying this effect was that Lys173 plays an essential role in catalyzing GTP hydrolysis. Specifically, the fact that the β-phosphate group of GDP interacts with Arg476, Arg478, and Arg580 led to the prediction that the γ-phosphate group of GTP needed to point away from the arginine pocket in order to avoid steric clash. Lys173 would presumably function to stabilize the transition-state for the GTP-hydrolytic reaction, because it is located close to the predicted position of the γ-phosphate group. An examination of the GTP hydrolysis-deficient Lys173Asn mutant in cells resulted in an interesting finding, namely, that the ability of TGase-2 to mediate signals from the α-adrenergic receptor to PLCδ was independent of its GTP-hydrolysis capability. This conclusion was demonstrated by the fact that the TGase-2 Lys173Asn mutant stimulated the production of similar levels of inositol phosphate (i.e., the product of PLCδ) as wild-type TGase-2 in COS-1 cells cotransfected with α_{1B}-adrenergic receptor. On the other hand, the Ser171Glu mutation in TGase-2, which disrupted GTP-binding capability, as well as the Arg580Lys mutation seriously impaired the production of inositol phosphates in COS-1 cells, indicating that the guanine nucleotide-binding activity of TGase-2 was necessary for mediating the signals from the α_1-adrenergic receptor [43].

D. ALLOSTERIC REGULATION OF TGase-2

1. Guanine Nucleotides Inhibit the Transamidation Activity of TGase-2

After the "open" structure of TGase-2 was determined, it was clear that the conformation of the enzyme active site did not exhibit an obvious change compared to the conformation of the active site when TGase-2 was in the "closed" state. Rather, the structure revealed that access to the active site was significantly hindered when TGase-2 is in the nucleotide-bound closed state, suggesting that guanine nucleotides serve as allosteric regulators of enzyme activity. This idea was first supported by the observation that the transamidation activity of rat TGase-2 in COS-1 cells was inhibited by the binding of guanine nucleotides [38]. Similarly, in vitro transamidation assays of human recombinant TGase-2 incubated with GDP or GTPγS, a nonhydrolyzable form of GTP, revealed a clear dose-dependent inhibition of transamidation activity by increasing concentrations of either GDP or GTPγS [44]. The structure for the closed state of TGase-2 shows that GDP binds to the first β-strand and adjoining loop within the first β-barrel domain and that this interaction helps to maintain TGase-2 in an enzymatically inactive state by stabilizing the strands and loops that block access to its transamidation catalytic site. As a result, guanine nucleotides not only play an essential role in mediating the actions of TGase-2 in α-adrenergic receptor-signaling, but also are important in the allosteric regulation of transamidation activity. Substitutions for Arg580, a critical residue for interacting with GDP as discussed above, abolished the ability of TGase-2 to bind guanine nucleotides and disabled their allosteric regulation of transamidation activity [42]. Interestingly, the overexpression in cells of TGase-2 mutants containing substitutions for the essential arginine, triggered cell death [23]. These results can not be simply explained by an inability of these TGase-2 mutants to exhibit guanine nucleotide-dependent regulation of their transamidation activity, because the TGase-2 Arg580Lys/Cys277Ala double mutant, which lacks both guanine nucleotide-binding capability and transamidation activity, caused cell death to the same level as the TGase-2 Arg580Lys mutant [23]. It was speculated that substitutions for Arg580 somehow affected the entire TGase-2 structure, rather than just influencing the guanine nucleotide-binding pocket. Consistent with this idea, native gel analysis revealed that substitutions for Arg580 resulted in TGase-2 adopting a semicompact conformation, which was clearly different from nucleotide-free TGase-2, GTP-bound TGase-2, or any other investigated TGase-2 mutant [45]. This unique conformation adapted by TGase-2 molecules containing substitutions at Arg580 might either selectively inactivate cellular proteins whose function were essential for cell survival or activate effector proteins that were sufficient to cause cell death. The ultimate answer awaits a high-resolution structural determination of a TGase-2 molecule containing a substitution for Arg580.

2. *Calcium Activates the Transamidation Activity of TGase-2*

Calcium ions serve as another crucial allosteric regulator of TGase-2. In contrast to the negative regulation of transamidation activity imparted by guanine nucleotides, Ca^{2+} strongly activates TGase-2's transamidation activity [46, 47]. Although Ca^{2+} was not cocrystalized with TGase-2 in either its open or closed structure, it is widely accepted that Ca^{2+} plays a key role in facilitating TGase's transition from a closed, inactive conformation to an open, active conformation. One piece of evidence that supports this claim came from small angle X-ray scattering (SAXS) experiments, which demonstrated a massive broadening of TGase-2's conformation in the presence of Ca^{2+}, as indicated by a dramatic 0.8-nm increase in the radius of its gyration after the binding of Ca^{2+} [35]. The Ca^{2+}-induced extension of TGase-2's structure was also suggested from proteolysis studies, in which the addition of Ca^{2+} promoted the further degradation of the large proteolytic fragments of TGase-2 into smaller peptides, thus suggesting that additional proteolytic sites on TGase-2 were exposed by binding Ca^{2+} [35].

On the basis of what we know about its open structure, the most dramatic change that occurs when TGase-2 becomes catalytically competent is the movement of the two *C*-terminal β-barrel domains, as opposed to the changes that occur to the central catalytic domain. Calcium binding was proposed to be the primary driving force of these large structural movements, which was supported by molecular dynamic simulation analysis. Specifically, the TGase-2 structure, when stimulated with Ca^{2+} for 50 ps, exhibited an unfolding event within the helix motif comprising Glu451 to Ala459, which is located adjacent to the hinge region between the first β-barrel domain and the catalytic domain. As a consequence, these two domains start to move away from each other, leaving a large gap between them. This conformation was believed to represent an initial snapshot of the entire transition that Ca^{2+}-stimulated TGase-2 undergoes from the closed state to the open state [35]. Given that Ca^{2+} is able to induce such a significant change within the TGase-2 structure, which leads to the activation of transamidation activity, it was of interest to learn more about the mechanistic basis of this important regulatory event. To develop a better understanding of how Ca^{2+} affects TGase-2 structure and activity, several groups, including our own, compared the sequence and structure of TGase-2 with other transglutaminase family members that bind Ca^{2+}, with the aim of identifying the Ca^{2+}-binding sites on TGase-2. Computer modeling of TGase-2 with human factor XIIIA pointed out three highly-conserved acidic regions, shared in both enzymes, which were assumed to represent Ca^{2+}-binding sites [35]. A similar comparison between TGase-2 and Ca^{2+}-bound TGase-3 was carried out by our group, where we proposed three pairs of Ca^{2+}-interacting residues within TGase-2 [44]. Introducing substitutions at each of these candidate Ca^{2+}-binding sites strongly compromised Ca^{2+}-induced transamidation activity, while having little effect on the guanine nucleotide-binding activity of TGase-2. This suggests that these changes caused a specific disruption of Ca^{2+}-initiated effects rather than causing a general disruption of protein folding. Nonetheless, mutations at the different putative Ca^{2+}-binding sites gave rise to distinct effects. Specifically, substitutions at Ca^{2+}-binding site 2 (Asp306Ala/Asn310Ala) had the most severe effects, resulting in no detectable transamidation activity at all Ca^{2+} and substrate concentrations that were examined, whereas changes at site 1 (Asn229Ala/Asp233Ala) and site 3 (Asn398Ala/Glu447Ala) shifted the dose-response curves for Ca^{2+} in the transamidation assays. The dramatic effects caused by substitutions at site 2 can be explained by the earlier structural studies [48] on TGase-3 showing that the binding of Ca^{2+} to what corresponds to site 2 caused the movement of a nearby loop and therefore opened the channel for substrates to access the active site. This argument perfectly matched the superimposing analysis of the open and closed structures for TGase-2, showing that Ca^{2+}-binding site 2 moved away with its connected loop (Figure 7A), thereby exposing the "tryptophan tunnel" that facilitated the interaction between the catalytic triad and transamidation substrates. The residues at Ca^{2+}-binding sites 1 and 3, on the other hand, did not exhibit obvious changes between the open and closed structures (Figure 7B), which probably explains their limited effect on Ca^{2+}-induced transamidation activity. In conclusion, site 2 was proposed to be the critical "calcium switch" that drives TGase-2 to the active state, whereas sites 1 and 3 might cooperate to assist in this process. However, this still does not explain all of the results regarding the actions of Ca^{2+} on TGase-2 function. For example, we have found that Ca^{2+} can weaken the binding of guanine nucleotides to TGase-2 by as much as 5-fold (unpublished data) as determined by the incorporation of the fluorescent nucleotide analog, BODIPY-GTPγS. However, it has not been possible to attribute this regulatory effect to any of the three Ca^{2+}-binding sites that have been identified thus far. At the same time, equilibrium dialysis experiments performed on erythrocyte transglutaminase suggested that it contained as many as six Ca^{2+}-binding sites [49]. Consequently, more efforts are needed to establish the number of Ca^{2+}-binding sites on TGase-2, their affinities, and the full spectrum of their possible regulatory effects.

E. INTERACTION BETWEEN FIBRONECTIN AND TGase-2

As early as 1988, fibronectin was already discovered as a binding partner for members of the transglutaminase family [50]. Native gel electrophoresis

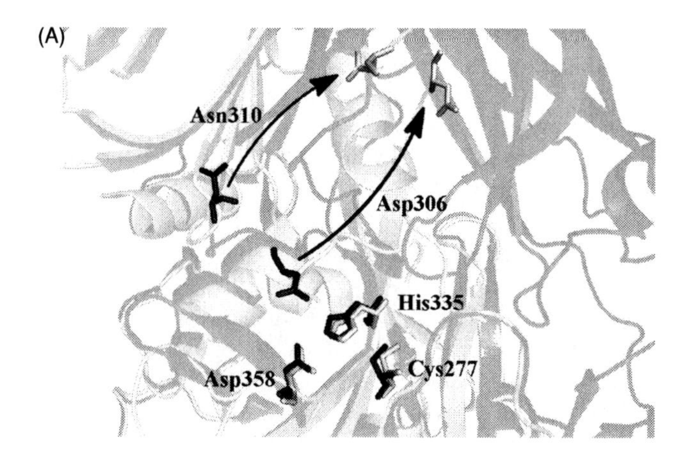

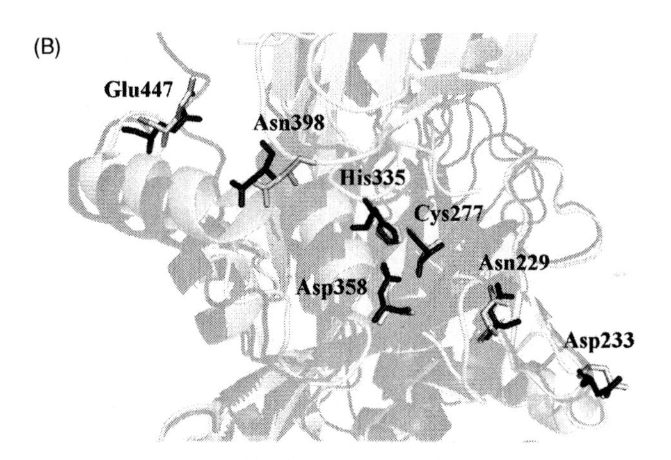

FIGURE 7. Comparison of the three Ca^{2+}-binding sites between inactive and active TGase-2. Ca^{2+}-binding site 1 (Asn229/Asp233), site 2 (Asp306/Asn310), and site 3 (Asn398/Glu447), as well as the catalytic triad of TGase-2, Cys277, His335, and Asp358, are shown in stick mode. All the other residues are shown in opaque ribbon mode. The inactive and active TGase-2 are displayed in dark gray and light gray, respectively, which are superimposed through the alignment of the *N*-terminal β-sandwich domain. (A) The movement of Ca^{2+}-binding site 2 (Asp306/Asn310) is indicated by arrows when TGase-2 changes from its inactive state to its active state. (B) Ca^{2+}-binding site 1 (Asn229/Asp233) and site 3 (Asn398/Glu447) exhibit few changes when the transition from the inactive TGase-2 to active TGase-2 occurs.

revealed that erythrocyte transglutaminase underwent a significant shift in mobility upon exposure to plasma, which contains a large amount of fibronectin. Depletion of fibronectin from plasma by gelatin affinity column chromatography compromised the mobility shift of erythrocyte transglutaminase on native gels. Later, it was found that TGase-2 exhibited a similar binding affinity for fibronectin in vitro, with a stoichiometry of binding of 2-mol TGase-2 (80 kD) per mol of fibronectin dimer (440 kD) [51]. This result matched very well with subsequent in vivo studies, where a 600-kDa TGase-2-containing complex was observed when eluting lysates from HeLa cervical carcinoma cells through a gel filtration column [52]. Although the cellular function of this binding complex was not yet fully understood, it was well accepted that the tight association of TGase-2 and fibronectin promoted wide-ranging effects on cell–matrix interactions, including the regulation of cell adhesion and migration, matrix turnover, and adhesion-dependent signaling [28, 53].

The multifunctional role of this complex drew attention to the question of how TGase-2 interacts with fibronectin. Efforts to address this issue led to a series of experiments focusing on the identification of the fibronectin-binding motif on TGase-2. First, controlled digestion of purified TGase-2 by endo-proteinase Lys-C revealed that the association with fibronectin relied on the 28-kDa *N*-terminal domain of TGase-2. Further digestion with another endo-proteinase enzyme, Glu-C, argued that the amino-terminal seven residues of TGase-2 were critical for fibronectin binding [54]. This conclusion was supported by in vivo studies, showing that an *N*-terminally truncated TGase-2 molecule failed to be deposited into the extracellular matrix [55]. An additional fibronectin-binding motif has been identified by GST-pull down assays, where different GST-tagged TGase-2 constructs were generated to examine their interactions with purified fibronectin. An extended hairpin region formed by the antiparallel β strands 5 and 6 within the *N*-terminal domain of TGase-2, ranging from amino acids 88 to 106, appears to contain the second fibronectin-binding motif [56]. Synthesized peptides corresponding to this region severely impaired the interaction of TGase-2 with fibronectin and decreased TGase-2-dependent cell adhesion and spreading. Two adjacent aspartic acid residues within this region, Asp94 and Asp97, seemed to be the critical contact points for fibronectin because substitution of these residues significantly inhibited the formation of the TGase-2-fibronectin complex. Although the direct interaction between TGase-2 and fibronectin in vitro seemed to be well understood, how these two proteins associated with each other in vivo was still unclear.

It has been shown that in some cell lines TGase-2 worked as a coreceptor for other cell surface receptors to better recognize matrix fibronectin, including β1 and β3 integrins, as well as the heparan sulfate proteoglycan, syndecan-4 [24, 57–60]. TGase-2 bridged the association of these cell surface receptors with fibronectin, intensified their ability to trigger cellular signaling activities, and promoted cell adhesion and migration. The coreceptor function of TGase-2 did not require its transamidation activity because the transamidation-deficient mutant of TGase-2 led to the same cellular effects as wild-type TGase-2 [24, 57]. However, in some other cases, the transamidation activity of TGase-2 did play a role because TGase-2 had the ability to cross-link fibronectin into nonreducible, high-molecular weight polymers. For instance, TGase-2 cross-linked fibronectin polymers on the surface of hepatocytes or osteoblasts, stimulating integrin-signaling and cell attachment [61, 62]. Even more interestingly, it was found that invading human glioblastomas demonstrated increasing levels and activities of TGase-2 together with a marked build-up of condensed fibronectin strands around tumor tissue, a distinguishing feature compared to normal brain stroma. Inhibition of TGase-2 activity in the extracellular matrix significantly affected the formation of fibronectin strands and sensitized brain tumors to the chemotherapeutic reagent BCNU [63]. Thus, the ability of TGase-2 to cross-link fibronectin in the extracellular environment may play an important role in the progression of brain cancer.

III. ROLE IN CELL SURVIVAL

A. THE ROLE OF TGase-2 AS A PROTECTIVE FACTOR IN CELLULAR DIFFERENTIATION

While there is now a growing appreciation for TGase-2 in the progression of various types of human cancer [27, 32, 33, 63–67], as little as nine years ago this was far from the case. In fact, at that time TGase-2 was widely considered as a protein that was important for either triggering the induction of apoptosis [18, 68–73] or for promoting cellular differentiation [14, 74–76], two cellular processes that are paradoxical with the development of the oncogenic state. The evidence implicating TGase-2 as a regulator of cell death stemmed largely from a line of research being conducted at that time aimed at understanding the molecular mechanisms that underlie the onset/progression of a class of neurodegenerative disorders, in particular Alzheimer's disease (AD), Huntington's disease (HD), and Parkinson's disease (PD), that are characterized by the presence of insoluble proteinaceous aggregates [77]. In each of these disease states, the size and number of protein aggregates detected in the brains of affected individuals tend to increase with the severity of the disease, and protein aggregation has been suggested to directly contribute to neuronal cell dysfunction and death [78, 79]. Interestingly, high levels of TGase-2 expression and its enzymatic transamidation activity have been identified as hallmarks of AD, HD, and PD [80–84]. Moreover, TGase-2 was shown to cross-link components of the protein aggregates associated with each of these diseases [81, 83, 85–87]. These findings, when coupled with the fact that inhibiting TGase-2-catalyzed transamidation activity suppresses protein aggregation and extends the lifespan of mouse models of HD and PD [85, 87, 88], provided some of the earliest and most intriguing data, suggesting that TGase-2 participated in the induction of cell death through the aberrant cross-linking of substrate proteins.

In addition to TGase-2 being upregulated in response to a variety of apoptosis-inducing cellular stresses (i.e., stressful conditions encountered by cells in neurodegenerative diseases), increases in its expression and activation were also shown to accompany cellular differentiation. For example, treatment of the human neuroblastoma cell line SH-SY5Y with the differentiation agent retinoic acid (RA) arrests their growth and stimulates the formation of neurite extensions [89, 90], making them a useful cell type to identify proteins that are important for neuronal differentiation. TGase-2 was identified as one such protein [14, 47, 74]. Not only was TGase-2 expression and transamidation activity found to be significantly enhanced in SH-SY5Y cells by RA stimulation, but inhibiting RA-induced TGase-2 expression using an antisense construct blocked the ability of RA to arrest cell cycle progression, as well as to promote neurite extension [74]. The study then went on to show that the ectopic expression of wild-type TGase-2 in SH-SY5Y cells, but not a transamidation-defective form in which a cysteine residue within the "catalytic triad" of the enzyme active site was changed to a valine (TGase-2 Cys277Val), was sufficient to cause these cells to spontaneously undergo differentiation. Thus, these findings demonstrated that induction of TGase-2 expression and activity in cells does not invariably result in cell death, but rather TGase-2 may in certain contexts also promote cellular differentiation.

Our laboratory's interest in the functional consequences of TGase-2 expression and activation started at about the same time that accumulating evidence, like those findings described above, were leading to the general belief that TGase-2 participated in biological processes that resulted in apoptosis and in a few instances limited cell growth through arresting the cell cycle. We were investigating the effects that the naturally occurring retinoid RA, as well as other synthetic retinoid analogs had on cell fate decisions. This question was particularly relevant at that time, since the growth inhibitory and differentiation activities associated with RA were prompting several research groups to examine the use of RA or RA-like synthetic derivatives as potential treatments for different types of human cancers ranging from prostate cancer to leukemia [91–94]. Our laboratory, as well as other groups, was able to show that treatment of the human leukemia cell line HL-60 with RA did indeed inhibit their growth [19, 95, 96]. However, when the same cell line was cultured in the presence of a synthetically derived analog that differs from RA by only the addition of a phenol ring to its carbon backbone (see Figure 8), and is referred to as *N*-(4-hydroxyphenol)retinamide (HPR or fenretinide), instead of inducing cell cycle arrest as is the case for RA, HPR elicited a potent apoptotic response in these cells [19, 97–99]. Likewise, HPR was also found to be much more effective than RA at inducing cell death in NIH-3T3 mouse fibroblasts [19], indicating that the abilities of these retinoids to differently influence cell fate decisions were not unique to a specific cell lineage.

How the highly related retinoid analogs RA and HPR gave rise to opposing cellular outcomes, such that RA promotes cell survival while HPR induces apoptosis, was puzzling and led us to consider possible mechanisms

Retinoic acid (RA)

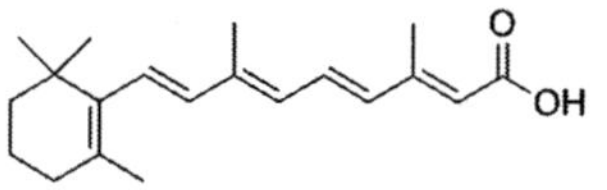

***N*-(4-hydroxyphenol)retinamide (HPR)**

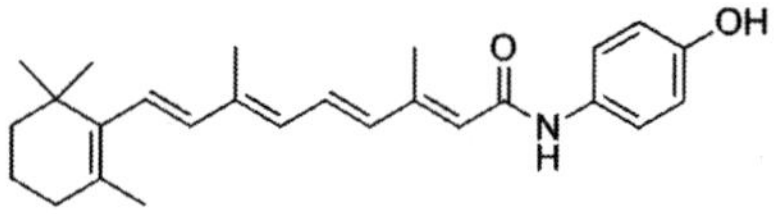

FIGURE 8. The chemical structures of retinoic acid (RA) and its synthetic derivative *N*-(4-hydroxyphenol)retinamide (HPR).

that could account for these effects. We obtained a clue as to how this may occur when we examined the effectiveness of HPR at inducing apoptosis in both HL-60 leukemia cells and NIH-3T3 fibroblasts that had first been stimulated with RA for different lengths of time [19]. While short-term RA pretreatments (up to 6 hours) caused little, if any, change in the ability of HPR to induce apoptosis in these cells, longer pretreatments with RA did have effects. For example, HL-60 and NIH-3T3 cells that had been preincubated with RA for two days prior to being challenged with HPR became highly resistant to the apoptotic-inducing actions of HPR. Since pretreating the cells with RA for extended lengths of time (on the order of days) was required in order to see the protective effects against HPR-induced apoptosis afforded by RA treatment, we began to favor the idea that RA-signaling might induce the expression of a distinct gene or set of genes that is crucial for ensuring cell viability during cellular differentiation.

While searching for potential candidate genes that were selectively upregulated by RA, we considered TGase-2, since the induction of its expression was already known to be tightly coupled to the effects of RA-signaling in HL-60 cells [95, 96], as well as some other cell types (i.e., SH-SY5Y cells) [14, 47, 74]. In unstimulated HL-60 leukemia cells and NIH-3T3 mouse fibroblasts, TGase-2 is expressed at very low or almost undetectable levels [19, 70, 100]. However, following the addition of RA to the cells, there was a progressive increase in the levels of TGase-2 expression, with maximal expression being reached in both cell lines between 2 and 3 days after RA treatment [19]. We then questioned how treating cells with HPR would influence TGase-2 expression. In contrast to RA, HPR was found to be completely ineffective at increasing TGase-2 expression levels beyond the nearly undetectable levels measured in unstimulated HL-60 or NIH-3T3 cells. These findings not only identified the induction of TGase-2 expression as a distinguishing feature between the signaling capabilities of RA and HPR, but they also opened the door to the possibility that TGase-2 may be important for mediating the ability of RA to protect cells from HPR-induced apoptosis. This indeed appears to be the case, as ectopically expressing TGase-2 in NIH-3T3 cells was as effective as RA treatment at protecting the cells from challenges with HPR. Perhaps even more telling of the potential significance that TGase-2 has in promoting cell survival came from another experiment in which its transamidation activity was blocked using monodansylcadaverine (MDC), a competitive inhibitor of the transamidation reaction, during the RA-induced differentiation of HL-60 cells. Rather than undergoing cell cycle arrest and stimulating the formation of mature granulocytes as occurs with RA treatment alone, RA-treated HL-60 cells cultured in the presence of MDC instead underwent apoptosis. These findings suggested for the first time that TGase-2 can serve as a protective factor, particularly in cells that are undergoing differentiation. In these instances, TGase-2 expression and its corresponding transamidation activity are induced by RA to ensure that cells that are exiting the cell cycle as part of the cellular differentiation process survive without undergoing apoptosis.

B. TGase-2 AND CHEMOTHERAPY RESISTANCE

In addition to establishing a novel and unexpected role for TGase-2 in maintaining the viability of HL-60 leukemia cells and NIH-3T3 fibroblasts undergoing RA-induced cellular differentiation [19], these initial findings also laid the foundation for subsequent investigations into whether TGase-2 might provide a protective effect to cells under a number of other conditions. One area of study that has been developing rapidly over the last few years is the relationship between TGase-2 and drug resistance in human cancer cells. Screening several different types of primary human tumor samples by immunohistochemistry for TGase-2 expression has revealed that its expression levels are increased in a significant percentage of breast, pancreatic, ovarian, and brain tumors [19, 63, 66, 67, 101–103]. Moreover, breast and pancreatic cancer cell lines that are resistant to chemotherapies also tend to have the highest levels of TGase-2 expression and transamidation activity [31, 64, 67, 104, 105], suggesting that TGase-2 may contribute to the drug-resistant phenotypes exhibited by some cancer cells. A case in point is the highly aggressive and chemotherapy-insensitive human breast carcinoma cell line MDA-MB231. These cells constitutively express high levels of TGase-2, and when extracts from the cells were assayed for transamidation activity, it was determined that TGase-2 was active [27, 31]. Knocking-down TGase-2 expression in these cancer cells using small interfering-RNAs (siRNAs) was found to be sufficient to sensitize them to the chemotherapeutic agent doxorubicin [27]. Likewise, treating MDA-MB231 cells with the transamidation inhibitor MDC also made the cells susceptible to doxorubicin-induced apoptosis [31], indicating that the cross-linking activity of TGase-2 is likely responsible for mediating its protective effects.

The notion that TGase-2 provides a survival advantage to human cancer cells was further strengthened by demonstrating that the prosurvival signaling capabilities of the epidermal growth factor (EGF)-receptor are dependent on TGase-2 in some cancer cell lines [31, 32]. The EGF receptor is a cell surface receptor tyrosine kinase that has been extensively investigated for its roles in promoting tumorigenesis [106–108]. Upon binding EGF, the EGF receptor becomes activated and initiates the activation of numerous intracellular signal transduction pathways that work in concert to directly influence cellular processes, as well as to induce the expression of a specific set of genes that promote cancer cell growth and survival. The interplay that exists between EGF-receptor signaling and TGase-2 in cancer cells was discovered when it was shown that the levels of TGase-2 expression and its corresponding transamidation activity were markedly enhanced in the human breast cancer cell line SKBR3 after EGF stimulation [31]. Consistent with the findings that RA-induced TGase-2 activation protects HL-60 leukemia cells and NIH-3T3 fibroblasts from apoptotic challenges [19], inhibiting the EGF-stimulated transamidation activity of TGase-2 using MDC also abolishes the ability of the EGF receptor to protect cells from doxorubicin-induced apoptosis, as well as from serum-deprivation-induced apoptosis [31, 32]. Although it is currently unclear how often and in which cancer cell types the EGF receptor elicits a survival response by upregulating TGase-2 expression and activation, two additional pieces of evidence suggest that this signaling mechanism is not limited to just SKBR3 breast cancer cells. First, it was shown that EGF stimulation was capable of potently inducing the expression and activation of TGase-2 in two additional breast cancer cell lines, namely (1) BT-20 and (2) MDA-MB468 cells [31]. Second, TGase-2 was identified in a screen as one of the most highly upregulated genes in a U87-MG brain tumor cell line that exogenously expresses a constitutively active mutant form of the EGF receptor [109]. Thus, the coupling of EGF-receptor signaling to the induction of TGase-2 expression and activation may be a relatively common occurrence in human cancers, particularly in those cancer cells whose oncogenic potential is dependent on EGF-receptor signaling activities. Taken together, these findings suggest that the upregulation of TGase-2 expression and activation in human tumors represent a critical step during cancer progression where tumor cells are transitioning to a high grade, drug-resistant phenotype.

C. HOW DOES TGase-2 PROMOTE CELL SURVIVAL?

In those cases where TGase-2 has been implicated as a protective factor in both normal and cancer cells, its enzymatic transamidation activity has been shown to be essential [19, 31, 32, 63, 64, 104, 109]. As a result of these findings, it should come as no surprise to learn that an important emphasis in the field has been to determine the identities of those proteins that are cross-linked by TGase-2 to elicit survival responses in cancer cells. While virtually hundreds of proteins have been shown to serve as putative substrates for the transamidation activity of TGase-2 using in vitro and high-throughput screening approaches [110, 111], determining which, if any, of these proteins are physiologically relevant substrates in different cellular contexts has proven to be far more challenging. However, recent results have highlighted two proteins that, when covalently modified by TGase-2, can have significant implications on cell viability. The first is caspase-3 (also referred to as CPP32, YAMA, and apopain) [112], an important component of the apoptotic-inducing machinery in the cell [113]. Caspase-3, as well as the other members of the caspase family, is a cysteine protease that is expressed in cells as an inactive, proform of the enzyme. However, when a cell commits to undergoing programed cell death or apoptosis, it initiates an irreversible activation of the caspase pathway that involves a series of cleavages that result in the conversion of inactive procaspases to active cysteine proteases. Once activated, caspase-3 catalyzes the specific cleavage of several key proteins that are essential for maintaining cell viability, resulting in cell death.

The findings showing that TGase-2 can cross-link caspase-3 as a way to promote cell survival came from a study that was investigating how a HCT-116 colon carcinoma cell line that lacked Bax expression was insensitive to the apoptotic-inducing effects of thapsigargin (THG), an inhibitor of the endoplasmic reticulum Ca^{2+}-ATPase [112]. The authors found by Western blot analysis that exposing Bax-deficient HCT-116 cells to THG for increasing lengths of time resulted in the progressive appearance of unusually high molecular weight forms of procaspase-3. Since the catalytic activity of TGase-2 has been shown to induce the oligomerization of several different proteins [81, 83, 85, 87, 114], the ability of TGase-2 to induce the oligomerization of procaspase-3 was examined as a possible explanation for the higher molecular weight forms of procaspase-3 detected in the cells treated with THG. Not only was it shown that THG potently induced TGase-2 expression and activation in the Bax-deficient HCT-116 colon carcinoma cells, but a recombinant form of procapase-3 could be cross-linked by TGase-2 in vitro [112]. This raised the question of whether blocking the ability of TGase-2 to cross-link proteins in THG-treated HCT-116 cells would influence the formation of the higher molecular weight species of procaspase-3, as well as sensitize these cells to THG. Interestingly, knocking-down TGase-2 expression using siRNA approaches or incubating HCT-116 cells with the inhibitor MDC caused a significant reduction in the ability of THG to induce the formation of the larger forms of procaspase-3. Coinciding with this effect, MDC treatment also made the cells susceptible to THG-induced apoptosis, indicating that TGase-2 affords a protective effect to HCT-116 colon carcinoma cells from THG-induced apoptosis by cross-linking an important component of the cell death machinery, namely caspase-3, into a nonfunctional oligomer.

The inhibitor of κBα (IκBα), a regulatory protein in the Rel/nuclear factor (NF)κB survival pathway, is the second example of a protein that can be cross-linked by TGase-2 to potentiate cell survival [66]. The central player in the NFκB survival pathway is NFκB itself, a transcription factor that can be rapidly and transiently activated by a variety of growth factors, and especially in response to harmful cellular stresses, to induce the expression of specific genes that promote cell survival, as well as cell growth [115, 116]. The transcriptional activity of NFκB in unstimulated cells is normally kept in check because NFκB is sequestered in the cytoplasm, where it is found in a complex with its negative regulator IκBα. For NFκB to become activated by extracellular stimuli, a complex sequence of events must take place that results in the phosphorylation of IκBα on two serine residues contained within its regulatory domain. This phosphorylated form of IκBα is then ubiquitinated and targeted for degradation in the proteosome. Once IκBα has been degraded, NFκB is free to translocate from the cytoplasm to the nucleus where it functions to regulate the expression of various genes. While many of the genes upregulated by NFκB promote cell survival and cell growth responses, NFκB also rapidly induces the expression of its own negative regulator IκBα, ensuring that NFκB activity is tightly controlled in normal cells.

Aberrant activation of the NFκB survival pathway has emerged as a hallmark of human cancers [116–119]. Moreover, inactivating NFκB by siRNA knockdown or using NFκB inhibitors has been shown to block the growth and survival advantages associated with numerous cancer cell lines [116, 120], suggesting that deregulation of the NFκB pathway may represent an important step during cancer progression. While searching for the mechanisms used by cancer cells to excessively stimulate NFκB activation, a connection with TGase-2 emerged. It was noted that a strong correlation existed between the levels of TGase-2 expression and the amount of constitutive NFκB activity detected in aggressive and drug-resistant forms of breast cancer cells (i.e., MDA-MB231 cells), malignant melanoma cells (i.e., A375 cells), and pancreatic cancer cells (i.e., Panc-28 cells), as well as in primary pancreatic tumors [66]. When the transamidation activity of TGase-2, associated with the Panc-28 pancreatic cancer cells, was inhibited by knocking-down TGase-2 using siRNAs or treating the cells with MDC, there was a corresponding reduction in NFκB activity, indicating that NFκB activation in Panc-28 cells is dependent on the catalytic function of TGase-2.

So how does the transamidation activity of TGase-2 stimulate NFκB activity? It appears to most likely involve the ability of TGase-2 to cross-link IκBα, the critical negative regulator of NFκB activity. In support of this idea, TGase-2 was found to induce the covalent oligomerization of IκBα both in vitro and within the Panc-28 cell line [66, 121]. Interestingly, the recombinant form IκBα that was cross-linked by TGase-2 was shown to bind NFκB more weakly than its non-cross-linked counterpart, suggesting that this TGase-2-induced modification of IκBα in cells functionally inactivates IκBα, resulting in the constitutive activation of NFκB [121]. In line with these findings, a separate study looking at how TGase-2 provides a protective effect to ovarian cancer cells challenged with the chemotherapeutic agent cisplatin also concluded that the TGase-2-stimulated chemo-resistance seen in these cancer cells was dependent on its ability to stimulate the activation of the NFκB survival pathway [104]. Thus, it appears that the survival effects of TGase-2 are coupled to NFκB activation in a number of cancer types.

Determining the mechanisms that give rise to drug resistance in cancer cells is extremely important for the development of more effective strategies to treat patients with advanced forms of cancer. The increasing amounts of evidence showing that TGase-2 expression and activation is frequently upregulated in human cancers as a means of protecting cells from harmful cellular stresses [31, 63, 64, 67, 104, 105] has provided significant insights into how cancer cells become insensitive to chemotherapies. Thus, targeting TGase-2 itself or its associated effects may prove to be an important strategy to treat cancer patients with advanced and/or chemotherapy-insensitive forms of the disease.

IV. THE ROLE OF TGase-2 IN CANCER CELL MIGRATION AND INVASION

Directional cell migration is a fundamental cellular process that is required for proper embryonic development, tissue homeostasis, and immune responses in animals [122, 123]. In order for a cell to efficiently migrate, it must coordinate the activation and distribution of signaling molecules with changes in its actin-cytoskeleton network and plasma membrane. Although directional cell migration is a highly complicated, cyclic process that is still far from being completely understood, the general principles underlying cell migration are known. To migrate, a cell must first take on a polarized morphology, such that it has a front or leading edge (for forward movement) and a back (for retraction). Actin polymerization-dependent protrusions are then extended from the leading edge and make contact with the extracellular matrix. These newly formed sites of cell adhesion function to anchor the protrusions, allowing the body of the cell to shift forward. Cell contraction at the back end of the cell, as an outcome of actin bundling and depolymerization, completes one cycle of the cell migration process. By repeating this sequence of events, a cell moves from one site to another.

Aside from its physiological relevance, cell migration is also a major factor in the development or progression of several different human pathologies, ranging from cardiovascular disease to cancer [122]. For example, during tumor progression a subset of the cancer cells found within the primary tumor mass often acquire the ability to move into or invade neighboring normal tissue. Tumors that exhibit this type of invasive behavior tend to be more aggressive in nature and correlate with a poor patient prognosis. Interestingly, TGase-2 has recently been implicated as an important contributor to the invasive phenotype of human cancer cells [27, 33, 65, 67, 103, 124, 125]. In this section, we will highlight some of the recent findings that have shed light on the emerging role for TGase-2 in promoting cancer cell migration and invasion.

One of the earliest indications that TGase-2 is associated with cancer cell migration and tumor dissemination came from proteomic screens performed on primary and metastatic human lung and breast tumors, where TGase-2 was identified as one of only eleven proteins that were consistently upregulated in metastatic tumors [126]. These findings were then followed up with a study that investigated the possible link between the constitutively expressed and activated TGase-2 found in the human breast cancer cell line MDA-MB231 and the excessive migration and invasive activity exhibited by these cells [27]. Experiments performed comparing the migration and invasive capability of the parental MDA-MB231 cells to a subline that was selected for its greatly reduced levels of TGase-2 expression, revealed that TGase-2 was required for MDA-MB231 cells to exhibit these aggressive properties. The authors then went on to show that TGase-2 was present at the plasma membrane of MDA-MB231 cells where it was bound to β1, β4, and β5 integrins, members of a family of cell surface receptors that primarily function to attach cells to specific components in the extracellular matrix (i.e., fibronectin) and to promote cell migration [127, 128]. TGase-2 has been shown to interact with specific members of the integrin family in a variety of normal and cancer cell types [24, 58, 103], suggesting that its ability to interact with proteins that function in cell attachment may likely contribute to its ability to stimulate cancer cell migration and invasion.

Signals that originate outside of the cell can play critical roles in directional cell motility. For example, simply adding growth factors like EGF to the culturing medium of normal fibroblasts or to HeLa cervical carcinoma cells can induce or enhance the ability of these cells to migrate [129, 130]. These findings, combined with the fact that accumulating evidence suggests that the EGF receptor contributes to tumor invasion and metastasis [131–133], have generated a considerable amount of interest in determining the molecular mechanisms that link EGF-receptor activation to the changes that cells must undergo in order to allow them to migrate and invade.

Our laboratory has recently implicated TGase-2 as an important mediator of EGF-stimulated cancer cell migration and invasive activity [33]. These studies stemmed from our long-standing interest in understanding how activation of the EGF receptor both initiates and orchestrates intracellular signaling events that give rise to cellular phenotypes that are characteristic of the oncogenic or transformed state. In this particular case, we set out to identify proteins that coupled the ability of the activated EGF receptor to give rise to an enhanced migratory and invasive capability in the human cervical carcinoma cell line HeLa. HeLa cell cultures exposed to EGF acquire the distinctive polarized cell morphology associated with actively migrating cells; with each cell having formed both a front or leading edge and a back. Since proteins that are localized to leading edges have consistently been shown to contribute to cell motility [122, 123], we began to look for proteins that accumulated at the leading edges of EGF-stimulated HeLa cells as a means to identify novel participants in cell migration. Because of the earlier findings showing that TGase-2 was vital for the aggressive phenotype exhibited by MDA-MB231 breast cancer cells [27], we considered TGase-2 as a possible candidate protein. Immunofluorescent studies carried out on untreated HeLa cells revealed that TGase-2 was primarily expressed throughout the cytoplasm of these cells [33]. However, EGF treatment caused a dramatic change in the subcellular localization of TGase-2, such that it could be readily detectable in the leading edges of cells. Moreover, we found that EGF promotes the activation of TGase-2 in HeLa cells, without having a significant effect on its expression, suggesting that TGase-2 might play an important role in EGF-promoted cell migration. Indeed, treating HeLa cells with the TGase-2 inhibitor MDC, or knocking-down TGase-2 expression by siRNA, potently inhibited the migration- and invasive-promoting effects of EGF. Interestingly, however, MDC treatment appeared to have no effect on serum-induced HeLa cell migration, raising the possibility that TGase-2 may not play a general role in cell migration, but rather is a specific component of the EGF-signaling pathway that promotes the motility of these cancer cells.

While these findings identify TGase-2 as a crucial regulator of EGF-stimulated cell migration and invasion, they also raise a number of additional questions. The foremost of these being, what are the proteins that are covalently modified by TGase-2 to promote cell migration and/or invasion? It is especially tempting to speculate that TGase-2 is recruited to the leading edges of cells in order to cross-link another protein or specific set of proteins that are also present at the leading edge during cell migration. The cross-linking of these proteins by TGase-2 would most likely affect their function in such a way as to allow EGF-stimulated cell migration to occur. However, we also recognize the possibility that TGase-2 might contribute to the development of the invasive/metastatic phenotype in cancer cells through additional mechanisms. In support of this idea, the aggressive nature of ovarian cancers was found to be coupled to the ability of TGase-2 to regulate gene expression [125]. More specifically, it was shown that TGase-2 stimulated the expression of the transcriptional repressor zinc finger E-box binding homeobox 1 (Zeb1) through an NFκB-dependent mechanism. Since Zeb1 functions to limit the expression of proteins that are important for maintaining tissue homeostasis (i.e., E-cadherin) [134], ovarian tumors that overexpress Zeb1 tend to disrupt tissue integrity, increasing the invasive and metastatic potential of these tumors.

V. ROLE IN ANCHORAGE-INDEPENDENT GROWTH

Activation of the EGF receptor has been shown to promote oncogenesis by stimulating both the growth and survival of cancer cells [107, 108]. The critical role of TGase-2 in EGF-stimulated cancer cell survival, which was demonstrated in a number of breast cancer cell lines as discussed above, prompted us to consider the possibility that TGase-2 might be essential for EGF-stimulated cancer cell growth. To investigate this issue, we took advantage of the fact that exposing SKBR3 breast cancer cells to EGF not only augmented their proliferative capacity and ability to form colonies in soft agar, but also resulted in the induction of TGase-2 expression and activation.

We found that by suppressing TGase-2 expression using siRNAs, or blocking transamidation activity using the competitive inhibitor MDC, potently inhibited the growth-promoting effects afforded to SKBR3 cells by EGF stimulation. Moreover, overexpressing TGase-2 in SKBR3 cells was sufficient to mimic the actions of EGF and strongly stimulate the growth of these cells in monolayer and under anchorage-independent conditions [32]. Similar results were obtained in other breast cancer cell lines that we examined. Thus, in addition to promoting chemoresistance and invasion, the induction of TGase-2 expression and activation in certain cancer cells contributes to another common hallmark of malignant transformation, namely aberrant cell growth [32, 65]. However, this conclusion prompted two follow-up questions: (1) How is TGase-2 upregulated by EGF stimulation and (2) how does it enhance cell proliferation? To answer the first question, we continued using the SKBR3 cell line as our model system and delineated the EGF-mediated signaling pathways that were responsible for inducing TGase-2 expression. For example, we found that individually activating signaling pathways known to be coupled to EGF stimulation by ectopically expressing activated forms of a variety of signaling proteins including Ras, Rac, RhoA, Cdc42, and PI 3-kinase [106] in SKBR3 cells did not result in the induction of TGase-2 expression. Surprisingly, coexpression of activated forms of Ras and Cdc42 in these cells was able to induce TGase-2 expression, demonstrating that the upregulation of TGase-2 upon EGF treatment did not involve just a single signaling pathway but rather required multiple signaling inputs. Although the combination of Ras and Rho GTPases appeared to be most effective at inducing TGase-2 expression, coactivation of different Rho GTPases in the absence of activated Ras (e.g., Cdc42 and RhoA) also exhibited the capability of upregulating TGase-2 in SKBR3 cells. Likewise, onco-Dbl, a guanine nucleotide exchange factor that activates both Cdc42- and RhoA-coupled signaling pathways [135], also induced an upregulation of TGase-2 expression. The fact that the upregulation of TGase-2 was caused by a combination of small GTPases led to the conclusion that a common effector(s) acting downstream of those different GTPases must be involved. Phosphoinositide 3-kinase (PI3K) turned out to be one such effector protein. We were able to show that PI3K was an important regulator of EGF-induced TGase-2 expression, and it was necessary for the induction of TGase-2 expression caused by coexpression of activated forms of Ras and Cdc42. Thus, it appears that PI3K can be activated by Cdc42 and/or Ras in these cells and that this activation event is essential for increases in TGase-2 expression. Although there was evidence suggesting that at least Cdc42 can directly stimulate PI3K activity in certain cellular contexts, we found that expressing a constitutively active form of PI3K along with either an activated form of Ras or Cdc42 in SKBR3 cells did not lead to increases in TGase-2 expression (unpublished data). These findings argued that while PI3K functioned as an essential signaling component in the induction of TGase-2 expression by the coactivation of Ras and Cdc42, additional Ras- and/or Cdc42-dependent signaling events were also needed. In fact, another downstream participant in the Ras- and Cdc42-dependent signaling pathways that contributed to the upregulation of TGase-2 was NFκB. The role of NFκB in the survival of breast cancer cells has been well established, and it is a downstream target of different Rho-family GTPases including RhoA, Cdc42, and Rac [136]. Indeed, blocking NFκB activity in SKBR3 cells using the small molecule inhibitor BAY 11-7082 completely abolished the upregulation of TGase-2 induced by either EGF treatment or cotransfection of activated Ras and Cdc42. Taken together, cosignaling events from multiple small GTPases, such as Ras and Cdc42, leading to the activation of PI3K and NFκB, provide a mechanism by which EGF upregulates TGase-2 in breast cancer cells (Figure 9). Among the most interesting questions that remain include why multiple signaling inputs are needed to upregulate TGase-2, and how does the activation of PI3K and NFκB lead to enhanced TGase-2 expression? Both of these questions are currently under investigation in our laboratory.

The findings described above demonstrated for the first time that the induction of TGase-2 expression and activation by EGF is both necessary and sufficient for mediating the growth-promoting actions of EGF in human breast cancer cells. How does increasing the levels of TGase-2 expression in SKBR3 cells exacerbate their oncogenic potential? We know that it most likely involves the ability to transamidate or cross-link specific protein substrates that are linked to cell growth processes. This is indicated by the fact that SKBR3 cells exogenously expressing the transamidation-defective TGase-2 (Cys277Val) mutant did not exhibit enhanced anchorage-independent growth, unlike the case for SKBR3 cells overexpressing wild-type TGase-2. These findings led us to begin searching for candidate transamidation substrates of TGase-2 whose function when altered in cells can lead to the induction of, or at least potentiate, cellular transformation. A potential candidate protein that meets these criteria and was recently shown to be cross-linked by TGase-2 is IκBα, a critical regulatory component in the NFκB-signaling cascade. The IκBα regulatory subunit specifically functions in this pathway by forming a complex with NFκB (p65/RelA or

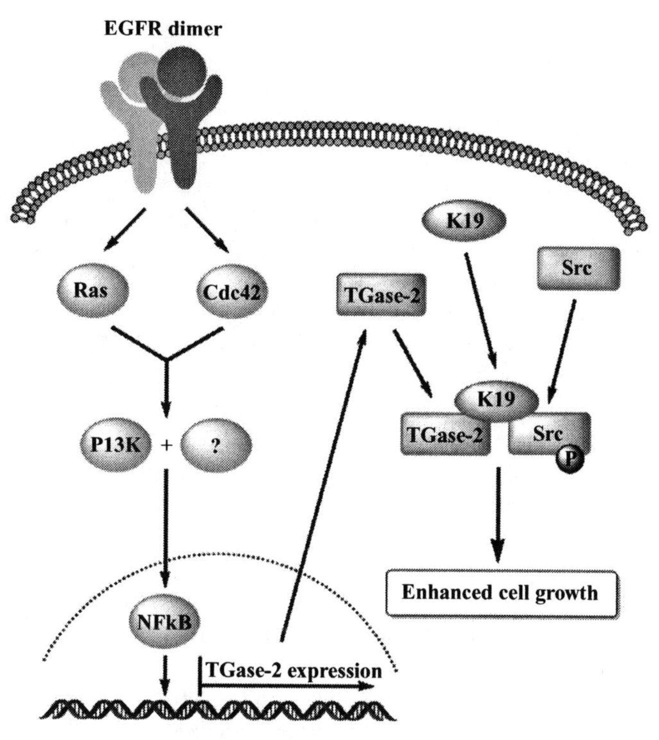

FIGURE 9. A schematic diagram of the signaling pathways used by the EGF receptor to induce TGase-2 expression in human cancer cells. Once upregulated, TGase-2 can form a complex with K19 and Src, resulting in enhanced Src kinase activity and aberrant cell growth.

p50/NFκB1) in the cytoplasm of cells, preventing it from translocating to the nucleus where it functions as a transcription factor [120]. As previously discussed, two independent groups recently found that TGase-2 activated NFκB by cross-linking IκBα and negating its regulatory function, thus allowing NFκB to enter the nucleus and influence the expression of genes that promote cell growth and survival [64, 66]. However, in SKBR3 cells overexpressing TGase-2, we could detect neither the presence of any cross-linked IκBα, nor the accumulation of NFκB in the nuclei of these cells beyond what was observed in control SKBR3 cells. Thus, it appeared that the ability of TGase-2 to promote the aberrant growth of SKBR3 cells was not through NFκB, but rather was mediated by another mechanism.

The fact that TGase-2 upregulation is crucial for EGF-mediated cell growth enhancement suggests that the signaling pathway(s) activated by TGase-2 to promote SKBR3 cell growth very likely overlaps EGF-coupled signaling cascades. As a result, we examined the activities of several signaling effectors, including ERK, AKT, JNK, P38, and S6K, which are frequently used to mediate signals from the EGF receptor to transcriptional factors that are responsible for stimulating cell proliferation [106, 137, 138]. However, none of these exhibited an increased activity in SKBR3 cells overexpressing TGase-2 as compared to control SKBR3 cells, demonstrating that TGase-2 promoted SKBR3 cell growth through a nontraditional signaling pathway. Since a large number of effectors involved in EGF-signaling cascades transmit signals via tyrosine phosphorylation events, we checked the tyrosine phosphorylation profile for lysates collected from SKBR3 cells overexpressing TGase-2 or vector. Surprisingly, a unique 60 kD protein exhibited a twofold increase in its tyrosine phosphorylation, which led us to consider whether TGase-2 promoted the phosphorylation and subsequent activation of Src, a well-established nonreceptor tyrosine kinase of $M_r \sim 60$ kDa. Indeed, we were able to show that SKBR3 cells overexpressing wild-type TGase-2 exhibited a more-than-twofold increase in Src activity compared to cells overexpressing vector or the transamidation-deficient TGase-2 (Cys277Val) mutant.

As one of the earliest identified oncogenes, a large amount of work has been done to support a role for Src in a number of aspects of malignant transformation, including cell proliferation, differentiation, motility, and survival [139, 140]. Our finding that TGase-2 upregulation led to Src activation gave a plausible explanation as to why TGase-2 also appears to be involved in multiple aspects of cancer progression. The need to further characterize the link between TGase-2 and Src raised the question of how TGase-2 activates Src in breast cancer cells. The classical model for Src activation starts with the recruitment of a phosphatase upon receiving upstream signals, which dephosphorylates a *C*-terminal tyrosine residue of Src (Tyr527) that normally forms an autoinhibitory loop with the SH2 domain of Src. Dephosphorylation of this particular tyrosine residue disrupts the intracellular inhibitory loop, exposing a critical tyrosine residue (Y416), whose phosphorylation eventually leads to the activation of Src [141]. However, this classical model is unable to provide a straightforward explanation of how TGase-2 activates Src because TGase-2 neither contains intrinsic phosphatase activity nor has it been reported to activate a phosphatase. On the other hand, several groups recently provided an unconventional mechanism for activating Src, namely through Src-binding proteins that displace the inhibitory loop away from the catalytic domain of Src, thereby driving it to assume its active conformation [142, 143]. We suspect that TGase-2 activates Src though this type of mechanism because the association of TGase-2 with Src was observed in multiple breast cancer cell lines. However, this association seems likely to be indirect due to the fact that recombinant TGase-2 did not exhibit any binding affinity toward GST-tagged Src, as demonstrated by in vitro GST pull-down assays.

Given that wild-type TGase-2 but not the transamidation-deficient mutant (Cys277Val) is capable of activating Src, the formation of a proposed Src-activating complex seems to require transamidation-competent TGase-2. This led to the possibility that the specific protein(s) mediating the TGase-2–Src interaction either can be transamidated by TGase-2 or interacts in the immediate vicinity of the transamidation catalytic site. To identify potential mediator protein(s), we immunoprecipitated TGase-2 from SKBR3 cells overexpressing either wild-type TGase-2 or the Cys277Val mutant, searching for targets that can be specifically precipitated with wild-type TGase-2 but not with the transamidation-defective mutant. As a result, two proteins in the molecular weight range of 40–45 kD were identified as specific candidate binding partners for wild-type TGase-2. One was identified to be actin, a known TGase-2-binding partner [70]. The other was the intermediate filament protein keratin-19. As a component of cytoskeleton-building blocks, the main function of the keratin family is believed to provide the mechanic support to maintain the desired cell shape [144, 145]. However, recent studies revealed that keratins were also able to bind and regulate the activities of signaling effectors, thereby functioning as scaffold proteins to modulate a number of cellular events including metabolic responses, cell migration, and cell growth [144–147]. For instance, keratin-10, a keratin-subtype highly expressed in the postmitotic, differentiating layers of epidermis, was shown to influence cell proliferation by directly interacting with AKT [148]. Another example came from studies in mouse skin keratinocytes, where keratin-17 activates the PI3K/AKT and mTOR/S6K pathway and promotes cell metabolism and growth though its ability to translocate the 14-3-3 protein from the nucleus to cytoplasm in a serum-dependent manner [149]. These findings prompted us to consider the possibility of keratin-19 as the mediator for TGase-2-Src interactions in breast cancer cells. This in fact turned out to be the case, as shown by the result that knocking-down keratin-19 expression seriously impaired the interaction between TGase-2 and Src in SKBR3 cells. Furthermore, suppressing keratin-19 expression greatly compromised the ability of TGase-2 to activate Src and its ability to promote the anchorage-independent growth of SKBR3 cells. Thus, our findings showing that keratin-19 bridges the interaction between TGase-2 and Src depicts a novel signaling pathway that is used by cancer cells to obtain a growth advantage (Figure 9). Further investigations ongoing in our laboratory are trying to understand how keratin-19 assists TGase-2 to activate Src and to identify additional participants involved in the formation or regulation of the TGase-2-Src-keratin-19 complex.

VI. CONCLUDING REMARKS/QUESTIONS FOR THE FUTURE

The combination of structure–function and cell biology-based studies performed on TGase-2 during the past 5–10 years have highlighted how this dual function GTP-binding protein/acyl transferase is much more than just a "house-keeping" enzyme that stabilizes membrane structure through its protein cross-linking capability. Instead, we now know, as highlighted in Figure 10, that TGase-2 plays interesting and in some cases, unanticipated roles in helping maintain cell survival under adverse stressful conditions and, when overexpressed in cancer cells, contributes strongly to the growth

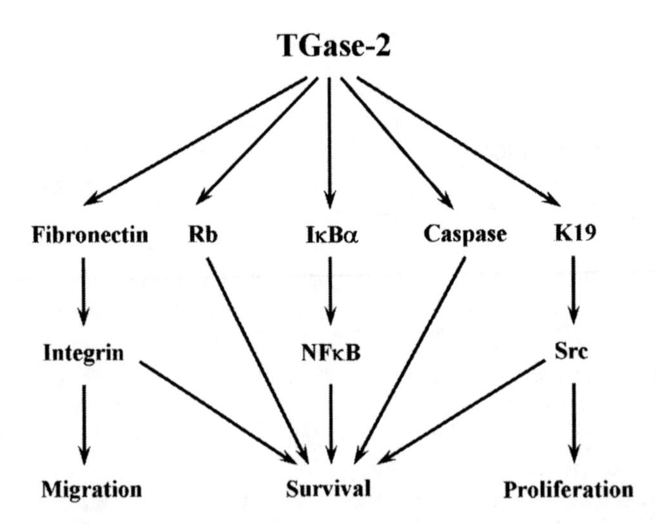

FIGURE 10. Diagram highlighting several of the mechanisms through which TGase-2 can trigger cell migration, survival, and growth.

and survival of these cells, as well as to their ability to migrate and exhibit invasive activity [19, 27, 32, 33, 63–65, 67, 105, 124, 125]. These roles have obvious implications for TGase-2 as a possible target for therapeutic intervention, but they also raise a number of intriguing questions that will likely be the focus of future research efforts. Among the most important of these questions will be to identify how the GTP-binding and transamidation activities of TGase-2 contribute to malignant transformation, and in particular, to the growth, migration, and metastatic capabilities of cancer cells. Assuming that the GTP-binding activity of TGase-2 will turn out to be important in the progression of certain cancers, the key issue will then become whether TGase-2 acts like a classical G-protein [40, 41], such that upon binding GTP, it engages and regulates the activity of an effector protein that is essential for one or more transforming activities. If so, we will want to know the identity of these effectors, how GTP-bound TGase-2 specifically regulates their activity, and in turn, how the GTP-binding/GTP-hydrolytic cycle of TGase-2 is regulated.

Likewise, we will want to know much more about how the transamidation activity of TGase-2 fits into different aspects of malignant transformation. In addressing this question, it will become important to develop the necessary methodologies to identify bona fide cellular transamidation substrates for TGase-2. A particularly intriguing candidate is keratin-19, because as discussed above, we have recently found that the binding of TGase-2 to keratin-19 is necessary for the TGase-2-mediated activation of the Src tyrosine kinase and that this activation is essential for the ability of TGase-2 to promote the transformation of certain breast cancer cells [32]. The mechanistic details underlying the molecular basis by which keratin-19 mediates the actions of TGase-2 on Src activity are still not understood. However, one attractive possibility is that the TGase-2-catalyzed cross-linking of keratin-19 enables it to serve as a scaffold to bind Src and help this tyrosine kinase to assume an activated conformational state. Once we better understand the nature of this interaction and whether the TGase-2-catalyzed cross-linking of keratin-19 in fact helps to promote the activation of Src, we will then want to know whether this accounts solely for the TGase-2-dependent stimulation of anchorage-dependent and independent cell growth in breast cancer cells, or if the activation of Src also contributes to the TGase-2-mediated promotional effects on cell migration and invasive activity. While the possibility that a single transamidation substrate accounts for many of the effects of TGase-2 on cancer cell progression would make for a simple picture, there are a number of reasons to suspect that the actions of TGase-2 are much more complex and are likely to be mediated through different substrates. Thus, it seems reasonable to anticipate that in different cellular contexts, and within specific types of cancers, TGase-2 will work in distinct ways to contribute to their transformed phenotypes.

In the future, it also will be important to begin to examine the question of just how the transamidation reaction is catalyzed within cells, given that the amount of Ca^{2+} required to detect activity in vitro (i.e., millimolar concentrations) greatly exceeds the concentration of free Ca^{2+} in cells [44]. Does this mean that the "true" transamidation substrates for TGase-2 require less Ca^{2+} for their reaction, and/or might the binding of these substrates significantly enhance the affinity of TGase-2 for its divalent metal ion cofactor? We also will need to contemplate the possibility that the apparent requirement for a functional transamidation active site in order for TGase-2 to mediate its biological functions might reflect the possibility that this enzyme active site in some cases actually functions as a binding domain for TGase-2-signaling partners. For example, one might imagine that in the absence of a donor amine-bearing substrate, a glutamine-containing acceptor molecule would form an acyl intermediate with the active site cysteine residue and thus undergo a binding interaction with TGase-2 that might be important for some aspect of cell growth or migration. Finally, it will be very interesting to see whether what we learn in the coming years regarding the roles of TGase-2 in malignant transformation has relevance to other disease states where TGase-2 has been suggested to have an involvement, in particular, neurodegenerative disorders like AD and HD [81, 83–85, 87, 88]. Will common transamidation substrates be involved or similar mechanisms underlie how TGase-2 plays a possible role in promoting or accentuating these disorders, compared to the roles it plays in cancer progression? All of this means that a good deal of work lies ahead for TGase-2 and some exciting opportunities will likely present themselves as we learn more about how this protein functions in normal cells and why it seems to be turning up in many pathological conditions.

REFERENCES

1. Folk, J. E. (1980) Transglutaminases, *Annu. Rev. Biochem. 49*, 517–531.
2. Greenberg, C. S., Birckbichler, P. J., and Rice, R. H. (1991) Transglutaminases: multifunctional cross-linking enzymes that stabilize tissues, *FASEB J. 5*, 3071–3077.
3. Fujita, K., Shibayama, K., Yamauchi, M., Kato, T., Ando, M., Takahashi, H., Iritani, K., Yoshimoto, N., and Nagata, Y. (1998) Alteration of enzymatic activities implicating

neuronal degeneration in the spinal cord of the motor neuron degeneration mouse during postnatal development, *Neurochem. Res. 23*, 557–562.

4. Tee, A. E., Marshall, G. M., Liu, P. Y., Xu, N., Haber, M., Norris, M. D., Iismaa, S. E., and Liu, T. (2009) The opposing effects of two tissue transglutaminase protein isoforms in neuroblastoma cell differentiation, *J. Biol. Chem.* [Epub ahead of print].
5. Aeschlimann, D., Kaupp, O., and Paulsson, M. (1995) Transglutaminase-catalyzed matrix cross-linking in differentiating cartilage: identification of osteonectin as a major glutaminyl substrate, *J. Cell Biol. 129*, 881–892.
6. Orlandi, A., Oliva, F., Taurisano, G., Candi, E., Di Lascio, A., Melino, G., Spagnoli, L. G., and Tarantino, U. (2009) Transglutaminase-2 differently regulates cartilage destruction and osteophyte formation in a surgical model of osteoarthritis, *Amino Acids 36*, 755–763.
7. Raghunath, M., Hopfner, B., Aeschlimann, D., Luthi, U., Meuli, M., Altermatt, S., Gobet, R., Bruckner-Tuderman, L., and Steinmann, B. (1996) Cross-linking of the dermo-epidermal junction of skin regenerating from keratinocyte autografts. Anchoring fibrils are a target for tissue transglutaminase, *J. Clin. Invest. 98*, 1174–1184.
8. Haroon, Z. A., Hettasch, J. M., Lai, T. S., Dewhirst, M. W., and Greenberg, C. S. (1999) Tissue transglutaminase is expressed, active, and directly involved in rat dermal wound healing and angiogenesis, *FASEB J. 13*, 1787–1795.
9. Verderio, E. A., Johnson, T., and Griffin, M. (2004) Tissue transglutaminase in normal and abnormal wound healing: review article, *Amino Acids 26*, 387–404.
10. Telci, D. and Griffin, M. (2006) Tissue transglutaminase (TG2)–a wound response enzyme, *Front. Biosci. 11*, 867–882.
11. Jones, R. A., Kotsakis, P., Johnson, T. S., Chau, D. Y., Ali, S., Melino, G., and Griffin, M. (2006) Matrix changes induced by transglutaminase 2 lead to inhibition of angiogenesis and tumor growth, *Cell Death Differ. 13*, 1442–1453.
12. Mehta, K. and Lopez-Berestein, G. (1986) Expression of tissue transglutaminase in cultured monocytic leukemia (THP-1) cells during differentiation, *Cancer Res. 46*, 1388–1394.
13. Aeschlimann, D., Wetterwald, A., Fleisch, H., and Paulsson, M. (1993) Expression of tissue transglutaminase in skeletal tissues correlates with events of terminal differentiation of chondrocytes, *J. Cell Biol. 120*, 1461–1470.
14. Singh, U. S., Pan, J., Kao, Y. L., Joshi, S., Young, K. L., and Baker, K. M. (2003) Tissue transglutaminase mediates activation of RhoA and MAP kinase pathways during retinoic acid-induced neuronal differentiation of SH-SY5Y cells, *J. Biol. Chem. 278*, 391–399.
15. Fesus, L., Thomazy, V., Autuori, F., Ceru, M. P., Tarcsa, E., and Piacentini, M. (1989) Apoptotic hepatocytes become insoluble in detergents and chaotropic agents as a result of transglutaminase action, *FEBS Lett. 245*, 150–154.
16. el Alaoui, S., Mian, S., Lawry, J., Quash, G., and Griffin, M. (1992) Cell cycle kinetics, tissue transglutaminase and programmed cell death (apoptosis), *FEBS Lett. 311*, 174–178.
17. Zhang, L. X., Mills, K. J., Dawson, M. I., Collins, S. J., and Jetten, A. M. (1995) Evidence for the involvement of retinoic acid receptor RAR alpha-dependent signaling

pathway in the induction of tissue transglutaminase and apoptosis by retinoids, *J. Biol. Chem. 270*, 6022–6029.

18. Oliverio, S., Amendola, A., Rodolfo, C., Spinedi, A., and Piacentini, M. (1999) Inhibition of "tissue" transglutaminase increases cell survival by preventing apoptosis, *J. Biol. Chem. 274*, 34123–34128.
19. Antonyak, M. A., Singh, U. S., Lee, D. A., Boehm, J. E., Combs, C., Zgola, M. M., Page, R. L., and Cerione, R. A. (2001) Effects of tissue transglutaminase on retinoic acid-induced cellular differentiation and protection against apoptosis, *J. Biol. Chem. 276*, 33582–33587.
20. Boehm, J. E., Singh, U., Combs, C., Antonyak, M. A., and Cerione, R. A. (2002) Tissue transglutaminase protects against apoptosis by modifying the tumor suppressor protein p110 Rb, *J. Biol. Chem. 277*, 20127–20130.
21. Liu, S., Cerione, R. A., and Clardy, J. (2002) Structural basis for the guanine nucleotide-binding activity of tissue transglutaminase and its regulation of transamidation activity, *Proc. Natl. Acad. Sci. U.S.A. 99*, 2743–2747.
22. Pinkas, D. M., Strop, P., Brunger, A. T., and Khosla, C. (2007) Transglutaminase 2 undergoes a large conformational change upon activation, *PLoS Biol. 5*, e327.
23. Datta, S., Antonyak, M. A., and Cerione, R. A. (2007) GTP-binding-defective forms of tissue transglutaminase trigger cell death, *Biochemistry 46*, 14819–14829.
24. Akimov, S. S., Krylov, D., Fleischman, L. F., and Belkin, A. M. (2000) Tissue transglutaminase is an integrin-binding adhesion coreceptor for fibronectin, *J. Cell Biol. 148*, 825–838.
25. Fâesèus, L. and Szondy, Z. (2005) Transglutaminase 2 in the balance of cell death and survival, *FEBS Lett. 579*, 3297–3302.
26. Antonyak, M. A., Jansen, J. M., Miller, A. M., Ly, T. K., Endo, M., and Cerione, R. A. (2006) Two isoforms of tissue transglutaminase mediate opposing cellular fates, *Proc. Natl. Acad. Sci. U.S.A. 103*, 18609–18614.
27. Mangala, L. S., Fok, J. Y., Zorrilla-Calancha, I. R., Verma, A., and Mehta, K. (2006) Tissue transglutaminase expression promotes cell attachment, invasion and survival in breast cancer cells, *Oncogene 26*, 2459–2470.
28. Zemskov, E. A., Janiak, A., Hang, J., Waghray, A., and Belkin, A. M. (2006) The role of tissue transglutaminase in cell-matrix interactions, *Front. Biosci. 11*, 1057–1076.
29. Antonyak, M. A., Boehm, J. E., and Cerione, R. A. (2002) Phosphoinositide 3-kinase activity is required for retinoic acid-induced expression and activation of the tissue transglutaminase, *J. Biol. Chem. 277*, 14712–14716.
30. Antonyak, M. A., McNeill, C. J., Wakshlag, J. J., Boehm, J. E., and Cerione, R. A. (2003) Activation of the Ras-ERK pathway inhibits retinoic acid-induced stimulation of tissue transglutaminase expression in NIH3T3 cells, *J. Biol. Chem. 278*, 15859–15866.
31. Antonyak, M. A., Miller, A. M., Jansen, J. M., Boehm, J. E., Balkman, C. E., Wakshlag, J. J., Page, R. L., and Cerione, R. A. (2004) Augmentation of tissue transglutaminase expression and activation by epidermal growth factor inhibit doxorubicin-induced apoptosis in human breast cancer cells, *J. Biol. Chem. 279*, 41461–41467.

32. Li, B., Antonyak, M. A., Druso, J. E., Cheng, L., Nikitin, A. Y., and Cerione, R. A. (2010) EGF potentiated oncogenesis requires a tissue transglutaminase-dependent signaling pathway leading to Src activation, *Proc. Natl. Acad. Sci. U.S.A. 107*, 1408–1413.

33. Antonyak, M. A., Li, B., Regan, A. D., Feng, Q., Dusaban, S. S., and Cerione, R. A. (2009) Tissue transglutaminase is an essential participant in the epidermal growth factor-stimulated signaling pathway leading to cancer cell migration and invasion, *J. Biol. Chem. 284*, 17914–17925.

34. Yee, V. C., Pedersen, L. C., Le Trong, I., Bishop, P. D., Stenkamp, R. E., and Teller, D. C. (1994) Three-dimensional structure of a transglutaminase: human blood coagulation factor XIII, *Proc. Natl. Acad. Sci. U.S.A. 91*, 7296–7300.

35. Casadio, R., Polverini, E., Mariani, P., Spinozzi, F., Carsughi, F., Fontana, A., Polverino de Laureto, P., Matteucci, G., and Bergamini, C. M. (1999) The structural basis for the regulation of tissue transglutaminase by calcium ions, *Eur. J. Biochem. 262*, 672–679.

36. Murthy, S. N., Iismaa, S., Begg, G., Freymann, D. M., Graham, R. M., and Lorand, L. (2002) Conserved tryptophan in the core domain of transglutaminase is essential for catalytic activity, *Proc. Natl. Acad. Sci. U.S.A. 99*, 2738–2742.

37. Iismaa, S. E., Holman, S., Wouters, M. A., Lorand, L., Graham, R. M., and Husain, A. (2003) Evolutionary specialization of a tryptophan indole group for transition-state stabilization by eukaryotic transglutaminases, *Proc. Natl. Acad. Sci. U.S.A. 100*, 12636–12641.

38. Nakaoka, H., Perez, D. M., Baek, K. J., Das, T., Husain, A., Misono, K., Im, M. J., and Graham, R. M. (1994) Gh: a GTP-binding protein with transglutaminase activity and receptor signaling function, *Science 264*, 1593–1596.

39. Lee, M. Y., Chung, S., Bang, H. W., Baek, K. J., and Uhm, D. (1997) Modulation of large conductance Ca^{2+}-activated K^+ channel by Galphah (transglutaminase II) in the vascular smooth muscle cell, *Pflugers Arch. 433*, 671–673.

40. Lambright, D. G., Noel, J. P., Hamm, H. E., and Sigler, P. B. (1994) Structural determinants for activation of the alpha-subunit of a heterotrimeric G protein, *Nature 369*, 621–628.

41. Sprang, S. R. and Coleman, D. E. (1998) Invasion of the nucleotide snatchers: structural insights into the mechanism of G protein GEFs, *Cell 95*, 155–158.

42. Begg, G. E., Holman, S. R., Stokes, P. H., Matthews, J. M., Graham, R. M., and Iismaa, S. E. (2006) Mutation of a critical arginine in the GTP-binding site of transglutaminase 2 disinhibits intracellular cross-linking activity, *J. Biol. Chem. 281*, 12603–12609.

43. Iismaa, S. E., Wu, M. J., Nanda, N., Church, W. B., and Graham, R. M. (2000) GTP binding and signaling by Gh/transglutaminase II involves distinct residues in a unique GTP-binding pocket, *J. Biol. Chem. 275*, 18259–18265.

44. Datta, S., Antonyak, M. A., and Cerione, R. A. (2006) Importance of Ca(2+)-dependent transamidation activity in the protection afforded by tissue transglutaminase against doxorubicin-induced apoptosis, *Biochemistry 45*, 13163–13174.

45. Begg, G. E., Carrington, L., Stokes, P. H., Matthews, J. M., Wouters, M. A., Husain, A., Lorand, L., Iismaa, S. E., and Graham, R. M. (2006) Mechanism of allosteric regulation of transglutaminase 2 by GTP, *Proc. Natl. Acad. Sci. U.S.A. 103*, 19683–19688.

46. Achyuthan, K. E. and Greenberg, C. S. (1987) Identification of a guanosine triphosphate-binding site on guinea pig liver transglutaminase. Role of GTP and calcium ions in modulating activity, *J. Biol. Chem. 262*, 1901–1906.
47. Zhang, J., Lesort, M., Guttmann, R. P., and Johnson, G. V. (1998) Modulation of the in situ activity of tissue transglutaminase by calcium and GTP, *J. Biol. Chem. 273*, 2288–2295.
48. Ahvazi, B., Kim, H. C., Kee, S. H., Nemes, Z., and Steinert, P. M. (2002) Three-dimensional structure of the human transglutaminase 3 enzyme: binding of calcium ions changes structure for activation, *EMBO J. 21*, 2055–2067.
49. Bergamini, C. M. (1988) GTP modulates calcium binding and cation-induced conformational changes in erythrocyte transglutaminase, *FEBS Lett. 239*, 255–258.
50. Lorand, L., Dailey, J. E., and Turner, P. M. (1988) Fibronectin as a carrier for the transglutaminase from human erythrocytes, *Proc. Natl. Acad. Sci. U.S.A. 85*, 1057–1059.
51. LeMosy, E. K., Erickson, H. P., Beyer, W. F., Jr., Radek, J. T., Jeong, J. M., Murthy, S. N., and Lorand, L. (1992) Visualization of purified fibronectin-transglutaminase complexes, *J. Biol. Chem. 267*, 7880–7885.
52. Singh, U. S. and Cerione, R. A. (1996) Biochemical effects of retinoic acid on GTP-binding protein/transglutaminases in HeLa cells. Stimulation of GTP-binding and transglutaminase activity, membrane association, and phosphatidylinositol lipid turnover, *J. Biol. Chem. 271*, 27292–27298.
53. Collighan, R. J. and Griffin, M. (2009) Transglutaminase 2 cross-linking of matrix proteins: biological significance and medical applications, *Amino Acids 36*, 659–670.
54. Jeong, J. M., Murthy, S. N., Radek, J. T., and Lorand, L. (1995) The fibronectin-binding domain of transglutaminase, *J. Biol. Chem. 270*, 5654–5658.
55. Gaudry, C. A., Verderio, E., Aeschlimann, D., Cox, A., Smith, C., and Griffin, M. (1999) Cell surface localization of tissue transglutaminase is dependent on a fibronectin-binding site in its N-terminal beta-sandwich domain, *J. Biol. Chem. 274*, 30707–30714.
56. Hang, J., Zemskov, E. A., Lorand, L., and Belkin, A. M. (2005) Identification of a novel recognition sequence for fibronectin within the NH2-terminal beta-sandwich domain of tissue transglutaminase, *J. Biol. Chem. 280*, 23675–23683.
57. Akimov, S. S. and Belkin, A. M. (2001) Cell-surface transglutaminase promotes fibronectin assembly via interaction with the gelatin-binding domain of fibronectin: a role in TGFbeta-dependent matrix deposition, *J. Cell Sci. 114*, 2989–3000.
58. Akimov, S. S. and Belkin, A. M. (2001) Cell surface tissue transglutaminase is involved in adhesion and migration of monocytic cells on fibronectin, *Blood 98*, 1567–1576.
59. Belkin, A. M., Tsurupa, G., Zemskov, E., Veklich, Y., Weisel, J. W., and Medved, L. (2005) Transglutaminase-mediated oligomerization of the fibrin(ogen) alphaC domains promotes integrin-dependent cell adhesion and signaling, *Blood 105*, 3561–3568.
60. Telci, D., Wang, Z., Li, X., Verderio, E. A., Humphries, M. J., Baccarini, M., Basaga, H., and Griffin, M. (2008) Fibronectin-tissue transglutaminase matrix rescues RGD-impaired cell adhesion through syndecan-4 and beta1 integrin co-signaling, *J. Biol. Chem. 283*, 20937–20947.

61. Barsigian, C., Stern, A. M., and Martinez, J. (1991) Tissue (type II) transglutaminase covalently incorporates itself, fibrinogen, or fibronectin into high molecular weight complexes on the extracellular surface of isolated hepatocytes. Use of 2-[(2-oxopropyl)thio] imidazolium derivatives as cellular transglutaminase inactivators, *J. Biol. Chem. 266*, 22501–22509.

62. Forsprecher, J., Wang, Z., Nelea, V., and Kaartinen, M. T. (2009) Enhanced osteoblast adhesion on transglutaminase 2-crosslinked fibronectin, *Amino Acids 36*, 747–753.

63. Yuan, L., Siegel, M., Choi, K., Khosla, C., Miller, C. R., Jackson, E. N., Piwnica-Worms, D., and Rich, K. M. (2007) Transglutaminase 2 inhibitor, KCC009, disrupts fibronectin assembly in the extracellular matrix and sensitizes orthotopic glioblastomas to chemotherapy, *Oncogene 26*, 2563–2573.

64. Kim, D. S., Park, S. S., Nam, B. H., Kim, I. H., and Kim, S. Y. (2006) Reversal of drug resistance in breast cancer cells by transglutaminase 2 inhibition and nuclear factor-kappaB inactivation, *Cancer Res. 66*, 10936–10943.

65. Hwang, J. Y., Mangala, L. S., Fok, J. Y., Lin, Y. G., Merritt, W. M., Spannuth, W. A., Nick, A. M., Fiterman, D. J., Vivas-Mejia, P. E., Deavers, M. T., Coleman, R. L., Lopez-Berestein, G., Mehta, K., and Sood, A. K. (2008) Clinical and biological significance of tissue transglutaminase in ovarian carcinoma, *Cancer Res. 68*, 5849–5858.

66. Mann, A. P., Verma, A., Sethi, G., Manavathi, B., Wang, H., Fok, J. Y., Kunnumakkara, A. B., Kumar, R., Aggarwal, B. B., and Mehta, K. (2006) Overexpression of tissue transglutaminase leads to constitutive activation of nuclear factor-kappaB in cancer cells: delineation of a novel pathway, *Cancer Res. 66*, 8788–8795.

67. Verma, A., Wang, H., Manavathi, B., Fok, J. Y., Mann, A. P., Kumar, R., and Mehta, K. (2006) Increased expression of tissue transglutaminase in pancreatic ductal adenocarcinoma and its implications in drug resistance and metastasis, *Cancer Res. 66*, 10525–10533.

68. Melino, G., Annicchiarico-Petruzzelli, M., Piredda, L., Candi, E., Gentile, V., Davies, P. J., and Piacentini, M. (1994) Tissue transglutaminase and apoptosis: sense and antisense transfection studies with human neuroblastoma cells, *Mol. Cell. Biol. 14*, 6584–6596.

69. Ou, H., Haendeler, J., Aebly, M. R., Kelly, L. A., Cholewa, B. C., Koike, G., Kwitek-Black, A., Jacob, H. J., Berk, B. C., and Miano, J. M. (2000) Retinoic acid-induced tissue transglutaminase and apoptosis in vascular smooth muscle cells, *Circ. Res. 87*, 881–887.

70. Nemes, Z., Jr., Adany, R., Balazs, M., Boross, P., and Fesus, L. (1997) Identification of cytoplasmic actin as an abundant glutaminyl substrate for tissue transglutaminase in HL-60 and U937 cells undergoing apoptosis, *J. Biol. Chem. 272*, 20577–20583.

71. Piacentini, M., Fesus, L., Farrace, M. G., Ghibelli, L., Piredda, L., and Melino, G. (1991) The expression of "tissue" transglutaminase in two human cancer cell lines is related with the programmed cell death (apoptosis), *Eur. J. Cell Biol. 54*, 246–254.

72. Piredda, L., Amendola, A., Colizzi, V., Davies, P. J., Farrace, M. G., Fraziano, M., Gentile, V., Uray, I., Piacentini, M., and Fesus, L. (1997) Lack of 'tissue' transglutaminase protein cross-linking leads to leakage of macromolecules from dying cells: relationship to development of autoimmunity in MRLIpr/Ipr mice, *Cell Death Differ. 4*, 463–472.

73. Piacentini, M., Farrace, M. G., Piredda, L., Matarrese, P., Ciccosanti, F., Falasca, L., Rodolfo, C., Giammarioli, A. M., Verderio, E., Griffin, M., and Malorni, W. (2002) Transglutaminase overexpression sensitizes neuronal cell lines to apoptosis by increasing mitochondrial membrane potential and cellular oxidative stress, *J. Neurochem. 81*, 1061–1072.

74. Tucholski, J., Lesort, M., and Johnson, G. V. (2001) Tissue transglutaminase is essential for neurite outgrowth in human neuroblastoma SH-SY5Y cells, *Neuroscience 102*, 481–491.

75. Condello, S., Caccamo, D., Curro, M., Ferlazzo, N., Parisi, G., and Ientile, R. (2008) Transglutaminase 2 and NF-kappaB interplay during NGF-induced differentiation of neuroblastoma cells, *Brain Res. 1207*, 1–8.

76. Chiocca, E. A., Davies, P. J., and Stein, J. P. (1989) Regulation of tissue transglutaminase gene expression as a molecular model for retinoid effects on proliferation and differentiation, *J. Cell. Biochem. 39*, 293–304.

77. Ross, C. A. and Poirier, M. A. (2004) Protein aggregation and neurodegenerative disease, *Nat. Med. 10(Suppl.)*, S10–S17.

78. Davies, S. W., Turmaine, M., Cozens, B. A., DiFiglia, M., Sharp, A. H., Ross, C. A., Scherzinger, E., Wanker, E. E., Mangiarini, L., and Bates, G. P. (1997) Formation of neuronal intranuclear inclusions underlies the neurological dysfunction in mice transgenic for the HD mutation, *Cell 90*, 537–548.

79. Scherzinger, E., Sittler, A., Schweiger, K., Heiser, V., Lurz, R., Hasenbank, R., Bates, G. P., Lehrach, H., and Wanker, E. E. (1999) Self-assembly of polyglutamine-containing huntingtin fragments into amyloid-like fibrils: implications for Huntington's disease pathology, *Proc. Natl. Acad. Sci. U.S.A. 96*, 4604–4609.

80. Lesort, M., Chun, W., Johnson, G. V., and Ferrante, R. J. (1999) Tissue transglutaminase is increased in Huntington's disease brain, *J. Neurochem. 73*, 2018–2027.

81. Karpuj, M. V., Garren, H., Slunt, H., Price, D. L., Gusella, J., Becher, M. W., and Steinman, L. (1999) Transglutaminase aggregates huntingtin into nonamyloidogenic polymers, and its enzymatic activity increases in Huntington's disease brain nuclei, *Proc. Natl. Acad. Sci. U.S.A. 96*, 7388–7393.

82. Kim, S. Y., Grant, P., Lee, J. H., Pant, H. C., and Steinert, P. M. (1999) Differential expression of multiple transglutaminases in human brain. Increased expression and cross-linking by transglutaminases 1 and 2 in Alzheimer's disease, *J. Biol. Chem. 274*, 30715–30721.

83. Junn, E., Ronchetti, R. D., Quezado, M. M., Kim, S. Y., and Mouradian, M. M. (2003) Tissue transglutaminase-induced aggregation of alpha-synuclein: Implications for Lewy body formation in Parkinson's disease and dementia with Lewy bodies, *Proc. Natl. Acad. Sci. U.S.A. 100*, 2047–2052.

84. Johnson, G. V., Cox, T. M., Lockhart, J. P., Zinnerman, M. D., Miller, M. L., and Powers, R. E. (1997) Transglutaminase activity is increased in Alzheimer's disease brain, *Brain Res. 751*, 323–329.

85. Dedeoglu, A., Kubilus, J. K., Jeitner, T. M., Matson, S. A., Bogdanov, M., Kowall, N. W., Matson, W. R., Cooper, A. J., Ratan, R. R., Beal, M. F., Hersch, S. M., and Ferrante, R. J.

(2002) Therapeutic effects of cystamine in a murine model of Huntington's disease, *J. Neurosci. 22*, 8942–8950.

86. Kahlem, P., Green, H., and Djian, P. (1998) Transglutaminase action imitates Huntington's disease: selective polymerization of Huntingtin containing expanded polyglutamine, *Mol. Cell 1*, 595–601.

87. Mastroberardino, P. G., Iannicola, C., Nardacci, R., Bernassola, F., De Laurenzi, V., Melino, G., Moreno, S., Pavone, F., Oliverio, S., Fesus, L., and Piacentini, M. (2002) 'Tissue' transglutaminase ablation reduces neuronal death and prolongs survival in a mouse model of Huntington's disease, *Cell Death Differ. 9*, 873–880.

88. Karpuj, M. V., Becher, M. W., Springer, J. E., Chabas, D., Youssef, S., Pedotti, R., Mitchell, D., and Steinman, L. (2002) Prolonged survival and decreased abnormal movements in transgenic model of Huntington disease, with administration of the transglutaminase inhibitor cystamine, *Nat. Med. 8*, 143–149.

89. Pahlman, S., Ruusala, A. I., Abrahamsson, L., Mattsson, M. E., and Esscher, T. (1984) Retinoic acid-induced differentiation of cultured human neuroblastoma cells: a comparison with phorbolester-induced differentiation, *Cell Differ. 14*, 135–144.

90. Preis, P. N., Saya, H., Nadasdi, L., Hochhaus, G., Levin, V., and Sadee, W. (1988) Neuronal cell differentiation of human neuroblastoma cells by retinoic acid plus herbimycin A, *Cancer Res. 48*, 6530–6534.

91. Sporn, M. B. and Roberts, A. B. (1983) Role of retinoids in differentiation and carcinogenesis, *Cancer Res. 43*, 3034–3040.

92. Breitman, T. R., Collins, S. J., and Keene, B. R. (1981) Terminal differentiation of human promyelocytic leukemic cells in primary culture in response to retinoic acid, *Blood 57*, 1000–1004.

93. Costa, A., Formelli, F., Chiesa, F., Decensi, A., De Palo, G., and Veronesi, U. (1994) Prospects of chemoprevention of human cancers with the synthetic retinoid fenretinide, *Cancer Res. 54*, 2032s–2037s.

94. Formelli, F., Barua, A. B., and Olson, J. A. (1996) Bioactivities of N-(4-hydroxyphenyl) retinamide and retinoyl beta-glucuronide, *FASEB J. 10*, 1014–1024.

95. Breitman, T. R., Selonick, S. E., and Collins, S. J. (1980) Induction of differentiation of the human promyelocytic leukemia cell line (HL-60) by retinoic acid, *Proc. Natl. Acad. Sci. U.S.A. 77*, 2936–2940.

96. Collins, S. J., Robertson, K. A., and Mueller, L. (1990) Retinoic acid-induced granulocytic differentiation of HL-60 myeloid leukemia cells is mediated directly through the retinoic acid receptor (RAR-alpha), *Mol. Cell. Biol. 10*, 2154–2163.

97. Delia, D., et al. (1993) N-(4-hydroxyphenyl)retinamide induces apoptosis of malignant hemopoietic cell lines including those unresponsive to retinoic acid, *Cancer Res. 53*, 6036–6041.

98. Dipietrantonio, A., Hsieh, T. C., and Wu, J. M. (1996) Differential effects of retinoic acid (RA) and N-(4-hydroxyphenyl) retinamide (4-HPR) on cell growth, induction of differentiation, and changes in p34cdc2, Bcl-2, and actin expression in the human promyelocytic HL-60 leukemic cells, *Biochem. Biophys. Res. Commun. 224*, 837–842.

99. Ponzoni, M., Bocca, P., Chiesa, V., Decensi, A., Pistoia, V., Raffaghello, L., Rozzo, C., and Montaldo, P. G. (1995) Differential effects of N-(4-hydroxyphenyl)retinamide and retinoic acid on neuroblastoma cells: apoptosis versus differentiation, *Cancer Res. 55*, 853–861.

100. Nagy, L., Thomazy, V. A., Chandraratna, R. A., Heyman, R. A., and Davies, P. J. (1996) Retinoid-regulated expression of BCL-2 and tissue transglutaminase during the differentiation and apoptosis of human myeloid leukemia (HL-60) cells, *Leuk. Res. 20*, 499–505.

101. Singer, C. F., Hudelist, G., Walter, I., Rueckliniger, E., Czerwenka, K., Kubista, E., and Huber, A. V. (2006) Tissue array-based expression of transglutaminase-2 in human breast and ovarian cancer, *Clin. Exp. Metastasis 23*, 33–39.

102. Hettasch, J. M., Bandarenko, N., Burchette, J. L., Lai, T. S., Marks, J. R., Haroon, Z. A., Peters, K., Dewhirst, M. W., Iglehart, J. D., and Greenberg, C. S. (1996) Tissue transglutaminase expression in human breast cancer, *Lab Invest. 75*, 637–645.

103. Satpathy, M., Cao, L., Pincheira, R., Emerson, R., Bigsby, R., Nakshatri, H., and Matei, D. (2007) Enhanced peritoneal ovarian tumor dissemination by tissue transglutaminase, *Cancer Res. 67*, 7194–7202.

104. Cao, L., Petrusca, D. N., Satpathy, M., Nakshatri, H., Petrache, I., and Matei, D. (2008) Tissue transglutaminase protects epithelial ovarian cancer cells from cisplatin-induced apoptosis by promoting cell survival signaling, *Carcinogenesis 29*, 1893–1900.

105. Verma, A., Guha, S., Wang, H., Fok, J. Y., Koul, D., Abbruzzese, J., and Mehta, K. (2008) Tissue transglutaminase regulates focal adhesion kinase/AKT activation by modulating PTEN expression in pancreatic cancer cells, *Clin. Cancer Res. 14*, 1997–2005.

106. Yarden, Y. and Shilo, B. Z. (2007) SnapShot: EGFR signaling pathway, *Cell 131*, 1018.

107. Nicholson, R. I., Gee, J. M., and Harper, M. E. (2001) EGFR and cancer prognosis, *Eur. J. Cancer 37(Suppl 4)*, S9–S15.

108. Mendelsohn, J. and Baselga, J. (2000) The EGF receptor family as targets for cancer therapy, *Oncogene 19*, 6550–6565.

109. Zhang, R., Tremblay, T. L., McDermid, A., Thibault, P., and Stanimirovic, D. (2003) Identification of differentially expressed proteins in human glioblastoma cell lines and tumors, *Glia 42*, 194–208.

110. Esposito, C. and Caputo, I. (2005) Mammalian transglutaminases. Identification of substrates as a key to physiological function and physiopathological relevance, *FEBS J. 272*, 615–631.

111. Facchiano, A. and Facchiano, F. (2009) Transglutaminases and their substrates in biology and human diseases: 50 years of growing, *Amino Acids 36*, 599–614.

112. Yamaguchi, H. and Wang, H. G. (2006) Tissue transglutaminase serves as an inhibitor of apoptosis by cross-linking caspase 3 in thapsigargin-treated cells, *Mol. Cell. Biol. 26*, 569–579.

113. Kurokawa, M. and Kornbluth, S. (2009) Caspases and kinases in a death grip, *Cell 138*, 838–854.

114. Hebert, S. S., Daviau, A., Grondin, G., Latreille, M., Aubin, R. A., and Blouin, R. (2000) The mixed lineage kinase DLK is oligomerized by tissue transglutaminase during apoptosis, *J. Biol. Chem. 275*, 32482–32490.

115. Lee, C. H., Jeon, Y. T., Kim, S. H., and Song, Y. S. (2007) NF-kappaB as a potential molecular target for cancer therapy, *Biofactors 29*, 19–35.

116. Basseres, D. S. and Baldwin, A. S. (2006) Nuclear factor-kappaB and inhibitor of kappaB kinase pathways in oncogenic initiation and progression, *Oncogene 25*, 6817–6830.

117. Biswas, D. K., Shi, Q., Baily, S., Strickland, I., Ghosh, S., Pardee, A. B., and Iglehart, J. D. (2004) NF-kappa B activation in human breast cancer specimens and its role in cell proliferation and apoptosis, *Proc. Natl. Acad. Sci. U.S.A. 101*, 10137–10142.

118. Nakshatri, H., Bhat-Nakshatri, P., Martin, D. A., Goulet, R. J., Jr., and Sledge, G. W., Jr. (1997) Constitutive activation of NF-kappaB during progression of breast cancer to hormone-independent growth, *Mol. Cell. Biol. 17*, 3629–3639.

119. Wang, W., Abbruzzese, J. L., Evans, D. B., Larry, L., Cleary, K. R., and Chiao, P. J. (1999) The nuclear factor-kappa B RelA transcription factor is constitutively activated in human pancreatic adenocarcinoma cells, *Clin. Cancer Res. 5*, 119–127.

120. Wu, J. T. and Kral, J. G. (2005) The NF-kappaB/IkappaB signaling system: a molecular target in breast cancer therapy, *J. Surg. Res. 123*, 158–169.

121. Lee, J., Kim, Y. S., Choi, D. H., Bang, M. S., Han, T. R., Joh, T. H., and Kim, S. Y. (2004) Transglutaminase 2 induces nuclear fact1or-kappaB activation via a novel pathway in BV-2 microglia, *J. Biol. Chem. 279*, 53725–53735.

122. Sahai, E. (2007) Illuminating the metastatic process, *Nat. Rev. Cancer 7*, 737–749.

123. Le Clainche, C. and Carlier, M. F. (2008) Regulation of actin assembly associated with protrusion and adhesion in cell migration, *Physiol. Rev. 88*, 489–513.

124. Joshi, S., Guleria, R., Pan, J., DiPette, D., and Singh, U. S. (2006) Retinoic acid receptors and tissue-transglutaminase mediate short-term effect of retinoic acid on migration and invasion of neuroblastoma SH-SY5Y cells, *Oncogene 25*, 240–247.

125. Shao, M., Cao, L., Shen, C., Satpathy, M., Chelladurai, B., Bigsby, R. M., Nakshatri, H., and Matei, D. (2009) Epithelial-to-mesenchymal transition and ovarian tumor progression induced by tissue transglutaminase, *Cancer Res. 69*, 9192–9201.

126. Jiang, D., Ying, W., Lu, Y., Wan, J., Zhai, Y., Liu, W., Zhu, Y., Qiu, Z., Qian, X., and He, F. (2003) Identification of metastasis-associated proteins by proteomic analysis and functional exploration of interleukin-18 in metastasis, *Proteomics 3*, 724–737.

127. Hynes, R. O. (2002) Integrins: bidirectional, allosteric signaling machines, *Cell 110*, 673–687.

128. Humphries, M. J., Travis, M. A., Clark, K., and Mould, A. P. (2004) Mechanisms of integration of cells and extracellular matrices by integrins, *Biochem. Soc. Trans. 32*, 822–825.

129. Zuo, X., Zhang, J., Zhang, Y., Hsu, S. C., Zhou, D., and Guo, W. (2006) Exo70 interacts with the Arp2/3 complex and regulates cell migration, *Nat. Cell Biol. 8*, 1383–1388.

130. Katz, M., Amit, I., Citri, A., Shay, T., Carvalho, S., Lavi, S., Milanezi, F., Lyass, L., Amariglio, N., Jacob-Hirsch, J., Ben-Chetrit, N., Tarcic, G., Lindzen, M., Avraham, R., Liao, Y. C., Trusk, P., Lyass, A., Rechavi, G., Spector, N. L., Lo, S. H., Schmitt, F.,

Bacus, S. S., and Yarden, Y. (2007) A reciprocal tensin-3-cten switch mediates EGF-driven mammary cell migration, *Nat. Cell Biol. 9*, 961–969.

131. Micallef, J., Taccone, M., Mukherjee, J., Croul, S., Busby, J., Moran, M. F., and Guha, A. (2009) Epidermal growth factor receptor variant III-induced glioma invasion is mediated through myristoylated alanine-rich protein kinase C substrate overexpression, *Cancer Res. 69*, 7548–7556.
132. Cui, X., Kim, H. J., Kuiatse, I., Kim, H., Brown, P. H., and Lee, A. V. (2006) Epidermal growth factor induces insulin receptor substrate-2 in breast cancer cells via c-Jun NH(2)-terminal kinase/activator protein-1 signaling to regulate cell migration, *Cancer Res. 66*, 5304–5313.
133. Ricono, J. M., Huang, M., Barnes, L. A., Lau, S. K., Weis, S. M., Schlaepfer, D. D., Hanks, S. K., and Cheresh, D. A. (2009) Specific cross-talk between epidermal growth factor receptor and integrin alphavbeta5 promotes carcinoma cell invasion and metastasis, *Cancer Res. 69*, 1383–1391.
134. Wellner, U., Schubert, J., Burk, U. C., Schmalhofer, O., Zhu, F., Sonntag, A., Waldvogel, B., Vannier, C., Darling, D., zur Hausen, A., Brunton, V. G., Morton, J., Sansom, O., Schuler, J., Stemmler, M. P., Herzberger, C., Hopt, U., Keck, T., Brabletz, S., and Brabletz, T. (2009) The EMT-activator ZEB1 promotes tumorigenicity by repressing stemness-inhibiting microRNAs, *Nat. Cell Biol. 11*, 1487–1495.
135. Cerione, R. A. and Zheng, Y. (1996) The Dbl family of oncogenes, *Curr. Opin. Cell Biol. 8*, 216–222.
136. Cammarano, M. S. and Minden, A. (2001) Dbl and the Rho GTPases activate NF kappa B by I kappa B kinase (IKK)-dependent and IKK-independent pathways, *J. Biol. Chem. 276*, 25876–25882.
137. Chen, Y. L., Law, P. Y., and Loh, H. H. (2005) Inhibition of PI3K/Akt signaling: an emerging paradigm for targeted cancer therapy, *Curr. Med. Chem. Anticancer Agents 5*, 575–589.
138. Roberts, P. J. and Der, C. J. (2007) Targeting the Raf-MEK-ERK mitogen-activated protein kinase cascade for the treatment of cancer, *Oncogene 26*, 3291–3310.
139. Summy, J. M. and Gallick, G. E. (2003) Src family kinases in tumor progression and metastasis, *Cancer Metastasis Rev. 22*, 337–358.
140. Ishizawar, R. and Parsons, S. J. (2004) c-Src and cooperating partners in human cancer, *Cancer Cell 6*, 209–214.
141. Parsons, S. J. and Parsons, J. T. (2004) Src family kinases, key regulators of signal transduction, *Oncogene 23*, 7906–7909.
142. Moarefi, I., LaFevre-Bernt, M., Sicheri, F., Huse, M., Lee, C. H., Kuriyan, J., and Miller, W. T. (1997) Activation of the Src-family tyrosine kinase Hck by SH3 domain displacement, *Nature 385*, 650–653.
143. Arias-Salgado, E. G., Lizano, S., Sarkar, S., Brugge, J. S., Ginsberg, M. H., and Shattil, S. J. (2003) Src kinase activation by direct interaction with the integrin beta cytoplasmic domain, *Proc. Natl. Acad. Sci. U.S.A. 100*, 13298–13302.
144. Stewart, M. (1990) Intermediate filaments: structure, assembly and molecular interactions, *Curr Opin. Cell Biol. 2*, 91–100.

145. Herrmann, H. and Aebi, U. (2004) Intermediate filaments: molecular structure, assembly mechanism, and integration into functionally distinct intracellular Scaffolds, *Annu. Rev. Biochem. 73*, 749–789.

146. Coulombe, P. A. and Wong, P. (2004) Cytoplasmic intermediate filaments revealed as dynamic and multipurpose scaffolds, *Nat. Cell Biol. 6*, 699–706.

147. Pallari, H. M. and Eriksson, J. E. (2006) Intermediate filaments as signaling platforms, *Sci. STKE 366*, 53.

148. Paramio, J. M., Segrelles, C., Ruiz, S., and Jorcano, J. L. (2001) Inhibition of protein kinase B (PKB) and PKCzeta mediates keratin K10-induced cell cycle arrest, *Mol. Cell Biol. 21*, 7449–7459.

149. Kim, S., Wong, P., and Coulombe, P. A. (2006) A keratin cytoskeletal protein regulates protein synthesis and epithelial cell growth, *Nature 441*, 362–365.

TRANSGLUTAMINASE 2 DYSFUNCTIONS IN THE DEVELOPMENT OF AUTOIMMUNE DISORDERS: CELIAC DISEASE AND TG2$^{-/-}$ MOUSE

ZSUZSA SZONDY
ILMA KORPONAY-SZABÓ
RÓBERT KIRÁLY
LÁSZLÓ FÉSÜS

CONTENTS

Advances in Enzymology and Related Areas of Molecular Biology, Volume 78.
Edited by Eric J. Toone.

I. INTRODUCTION

A. HISTORICAL BACKGROUND

The name transglutaminase was assigned by Heinrich Waelsch in 1959 [1] to an enzymatic activity in guinea pig liver, which catalyzes the incorporation of primary amines into proteins. It took almost a decade to learn that blood coagulation FXIII is also a transglutaminase [2] forming ε(γ-glutamyl)lysine cross-links in fibrin [3–5]. It soon has been revealed that the basic catalytic mechanism of all transglutaminases is the same [6], involving deamidation of the γ-carboxamide groups of specific protein-bound glutamine residues and the exchange of any of a wide variety of primary amines, water, or the ε-amino group of a lysine residue in a protein to ammonia in a Ca^{2+}-dependent manner. In mammalian genomes, eight transglutaminases are encoded [7] and the specific physiological functions of some of these are straightforward. For example, FXIII cross-links fibrin in the final stage of blood coagulation and TG1 forms the cornified envelope during terminal differentiation of keratinocytes. However, the biological role of other transglutaminases is still not fully determined and this is particularly true for the much studied and ubiquitously distributed TG2, which has been also called liver or tissue transglutaminase [8, 9]. This transglutaminase has acquired additional biochemical functions during evolution. It can work as a signal transduction protein utilizing GTP [10]. It is secreted from cells using an unconventional mechanism and binds fibronectin [11], integrins [12], and syndecan 4 [13] on the cell surface. It may have even protein disulphide [14] and protein kinase [15] activity. Although a large number of TG2 glutamine substrates, binding partners, and possible functions, both in various cell compartments and outside of cells, have been described or suggested, the physiological significance of TG2 is still an enigma. Most of the uncertainties are related to the lack of an apparent phenotype in TG2 knockout mice and the methodological difficulties to confirm cell culture results in in vivo studies. However, much attention has been paid to the possible role of TG2 in pathological phenomena with the obvious intention of finding new elements in the pathomechanism of various diseases and using them for the development of novel diagnostic and therapeutic tools. The accumulated pathologic data has also served, in an indirect way, to understand more the biology of TG2 as it is demonstrated by listing typical examples of TGase pathologies in the next section.

B. TRANSGLUTAMINASE 2 PATHOLOGIES

1. Gain of Function

Under physiological conditions, the two dominant activities of TG2 are tightly regulated by the presence of GDP/GTP or rising Ca^{2+} concentrations, which regulate its G protein-type signaling function (inhibited by Ca^{2+}) and transglutaminase activity (inhibited by GTP), respectively, in an opposing way [8]. Therefore, pathologic levels of intracellular Ca^{2+}, inappropriate appearance of TG2 in the extracellular space, or the unusual appearance of protein substrates may lead to unwanted protein modifications and nonphysiological reactions. One of the most striking examples is celiac disease, a major topic of this review, in which TG2-deamidated gluten peptides can trigger an immune reaction involving gliadin-reactive T cell and mucosal destruction in some of the individuals with specific HLA settings. Furthermore, TG2 becomes the major autoantigen during the course of the disease [16]. In ageing and stressed neurons, activated TG2 contributes to the accumulation of inclusion bodies with cross-linked protein polymers formed particularly from α-synuclein, ubiquitin, and heat shock proteins; when the proteasomal and autophagic protein degradation pathways cannot cope with the inclusions, the cells die in an increasing number, determining progression of Alzheimer's, Pick, or other neurodegenerative diseases [17]. Similar phenomena may take place during Mallory body formation in steatohepatitis [18]. Age-related cataract formation may involve TG2-mediated cross-linking of crystallins as a result of oxidative stress [19]. Formation of nondissociable angiotensin II type 2 receptor oligomers by TG2 contributes to the pathogenesis of severe Alzheimer's disease [20]. When TG2 is released from damaged cells in diseased liver or kidney, it may initiate fibrosis by cross-linking extracellular proteins [21, 22]. TG2 may enhance inflammatory reactions by activating phospholipase A2 [23, 24], freeing NFκB via cross-linking IκB [25, 26], or through increased degradation of PPAR-γ, a negative regulator of inflammation [27]. Overexpression of TG2 in the heart of a transgenic mouse model leads to hypertrophy, fibrosis, and increased apoptosis through enhanced transglutaminase activity as well as signaling with cyclooxygenase-2 activation and increased lipid peroxidation [28]. High amount of TG2 in cancer cells is associated with drug resistance and increased metastatic potential [29–31].

2. *Loss of Function*

Surprisingly, when knockout mouse models of TG2 were produced, they did not reveal any obvious phenotypes [32, 33]. However, when these mice have aged, autoimmune symptoms developed [34] mainly because of the deficient clearance of apoptotic cells, as discussed in the third part of this review. Also, symptoms of the MODY type of diabetes could be observed [35] and explained by missing the regulatory contribution of TG2 to insulin secretion. $TG2^{-/-}$ fibroblasts show decreased adherence to extracellular matrix and the knockout mice have impaired wound healing [36]. Several neutrophil functions, including migration and superoxide production, are compromised in the absence of TG2 [37]. Consistent with the proinflammatory role of TG2, the knockout mice were less sensitive to endotoxic shock [38]. TG2 has a protective role against some form of cell death since its absence sensitizes hepatocytes to Fas-dependent liver damage and cardiomyocytes to reperfusion injury [39, 40]. In $TG2^{-/-}$ mice, there is a reduction in inward vascular remodeling in response to reduced blood flow [41] probably because TG2-mediated dimerization of TG2 does not occur. These examples clearly show that the lack of TG2 has significant biological consequences that appear either during ageing or after various stress or pathologic conditions. Though no systematic investigation has been carried out, it should be also considered that TG2 knockout mice may adapt to life by turning on the expression of other transglutaminase isoforms as a compensatory mechanism.

II. TRANSGLUTAMINASE 2 AND CELIAC DISEASE

A. AUTOIMMUNITY AGAINST TG2

In some circumstances, autoimmunity can evolve against TG2 itself. Specifically, TG2 is a major autoantigen in celiac disease (also called as celiac sprue or gluten-sensitive enteropathy), and the antibodies targeting the enzyme are among the most characteristic features of this condition [42, 43]. Celiac anti-TG2 antibodies have some distinct and peculiar properties. They are induced by gluten, a complex of exogenous food proteins contained in wheat, rye, and barley. The autoantibody production continues as long as the patient regularly ingests gluten, but subsides when patients are placed on a gluten-free diet. This on–off feature is reproducible in unlimited number of occasions, which is often the case with subjects who transgress their gluten-free diet. This type of gluten intolerance lasts with very rare exceptions lifelong and is strongly inheritable, presumably because of congenital but yet unknown defects in handling gluten peptides. Further, celiac anti-TG2 antibodies are also found in dermatitis herpetiformis, where gluten induces a blistering skin disease in addition to the enteropathy. Typical celiac antibodies are produced in the gut mucosa [44], belong to IgA class, and target extracellular TG2 [45], as seen by their binding to frozen tissue sections containing high amounts of TG2 in reticulin fibers or in the endomysial structures of smooth muscles (endomysial antibody test). The endomysial-type TG2 antibodies are very reliable disease markers of the gluten-sensitivity disorders, also suitable for clinical diagnostic use and for celiac disease screening in asymptomatic populations.

These features are in sharp contrast with those of antibodies against TG2 seen in other disease conditions. Among others, tissue destruction and apoptosis can lead to various TG2 antibodies (IgA or IgG) with the mechanisms described later in context of apoptosis. Antibodies to TG2 were detected in human inflammatory and autoimmune diseases, like systemic lupus erythematodes, Sjögren's syndrome, psoriasis, ulcerative colitis, Crohn's disease, chronic liver diseases, and cirrhosis [46–50], as well as in autoimmunity-prone MRL*lpr/lpr* mice [51], but their pathologic significance is less clear. Further, low levels of serum anti-TG2 antibodies were demonstrated in tumor patients [52] and after myocardial injury [53]. In some cases, heat treatment of serum samples may also result in anti-TG2 ELISA reactivity [54]. Recently, temporary production of anti-TG2 antibodies was described in children even with minor infective diseases, which was self-limited and often belonged to only the IgG class [55]. However, these other antibodies do not bind to the endomysium and are independent of gluten-sensitivity. In the further description, we shall focus on the pathologic significance of TG2 in celiac disease.

B. PATHOLOGIC SIGNIFICANCE OF TG2 IN CELIAC DISEASE

1. Deamidation of Gliadin

In celiac disease, gliadin peptides derived from the ethanol-soluble toxic fraction of gluten evoke a specific T lymphocyte response in the gut, which then leads to chronic inflammation and tissue destruction characterized by villous atrophy, crypt hyperplasia, and elevation of intraepithelial lymphocytes [42]. Gliadin peptides also activate the innate immune system in multiple ways and can induce direct cytotoxicity and apoptosis [56–58].

Intraepithelial lymphocytes act as effectors of innate immunity to exclude stressed or damaged epithelial cells expressing MIC-A [59].

Gliadin peptides are rich in prolines and glutamines and are good substrates for TG2 in vitro. These peptides can be either incorporated into cross-linked complexes or can be deamidated, depending on the reaction environment and the amounts of available amine substrates. Some of the gliadin peptides can be cross-linked to TG2 itself on its defined lysine residues [60]. Deamidation seems to be determined by acidic pH, concentration of the substrates, and by the amino acid sequence of the peptides. Deamidation also can be indirect and secondary by hydrolysis of the isopeptide bonds [61]. Interestingly, anti-TG2 antibodies derived from celiac patients could enhance both the cross-linking and deamidating activity of purified recombinant TG2 [62], and a sufficient cross-linking of amines was also demonstrated in the presence of the antibodies with gliadin peptides [63]. Both reactions were modified by the presence of other proteins or cell lysate constituents that may result in variable outcomes in tissue conditions or on the surface of cells. This can explain the contradictory data generated in different experimental settings, and in particular, the decrease of TG2 activity when celiac antibodies were added to TG2 from cell lysate [62, 64].

Deamidation of gliadin creates peptides with enhanced immunogenicity. It has long been known that celiac disease occurs almost exclusively in subjects carrying MHC class II molecules HLA-DQ2 or DQ8, and gliadin-specific T lymphocytes are DQ restricted [65, 66]. However, these DQ heterodimers require negatively charged peptides at some well-defined positions for the binding to their antigen-presenting grooves [67]. The enzymatic action of TG2 can introduce such negative charges into gliadin peptides by deamidation of certain glutamines to glutamic acid on the basis of their spacing to proline residues [68]. This deamidation pattern perfectly matches the requirements to bind to DQ2 or DQ8. Alternatively, negative charge on the T cell receptor complementary regions may favor the complexing with native gliadin peptides [69]. Typically, multiple gliadin peptides, both native and deamidated, can induce immune response in celiac individuals. Also immunization of mice transgenic for human DQ8 with native gliadin peptides led to T lymphocytes that reacted to both native and deamidated gliadin peptides, and showed a stronger (heteroclitic) response to the deamidated forms. But none of the activated clones recognized exclusively the deamidated peptide, and notably, a heteroclitic response was never observed after the immunization with the deamidated peptide [69].

2. *Cell Surface TG2*

It is still unclear where the deamidation of gliadin peptides takes place during the development of celiac disease and whether this process would differ in nonceliac and celiac subjects. Interestingly, celiac patients produce antibodies against deamidated gliadin sequences, whereas healthy people do not [70]. Certain 12-mer, 19-mer, and 33-mer gliadin peptides survive quite long within the gastrointestinal tract because of their high proline content [71]. The 33-mer peptide contains multiple peptide sequences that can be recognized in their deamidated forms by gliadin-specific T lymphocytes [72], and the recognition does not require any further processing within antigen presenting cells [73]. The peptic-tryptic digest of gliadin obtained from wheat flour by the incubation with pepsin and trypsin is still toxic for celiac patients in vivo [74, 75], as well as is frequently used in vitro to investigate the reaction of lymphocytes or that of small bowel biopsy fragments in organ culture. Human digestive enzymes have difficulties in breaking further down the peptides with a penultimate proline, although dipeptidyl-peptidase IV (DPP-IV) may help other proteases [71, 76] to detoxify gluten and does not appear to be deficient in celiac patients. Human prolyl-endopeptidase (PEP), the human counterpart of potent gluten-degrading microbial enzymes recently proposed for treating celiac disease, is produced in high quantities in both normal and celiac small bowel epithelial cells [77, 78]. However, the human PEP enzyme cannot degrade the 33-mer peptide, which rather acts as an inhibitor of it. In addition, PEP is a cytoplasmic enzyme that does not reach the surface brush border layer [78]. Further, it has been shown that gliadin peptides can cross the epithelial barrier [77] and could be available in the tunica propria where they can meet TG2. Mainly nondeamidated gliadin peptides could be traced within the epithelial cells using epitope-specific monoclonal antibodies [79].

Although direct evidence is still lacking that TG2 in fact really deamidates gliadin in vivo, it has been demonstrated that gliadin peptides can bind to tissue sections where transglutaminases (TG1 or TG2) are present. In very old times, gliadin-specific serum antibodies were detected by a simple immunofluorescent method incubating frozen sections of esophagus or skin with gliadin solution and then adding serum antibodies [80]. These pictures show the binding of gliadin to the intercellular spaces of the epithelial layer, which is rich in TG1 but where TG2 is not recognizable. Later studies confirmed the role of TG2 in binding of gliadin to the extracellular matrix [81], which was Ca^{2+}-dependent and could be inhibited by putrescin or excess amounts of purified TG2. This and other studies also showed the colocalization of TG2 and externally added gliadin on reticulin fibers by double staining in confocal microscopy [81] or in immunofluorescent in vitro incorporation studies with tissue sections [82, 83]. These structures are close to the basolateral membrane of enterocytes. The immunodominant gliadin peptides may reach these compartments after their sorting to late endosomes [79], whereas other gliadin sequences, like the so-called "toxic" 31–49 alpha-gliadin fragment to which T lymphocytes do not react, may be differently transported to early endosomes and escape antigen presentation at the basolateral membrane [79]. Gliadin may also bind to surface TG2 expressed on lymphocytes and antigen presenting cells, like intestinal dendritic cells (DC). Several studies addressed this issue [84, 85], suggesting that gliadin peptides could undergo deamidation and subsequent loading onto surface HLA-DQ molecules even in extracellular compartments, but these results were later questioned because the antibody used to detect surface TG2 was subsequently found to cross-react with CD44 [86]. However, we confirmed the presence of surface TG2 on blood-derived DC and monocytes also by other commercial monoclonal TG2-specific antibody, TG100 [87]. This surface TG2 is catalytically active and upregulated by inflammatory signals, like LPS. Further, TG100 antibody showed surface TG2 expression on several other cells in the tunica propria of intestinal mucosa, and these cells were more abundant in celiac patients than in controls [87]. These results, taken together, strongly support the assumption that gliadin may meet TG2 before antigen presentation and deamidation can take place in the close vicinity of antigen presenting cells. Such a deamidation can also occur within the lumen of the gastrointestinal tract mediated by microbial transglutaminases often used as food additives [88] or in nonenzymatic ways by stomach acidity [89].

3. Mechanism of Neoantigen Formation

Involvement of TG2 in handling gliadin peptides can be implicated in multiple ways in the generation of autoantigenicity and antibody production against TG2. In particular, close contact of TG2 with gliadin, used as substrate, is thought to be the main cause why TG2 becomes an autoantigen. However, it must be noted that TG2-specific antibodies are not produced in other food intolerance diseases, like cow's milk enteropathy often leading to malabsorption syndrome and villous atrophy similar to celiac disease in young children [90]. This is very interesting because casein and its derivatives are well-recognized substrates of TG2 as well. According to one model, cross-linking of gliadin peptides to TG2 may generate neo-epitopes or can disturb the physiologic opening–closing cycle of TG2 and thereby exposing normally not accessible surface parts. In fact, potent active site-directed TG2 inhibitors were designed and developed by modifying certain gliadin peptides. These inhibitors can trap TG2 in a permanently open-extended conformation as shown by crystallography [91]. This large conformational change of TG2 was implicated in the induction of autoimmunity. Interestingly, complexing recombinant TG2 with these or similar inhibitors did not lead to enhanced reactivity of celiac antibodies [92]. It thus seems that epitope determinants in the hidden TG2 regions are not dominating over the epitopes already available for the antibodies in the resting form. Further, in contrast to inhibitors that "freeze" TG2 in the open form, gliadin peptides are continuously incorporated into high molecular weight complexes or deamidated with a constant rate during in vitro enzymatic assays, which means that most gliadin peptides are released from the enzyme's active site after some time.

Crosslinking of gliadin peptides to TG2 or other proteins may also generate neo-epitope formation in indirect ways. In fact, cross-linking to proteins and deamidation of gliadin occurred simultaneously when gliadin peptides were added to complex biological systems or tissue sections [83]. Citrullination (deimination of arginine residues) of various matrix proteins generates potent autoantigens in rheumatoid arthritis [93, 94] because change of even a low proportion of arginins may result in the deformation of sensitive proteins [95]. A similar process comprising deamidation of glutamine residues in proteins with cross-linked gliadin can enhance gliadin toxicity towards various extracellular matrix proteins, e.g., fibronectin, collagens, or TG2 itself [82]. However, such a process leads to antibodies against these proteins in rheumatoid patients with variable and much lower (around 30–40%) frequency [96, 97] than the prevalence of anti-TG2 antibodies in uncomplicated celiac disease. Antibodies against TG2 are found in virtually all celiac patients, whereas antibodies against fibronectin are exceptional [45, 98, 99] and purified antideamidated gliadin antibodies do not bind to tissues in any well-defined pattern that would indicate a clinically meaningful reaction with any of these matrix proteins [100]. Further, expression of matrix proteins is not altered during the course of celiac disease [101]. Celiac subjects with selective humoral IgA deficiency produce anti-TG2 circulating antibodies in IgG class [102], whereas IgM class anti-TG2 is found locally in the gut [103]. In addition, in seronegative celiac patients, anti-TG2 antibodies bound to TG2 locally in gut tissues can be found [104]. There was only one very old case report on celiac-type malabsorption in a male patient with agammaglobulinemia [105], but later evolution of this particular case showed incompatibilities (unpublished information from the authors).

Although some neo-epitopes may have role in the initiation of the autoimmune reaction, celiac anti-TG2 antibodies are mainly characterized by the capability to bind even normal TG2 in nondiseased tissues. Moreover, the "good" shape of the protein is required for a sensitive and specific reaction to occur [106]. Indirect immunofluorescent studies have long been used to detect celiac-specific antibodies in clinical laboratories, and these assays use various normal animal and human fetal tissues not exposed to food gliadin before. For example, the preferred substrate for the determination of endomysial antibodies is human umbilical cord tissue [107] and umbilical cord sections from premature babies work even better. Also recombinant TG2 antigens expressed in various cells systems or bacteria are good antigens often used in various clinical immunoassay formats for clinical diagnosis. These facts demonstrate that the presence of gliadin determinants is not needed for celiac autoantibodies for their binding, and TG2 alone is sufficient. No difference in antigenicity could be found between normal TG2 and the TG2 of the patients in antibody-binding studies [108]. Recognition of the own erythrocyte transglutaminase released by hemolysis of capillary blood samples constitutes a potent and rapid method to detect plasma anti-TG2 antibodies in the same samples [109], but antibodies of different celiac patients are capable to recognize TG2 of other patients as well. Further, no pathologic alterations in the TG2 gene could be found in celiac patients by sequence [110] and polymorphism analysis [111, 112], and neither celiac-type small bowel alterations nor anti-TG2 antibodies develop in $TG2^{-/-}$ mice [45].

However, it is still unclear whether unprocessed TG2 in its native form is the primary autoantigen in celiac disease or TG2 requires some post-translational modification to become an autoantigen. The extracellular TG2 form, i.e., the one contacted before with Ca^{2+}, is well recognized, while the intracellular protein is not. The role of Ca^{2+} was controversial in clinical assays. While it was evident that guinea pig TG2 needed Ca^{2+} activation to work properly in diagnostic assays to detect celiac antibodies [113], the results with human recombinant TG2 proteins were more sensitive and satisfactory even without a pretreatment with Ca^{2+}. The commercial guinea pig TG2 is purified from liver under conditions to prevent premature enzymatic activation, but the recombinant proteins may have had some contact with

Ca^{2+} already during purification. We indeed have shown that human TG2 possesses a strong Ca^{2+}-binding site (amino acids 229–233) that contains already one bound Ca^{2+} originating from the expression system and this cannot be removed by even extensive treatment with EDTA [114]. Further, we also showed that the binding epitope of celiac antibodies is related to Ca^{2+} binding, as mutation within one of the additional weak Ca^{2+} binding sites of TG2 (amino acids 151–158) resulted in profoundly decreased binding of celiac autoantibodies to TG2 [114].

Calreticulin (CRT) acts as a regulatory subunit for the G protein form of TG2 and can be found on the cell surface as well. CRT also has some sequence homology with gliadin peptides [115]. In biochemical assays, we found that TG2-CRT complexes are less recognized by celiac antibodies than either proteins alone, and gliadin peptides were able to dose-dependently displace CRT from these complexes, unmasking at the same time the epitopes recognized by celiac antibodies [116].

For the explanation of antibody production against normal TG2, a hapten-carrier model has been proposed [117]. According to this, gliadin-specific helper T lymphocytes provide the driving force to the antibody production, which can well explain why the anti-TG2 antibodies are eliminated after gluten is removed from the patient's diet. The presence of gliadin-specific T lymphocytes is well documented [66], while TG2-specific T cells were not yet found. According to the hapten-carrier model, gliadin-TG2 complexes bind to TG2-specific B cells, where both gliadin and TG2 fragments can be presented on DQ2. The presented gliadin-DQ2 complexes can activate gliadin-specific T-cells, which provide help for the TG2-specific B cells to produce anti-TG2 antibodies and for gliadin-specific B cells to produce antigliadin antibodies [117]. This hapten-carrier model also provides explanation for epitope spreading. Epitope mapping of TG2 antibodies with TG2 fragments has revealed that in addition to the first part of the core domain recognized as one main epitope, additional targeting of the *N*-terminal and *C*-terminal parts of TG2 occurs [118–120]. Further, some patients elaborate antibodies that also can react with homologous other transglutaminases, like TG3 [121] and TG6 [122]. The targeting of TG3 is associated with a skin manifestation of gluten sensitivity, dermatitis herpetiformis. In this condition, antibody complexes are also deposited in the subepidermal parts of the skin. These antibody complexes contain both patient IgA and TG3, which are normally not found in this localization, only in the upper layers of the skin. Dermatitis herpetiformis is usually associated with a celiac-type enteropathy in 90% of patients, and in these subjects, antibodies against

TG2 are also found. In the others, antibodies against TG3 can be the only antibodies. The antibodies against TG6 were implicated in the neural manifestations of celiac disease. Also in this group, some patients have both central or peripheral nervous system symptoms and classical celiac disease with anti-TG2 antibodies. The main neural symptom is a slowly developing sporadic cerebellar ataxia due to the loss of Purkinje cells. This ataxia can be responsive to gluten withdrawal, and patients may somewhat improve clinically, even if a complete restitution is not achieved. Some ataxia patients do not have villous atrophy in their small bowel, and at least in one-third, serum is negative for anti-TG2 antibodies. However, some of these cases may react with purified TG6 proteins [122]. In addition, antibodies against TG6 were found in brain parenchymal tissues of ataxia patients in the same study. Gluten ataxia has many relations also to TG2 and TG2-specific antibodies. TG2-specific antibodies deposit in brain vessels, mostly in those parts (cerebellum, brainstem) that are more involved, and ataxia patients also may have anti-TG2 antibodies in their small bowel even in the absence of circulating antibodies [123].

As a modification of the original hapten-carrier model, recent investigations identified that the antibody response against deamidated gliadin peptides is more specific for celiac disease than antibodies against native gliadin sequences [70]. Interestingly, some of the targeted gliadin peptides show a three-dimensional homology with TG2. TG2-specific monoclonal antibodies can cross-react with deamidated gliadin and antibodies purified with deamidated gliadin peptides reacted with TG2 [100]. These monoclonal antibodies are specific for the core domain or *C*-terminal part of TG2 and were produced in mice independently of celiac pathology. The findings raise the possibility that a molecular mimicry between gliadin sequences and TG2 may also contribute to autoimmunity. Although the primary amino acid sequences of gliadin and TG2 are different, deamidated peptides may act as mimotopes of TG2. Similarly, celiac patient antibodies purified with a rotavirus-homolog peptide sequence cross-reacted with TG2 and also with other self-proteins. These proteins included, among many others, desmoglein-1, myotubularin-related protein 2, heat shock protein 60, and toll-like receptor 4 [124]. Other results show that toll-like receptors 2 and 4 are constitutively upregulated in celiac patients both before and after diet [125]. These findings indicate that induction of some anti-TG2 antibodies may also be the result of a combined hapten-carrier and molecular mimicry reaction. Microorganisms, like rotavirus or other gastrointestinal infections, have been implicated in the induction of celiac disease [126] by creating an inflammatory environment that promotes increased permeability for gliadin peptides and results in enhanced reaction of the innate immune system. This can initiate an intestinal damage that is then perpetuated by autoimmune reactions. It is also possible that gliadin or other microbial antigens act as homolog of a self-protein that is deranged or defective in celiac patients, which has not induced a proper tolerance in fetal life. We propose that such a mechanism would also explain the familiar occurrence and heredity of gluten intolerance.

C. BIOLOGICAL EFFECTS OF CELIAC ANTIBODIES

Endomysial antibodies and antibodies to TG2 have long been considered as a kind of innocent bystander epiphenomenon in celiac disease. In the classical disease model, gliadin-specific T cells alone were implicated as responsible for the induction of epithelial cell damage, inflammation, small bowel villous atrophy, and clinical symptoms. Later it became evident that some features of the local immune reaction in the gut are not T cell mediated like early inflammatory changes and epithelial cell damage. In fact, the master cytokine of these responses is IL-15 and originates from the reaction of the innate immune system. Also increased number of intraepithelial lymphocytes was once held as indispensable for the celiac tissue destruction. Now, it was shown that intraepithelial lymphocytes represent rather a defense mechanism of the innate immunity, eliminating damaged epithelial cells by the engagement of NKG2D receptors on intraepithelial lymphocytes with the increased epithelial expression of MIC-A [59, 127]. Further, a recent paper shows that gliadin-specific CD4+ T lymphocytes did not induce celiac lesions in human DR3DQ2 mice model [128]. On the other hand, evidence is accumulating about the role of antibodies in the pathologic modification of TG2 functions. Serum antibodies to reticulin structures were described as early as 1971 [129] and later these antibodies were shown to be identical with the endomysial and anti-TG2 antibodies [45, 130]. Antibody positivity can reveal clinically silent celiac disease with villous atrophy as a noninvasive biomarker [131]. Further, antibodies may herald forthcoming disease already in the early stage when small bowel villous atrophy is not yet present [132, 133] and can be associated with extraintestinal disease manifestations of gluten-sensitivity that often manifest without a small bowel lesion (e.g., gluten ataxia). Even before the identification of the target of celiac antibodies, it was demonstrated that celiac IgA antibodies block the differentiation of intestinal epithelial cells in a 3-D tissue culture model

[134, 135] by interfering with the activation of transforming growth factor (TGF)-β. Increasing knowledge on TG2 biology has helped identify additional effects that can be TG2-mediated. It became evident that celiac anti-TG2 antibodies are functional and can also bind to their autoantigen in vivo during disease development [104, 136]. In celiac patients, this binding is well seen in small biopsy specimens below the basement membrane of the epithelial layer, around the crypts and submucosal glands, in the endomysial sheet of the tunica muscularis mucosae and around small vessels. Anti-TG2 antibodies were also found deposited in other tissues of celiac patients, particularly in the liver, muscles, kidney, lymph nodes, and brain [136]. The deposited antibodies were seen on vessel structures, reticulin fibers, or muscle structures where a concomitant expression of TG2 was also detected. TG2-specificity and functionality was demonstrated by the specific binding of external TG2 by the deposited antibody [104].

The initial assumption was that celiac antibodies disturb the function of TG2 by blocking its transglutaminase catalytic activity [137] via a presumed binding to the catalytic triad [138]. This would be a protective effect in regard of the deamidation of gliadin peptides, but also would diminish extracellular matrix cross-linking and stability. This effect was similar to the effect of function blocking monoclonal antibody CUB7402 when TG2 from cell lysate was used [55, 64]. However, other results show that incorporation and cross-linking of peptides also continues in the presence of antibodies [63] and when using purified recombinant TG2 bound to fibronectin, it can be even faster and longer lasting than without patient antibodies [62]. Taking into account these results and other observations showing that celiac antibodies can well bind to TG2 in ELISA even when the active site of the enzyme is occupied by an irreversible inhibitor [92], it seems improbable that celiac antibodies would in fact bind to the active site. Further, subepithelial TG2 can incorporate amines into extracellular matrix proteins in situ in tissue sections of celiac patients in a distribution pattern consistent with in vivo bound antibodies [137], whereas other experiments show that celiac antibodies themselves are not catalytic and do not exhibit per se transglutaminase activity in vitro [62]. As shown by active-site labeling in experimental animals, this extracellular TG2 is not constitutively active under normal conditions, but inflammation may lead to enhanced cross-linking in the extracellular matrix [139]. Relation of celiac antibody binding to a Ca^{2+}-binding site of TG2 [114] could be an explanation for the enhancing effect and possible gain of function. It is possible that the particular structure of the enzyme and its change during the catalytic cycle explain this property; it has already been very early on known that immunization of rabbits with TG2 can lead to two diverse populations of antibodies: (1) enhancing transamidation and (2) blocking transamidation [140].

TG2-specific antibodies derived from celiac patients inhibit several steps of angiogenesis in endothelial cells in culture, like tube formation, sprouting, and migration of endothelial and mesenchymal cells [141], and these negative effects can be reversed by TG2-specific inhibitors [142]. Further dissection of the pathomechanism has shown that effects of the antibodies are mediated by the increased TG2 activity on the cell surface, leading to increased vascular permeability and the upregulation of RhoA signaling pathway [92], which was different from the effects seen with the TG2 function blocking CUB 7402 antibody. Vessels may be preferential targets of celiac antibodies. They express high levels of TG2 and attract antibody binding in vivo both in the small bowel and elsewhere in the body. Strikingly, antibodies were also found to deposit in brain vessels in gluten ataxia [123], which can lead to increased permeability and various other problems with the blood–brain barrier. Loss of Purkinje cells is a characteristic feature in gluten ataxia. These big neurons are exceptionally sensitive to hypoxia due to the special arrangement of cerebellar vasculature, which makes highly probable that deposition of antibodies in brain vessels may also induce disturbances in blood supply. During the progression of gluten ataxia, usually a universal cortical and cerebellar atrophy develops, which further supports a possible role of small vessel-related diffuse ischemia in the development of celiac-related brain lesions. The vasculature could be equally important in the small bowel where high turnover of cells increases demands. The mucosal vessels are embedded in a fibrous sheet below the epithelial layer in which myofibroblasts move upwards during the differentiation of epithelial cells. Further, blood and lymphatic vessels contribute to the erectile force of villi. Irregular mucosal arrangement of vessels is readily seen with stereomicroscope [143] and well-known from early pathology studies [144] of celiac small bowel samples exhibiting villous atrophy. These studies also show that the number of mucosal vessels is diminished, and they also exhibit an abnormal ratio of endothelial/muscular components. Further, we recently observed that in long-standing active disease without proper dietary treatment, the deeper mucosal layer containing less vessels and TG2 is expanded. This could be a protective downregulation of TG2 autoantigenic targets in the small intestinal mucosa, explaining why many patients with seropositivity and manifest mucosal atrophy do not experience heavy symptoms.

Studies with mesenchymal cells found that celiac antibodies also affect the migration of fibroblasts and lead to enhanced matrix degradation [145]. In addition, celiac patient-derived antibodies expressed in and purified from phages caused shift in cell proliferation, inducing the entry in S phase [146]. This effect was extracellular signal-regulated kinase (ERK) dependent and presumably mediated by altered integrin-mediated signaling, while extracellular TG2 activity was not inhibited and apoptosis was not increased. In fact, enzymatic activity of TG2 might not be important in cellular processes mediated mainly by protein–protein interactions in which extracellular TG2 can be involved. Another constant finding throughout most studies was the alteration of the cytoskeleton in various cell types [141, 146]. Celiac antibodies also caused moderate degree of apoptotic features via mitochondrial pathway in cultured neuroblastoma cells [147]. These comprised the translocation of Bax into mitochondria and the cytoplasmatic release of cytochrome c at higher level than seen with control sera, but alterations were less expressed than those caused by celiac sera originating from patients with additional neurological disorders and antineuronal antibodies. Another fraction of celiac-related antibodies, which shows homology with the a rotavirus peptide, induced via cross-reaction to toll-like receptor 4 monocyte activation with elevated CD83 and CD40 expression and cytokine production [124].

It is less investigated whether celiac antibodies also may influence the GTP-ase function of TG2. Celiac antibodies decrease in vitro the hydrolysis of GTP [62], but it is a question whether the antibodies could penetrate into cells to exert such an effect in vivo. Under some conditions, both IgA and IgG class antibodies can be endocytosed or transported via hepatocytes in other autoimmune conditions [148]. It also has been demonstrated that antigliadin antibodies can enter intestinal epithelial cells linked to the transferrin receptor [77]. In preliminary experiments with endothelial cells in culture, we observed the accumulation of monoclonal anti-TG2 antibodies in various cell organelles (unpublished results).

There are only limited data regarding the effect of antibodies in more complex biological systems than cells. Unfortunately, there are no full animal models for celiac disease. In particular, none of the proposed models produces antibodies against TG2. Celiac antibodies expressed in mice generated high level of anti-idiotypic response [149] that led to the clearance of antibodies from the circulation and the mice remained asymptomatic. There are some indirect clinical observations showing that passive transfer of maternal antibodies into newborns may indeed cause symptoms [150]. Further, ataxia-like symptoms were elicited in mice by the intraventricular injection of patient serum from celiac cases with neurological symptoms or of TG2-specific antibodies expressed in phage display [151]. Effect of antibodies in patients may also be dependent on certain genetic predisposing factors. The clinical manifestations seen in celiac patients are highly variable ranging from severe gastrointestinal problems to lifelong latency [42]. In addition to HLA, an increasing number of non-HLA genetic polymorphisms associated with celiac disease are described with at least 40 loci identified in a large genome-wide study [152]. Most of these genes are related to the immune system or might influence immune gene expression and can determine the symptomatic outcome of the disease process in an individual patient.

Various biologic effects of gliadin may be antagonized at the level of antibodies, and these novel strategies may offer in the future additional therapeutic tools for celiac disease. At present, a lifelong gluten-free diet is the only effective therapy. TG2 inhibitors may block the deamidation of gliadin peptides and may diminish antigen presentation, but as reaction of gliadin-specific T cells is not restricted to the deamidated forms, this approach may not fully prevent the effects of natural gliadin ingestion. Further, blocking surface TG2 by a specific antibody was found to interfere with the toxic action of the 31–49 gluten peptide [84]. Interference with the enhancing effect of celiac antibodies on transglutaminase activity might be achieved by suitable inhibitors or other competing compounds, provided they are related to a limited number of target epitopes. However, it is difficult to foresee whether such an approach would interfere with all actions of antibodies, as target epitopes may vary.

III. TRANSGLUTAMINASE 2 AND SLE LIKE SYNDROMES

A. APOPTOSIS

Apoptosis, in contrast to necrosis (cell disintegration), does not provoke proinflammatory responses and plays a fundamental role in almost all physiological processes [153, 154]. Apoptotic cells are characterized by rapid shrinkage of the cytoplasm, nuclear condensation, membrane budding, and formation of apoptotic bodies. Importantly, membrane integrity during apoptosis is maintained, preventing liberation of intracellular histotoxic contents, thereby limiting the potential for propagation and exaggeration of inflammatory processes. To avoid cell disintegration (secondary necrosis), apoptotic cells must be efficiently and rapidly removed by macrophages [155], DC

[156], or in certain circumstances other cell types such as endothelial cells [157], vascular smooth muscle cells [158], and fibroblasts [159]. Moreover, clearance of apoptotic cells results in diminished proinflammatory mediator release and an augmented secretion of anti-inflammatory cytokines, TGF-β and IL-10 [160, 161], whereas uptake of necrotic cells, including secondarily necrotic cells derived from nonengulfed apoptotic cells, stimulates release of proinflammatory mediators (TNFα, IL-1β and NO) [160]. It is now widely accepted that failed clearance of apoptotic cells and consequent secondary necrosis may lead to the development of diseases, such as systemic lupus erythematosus (SLE) [162], type II diabetes [163], cystic fibrosis [164], chronic obstructive pulmonary disease [165], and atherosclerosis [166]. We will discuss all these events in the context of the loss of TG2 function.

B. PHAGOCYTOSIS OF APOPTOTIC CELLS

1. Recognition of Apoptotic Cells by Macrophages

Removal of apoptotic cells usually involves three central elements: (1) attraction of phagocytes via soluble "find me" signals, (2) recognition and phagocytosis via cell surface presenting "eat me" and removing of "don't eat me" signals, and (3) induction of release of anti-inflammatory, while active inhibition of production of proinflammatory cytokines.

Normal, healthy cells present signals that prevent binding to phagocytes or inhibit the engulfment activity. One of these "don't eat me" signals is CD47, which is known to be an integrin-associated protein. It is widely distributed on living cells and can act as a ligand for the heavily glycosylated, immunoreceptor tyrosine-based inhibitory motif-containing, signal regulatory protein α on the phagocyte [167].

In the initial part of apoptosis, dying cells stop presenting these signals and start to express surface markers, which are positive signals for their engulfment. In addition, they release soluble molecules that act as chemoattractant for the phagocytes. Although "find me" signals are poorly characterized, recent studies demonstrated that apoptotic bodies secrete in a caspase-3-dependent manner the phospholipid lysophosphatidylcholine [168] and nucleotides [169] that act on macrophage cell surface receptors to direct movement. In addition, split human tyrosyl-tRNA synthetase [170] and thrombospondin-1 (TSP) [171] were also recognized as attraction signals.

Once phagocytes reach their prey, their phagocytic receptors recognize specific "eat-me" signals on the apoptotic cells, through either direct apoptotic cell–phagocyte interactions or serum opsonizing proteins that bridge apoptotic ligands and the phagocyte receptors [172, 173]. The redistribution of phosphatidylserine (PS) on the surface of the apoptotic cell is the best characterized mark that distinguishes cellular life from death. Cellular plasma membranes maintain an asymmetric distribution of lipid molecules in the bilayer: almost all of the phospholipids that contain terminal primary amino groups, phosphatidylethanolamine, and PS are localized on the cytoplasmic surface. During apoptosis, cellular mechanisms that maintain PS asymmetry become inactivated and PS concentrates in patches on the outer leaflet [173]. In addition to PS, other molecules normally confined to the cytosolic side of the cell membrane also appear on the cell surface, including annexin I that colocalizes with PS [174], and the endoplasmic reticulum protein CRT whose levels increase on the cell surface during programed cell death by unknown mechanisms [167].

Some other "eat me signals" signals are shared with molecular patterns on the surface of microbial pathogens and have been termed apoptotic cell-associated molecular patterns (ACAMPs) [175]. ACAMPs are composed of oxidized and covalently modified lipids and proteoglycans on the apoptotic cell that interact with pattern recognition receptors (PRRs) in a similar manner to the recognition of bacterial pathogens by the innate immune system [176]. However, a big number of bridging molecules possesses rather poorly characterized "eat me signals" such as complement proteins C1q and C3b/bi, collectins, or the lung surfactant proteins-A and -D.

During the past decade, a large number of receptors that participate in cooperation in the phagocytosis of apoptotic cells have been identified. Many of these receptors recognize PS or its oxidized form directly or indirectly using bridging molecules such as milk fat globule EGF-factor 8 (MFG-E8), growth arrest-specific-6 (Gas6), protein S, or TSP-1. MFG-E8, Gas6, and protein S are secreted proteins characterized by posttranslational γ-carboxylation modification of glutamic acid residues that bind PS. Phagocytic receptors, $\alpha_v\beta_5$, and $\alpha_v\beta_3$ integrins bind an RGD motif in the EGF-like domain of MFG-E8 [177]. On the other hand, Gas6 [178] and protein S [179] are ligands for Mer family tyrosine kinases and activate them by inducing receptor dimerization and *trans*-phosphorylation [180]. TSP-1 is a multifunctional, 450-kDa adhesive glycoprotein, which facilitates cell adhesion with interacting fibronectin or fibrinogen. In phagocytosis, a model was proposed in which the alpha$_v\beta_3$ integrin and CD36 on phagocytes cooperate to form a high-affinity binding site for TSP-1, which binds to the oxidized PS on the apoptotic cells [181].

In addition to receptors that recognize PS via bridging molecules, simultaneous efforts by three separate laboratories, using very different strategies, have recently identified three directly binding PS receptors (PSR), suggesting that PS-binding bridge molecules is not the only way of PS recognition. T cell immunoglobulin- and mucin-domain-containing molecule (Tim4) was identified in a screen for engulfment-blocking monoclonal antibodies raised against naive peritoneal macrophages [182]. Meanwhile, brain-specific angiogenesis inhibitor 1 (BAI1) was identified as an interacting partner of engulfment-promoting protein ELMO, which associates and acts with Dock180 as a bipartite guanine nucleotide exchange factor for the small GTPase Rac [183]. The third PSR is stabilin-2, a multifunctional receptor binding a large array of ligands, and perhaps best known for its scavenger receptor and endocytic functions [184]. Intriguingly, each of these receptors seems to use different sequence structures to recognize PS and these may be different again from such recognition domains in the PS-binding bridge molecules or the scavenger receptors (SRs). While both BAI1 and stabilin-2 actively participate in the cellular uptake, Tim4 was recognized as a tethering receptor [185]

Since SRs can recognize diverse classes of molecules, including lipids and modified proteins; they also can detect apoptosis-related changes on cells. CD36 was one of the first macrophage receptors to be involved in recognition of apoptotic cells [181]. Besides binding TSP-1, it can also recognize oxidized PS [186]. Other SRs potentially involved in apoptotic cell clearance include SRB1, SRA, LOX-1, CD68, CD91, and CD14.

The low-density lipoprotein receptor-related protein 1 (LRP1) or CD91 is a multifunctional scavenger and signaling receptor. It interacts with CRT, which is highly expressed on the surface of apoptotic cells [167]. Mammalian ATP-binding cassette (ABCA7), which also enhances phagocytosis of apoptotic cells by macrophages, in response to apoptotic cells, moves to the macrophage cell surface and colocalizes with the LRP1 in phagocytic cups. ABCA7 appears to facilitate the cell-surface localization of LRP1 and associated signaling via ERK [187].

CD14 is found as a plasma-membrane-anchored molecule (mCD14) on the surface of monocytes, macrophages, and granulocytes as well as in soluble form (sCD14) in plasma. This molecule previously was much better known as a receptor for bacterial lipopolysaccharide. Apoptotic cells are docking through CD14 receptor and the macrophage CD14 mediates clearance of apoptotic cells without indicating inflammation [188]. The CD14 clearance pathway more probably can interact with common or

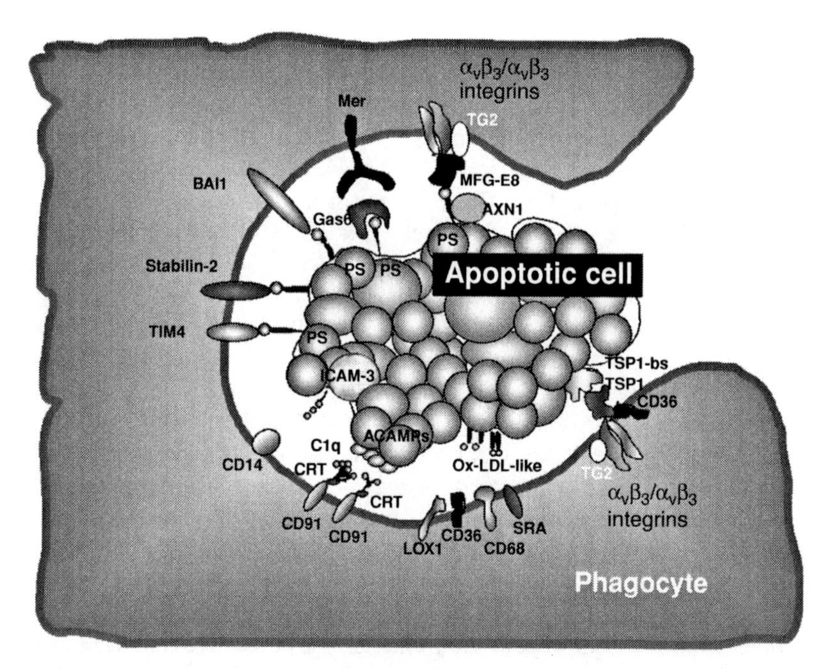

FIGURE 1. Macrophage cell surface receptors participating in the uptake of apoptotic cells. ACAMP, apoptotic cell-associated molecular pattern; AXN1, annexin 1; C1q, complement factor C1q; CRT, calreticulin; LOX-1, lectin-like oxidized low-density lipoprotein receptor-1; MFG-E8, milk fat globule EGF-factor 8; ox-LDL, oxidized low density lipoprotein; TG2, transglutaminase 2; TSP-1 thrombspondin-1; TSP-1 bs, TSP-1 binding sites.

multiple apoptosis-associated plasma-membrane structures, most probably ACAMPs.

The apoptotic signals and their recognition receptors are summarized in Figure 1.

2. *Uptake of Apoptotic Cells by Macrophages*

Independent of the ingested cell (apoptotic or necrotic) and the receptors involved, a number of common principles have been established: following receptor-mediated particle recognition actin polymerization occurs at the site of ingestion and the particle is internalized via an actin-dependent mechanism. After internalization, actin is shed from the phagosome and the phagosome matures by a series of fusion and fission events with components of the endocytic pathway culminating in the mature phagolysosome.

However, the signal pathways involved in uptake of apoptotic cells by either macrophages or nonprofessional phagocytes appear to be unique and very highly conserved evolutionarily, involving genes working within two parallel signaling pathways promoting cytoskeletal reorganization via the activation of the low molecular weight GTPase Rac1, which is obligatorily required for the uptake [189]. The first pathway is initiated by LRP-1(CD91) or stabilin-2 as receptors, and the adaptor protein GULP, and regulates Rac1 activity by yet unknown mechanisms [190, 191]. The second pathway is mediated via the 180-kDa protein downstream of chicken tumor virus no. 10 regulator kinase II (Dock180) and the engulfment and migration protein (ELMO), which form together an unconventional two-part guanine nucleotide exchange factor for Rac1 [192]. Dock180 contains an *N*-terminal SH3 domain, a large central region with a DOCKER domain and a *C*-terminal proline-rich region that likely binds the first SH3 domain of CrkII. ELMO, on the other hand, contains a PH domain, which can target interacting proteins to cell membranes containing phosphorylated inositides. ELMO1 can indeed facilitate the recruitment of Dock180 to cell membranes [193], and phagocytosis of apoptotic cells depends on the activation of phosphatidylinositol-3-OH kinase (PI-3kinase) [194]. Various upstream components exist for Rac activation. One scheme of the upstream regulation of the Dock180/ELMO/Rac signaling is the following: engagement of a ligand with the receptor leads to the tyrosine phosphorylation of that receptor or of a receptor-associated protein that interacts with the SH2 domain of CrkII. An SH3 domain in CrkII binds the *C*-terminal proline-rich region of Dock180 and recruits it to cell membranes. The proline-rich region of ELMO binds the *N*-terminal SH3 domain of Dock180 [193], which then promotes the GTP loading of Rac [192]. Integrin receptors are candidate regulators of this pathway [195]. Another upstream pathway of Rac identified in both mammalian cells and worms involves RhoG and the guanine nucleotide exchange factor TRIO [196], and this pathway also seems to be $\alpha v\beta 3$ integrin-regulated [197].

Mer activation by Gas-6 induces a postreceptor signaling cascade involving Src-mediated tyrosine phosphorylation of FAK on Tyr^{861}, the recruitment of FAK^{Tyr861} to the $\alpha v\beta 5$ integrin, and increased formation of the $p130^{CAS}$/CrkII/Dock180 complex to activate Rac1. Gas6 failed to stimulate phagocytosis in $\beta 5$-deficient phagocytes, which indicates that Mer is directly and functionally linked to the integrin pathway [198].

The CD36 signaling pathway and its contribution to the integrin signaling has not been studied in details during the phagocytosis of apoptotic cells, but because of its known components [199], a scheme can be suggested.

According to this scheme, CD36 activates the Src family tyrosine kinase Fyn, which can phosphorylate the substrate domain of the p130Cas, which can then transmit the signal to Dock180 and Rac [200].

The recently identified PS receptor, BAI1, can also form a trimeric complex with ELMO and Dock180 leading to the activation of Rac [183]. All these data indicate that many PS-dependent signals seem to converge on the Dock/ELMO downstream pathway and "synapse" to amplify internalization signals for the phagocytosis of apoptotic cells. These pathways are summarized in Figure 2.

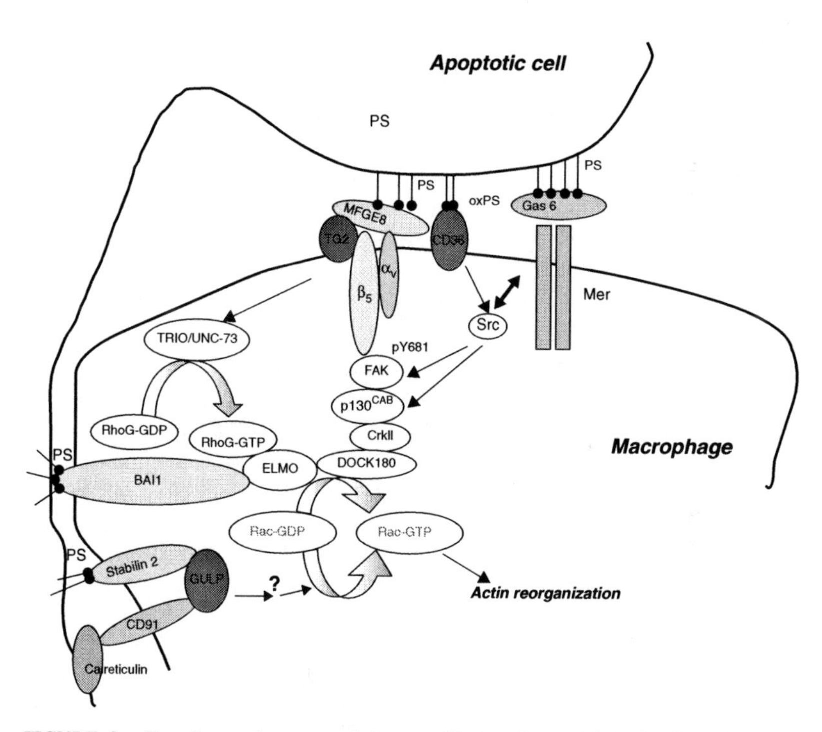

FIGURE 2. Signaling pathways regulating engulfment of apoptotic cells. GTP loading of the small GTPase Rac-1 plays a central role in initiation of actin reorganization leading to the apoptotic cell uptake. Rac-1 can be activated via the GULP pathway by unknown mechanisms or by the formation and activation of the bipartite guanine exchange factor DOCK180/ELMO. Activation of DOCK180/ELMO can be induced directly by the BAI1 receptor or via two main signaling pathways, which involve either GTP bound RhoG, or the CRKII/pCas complex. In addition, for full activation, the DOCK180/ELMO complex has to translocate to the membrane 3-phosphoinositide sites formed by the phosphoinositide-3-kinase activated during the phagocytosis process. (See insert for color representation.)

3. *Inhibition of Proinflammatory Cytokine Formation by Apoptotic Cells*

Although the identification of surface molecules involved in the recognition of apoptotic cells by macrophages has been the subject of sustained research efforts in recent years, much remains to be learned about the effect of apoptotic cell phagocytosis on macrophage function. This represents an intriguing issue, as phagocytosis of a variety of targets normally triggers a battery of proinflammatory responses in macrophages. In sharp contrast, ingestion of apoptotic cells by macrophages induces an anti-inflammatory phenotype. Apoptotic cells do not simply fail to provide proinflammatory signals, but they actively interfere with the inflammatory program, as preincubation with apoptotic cells strongly suppresses the lipopolysacharide-induced inflammatory response of macrophages [160, 161, 201]. Conversely, macrophages secrete increased amounts of the anti-inflammatory TGF-β and IL-10 under the same conditions [160, 161, 202]. Active inhibition of proinflammatory cytokine production of macrophages by apoptotic cells might play a crucial role in the resolution of acute inflammations, where dying neutrophils will alter the phenotype of inflammatory macrophages and promote their clearance function.

How the anti-inflammatory phenotype and the production and release of TGF-β and IL-10 are regulated was not investigated so far, but recently several reports suggested that receptors, such as Mer, C1q receptor, or stabilin2 involved in the phagocytosis of apoptotic cells, play a determining role in the phenomenon [160, 184, 203, 204]. Collectively, these studies suggest that macrophages not only contribute to the resolution of inflammation through apoptotic cell removal, but also by actively suppressing inflammatory mediator production.

C. FAILURE IN THE CLEARANCE IN APOPTOTIC CELLS AND AUTOIMMUNITY

Increasing evidence suggest that failure in the clearance of apoptotic cells can lead to the development of autoimmunity. First, development of SLE-like autoimmunity was observed in many of the mouse models, in which phagocytosis of apoptotic cells was impaired [205–209]. These mice develop splenomegaly, spontaneous antibodies against nuclear antigens, as well as IgG deposits in the kidney, leading to glomerulonephritis, a common complication of SLE. Second, in a number of human SLE patients, impaired phagocytosis of apoptotic cells can be detected, though in most of these cases, it is not clear whether the impaired phagocytosis of apoptotic cells is the cause or the consequence of the disease [210]. In addition, failure in the clearance of apoptotic cells was observed in some other diseases as well, which have a chronic inflammatory component, such as type II diabetes [163], cystic fibrosis [164], chronic obstructive pulmonary disease [165], and atherosclerosis [166].

In the absence of proper clearance of apoptotic cells, several mechanisms have been attributed to the development of autoimmunity. Apoptotic cells, which are not taken up, undergo secondary necrosis, representing a source of proinflammatory stimuli and potentially immunogenic autoantigens. Nucleosomal proteins released from DNA represent one of the most abundant classes of autoantigens found in patients with autoimmune disorders [211]. Additionally, delayed uptake of apoptotic cells may generate caspase-derived neoantigenic peptides that are also capable of triggering autoimmune responses [212]. If apoptotic cells are not properly cleared, they become accessible to tissue or follicular DC, which play a key role in presenting antigens and activating the immune system. The capacity of DC to induce immune activation to a given antigen is strongly induced by proinflammatory signals, while the presence of the anti-inflammatory cytokines such as TGF-β or IL-10 promotes peripheral tolerance [213]. Altered capability of apoptotic cells to inhibit proinflammatory cytokine production itself may be important in the development of autoimmunity, as in the mouse models, where the impaired clearance of apoptotic cells was not accompanied by altered proinflammatory production upon apoptotic cell uptake, autoimmunity was not observed [188]. The proinflammatory milieu can be further propagated in the later phases of the disease, when autoantibodies opsonizing apoptotic corpses appear, which may alter the type of cellular uptake by promoting Fc receptor-mediated phagocytosis of apoptotic cells with proinflammatory consequences [214].

D. TRANSGLUTAMINASE 2, CLEARANCE OF APOPTOTIC CELLS, AND AUTOIMMUNITY

Increasing evidence suggests that TG2, a multifunctional protein [9], is strongly coupled to the above processes and the loss of it will enhance the sensitivity to the development of autoimmunity. Eight distinct enzymatically active transglutaminases have so far been described [8]. TG2 is very unique among the TGase family members, because in addition to catalyzing the formation of protein cross-links, it also possesses GTPase activity (and acts as a G protein) [10], protein disulfide isomerase activity [14], and protein kinase activity [215]. In apoptotic cells, TG2 translocates to the mitochondria and contributes to the cell death decision using its BH3-only domain [216]. Although TG2 is localized predominantly in the cytoplasm, substantial amounts of the protein is present also on the surface in various cell types and in the extracellular matrix [217, 218], despite the fact that TG2 has no leader sequence, hydrophobic domains, or posttranslational modifications for targeting the endoplasmic reticulum or Golgi apparatus. In addition, TG2 also interacts with integrins of the β_1 and β_3 subfamilies, and integrin/TG2 complexes are detected inside the cell during biosynthesis and accumulate as coreceptors on the cell surface [219, 220]. TG2 can also bind to the major extracellular protein fibronectin [221] and cooperate with integrins in cell adhesion through either a direct noncovalent interaction with the β_1 and β_3 integrin subunits or formation of stable ternary complexes with integrins and fibronectin [219, 220]. These interactions induce integrin clustering and modify integrin signaling [222].

TG2 was first shown by us to be coupled to the in vivo apoptosis program based on the observation that the protein appeared very often in tissues, where continuous cell death program was going on, including the intestine or the thymus [217, 223]. This observation then was confirmed by others widening the range of tissues, in which TG2 appeared during enhanced apoptosis [224–228]. In addition, the enzyme was induced in various in vivo apoptosis models [229–232], and the regulation seemed to be transcriptional [233]. However, at least in the case of thymocytes apoptotic signals alone were not sufficient to induce the expression of the enzyme, because these cells efficiently died by apoptosis in vitro, and still the induction of TG2 was not observed [233]. This observation suggested that tissue factors present in the tissue environment contribute to the induction of the enzyme. Indeed, our recent partially published studies suggest that retinoids produced by thymic epithelial cells [234] and TGF-β produced by macrophages engulfing apoptotic cells [34] contribute to the induction of the protein in apoptotic thymocytes, as the TG2 promoter contains both retinoid [235] and TGF-β response elements [236].

Though TG2 can contribute to the apoptosis by using its BH3-only domain [216] or its protein kinase activity [215], early studies suggested that intracellular increases in the cytosolic Ca^{2+} concentrations during the apoptosis program will trigger the cross-linking activity of TG2 [229, 232]. Indeed, in vivo apoptosis induction and appearance of the enzyme was accompanied by increases in the levels of $\varepsilon(\gamma$-glutamyl)lysine dipeptide [232]. Covalent $\varepsilon(\gamma$-glutamyl)lysine cross-links are formed between polypeptide chains of proteins only when TG2 is active within cells. Since proteases do not cleave the $\varepsilon(\gamma$-glutamyl)lysine cross-link formed by TG2 [237], the amount of isodipeptide released into blood circulation following phagocytosis and degradation of apoptotic bodies reflected the rate of apoptosis in tissues [238]. Though TG2 could contribute to the initiation of apoptosis by cross-linking the retinoblastoma protein [239] or the adenine nucleotide translocator 1 [240], and at the beginning, the enzyme was considered as a determining player in the effector phase of the apoptosis [241], it has been proposed that the main role of the cross-linked proteins is to prevent the leakage of the harmful cell content, thus preventing tissue damage and inflammation [51]. One of the proteins identified as an abundant TG2 substrate was actin [242]. This proposal indicated already very early that an important role of TG2 during apoptosis is to maintain the tissue homeostasis.

In addition, two further studies demonstrated that TG2 expressed by apoptotic cells contribute to the proper clearance of apoptotic cells. First, TG2 in apoptotic cells was shown to cross-link the S19 ribonuclear protein that acted as chemotactic factor for macrophages [243]. Second, TG2 promoted PS exposure and thus early recognition of dying red blood cells [244]. After all these observations, it was not a surprise that TG2 null mice developed autoimmunity with a dominance in females and is characterized by splenomegaly, antinuclear and antitissue antibodies, and glomerulonephritis induced by IgG deposits [34]. In addition to the evidence of the existence of autoreactive B cells, we found IgG type anti-IgG antibodies in these mice, the production of which requires the function of autoreactive $CD4^+$ T cells. Very unusually, apoptosis in these mice was associated with an inflammatory response, as evidenced by the presence of large infiltrates of blood cells in the surrounding tissue. Surprisingly, however, these mice did not show a clear apoptosis phenotype [32, 33]. Instead accumulation of apoptotic cells could be detected in tissues, when excess apoptosis was induced, indicating failure in the clearance of apoptotic cells [34]. However, the failure in the clearance seemed to be related to the macrophages, rather than to the apoptotic cells, and affected only the removal of apoptotic cells.

Macrophages have been known for a long time to express TG2, and also that inflammation induced its levels [245, 246]. It has been shown that 10% of the total TG2 is expressed on the cell surface of macrophages, where it acts as an integrin coreceptor for cell adhesion and migration of these cells [220]. Interestingly, two receptors participating in the phagocytosis of apoptotic cells have been connected to the TG2 cell surface traffic: (1) integrin β_3/TG2 complexes are detected inside the cell during biosynthesis and accumulate together as coreceptors on the cell surface [219, 220], while (2) LRP/CD91 is responsible for the removal of TG2 from the cell surface [247].

Detailed analysis of the phagocytosis deficiency has shown that TG2 is also required for proper phagocytosis in the context of integrin β_3. In the absence of TG2, integrin β_3 could not accumulate in the phagocytic cup, and the integrin β_3 signaling via RhoG leading to Rac1 activation was found to be impaired during phagocytosis of apoptotic cells [248, 249]. In addition, activation of the PI-3kinase required for the proper localization of the Dock180/ELMO complex also seemed to be impaired. To compensate the defect, TG2 null macrophages induced the expression of integrin β_3 and that of RhoG, resulting in increased basal levels of GTP-bound Rac1 and enhanced motility, as motility is also coupled to the integrin/RhoG pathway [196], but seems to be less affected by the loss of TG2. In the absence of TG2, macrophages had difficulties in initiating the formation of the phagocyte portals, and even those, which could initiate the formation of the portal, showed a much slower engulfment [248]. So we proposed that the main role of TG2 in phagocytosis is to stabilize phagocyte portals via regulating integrin clustering. How can TG2 achieve this?

Previous studies have shown that in the context of cell adhesion and migration, TG2 can promote the function of integrin receptors by simultaneously binding to integrin and its ligand fibronectin forming a ternary complex [219–221]. Similarly, we found that TG2 also interacts with MFG-E8, which is the integrin β_3/β_5 ligand during phagocytosis, indicating that a ternary complex formation might also be needed for proper phagocytosis. TG2 may exist in two opposing conformations: (1) in an open conformation, in which its cross-linking function is active, and (2) in a closed conformation, which is stabilized by guanine nucleotide binding [91]. Our not yet published studies indicate that MFG-E8 binding by TG2 is dependent on the conformation of the protein, preferring the guanine nucleotide-bound form. The binding involves the catalytic core domain and the first β-barrel domain, which twists away from the core domain during activation. These data are in good correlation with those, which indicate that TG2 must bind guanine nucleotides to promote phagocytosis [248, 250]. In addition, studies, in which we tested the compensation in the expression of various phagocytic receptors in macrophages lacking a phagocytic receptor, indicate that integrin β_3 and TG2 might form a functional unit with CD36, as their expression was found to be up- or downregulated in each type of macrophages simultaneously. Whether cell surface TG2 itself can interact with the apoptotic cells is not clear yet. We have not found evidence for PS binding [248]; however, the protein is known to interact with both CRT [251] and annexin I [252] appearing on the cell surface of apoptotic cells.

Similar to many other phagocytic receptor mutants, the loss of which lead to autoimmunity, TG2 null macrophages also showed an altered proinflammatory cytokine production, when exposed to apoptotic cells [253]. Besides affecting directly the production of proinflammatory cytokines as a coreceptor, this observation might be related to the fact that macrophage cell surface TG2 contributes to the latent TGF-β activation [254]. Using TG2 null macrophages, we have confirmed that these cells cannot produce active TGF-β. The activation of latent TGF-β is rather complex and the exact role of TG2 is not clear in it. Though in the initial studies it was proposed that cross-linking activity of TG2 is required, it is interesting to note that integrins [255], CD36 [256], and TSP [257] all play a role in it.

Lack of TGF-β activation by macrophages might have several consequences, which might contribute to the development of autoimmunity. TGF-β will not promote TG2 expression in macrophages and in apoptotic cells, and it will not inhibit proinflammatory cytokine formation by macrophages during engulfment of apoptotic cells. What is more, in the absence of proper TGF-β formation, the development of regulatory T cells, which inhibit the development of autoimmunity to a given antigen, might also be less effective [258].

TG2 null macrophages also show enhanced proinflammatory cytokine production as compared to their wild-type counterparts, when exposed to bacterial LPS. Though previous studies indicated that TGF-β produced by LPS stimulation might play a feedback role in the regulation of LPS-induced proinflammatory cytokine production [259], our data indicate that the compensatory increase in integrin β3 expression leads to an enhanced src kinase activation, which sensitizes NFκB. As a result, these cells also express higher levels of NFκB in the nucleus under basal conditions and respond to LPS stimulation more vigorously [260].

Inducing apoptosis in the thymus by injecting anti-T cell receptor antibodies or glucocorticoids or by γ-irradiation results in the loss of the 80% of thymocytes in the mouse thymus within 48 hours [232], still apoptotic cells are cleared efficiently without inducing inflammation. Thus macrophages must sense the amount of apoptotic cells and respond to the increased demand by an enhanced phagocytic capacity. Indeed, a recent report indicates that one of such mechanisms involves the Liver X receptor (LXR) signaling pathway, which might sense the amount of cholesterol taken up within the apoptotic cells [209]. Following apoptotic cell uptake, the expression of Mer was found to be induced in a direct LXR-dependent manner, while induction of TG2 expression involved the LXR-dependent induction of RARα [261]. Macrophages from LXR null mice showed Mer- and TG2-dependent impairment in the phagocytosis of apoptotic cells, and the mice developed autoimmunity similar to the one found in TG2 null or Mer null mice [209].

A number of reports indicate that the frequency of atherosclerosis is enhanced in autoimmune patients [262]. In addition, it was also observed that phagocytosis of apoptotic cells is very often impaired in atherosclerotic lesions [166]. Interestingly, bone marrow transplantation of macrophages showing defect in TG2 or Mer can promote the development of atherosclerosis on an $ApoE^{-/-}$ or an LDL receptor null background [263]. These data indicate that impaired phagocytosis and altered proinflammatory cytokine production in autoimmune diseases might contribute to the increased sensitivity of these patients to the development of atherosclerosis.

Could then therapies that enhance the expression of TG2 and the phagocytosis of apoptotic cells attenuate the development of autoimmunity or of atherosclerosis? Interestingly, ligation of the LXR receptor by synthetic ligands, which upregulated TG2 in macrophages, could attenuate both processes in animal models [209, 264]. Altogether, these data indicate that TG2 plays a determining role in the regulation of phagocytosis of apoptotic cells and in the downregulation of inflammation, and pharmacological induction of its expression in macrophages might contribute to the therapies of autoimmune and atherosclerotic diseases.

IV. CONCLUDING REMARKS

There are diverse pathomechanisms and manifestations of autoimmune diseases. It is still quite unusual that, as it has been detailed above, both gain of function and loss of function disturbances of transglutaminase 2 physiology may lead to autoimmunity. However, the molecular background and the consequences of the two TG2 dysfunctions differ in several respects. First, in celiac disease, TG2 is both the initiator (when gluten peptides appear) and the target (in organs where it is normally present extracellularly) of the immune reaction. Lack of TG2 in mice, on the other hand, results in a classical autoimmune disease gradually developing against various tissue components, suggesting disturbance in a general tissue phenomenon, which is in clearance of apoptotic cells. Celiac disease appears in about 1% of the human population and it is not related to single nucleotide polymorphisms (SNPs) or mutations of the TG2 gene [110–112]; it appears that the normally available TG2 reacts to an alimentary constituent in susceptible individuals who have a so far unrevealed sensitizing condition. Autoimmune disease as a result of deficient TG2 activity has not been observed in humans. In fact, although there are more than 300 TG2 SNPs in the normal populations (all available public databases), the frequency of those determined so far is quite low and there is no report of their association with any disease. Again, it is possible that systematic studies will reveal that deficient activity of TG2 developed in a nongenetic way may contribute to the progression of human autoimmune disease. Beyond contributing to better understanding of autoimmune diseases, TG2 may also offer novel therapeutic possibilities. In celiac diseases, specific inhibitors of TG2 [31] and prevention or elution of the deposition of anti-TG2 antibodies to tissues may provide therapeutic benefits. Replacement or upregulation of TG2 may also have a place in future treatment protocols in human autoimmune syndromes.

ACKNOWLEDGMENTS

This work was supported by Hungarian Scientific Research Funds (OTKA NI 67877, K 77587, K 61868), Hungarian Ministry of Health, TAMOP 4.2.1/B-09/1/KONV-2010-0007 project (implemented through the New Hungary Development Plan, Co-Financial by the European Social Fund) and EU (MRTN-CT-2006-036032, MRTN-CT 2006-035624, LSHB-CT-2007-037730).

REFERENCES

1. Mycek, M. J., Clarke, D. D., Neidle, A., and Waelsch, H. (1959) Amine incorporation into insulin as catalised by transglutaminase, *Arch. Biochem. Biophys. 84*, 528–540.
2. Loewy, A. G. (1968) Mechanism of action of factor XIII, *Thromb. Diath. Haemorrh. Suppl. 28*, 1–12; discussion, 23–54.
3. Pisano, J. J., Finlayson, J. S., and Peyton, M. P. (1968) Cross-link in fibrin polymerized by factor 13: epsilon-(gamma-glutamyl)lysine, *Science 160*, 892–893.
4. Matacić, S. and Loewy, A. G. (1968) The identification of isopeptide crosslinks in insoluble fibrin, Biochem. Biophys. Res. Commun. *30*, 356–362.
5. Lorand, L., Downey, J., Gotoh, T., Jacobsen, A., and Tokura, S. (1968) The transpeptidase system which crosslinks fibrin by gamma-glutamyle-episilon-lysine bonds, *Biochem. Biophys. Res. Commun. 31*, 222–230.
6. Folk J. E. and Chung, S. I. (1973) Molecular and catalytic properties of transglutaminases, *Adv. Enzymol. Relat. Areas Mol. Biol. 38*, 109–191.
7. Aeschlimann, D., Koeller, M. K., Allen-Hoffmann, B. L., and Mosher, D. F. (1998) Isolation of a cDNA encoding a novel member of the transglutaminase gene family from human keratinocytes. Detection and identification of transglutaminase gene

products based on reverse transcription-polymerase chain reaction with degenerate primers, *J. Biol. Chem. 273*, 3452–3460.

8. Lorand, L. and Graham, R. M. (2003) Transglutaminases: crosslinking enzymes with pleiotropic functions, *Nat. Rev. Mol. Cell Biol. 4*, 140–156.
9. Fesus, L. and Piacentini, M. (2002) Transglutaminase 2: an enigmatic enzyme with diverse functions, *Trends Biochem. Sci. 27*, 534–539.
10. Nakaoka, H., Perez, D. M., Baek, K. J., Das, T., Husain, A., Misono, K., Im, M. J., and Graham, R. M. (1994) Gh: a GTP-binding protein with transglutaminase activity and receptor signaling function, *Science 264*, 1593–1596.
11. Radek, J. T., Jeong, J. M., Murthy, S. N., Ingham, K. C., and Lorand, L. (1993) Affinity of human erythrocyte transglutaminase for a 42-kDa gelatin-binding fragment of human plasma fibronectin, *Proc. Natl. Acad. Sci. U.S.A. 90*, 3152–3156.
12. Zemskov, E. A., Janiak, A., Hang, J., Waghray, A., and Belkin, A. M. (2006) The role of tissue transglutaminase in cell-matrix interactions, *Front. Biosci. 11*, 1057–1076.
13. Telci, D., Wang, Z., Li, X., Verderio, E. A., Humphries, M. J., Baccarini, M., Basaga, H., and Griffin, M. (2008) Fibronectin-tissue transglutaminase matrix rescues RGD-impaired cell adhesion through syndecan-4 and beta1 integrin co-signaling, *J. Biol. Chem. 283*, 20937–20947.
14. Hasegawa, G., Suwa, M., Ichikawa, Y., Ohtsuka, T., Kumagai, S., Kikuchi, M., and Saito, Y. (2003) A novel function of tissue-type transglutaminase: protein disulphide isomerase, *Biochem. J. 373*, 793–803.
15. Mishra, S. and Murphy, L. J. (2004) Tissue transglutaminase has intrinsic kinase activity: identification of transglutaminase 2 as an insulin-like growth factor-binding protein-3 kinase, *J. Biol. Chem. 279*, 23863–23868.
16. Dieterich, W., Ehnis, T., Bauer, M., Donner, P., Volta, U., Riecken, E. O., and Schuppan, D. (1997) Identification of tissue transglutaminase as the autoantigen of celiac disease, *Nat. Med. 3*, 797–801.
17. Nemes, Z., Devreese, B., Steinert, P. M., Van Beeumen, J., and Fésüs, L. (2004) Cross-linking of ubiquitin, HSP27, parkin, and alpha-synuclein by gamma-glutamyl-epsilon-lysine bonds in Alzheimer's neurofibrillary tangles, *FASEB J. 18*, 1135–1137.
18. Strnad, P., Harada, M., Siegel, M., Terkeltaub, R. A., Graham, R. M., Khosla, C., and Omary, M. B. (2007) Transglutaminase 2 regulates mallory body inclusion formation and injury-associated liver enlargement, *Gastroenterology 132*, 1515–1526.
19. Shin, D. M., Jeon, J. H., Kim, C. W., Cho, S. Y., Lee, H. J., Jang, G. Y., Jeong, E. M., Lee, D. S., Kang, J. H., Melino, G., Park, S. C., and Kim, I. G. (2008) TGFbeta mediates activation of transglutaminase 2 in response to oxidative stress that leads to protein aggregation, *FASEB J. 22*, 2498–2507.
20. AbdAlla, S., Lother, H., el Missiry, A., Langer, A., Sergeev, P., el Faramawy, Y., and Quitterer, U. (2009) Angiotensin II AT2 receptor oligomers mediate G-protein dysfunction in an animal model of Alzheimer disease, *J. Biol. Chem. 284*, 6554–6565.
21. Nardacci, R., Lo Iacono, O., Ciccosanti, F., Falasca, L., Addesso, M., Amendola, A., Antonucci, G., Craxì, A., Fimia, G. M., Iadevaia, V., Melino, G., Ruco, L., Tocci, G., Ippolito, G., and Piacentini, M. (2003) Transglutaminase type II plays a protective role in hepatic injury, *Am. J. Pathol. 162*, 1293–1303.

22. Skill, N. J., Johnson, T. S., Coutts, I. G., Saint, R. E., Fisher, M., Huang, L., El Nahas, A. M., Collighan, R. J., and Griffin, M. (2004) Inhibition of transglutaminase activity reduces extracellular matrix accumulation induced by high glucose levels in proximal tubular epithelial cells, *J. Biol. Chem. 279*, 47754–47762.

23. Cordella-Miele, E., Miele, L., and Mukherjee, A. B. (1990) A novel transglutaminase-mediated post-translational modification of phospholipase A2 dramatically increases its catalytic activity, *J. Biol. Chem. 265*, 17180–17188.

24. Sohn, J, Kim, T. I., Yoon, Y. H., Kim, J. Y., and Kim, S. Y. (2003) Novel transglutaminase inhibitors reverse the inflammation of allergic conjunctivitis, *J. Clin.* Invest. *111*, 121–128.

25. Kuncio, G. S., Tsyganskaya, M., Zhu, J., Liu, S. L., Nagy, L., Thomazy, V., Davies, P. J., and Zern, M. A. (1998) TNF-alpha modulates expression of the tissue transglutaminase gene in liver cells, *Am. J. Physiol. 274*, G240–G245.

26. Lee, J., Kim, Y. S., Choi, D. H., Bang, M. S., Han, T. R., Joh, T. H., and Kim, S. Y. (2004) Transglutaminase 2 induces nuclear factor-kappaB activation via a novel pathway in BV-2 microglia, *J. Biol. Chem. 279*, 53725–53735.

27. Maiuri, L., Luciani, A., Giardino, I., Raia, V., Villella, V. R., D'Apolito, M., Pettoello-Mantovani, M., Guido, S., Ciacci, C., Cimmino, M., Cexus, O. N., Londei, M., and Quaratino, S. (2008) Tissue transglutaminase activation modulates inflammation in cystic fibrosis via PPARgamma down-regulation, *J. Immunol. 180*, 7697–7705.

28. Zhang, Z., Vezza, R., Plappert, T., McNamara, P., Lawson, J. A., Austin, S., Praticò, D., Sutton, M. S., and FitzGerald, G. A. (2003) COX-2-dependent cardiac failure in Gh/tTG transgenic mice, *Circ. Res. 92*, 1153–1161.

29. Verma, A., Wang, H., Manavathi, B., Fok, J. Y., Mann, A. P., Kumar, R., and Mehta, K. (2006) Increased expression of tissue transglutaminase in pancreatic ductal adenocarcinoma and its implications in drug resistance and metastasis, *Cancer Res. 66*, 10525–10533.

30. Mangala, L. S., Fok, J. Y., Zorrilla-Calancha, I. R., Verma, A., and Mehta, K. (2007) Tissue transglutaminase expression promotes cell attachment, invasion and survival in breast cancer cells, *Oncogene 26*, 2459–2470.

31. Yuan, L., Siegel, M., Choi, K., Khosla, C., Miller, C. R., Jackson, E. N., Piwnica-Worms, D., and Rich, K. M. (2007) Transglutaminase 2 inhibitor, KCC009, disrupts fibronectin assembly in the extracellular matrix and sensitizes orthotopic glioblastomas to chemotherapy, *Oncogene 26*, 2563–2573.

32. De Laurenzi, V. and Melino, G. (2001) Gene disruption of tissue transglutaminase, *Mol. Cell Biol. 21*, 148–155.

33. Nanda, N., Iismaa, S. E., Owens, W. A., Husain, A., Mackay, F., and Graham, R. M. (2001) Targeted inactivation of Gh/tissue transglutaminase II, *J. Biol. Chem. 276*, 20673–20678.

34. Szondy, Z., Sarang, Z., Molnar, P., Nemeth, T., Piacentini, M., Mastroberardino, P. G., Falasca, L., Aeschlimann, D., Kovacs, J., Kiss, I., Szegezdi, E., Lakos, G., Rajnavolgyi, E., Birckbichler, P. J., Melino, G., and Fesus, L. (2003) Transglutaminase $2^{-/-}$ mice reveal a phagocytosis-associated crosstalk between macrophages and apoptotic cells, *Proc. Natl. Acad. Sci. U.S.A. 100*, 7812–7817.

35. Bernassola, F., Federici, M., Corazzari, M., Terrinoni, A., Hribal, M. L., De Laurenzi, V., Ranalli, M., Massa, O., Sesti, G., McLean, W. H., Citro, G., Barbetti, F., and Melino, G. (2002) Role of transglutaminase 2 in glucose tolerance: knockout mice studies and a putative mutation in a MODY patient, *FASEB J. 16*, 1371–1378.

36. Stephens, P., Grenard, P., Aeschlimann, P., Langley, M., Blain, E., Errington, R., Kipling, D., Thomas, D., and Aeschlimann, D. (2004) Crosslinking and G-protein functions of transglutaminase 2 contribute differentially to fibroblast wound healing responses, *J. Cell. Sci. 117*, 3389–3403.

37. Balajthy, Z., Csomós, K., Vámosi, G., Szántó, A., Lanotte, M., and Fésüs, L. (2006) Tissue-transglutaminase contributes to neutrophil granulocyte differentiation and functions, *Blood 108*, 2045–2054.

38. Falasca, L., Farrace, M. G., Rinaldi, A., Tuosto, L., Melino, G., and Piacentini, M. (2008) Transglutaminase type II is involved in the pathogenesis of endotoxic shock, *J. Immunol. 180*, 2616–2624.

39. Sarang, Z., Molnár, P., Németh, T., Gomba, S., Kardon, T., Melino, G., Cotecchia, S., Fésüs, L., and Szondy, Z. (2005) Tissue transglutaminase (TG2) acting as G protein protects hepatocytes against Fas-mediated cell death in mice, *Hepatology 42*, 578–587.

40. Szondy, Z., Mastroberardino, P. G., Váradi, J., Farrace, M. G., Nagy, N., Bak, I., Viti, I., Wieckowski, M. R., Melino, G., Rizzuto, R., Tósaki, A., Fesus, L., and Piacentini, M. (2006) Tissue transglutaminase (TG2) protects cardiomyocytes against ischemia/reperfusion injury by regulating ATP synthesis, *Cell Death Differ. 13*, 1827–1829.

41. Bakker, E. N., Pistea, A., Spaan, J. A., Rolf, T., de Vries, C. J., van Rooijen, N., Candi, E., and VanBavel, E. (2006) Flow-dependent remodeling of small arteries in mice deficient for tissue-type transglutaminase: possible compensation by macrophage-derived factor XIII, *Circ. Res. 99*, 86–92.

42. Green, P. H. and Cellier, C. (2007) Celiac disease, *N. Engl. J. Med. 357*, 1731–1743.

43. Schuppan, D., Junker, Y., and Barisani, D. (2009) Celiac disease: from pathogenesis to novel therapies, *Gastroenterology 137*, 1912–1933.

44. Marzari, R., Sblattero, D., Florian, F., Tongiorgi, E., Not, T., Tommasini, A., Ventura, A., and Bradbury, A. (2001) Molecular dissection of tissue transglutaminase autoantibody response in celiac disease, *J. Immunol. 166*, 4170–4176.

45. Korponay-Szabó, I. R., Laurila, K., Szondy, Z., Halttunen, T., Szalai, Z., Dahlbom, I., Rantala, I., Kovács, J. B., Fésüs, L., and Mäki, M. (2003) Missing endomysial and reticulin binding of coeliac antibodies in transglutaminase 2 knockout tissues, *Gut 52*, 199–204.

46. Villalta, D., Bizzaro, N., Tonutti, E., and Tozzoli, R. (2002) IgG anti-transglutaminase autoantibodies in systemic lupus erythematosus and Sjögren syndrome, *Clin. Chem. 48*, 1133.

47. Bizzaro, N., Villalta, D., Tonutti, E., Doria, A., Tampoia, M., Bassetti, D., and Tozzoli, R. (2003) IgA and IgG tissue transglutaminase antibody prevalence and clinical significance in connective tissue diseases, inflammatory bowel disease, and primary biliary cirrhosis, *Dig. Dis. Sci. 48*, 2360–2365.

48. Luft, L. M., Barr, S. G., Martin, L. O., Chan, E. K., and Fritzler, M. J. 2003) Autoantibodies to tissue transglutaminase in Sjögren's syndrome and related rheumatic diseases, *J. Rheumatol. 30*, 2613–2619.

49. Bizzaro, N., Tampoia, M., Villalta, D., Platzgummer, S., Liguori, M., Tozzoli, R., and Tonutti, E. (2006) Low specificity of anti-tissue transglutaminase antibodies in patients with primary biliary cirrhosis, *J. Clin. Lab. Anal. 20*, 184–189.

50. Damasiewicz-Bodzek, A. and Wielkoszyński, T. (2008) Serologic markers of celiac disease in psoriatic patients, *J. Eur. Acad. Dermatol. Venereol. 22*, 1055–1061.

51. Piredda, L., Amendola, A., Colizzi, V., Davies, P. J., Farrace, M. G., Fraziano, M., Gentile, V., Uray, I., Piacentini, M., and Fesus, L. (1997) Lack of 'tissue' transglutaminase protein cross-linking leads to leakage of macromolecules from dying cells: relationship to development of autoimmunity in MRLIpr/Ipr mice, *Cell Death Differ. 4*, 463–472.

52. Sárdy, M., Csikós, M., Geisen, C., Preisz, K., Kornseé, Z., Tomsits, E., Töx, U., Hunzelmann, N., Wieslander, J., Kárpáti, S., Paulsson, M., and Smyth, N. (2007) Tissue transglutaminase ELISA positivity in autoimmune disease independent of gluten-sensitive disease, *Clin. Chim. Acta 376*, 126–135.

53. Di Tola, M., Barillà, F., Trappolini, M., Palumbo, H. F., Gaudio, C., and Picarelli, A. (2008) Antitissue transglutaminase antibodies in acute coronary syndrome: an alert signal of myocardial tissue lesion? *J. Intern. Med. 263*, 43–51.

54. Zöller-Utz, I. M., Esslinger, B., Schulze-Krebs, A., and Dieterich, W. (2009) Natural Hidden Autoantibodies to Tissue Transglutaminase Cross-React with Fibrinogen, *J. Clin. Immunol. 30*, 204–212.

55. Ferrara, F., Quaglia, S., Caputo, I., Esposito, C., Lepretti, M., Pastore, S., Giorgi, R., Martelossi, S., Dal Molin, G., Di Toro, N., Ventura, A., and Not, T. (2010) Anti-transglutaminase antibodies in non-coeliac children suffering from infectious diseases, *Clin. Exp. Immunol. 159*, 217–223.

56. Giovannini, C., Matarrese, P., Scazzocchio, B., Varí, R., D'Archivio, M., Straface, E., Masella, R., Malorni, W., and De Vincenzi, M. (2003) Wheat gliadin induces apoptosis of intestinal cells via an autocrine mechanism involving Fas-Fas ligand pathway, *FEBS. Lett. 540*, 117–124.

57. Elli, L., Dolfini, E., and Bardella, M. T. (2003) Gliadin cytotoxicity and in vitro cell cultures, *Toxicol. Lett. 146*, 1–8.

58. Sakly, W., Thomas, V., Quash, G., and El Alaoui, S. (2006) A role for tissue transglutaminase in alpha-gliadin peptide cytotoxicity, *Clin. Exp. Immunol. 146*, 550–558.

59. Hue, S., Mention, J. J., Monteiro, R. C., Zhang, S., Cellier, C., Schmitz, J., Verkarre, V., Fodil, N., Bahram, S., Cerf-Bensussan, N., and Caillat-Zucman, S. (2004) A direct role for NKG2D/MICA interaction in villous atrophy during celiac disease, *Immunity 21*, 367–377.

60. Fleckenstein, B., Qiao, S. W., Larsen, M. R., Jung, G., Roepstorff, P., and Sollid. L. M. (2004) Molecular characterisation of covalent complexes between tissue transglutaminase and gliadin peptides, *J. Biol. Chem. 279*, 17607–17616.

61. Stamnaes, J., Fleckenstein, B., and Sollid, L. M. (2008) The propensity for deamidation and transamidation of peptides by transglutaminase 2 is dependent on substrate affinity and reaction conditions, *Biochim. Biophys. Acta 1784*, 1804–1811.

62. Kiraly, R., Vecsei, Z., Demenyi, T., Korponay-Szabo, I. R., and Fesus, L. (2006) Coeliac autoantibodies can enhance transamidating and inhibit GTPase activity of tissue transglutaminase: Dependence on reaction environment and enzyme fitness, *J. Autoimmun. 26*, 278–287.

63. Dieterich, W., Trapp, D., Esslinger, B., Leidenberger, M., Piper, J., Hahn, E., and Schuppan, D. (2003) Autoantibodies of patients with celiac disease are insufficient to block tissue transglutaminase activity, *Gut 52*, 1562–1566.

64. Esposito, C., Paparo, F., Caputo, I., Rossi, M., Maglio, M., Sblattero, D., Not, T., Porta, R., Auricchio, S., Marzari, R., and Troncone, R. (2002) Anti-tissue transglutaminase antibodies from coeliac patients inhibit transglutaminase activity both in vitro and in situ, *Gut 51*, 177–181.

65. Sollid, L. M., Markussen, G., Ek, J., Gjerde, H., Vartdal, F., and Thorsby, E. (1989) Evidence for a primary association of celiac disease to a particular HLA-DQ alpha/beta heterodimer, *J. Exp. Med. 169*, 345–350.

66. Lundin, K. E., Scott, H., Hansen, T., Paulsen, G., Halstensen, T. S., Fausa, O., Thorsby, E., and Sollid, L. M. (1993) Gliadin-specific, HLA-DQ(alpha 1*0501,beta 1*0201) restricted T cells isolated from the small intestinal mucosa of celiac disease patients, *J. Exp. Med. 178*, 187–196.

67. Kim, C. Y., Quarsten, H., Bergseng, E., Khosla, C., and Sollid, L. M. (2004) Structural basis for HLA-DQ2-mediated presentation of gluten epitopes in celiac disease, *Proc. Natl. Acad. Sci. U.S.A. 101*, 4175–4179.

68. Vader, L. W., de Ru, A., Van Der Wal, Y., Kooy, Y. M., Benckhuijsen, W., Mearin, M. L., Drijfhout, J. W., van Veelen, P., and Koning, F. (2002) Specificity of tissue transglutaminase explains cereal toxicity in celiac disease, *J. Exp. Med. 195*, 643–649.

69. Hovhannisyan, Z., Weiss, A., Martin, A., Wiesner, M., Tollefsen, S., Yoshida, K., Ciszewski, C., Curran, S. A., Murray, J. A., David, C. S., Sollid, L. M., Koning, F., Teyton, L., and Jabri, B. (2008) The role of HLA-DQ8 beta57 polymorphism in the anti-gluten T-cell response in coeliac disease, *Nature 456*, 534–538.

70. Lewis, N. R. and Scott, B. B. (2010) Meta-analysis: deamidated gliadin peptide antibody and tissue transglutaminase antibody compared as screening tests for coeliac disease, *Aliment. Pharmacol. Ther. 31*, 73–81.

71. Hausch, F., Shan, L., Santiago, N. A., Gray, G. M., and Khosla, C. (2002) Intestinal digestive resistance of immunodominant gliadin peptides, *Am. J. Physiol. Gastrointest. Liver Physiol. 283*, G996–G1003.

72. Shan, L., Molberg, Ø., Parrot, I., Hausch, F., Filiz, F., Gray, G. M., Sollid, L. M., and Khosla, C. (2002) Structural basis for gluten intolerance in celiac sprue, *Science 297*, 2275–2279.

73. Qiao, S. W., Bergseng, E., Molberg, Ø., Xia, J., Fleckenstein, B., Khosla, C., and Sollid, L. M. (2004) Antigen presentation to celiac lesion-derived T cells of a 33-mer gliadin peptide naturally formed by gastrointestinal digestion, *J. Immunol. 173*, 1757–1762.

74. Anderson, C. M., French, J. M., Sammons, H. G., Frazer, A. C., Gerrard, J. W., and Smellie, J. M. (1952) Coeliac disease; gastrointestinal studies and the effect of dietary wheat flour, *Lancet 1*, 836–842.

75. Alvey, C., Anderson, C. M., and Freeman, M. (1957) Wheat gluten and coeliac disease, *Arch. Dis. Child. 32*, 434–437.

76. Ehren, J., Morón, B., Martin, E., Bethune, M. T., Gray, G. M., and Khosla, C. (2009) A food-grade enzyme preparation with modest gluten detoxification properties, *PLoS. One. 4*, e6313.

77. Matysiak-Budnik, T., Candalh, C., Cellier, C., Dugave, C., Namane, A., Vidal-Martinez, T., Cerf-Bensussan, N., and Heyman, M. (2005) Limited efficiency of prolyl-endopeptidase in the detoxification of gliadin peptides in celiac disease, *Gastroenterology 129*, 786–796.

78. Garcia-Horsman, J. A., Venäläinen, J. I., Lohi, O., Auriola, I. S., Korponay-Szabo, I. R., Kaukinen, K., Mäki, M., and Männistö, P. T. (2007) Deficient activity of mammalian prolyl oligopeptidase on the immunoactive peptide digestion in coeliac disease, *Scand. J. Gastroenterol. 42*, 562–571.

79. Zimmer, K. P., Fischer, I., Mothes, T., Weissen-Plenz, G., Schmitz, M., Wieser, H., Mendez, E., Buening, J., Lerch, M. M., Ciclitira, P. C., Weber, P., and Naim, H. Y. (2009) Endocytotic segregation of gliadin peptide 31–49 in enterocytes, *Gut* 2009 Aug 3. [Epub ahead of print].

80. Chorzelski, T. P., Sulej, J., Tchorzewska, H., Jablonska, S., Beutner, E. H., and Kumar, V. (1983) IgA class endomysium antibodies in dermatitis herpetiformis and coeliac disease, *Ann. N. Y. Acad. Sci. 420*, 325–334.

81. Uhlig, H., Osman, A. A., Tanev, I. D., Viehweg, J., and Mothes, T. (1998) Role of tissue transglutaminase in gliadin binding to reticular extracellular matrix and relation to coeliac disease autoantibodies, *Autoimmunity 28*, 185–195.

82. Dieterich, W., Esslinger, B., Trapp, D., Hahn, E., Huff, T., Seilmeier, W., Wieser, H., and Schuppan, D. (2006) Cross linking to tissue transglutaminase and collagen favours gliadin toxicity in coeliac disease, *Gut 55*, 478–484.

83. Skovbjerg, H., Anthonsen, D., Knudsen, E., and Sjöström, H. (2008) Deamidation of gliadin peptides in lamina propria: implications for celiac disease, *Dig. Dis. Sci. 53*, 2917–2924.

84. Maiuri, L., Ciacci, C., Ricciardelli, I., Vacca, L., Raia, V., Rispo, A., Griffin, M., Issekutz, T., Quaratino, T., and Londei, M. (2005) Unexpected Role of Surface Transglutaminase Type II in Celiac Disease, *Gastroenterology 129*, 1400–1413.

85. Ráki, M., Schjetne, K. W., Stamnaes, J., Molberg, Ø., Jahnsen, F. L., Issekutz, T. B., Bogen, B., and Sollid, L. M. (2007) Surface expression of transglutaminase 2 by dendritic cells and its potential role for uptake and presentation of gluten peptides to T cells, *Scand. J. Immunol. 65*, 213–220.

86. Stamnaes, J., Fleckenstein, B., Lund-Johansen, F., and Sollid, L. M. (2008) The monoclonal antibody 6B9 recognizes CD44 and not cell surface transglutaminase 2, *Scand. J. Immunol. 68*, 534–542.

87. Hodrea, J., Demény, M. A., Majai, G., Sarang, Z., Korponay-Szabó, I. R., and Fésüs, L. (2010) Transglutaminase 2 is expressed and active on the surface of human monocyte-derived dendritic cells and macrophages, *Immunol. Lett.* 2010 Feb 9. [Epub ahead of print]

88. Skovbjerg, H., Norén, O., Anthonsen, D., Moller, J., and Sjöström, H. (2002) Gliadin is a good substrate of several transglutaminases: possible implication in the pathogenesis of coeliac disease, *Scand. J. Gastroenterol.* *37*, 812–817.

89. Sjöström, H., Lundin, K. E., Molberg, O., Körner, R., McAdam, S. N., Anthonsen, D., Quarsten, H., Norén, O., Roepstorff, P., Thorsby, E., and Sollid, L. M. (1998) Identification of a gliadin T-cell epitope in coeliac disease: general importance of gliadin deamidation for intestinal T-cell recognition, *Scand. J. Immunol.* *48*, 111–115.

90. Dahlbom, I., Korponay-Szabó, I. R., Kovács, J. B., Szalai, Z., Mäki, M., and Hansson, T. (2010) Prediction of clinical and mucosal severity of coeliac disease and dermatitis herpetiformis by quantification of IgA/IgG serum antibodies to tissue transglutaminase, *J. Pediatr. Gastroenterol. Nutr.* *50*, 140–146.

91. Pinkas, D. M., Strop, P., Brunger, A. T., and Khosla, C. (2007) Transglutaminase 2 undergoes a large conformational change upon activation, *PLoS. Biol.* *5*, e327.

92. Myrsky, E., Caja, S., Simon-Vecsei, Z., Korponay-Szabo, I. R., Nadalutti, C., Collighan, R., Mongeot, A., Griffin, M., Mäki, M., Kaukinen, K., and Lindfors, K. (2009) Celiac disease IgA modulates vascular permeability in vitro through the activity of transglutaminase 2 and RhoA, *Cell. Mol. Life. Sci.* *66*, 3375–3385.

93. Molberg, Ø., and Sollid, L. M. (2006) A gut feeling for joint inflammation - using coeliac disease to understand rheumatoid arthritis, *Trends. Immunol.* *27*, 188–194.

94. Stenberg, P., Roth, B., and Wollheim, F. A. (2009) Peptidylarginine deiminases and the pathogenesis of rheumatoid arthritis: A reflection of the involvement of transglutaminase in coeliac disease, *Eur. J.* Intern. Med. *20*, 749–755.

95. Tarcsa, E., Marekov, L. N., Mei, G., Melino, G., Lee, S. C., and Steinert, P. M. (1996) Protein unfolding by peptidylarginine deiminase. Substrate specificity and structural relationships of the natural substrates trichohyalin and filaggrin, *J. Biol. Chem.* *271*, 30709–30716.

96. Picarelli, A., Di Tola, M., Sabbatella, L., Vetrano, S., Anania, M. C., Spadaro, A., Sorgi, M. L., and Taccari, E. (2003) Anti-tissue transglutaminase antibodies in arthritic patients: a disease-specific finding? *Clin. Chem.* *49*, 2091–2094.

97. Roth, E. B., Stenberg, P., Book, C., and Sjöberg, K. (2006) Antibodies against transglutaminases, peptidylarginine deiminase and citrulline in rheumatoid arthritis—new pathways to epitope spreading, *Clin. Exp. Rheumatol.* *24*, 12–18.

98. Unsworth, D. J., Scott, D. L., Walton, K. W., Walker-Smith, J. A., and Holborow, E. J. (1984) Failure of R1 type anti-reticulin antibody to react with fibronectin, collagen type III or the non-collagenous reticulin component (NCRC), *Clin. Exp. Immunol.* *57*, 609–613.

99. Korponay-Szabó, I. R., Raivio, T., Laurila, K., Opre, J., Király, R., Kovács, J. B., Kaukinen, K., Fésüs, L., and Mäki, M. (2005) Coeliac disease case finding and diet monitoring by point-of-care testing, *Aliment. Pharmacol. Ther.* 22, 729–737.

100. Korponay-Szabó, I. R., Vecsei, Z., Király, R., Dahlbom, I., Chirdo, F., Nemes, E., Fésüs, L., and Mäki, M. (2008) Deamidated gliadin peptides form epitopes that transglutaminase antibodies recognize, *J. Pediatr. Gastroenterol. Nutr.* *46*, 253–261.

101. Korhonen, M., Ormio, M., Burgeson, R. E., Virtanen, I., and Savilahti, E. (2000) Unaltered distribution of laminins, fibronectin, and tenascin in celiac intestinal mucosa, *J. Histochem. Cytochem. 48*, 1011–1020.

102. Korponay-Szabó, I. R., Dahlbom, I., Laurila, K., Koskinen, S., Woolley, N., Partanen, J., Kovács, J. B., Mäki, M., and Hansson, T. (2003) Elevation of IgG antibodies against tissue transglutaminase as a diagnostic tool for coeliac disease in selective IgA deficiency, *Gut 52*, 1567–1571.

103. Borrelli, M., Maglio, M., Agnese, M., Paparo, F., Gentile, S., Colicchio, B., Tosco, A., Auricchio, R., and Troncone, R. (2009) High density of intraepithelial gammadelta lymphocytes and deposits of immunoglobulin (Ig)M anti-tissue transglutaminase antibodies in the jejunum of coeliac patients with IgA deficiency, *Clin. Exp. Immunol.* 2009 Dec 21. [Epub ahead of print]

104. Salmi, T. T., Collin, P., Korponay-Szabó, I. R., Laurila, K., Partanen, J., Huhtala, H., Király, R., Lorand, L., Reunala, T., Mäki, M., and Kaukinen, K. (2006) Endomysial antibody-negative coeliac disease: clinical characteristics and intestinal autoantibody deposits, *Gut 55*, 1746–1753.

105. Webster, A. D., Slavin, G., Shiner, M., Platts-Mills, T. A., and Asherson, G. L. (1981) Coeliac disease with severe hypogammaglobulinaemia, *Gut 22*, 153–157.

106. Naiyer, A. J., Hernandez, L., Ciaccio, E. J., Papadakis, K., Manavalan, J. S., Bhagat, G., and Green, P. H. (2009) Comparison of commercially available serologic kits for the detection of celiac disease, *J. Clin. Gastroenterol. 43*, 225–232.

107. Ladinser, B., Rossipal, E., and Pittschieler, K. (1994) Endomysium antibodies in coeliac disease: an improved method, *Gut 35*, 776–778.

108. Mäki, M. and Korponay-Szabo, I. R. (2002) Methods and means for detecting gluten-induced diseases. Patent application PCT/FI02/00340, International publication number W002/086509 A19.

109. Raivio, T., Kaukinen, K., Nemes, E., Laurila, K., Collin, P., Kovács, J. B., Mäki, M., and Korponay-Szabó, I. R. (2006) Self transglutaminase-based rapid coeliac disease antibody detection by a lateral flow method, *Aliment. Pharmacol. Ther. 24*, 147–154.

110. Popat, S., Hogberg, L., McGuire, S., Green, H., Bevan, S., Stenhammar, L., and Houlston, R. S. (2001) Germline mutations in TGM2 do not contribute to coeliac disease susceptibility in the Swedish population, *Eur. J. Gastroenterol. Hepatol. 13*, 1477–1479.

111. Aldersley, M. A., Hamlin, P. J., Jones, P. F., Markham, A. F., Robinson, P. A., and Howdle, P, D. (2000) No polymorphism in the tissue transglutaminase gene detected in coeliac disease patients, *Scand. J. Gastroenterol. 35*, 61–63.

112. van Belzen, M. J., Mulder, C. J., Pearson, P. L., Houwen, R. H., and Wijmenga, C. (2001) The tissue transglutaminase gene is not a primary factor predisposing to celiac disease, *Am. J. Gastroenterol. 96*, 3337–3340.

113. Sulkanen, S., Halttunen, T., Laurila, K., Kolho, K. L., Korponay-Szabó, I. R., Sarnesto, A., Savilahti, E., Collin, P., and Mäki, M. (1998) Tissue transglutaminase autoantibody enzyme-linked immunosorbent assay in detecting celiac disease, *Gastroenterology 115*, 1322–1328.

114. Király, R., Csosz, E., Kurtán, T., Antus, S., Szigeti, K., Simon-Vecsei, Z., Korponay-Szabó, I. R., Keresztessy, Z., and Fésüs, L. (2009) Functional significance of five

noncanonical Ca2+-binding sites of human transglutaminase 2 characterized by site-directed mutagenesis, *FEBS J. 276*, 7083–7096.

115. Tucková, L., Karská, K., Walters, J. R., Michalak, M., Rossmann, P., Krupicková, S., Verdu, E. F., Saalman, R., Hanson, L. A., and Tlaskalová-Hogenová, H. (1997) Anti-gliadin antibodies in patients with celiac disease cross-react with enterocytes and human calreticulin, *Clin. Immunol. Immunopathol. 85*, 289–296.
116. Korponay-Szabó, I. R., Halttunen, T., Laurila, K., Sulkanen, S., Szalai, Z., Dahlbom, I., Kovács, J. B., and Mäki, M. (2000) Calreticulin-bound tissue transglutaminase is less antigenic for celiac autoantibodies, *J. Pediatr. Gastr. Nutr. 31*(Suppl3), S9.
117. Sollid, L. M., Molberg, O., McAdam, S., and Lundin, K. E. (1997) Autoantibodies in coeliac disease: tissue transglutaminase-guilt by association? *Gut 41*, 851–852.
118. Seissler, J., Wohlrab, U., Wuensche, C., Scherbaum, W. A., and Boehm, B. O. (2001) Autoantibodies from patients with coeliac disease recognise distinct functional domains of the autoantigen tissue transglutaminase, *Clin. Exp. Immunol. 125*, 216–221.
119. Sblattero, D., Florian, F., Azzoni, E., Zyla, T., Park, M., Baldas, V., Not, T., Ventura, A., Bradbury, A., and Marzari, R. (2002) The analysis of the fine specificity of celiac disease antibodies using tissue transglutaminase fragments, *Eur. J. Biochem. 269*, 5175–5181.
120. Nakachi, K., Powell, M., Swift, G., Amoroso, M. A., Ananieva-Jordanova, R., Arnold, C., Sanders, J., Furmaniak, J., and Rees Smith, B. (2004) Epitopes recognised by tissue transglutaminase antibodies in coeliac disease, *J. Autoimmun. 22*, 53–63.
121. Sárdy, M., Kárpáti, S., Merkl, B., Paulsson, M., and Smyth, N. (2002) Epidermal transglutaminase (TGase 3) is the autoantigen of dermatitis herpetiformis, *J. Exp. Med. 195*, 747–757.
122. Hadjivassiliou, M., Aeschlimann, P., Strigun, A., Sanders, D. S., Woodroofe, N., and Aeschlimann, D. (2008) Autoantibodies in gluten ataxia recognize a novel neuronal transglutaminase, *Ann. Neurol. 64*, 332–343.
123. Hadjivassiliou, M., Mäki, M., Sanders, D. S., Williamson, C. A., Grünewald, R. A., Woodroofe, N. M., and Korponay-Szabó, I. R. (2006) Autoantibody targeting of brain and intestinal transglutaminase in gluten ataxia, *Neurology 66*, 373–377.
124. Zanoni, G., Navone, R., Lunardi, C., Tridente, G., Bason, C., Sivori, S., Beri, R., Dolcino, M., Valletta, E., Corrocher, R., and Puccetti, A. (2006) In celiac disease, a subset of autoantibodies against transglutaminase binds toll-like receptor 4 and induces activation of monocytes, *PLoS. Med. 3*, e358.
125. Szebeni, B., Veres, G., Dezsofi, A., Rusai, K., Vannay, A., Bokodi, G., Vásárhelyi, B., Korponay-Szabó, I. R., Tulassay, T., and Arató, A. (2007) Increased mucosal expression of Toll-like receptor (TLR)2 and TLR4 in coeliac disease, *J. Pediatr. Gastroenterol. Nutr. 45*, 187–193.
126. Stene, L. C., Honeyman, M. C., Hoffenberg, E. J., Haas, J. E., Sokol, R. J., Emery, L., Taki, I., Norris, J. M., Erlich, H. A., Eisenbarth, G. S., and Rewers, M. (2006) Rotavirus infection frequency and risk of celiac disease autoimmunity in early childhood: a longitudinal study, *Am. J. Gastroenterol. 101*, 2333–2340.
127. Meresse, B., Chen, Z., Ciszewski, C., Tretiakova, M., Bhagat, G., Krausz, T. N., Raulet, D. H., Lanier, L. L., Groh, V., Spies, T., Ebert, E. C., Green, P. H., and Jabri, B. (2004) Coordinated induction by IL15 of a TCR-independent NKG2D signaling pathway

converts CTL into lymphokine-activated killer cells in celiac disease, *Immunity 21*, 357–366.

128. de Kauwe, A. L., Chen, Z., Anderson, R. P., Keech, C. L., Price, J. D., Wijburg, O., Jackson, D. C., Ladhams, J., Allison, J., and McCluskey, J. (2009) Resistance to celiac disease in humanized HLA-DR3-DQ2-transgenic mice expressing specific anti-gliadin CD4+ T cells, *J. Immunol. 182*, 7440–7450.

129. Seah, P. P., Fry, L., Rossiter, M. A., Hoffbrand, A. V., and Holborow, E. J. (1971) Anti-reticulin antibodies in childhood coeliac disease, *Lancet 2*, 681–682.

130. Korponay-Szabó, I. R., Sulkanen, S., Halttunen, T., Maurano, F., Rossi, M., Mazzarella, G., Laurila, K., Troncone, R., and Mäki, M. (2000) Tissue transglutaminase is the target in both rodent and primate tissues for celiac disease-specific autoantibodies, J. Pediatr. Gastroenterol. Nutr. *31*, 520–527.

131. Mäki, M., Mustalahti, K., Kokkonen, J., Kulmala, P., Haapalahti, M., Karttunen, T., Ilonen, J., Laurila, K., Dahlbom, I., Hansson, T., Höpfl, P., and Knip, M. (2003) Prevalence of Celiac disease among children in Finland, *N. Engl. J. Med. 348*, 2517–2524.

132. Salmi, T. T., Collin, P., Järvinen, O., Haimila, K., Partanen, J., Laurila, K., Korponay-Szabo, I. R., Huhtala, H., Reunala, T., Mäki, M., and Kaukinen, K. (2006) Immunoglobulin A autoantibodies against transglutaminase 2 in the small intestinal mucosa predict forthcoming coeliac disease, *Aliment. Pharmacol. Ther. 24*, 541–552.

133. Kurppa, K., Collin, P., Viljamaa, M., Haimila, K., Saavalainen, P., Partanen, J., Laurila, K., Huhtala, H., Paasikivi, K., Mäki, M., and Kaukinen, K. (2009) Diagnosing mild enteropathy celiac disease: a randomized, controlled clinical study, *Gastroenterology 136*, 816–823.

134. Halttunen, T., Marttinen, A., Rantala, I., Kainulainen, H., and Mäki, M. (1996) Fibroblasts and transforming growth factor beta induce organization and differentiation of T84 human epithelial cells, *Gastroenterology 111*, 1252–1262.

135. Halttunen, T., and Maki, M. (1999) Serum immunoglobulin A from patients with celiac disease inhibits human T84 intestinal crypt epithelial cell differentiation, *Gastroenterology 116*, 566–572.

136. Korponay-Szabó, I. R., Halttunen, T., Szalai, Z., Laurila, K., Király, R., Kovács, J. B., Fésüs, L., and Mäki, M. (2004) In vivo targeting of intestinal and extraintestinal transglutaminase 2 by coeliac autoantibodies, *Gut 53*, 641–648.

137. Esposito, C., Paparo, F., Caputo, I., Porta, R., Salvati, V. M., Mazzarella, G., Auricchio, S., and Troncone, R. (2003) Expression and enzymatic activity of small intestinal tissue transglutaminase in celiac disease, *Am. J. Gastroenterol. 98*, 1813–1820.

138. Byrne, G., Ryan, F., Jackson, J., Feighery, C., and Kelly, J. (2007) Mutagenesis of the catalytic triad of tissue transglutaminase abrogates coeliac disease serum IgA autoantibody binding, *Gut 56*, 336–341.

139. Siegel, M., Strnad, P., Watts, R. E., Choi, K., Jabri, B., Omary, M. B., and Khosla, C. (2008) Extracellular transglutaminase 2 is catalytically inactive, but is transiently activated upon tissue injury, *PLoS. One 3*, e1861.

140. Fesus, L. and Laki, K. (1977) Two antigenic sites of tissue transglutaminase, *Biochemistry 16*, 4061–4066.

141. Myrsky, E., Kaukinen, K., Syrjänen, M., Korponay-Szabó, I. R., Mäki, M., and Lindfors, K. (2008) Coeliac disease-specific autoantibodies targeted against transglutaminase 2 disturb angiogenesis, *Clin. Exp. Immunol. 152*, 111–119.

142. Caja, S., Myrsky, E., Korponay-Szabo, I. R., Nadalutti, C., Sulic, A. M., Lavric, M., Sblattero, D., Marzari, R., Collighan, R., Mongeot, A., Griffin, M., Mäki, M., Kaukinen, K., and Lindfors, K. (2010) Inhibition of transglutaminase 2 enzymatic activity ameliorates the anti-angiogenic effects of coeliac disease autoantibodies, *Scand. J. Gastroenterol.* 2010 Jan 22. [Epub ahead of print]

143. Myrsky, E., Syrjänen, M., Korponay-Szabo, I. R., Mäki, M., Kaukinen, K., and Lindfors, K. (2009) Altered small-bowel mucosal vascular network in untreated coeliac disease, *Scand. J. Gastroenterol. 44*, 162–167.

144. Cooke, W. T. and Holmes, G. K. T. (1984) *Coeliac Disease*. Churchill Livingstone, New York.

145. Halttunen, T. (2000) *Biological Functions of Coeliac Disease Autoantibodies*. Tampere University Press, Tampere.

146. Barone, M. V., Caputo, I., Ribecco, M. T., Maglio, M., Marzari, R., Sblattero, D., Troncone, R., Auricchio, S., and Esposito, C. (2007) Humoral immune response to tissue transglutaminase is related to epithelial cell proliferation in celiac disease, *Gastroenterology 132*, 1245–1253.

147. Cervio, E., Volta, U., Verri, M., Boschi, F., Pastoris, O., Granito, A., Barbara, G., Parisi, C., Felicani, C., Tonini, M., and De Giorgio, R. (2007) Sera of patients with celiac disease and neurologic disorders evoke a mitochondrial-dependent apoptosis in vitro, *Gastroenterology 133*, 195–206.

148. Malmborg, A. C., Shultz, D. B., Luton, F., Mostov, K. E., Richly, E., Leung, P. S., Benson, G. D., Ansari, A. A., Coppel, R. L., Gershwin, M. E., and Van de Water, J. (1998) Penetration and co-localization in MDCK cell mitochondria of IgA derived from patients with primary biliary cirrhosis, *J. Autoimmun. 11*, 573–580.

149. Di Niro, R., Sblattero, D., Florian, F., Stebel, M., Zentilin, L., Giacca, M., Villanacci, V., Galletti, A., Not, T., Ventura, A., and Marzari, R. (2008) Anti-idiotypic response in mice expressing human autoantibodies, *Mol. Immunol. 45*, 1782–1791.

150. Korponay-Szabó, I. R., Toth, B., Nemes, E., Kiraly, R., Vecsei, Z., Juuti-Uusitalo, K., Csipo, A., Gyimesi, J., Fesus, L., and Maki, M. (2009) *Passive Transfer Model of Coeliac Autoimmunity in Newborns from Mothers with Coeliac Disease and Transglutaminase Antibodies*, Abstract, ESPGHAN 2009, Budapest, June 3–6.

151. Boscolo, S., Sarich, A., Lorenzon, A., Passoni, M., Rui, V., Stebel, M., Sblattero, D., Marzari, R., Hadjivassiliou, M., and Tongiorgi, E. (2007) Gluten ataxia: passive transfer in a mouse model, *Ann. N Y. Acad. Sci. 1107*, 319–328.

152. Dubois, P. C., Trynka, G., Franke, L., Hunt, K. A., Romanos, J., Curtotti, A., Zhernakova, A., Heap, G. A., Adány, R., Aromaa, A., Bardella, M. T., Van Den Berg, L. H., Bockett, N. A., de la Concha, E. G., Dema, B., Fehrmann, R. S., Fernández-Arquero, M., Fiatal, S., Grandone, E., Green, P. M., Groen, H. J., Gwilliam, R., Houwen, R. H., Hunt, S. E., Kaukinen, K., Kelleher, D., Korponay-Szabo, I., Kurppa, K., Macmathuna, P., Mäki, M., Mazzilli, M. C., McCann, O. T., Mearin, M. L., Mein, C. A., Mirza, M. M., Mistry, V., Mora, B., Morley, K. I., Mulder, C. J., Murray, J. A., Núñez, C., Oosterom, E., Ophoff,

R. A., Polanco, I., Peltonen, L., Platteel, M., Rybak, A., Salomaa, V., Schweizer, J. J., Sperandeo, M. P., Tack, G. J., Turner, G., Veldink, J. H., Verbeek, W. H., Weersma, R. K., Wolters, V. M., Urcelay, E., Cukrowska, B., Greco, L., Neuhausen, S. L., McManus, R., Barisani, D., Deloukas, P., Barrett, J. C., Saavalainen, P., Wijmenga, C., and van Heel, D. A. (2010) Multiple common variants for celiac disease influencing immune gene expression, *Nat. Genet.* 2010 Feb 28. [Epub ahead of print]

153. Haslett, C. (1999) Granulocyte apoptosis and its role in the resolution and control of lung inflammation, *Am. J. Respir. Crit. Care Med. 160*, S5–S11.

154. Wyllie, A. H., Kerr, J. F., and Currie, A. R. (1980) Cell death: the significance of apoptosis, *Int. Rev. Cytol. 68*, 251–306.

155. Savill, J. S., Wyllie, A. H., Henson, J. E., Walport, M. J., Henson, P. M., and Haslett, C. (1989) Macrophage phagocytosis of aging neutrophils in inflammation. Programmed cell death in the neutrophil leads to its recognition by macrophages, *J. Clin. Invest. 83*, 865–875.

156. Albert, M. L., Pearce, S. F., Francisco, L. M., Sauter, B., Roy, P., Silverstein, R. L., and Bhardwaj, N. (1998) Immature dendritic cells phagocytose apoptotic cells via alphavbeta5 and CD36, and cross-present antigens to cytotoxic T lymphocytes, *J. Exp. Med. 188*, 1359–1368.

157. Dini, L., Lentini, A., Diez, G. D., Rocha, M., Falasca, L., Serafino, L., and Vidal-Vanaclocha, F. (1995) Phagocytosis of apoptotic bodies by liver endothelial cells, *J. Cell Sci. 108*, 967–973.

158. Bennett, M. R., Gibson, D. F., Schwartz, S. M., and Tait, J. F. (1995) Binding and phagocytosis of apoptotic vascular smooth muscle cells is mediated in part by exposure of phosphatidylserine, *Circ. Res. 77*, 1136–1142.

159. Hall, S. E., Savill, J. S., Henson, P. M., and Haslett, C. (1994) Apoptotic neutrophils are phagocytosed by fibroblasts with participation of the fibroblast vitronectin receptor and involvement of a mannose/fucose-specific lectin, *J. Immunol. 153*, 3218–3227.

160. Voll, R. E., Herrmann, M., Roth, E. A., Stach, C., Kalden, J. R., and Girkontaite, I. (1997) Immunosuppressive effects of apoptotic cells, *Nature 390*, 350–351.

161. Fadok, V. A., Bratton, D. L., Konowal, A., Freed, P. W., Wescott, J. Y., and Henson, P. M. (1998) Macrophages that have ingested apoptotic cells in vitro inhibit proinflammatory cytokine production through autocrine/paracrine mechanisms involving TGFβPGE$_2$ and PAF, *J. Clin.* Invest. *101*, 890–898.

162. Ren Y, Tang, J., Mok, M. Y., Chan, A. W., Wu, A., and Lau, C. S. (2003) Increased apoptotic neutrophils and macrophages and impaired macrophage phagocytic clearance of apoptotic neutrophils in systemic lupus erythematosus. Arthritis Rheum. *48*, 2888–2897.

163. O'Brien, B. A., Fieldus, W. E., Field, C. J., and Finegood, D. T. (2002) Clearance of apoptotic beta-cells is reduced in neonatal autoimmune diabetes-prone rats, *Cell Death Differ. 9*, 457–464.

164. Vandivier, R. W., Fadok, V. A., Hoffmann, P. R., Bratton, D. L., Penvari, C., Brown, K. K., Brain, J. D., Accurso, F. J., and Henson, P. M. (2002) Elastase-mediated phosphatidylserine receptor cleavage impairs apoptotic cell clearance in cystic fibrosis and bronchiectasis, *J. Clin. Invest. 109*, 661–670.

165. Hodge, S., Hodge, G., Scicchitano, R., Reynolds, P. N., and Holmes, M. (2003) Alveolar macrophages from subjects with chronic obstructive pulmonary disease are deficient in their ability to phagocytose apoptotic airway epithelial cells, *Immunol. Cell. Biol. 81*, 289–296.

166. Schrijvers, D. M., De Meyer, G. R., Kockx, M. M., Herman. A. G., and Martinet, W. (2005) Phagocytosis of apoptotic cells by macrophages is impaired in atherosclerosis, *Arterioscler. Thromb. Vasc. Biol. 25*, 1256–1261.

167. Gardai, S. J., McPhillips, K. A., Frasch, S. C., Janssen, W. J., Starefeldt, A., Murphy-Ullrich, J. E., Bratton, D. L., Oldenborg, P. A., Michalak, M., and Henson, P. M. (2005) Cell-surface calreticulin initiates clearance of viable or apoptotic cells through trans-activation of LRP on the phagocyte, *Cell 123*, 321–334.

168. Lauber, K., Bohn, E., Kröber, S. M., Xiao, Y. J., Blumenthal, S. G., Lindemann, R. K., Marini, P., Wiedig, C., Zobywalski, A., Baksh, S., Xu. Y., Autenrieth, I. B., Schulze-Osthoff, K., Belka, C., Stuhler, G., and Wesselborg, S. (2003) Apoptotic cells induce migration of phagocytes via caspase-3-mediated release of a lipid attraction signal, *Cell 113*, 717–730.

169. Elliott, M. R., Chekeni, F. B., Trampont, P. C., Lazarowski, E. R., Kadl, A., Walk, S. F., Park, D., Woodson, R. I., Ostankovic, M., Sharma, P., Lysiak, J. J., Harden, T. K., Leitinger, N., and Ravichandran, K. S. (2009) Nucleotides released by apoptotic cells act as a find-me signal to promote phagocytic clearance, *Nature 461*, 282–286.

170. Wakasugi, K. and Schimmel, P. (1999) Highly differentiated motifs responsible for two cytokine activities of a split human tRNA synthetase, *J. Biol. Chem. 274*, 23155–23159.

171. Moodley, Y., Rigby, P., Bundell, C., Bunt, S., Hayashi, H., Misso, N., McAnulty, R., Laurent, G., Scaffidi, A., Thompson, P., and Knight, D. (2003) Macrophage recognition and phagocytosis of apoptotic fibroblasts is critically dependent on fibroblast-derived thrombospondin 1 and CD36, *Am. J. Pathol. 162*, 771–779.

172. Fadok, V. A. and Henson, P. M. (2003) Apoptosis: giving phosphatidylserine recognition an assist–with a twist, *Curr. Biol. 13*, R655–R657.

173. Henson, P. M., Bratton, D. L., and Fadok, V. A. (2001) Apoptotic cell removal, *Curr. Biol. 11*, R795–R805.

174. Arur, S., Uche, U. E., Rezaul, K., Fong, M., Scranton, V., Cowan, A. E., Mohler, W., and Han, D. K. (2003) Annexin I is an endogenous ligand that mediates apoptotic cell engulfment, *Dev. Cell. 4*, 587–598.

175. Savill, J., Dransfield, I., Gregory, C., and Haslett, C. (2002) A blast from the past: clearance of apoptotic cells regulates immune responses, *Nat. Rev. Immunol. 2*, 965–975.

176. Stuart, L. M. and Ezekowitz, R. A. (2005) Phagocytosis: elegant complexity, *Immunity 22*, 539–550.

177. Hanayama, R., Tanaka, M., Miwa, K., Shinohara, A., Iwamatsu, A., and Nagata, S. (2002) Identification of a factor that links apoptotic cells to phagocytes, *Nature 417*, 182–187.

178. Chen, J., Carey, K., and Godowski, P. J. (1997) Identification of Gas6 as a ligand for Mer, a neural cell adhesion molecule related receptor tyrosine kinase implicated in cellular transformation, *Oncogene 14*, 2033–2039.

179. Hall, M. O., Obin, M. S., Heeb, M. J., Burgess, B. L., and Abrams, T. A. (2005) Both protein S and Gas6 stimulate outer segment phagocytosis by cultured rat retinal pigment epithelial cells, *Exp. Eye Res. 81*, 581–591.

180. Ling, L., Templeton, D., and Kung, H. J. (1996) Identification of the major autophosphorylation sites of Nyk/Mer, an NCAM-related receptor tyrosine kinase, *J. Biol. Chem.* 271, 18355–18362.

181. Savill, J. S., Hogg, N., Ren, Y., and Haslett, C. (1992) Thrombospondin cooperates with CD36 and the vitronectin receptor in macrophage recognition of neutrophils undergoing apoptosis, *J. Clin. Invest. 90*, 1513–1522.

182. Miyanishi, M., Tada, K., Koike, M., Uchiyama, Y., and Nagata, S. (2007) Identification of Tim4 as a phosphatidylserine receptor, *Nature 450*, 435–439.

183. Park, D., Tosello-Trampont, A. C., Elliott, M. R., Lu, M., Haney, L. B., Ma, Z., Klibanov, A. L., Mandell, J. W., and Ravichandran, K. S. (2007) BAI1 is an engulfment receptor for apoptotic cells upstream of the ELMO/Dock180/Rac module, *Nature* 450, 430–434.

184. Park, S. Y., Jung, M. Y., Kim, H. J., Lee, S. J., Kim, S. Y., Lee, B. H., Kwon, T. H., Park, R. W., and Kim, I. S. (2008) Rapid cell corpse clearance by stabilin-2, a membrane phosphatidylserine receptor, *Cell Death Differ. 15*, 192–201.

185. Park, D., Hochreiter-Hufford, A., and Ravichandran, K. S. (2009) The phosphatidylserine receptor TIM-4 does not mediate direct signaling, *Curr. Biol. 19*, 346–351.

186. Greenberg, M. E., Sun, M., Zhang, R., Febbraio, M., Silverstein, R., and Hazen, S. L. (2006) Oxidized phosphatidylserine-CD36 interactions play an essential role in macrophage-dependent phagocytosis of apoptotic cells, *J. Exp. Med. 203*, 2613–2625.

187. Jehle, A. W., Gardai, S. J., Li, S., Linsel-Nitschke, P., Morimoto, K., Janssen, W. J., Vandivier, R. W., Wang, N., Greenberg, S., Dale, B. M., Qin, C., Henson, P. M., and Tall, A. R. (2006) ATP-binding cassette transporter A7 enhances phagocytosis of apoptotic cells and associated ERK signaling in macrophages, *J. Cell. Biol. 174*, 547–556.

188. Devitt, A., Parker, K. G., Ogden, C. A., Oldreive, C., Clay, M. F., Melville, L. A., Bellamy, C. O., Lacy-Hulbert, A., Gangloff, S. C., Goyert, S. M., and Gregory, C. D. (2004) Persistence of apoptotic cells without autoimmune disease or inflammation in $CD14^{-/-}$ mice, *J. Cell. Biol. 167*, 1161–1170.

189. Kinchen, J. M., Cabello, J., Klingle, D., Wong, K., Freichtinger, R., Schnabel, H., Schnabel, R., and Hengartner, M. O. (2005) Two pathways converge at CED-10 to mediate actin rearrangement and corpse removal in C. elegans, *Nature 43*, 93–99.

190. Su, H. P., Nakada-Tsukui, K., Tosello-Trampont, A. C., Li, Y., Bu, G., Henson, P. M., and Ravichandran, K. S. (2002) Interaction of CED-6/GULP, an adapter protein involved in engulfment of apoptotic cells with CED-1 and CD91/low density lipoprotein receptor-related protein (LRP), *J. Biol. Chem. 277*, 1772–11779.

191. Park, S. Y., Kang, K. B., Thapa, N., Kim, S. Y., Lee, S. J., and Kim, I. S. (2008) Requirement of adapter protein GULP-6 for stabilin-2 mediated cell corpse engulfment, *J. Biol. Chem. 283*, 10593–10600.

192. Brugnera, E., Haney, L., Grimsley, C., Lu, M., Walk, S. F., Tosello-Tampont, A. C., Macara, I. G., Madhani, H., Fink, G. R., and Ravichandran, K. S. (2002) Unconventional Rac-GEF activity is mediated through the Dock180-ELMO complex, *Nat. Cell Biol. 4*, 574–582.

193. Gumienny, T. L., Brugnera, E., Tosello-Trampont, A. C., Kinchen, J. M., Haney, L. B., Nishiwaki, K., Walk, S. F., Nemergut, M. E., Macara, I. G., Francis, R., Schedl, T., Qin, Y., Van Aelst, L., Hengartner, M. O., and Ravichandran, K. S. (2001) CED-12/ELMO, a novel member of the CrkII/Dock180/Rac pathway, is required for phagocytosis and cell migration, *Cell 107*, 27–41.

194. Leverrier, Y., Okkenhaug, K., Sawyer, C., Bilancio, A., Vanhaesebroeck, B., and Ridley, A. J. (2003) Class I phosphoinositide 3-kinase p110beta is required for apoptotic cell and Fcgamma receptor-mediated phagocytosis by macrophages, *J. Biol. Chem. 278*, 38437–38442.

195. Albert, M. L., Kim, J. I., and Birge, R. B. (2000) alphavbeta5 integrin recruits the CrkII-Dock180-rac1 complex for phagocytosis of apoptotic cells, *Nat. Cell Biol. 2*, 899–905.

196. deBakker, C. D., Haney, L. B., Kinchen, J. M., Grimsley, C., Lu, M., Klingele, D., Hsu, P. K., Chou, B. K., Cheng, L. C., Blangy, A., Sondek, J., Hengartner, M. O., Wu, Y. C., and Ravichandran, K. S. (2004) Phagocytosis of apoptotic cells is regulated by a UNC-73/TRIO-MIG-2/RhoG signaling module and armadillo repeats of CED-12/ELMO, *Curr. Biol. 29*, 2208–2216.

197. Nakaya, M., Tanaka, M., Okabe, Y., Hanayama, R., and Nagata, S. (2006) Opposite effects of rho family GTPases on engulfment of apoptotic cells by macrophages, *J. Biol. Chem. 281*, 8836–8842.

198. Wu, Y., Singh, S., Georgescu, M. M., and Birge, R. B. (2005) A role for Mer tyrosine kinase in alphavbeta5 integrin-mediated phagocytosis of apoptotic cells, *J. Cell Sci. 118*, 539–553.

199. Stuart, L. M., Bell, S. A., Stewart, C. R., Silver, J. M., Richard, J., Goss, J. L., Tseng, A. A., Zhang, A., El Khoury, J. B., and Moore, K, J. (2007) CD36 signals to the actin cytoskeleton and regulates microglial migration via a p130Cas complex, *J. Biol. Chem. 282*, 27392–27401.

200. Kiyokawa, E., Hashimoto, Y., Kurata, T., Sugimura, H., and Matsuda, M. (1998) Evidence that DOCK180 up-regulates signals from the CrkII-p130(Cas) complex, *J. Biol. Chem. 273*, 24479–24484.

201. Cvetanovic, M. and Ucker, D. S. (2004) Innate immune discrimination of apoptotic cells: Repression of proinflammatory macrophage transcription is coupled to specific recognition, *J. Immunol. 172*, 880–889.

202. McDonald, P. P., Fadok, V. A., Bratton, D., and Hanson, P. M. (1999) Transcriptional and translational regulation of inflammatory mediator production by endogenous TGF- β in macrophages that have ingested apoptotic cells, *J. Immunol. 163*, 6164–6172.

203. Fraser, D. A., Arora, M., Bohlson, S. S., Lozano, E., and Tenner, A. J. (2007) Generation of NFkappaB complexes and phosphorylated cAMP response element-binding protein correlates with the anti-inflammatory activity of complement protein C1q in human monocytes, *J. Biol. Chem. 282*, 7360–7367.

204. Sen, P., Wallet, M. A., Yi, Z., Huang, Y., Henderson, M., Mathews, C. E., Earp, H. S., Matsushima, G., Baldwin, A. S., and Tisch, R. M. (2007) Apoptotic cells induce Mer tyrosine kinase-dependent blockade of NF-κB activation in dendritic cells, *Blood 109*, 653–660.

205. Mukundan, L., Odegaard, J. I., Morel, C. R., Heredia, J. E., Mwangi. J. W., Ricardo-Gonzalez, R. R., Goh, Y. P., Eagle, A. R., Dunn, S. E., Awakuni, J. U., Nguyen, K. D., Steinman. L., Michie, S. A., and Chawla, A. (2009) PPAR-delta senses and orchestrates clearance of apoptotic cells to promote tolerance, *Nat. Med. 15*, 1266–1272.

206. Taylor, P. R., Carugati, A., Fadok, V. A., Cook, H. T., Andrews, M., Carroll, M. C., Savill, J. S., Henson, P. M., Botto, M., and Walport, M. J. (2000) A hierarchical role for classical pathway complement proteins in the clearance of apoptotic cells in vivo, *J. Exp. Med. 192*, 359–366.

207. Lu, Q. and Lemke, G. (2001) Homeostatic regulation of the immune system by receptor tyrosine kinases of the Tyro 3 family, *Science 293*, 306–311.

208. Asano, K., Miwa, M., Miwa, K., Hanayama, R., Nagase, H., Nagata, S., and Tanaka, M. (2004) Masking of phosphatidylserine inhibits apoptotic cell engulfment and induces autoantibody production in mice, *J. Exp. Med. 200*, 459–467.

209. A-Gonzalez, N., Bensinger, S. J., Hong, C., Beceiro, S., Bradley, M. N., Zelcer, N., Deniz, J., Ramirez, C., Díaz, M., Gallardo, G., de Galarreta, C. R., Salazar, J., Lopez, F., Edwards, P., Parks, J., Andujar, M., Tontonoz, P., and Castrillo, A. (2009) Apoptotic cells promote their own clearance and immune tolerance through activation of the nuclear receptor LXR, *Immunity 31*, 245–258.

210. Munoz, L. E., Gaipl, U. S., Franz, S., Sheriff, A., Voll, R. E., Kalden, J. R., and Herrmann, M. (2005) SLE - a disease of clearance deficiency? *Rheumatology (Oxford) 44*, 1101–1107.

211. Baumann, I., Kolowos, W., Voll, R. E., Manger, B., Gaipl, U., Neuhuber, W. L., Kirchner, T., Kalden, J. R., and Herrmann, M. (2002) Impaired uptake of apoptotic cells into tingible body macrophages in germinal centers of patients with systemic lupus erythematosus, *Arthritis Rheum. 46*, 191–201.

212. Casiano, C. A. and Tan, E. M. (1996) Recent developments in the understanding of antinuclear autoantibodies, *Int. Arch. Allergy Immunol. 111*, 308–313.

213. Manderson, A. P., Botto, M., and Walport, M. J. (2004) The role of complement in the development of systemic lupus erythematosus, *Annu. Rev. Immunol. 22*, 431–456.

214. Price, B. E., Rauch, J., Shia, M. A., Walsh, M. T., Lieberthal, W., Gilligan, H. M., O'Laughlin, T., Koh, J. S., and Levine, J. S. (1996) Anti-phospholipid autoantibodies bind to apoptotic, but not viable, thymocytes in a beta 2-glycoprotein I-dependent manner, *J. Immunol. 157*, 2201–2208.

215. Mishra, S., Melino, G., and Murphy, L. J. (2007) Transglutaminase 2 kinase activity facilitates protein kinase A-induced phosphorylation of retinoblastoma protein, *J. Biol. Chem. 282*, 18108–18115.

216. Rodolfo, C., Mormone, E., Matarrese, P., Ciccosanti, F., Farrace, M. G., Garofano, E., Piredda, L., Fimia, G. M., Malorni, W., and Piacentini, M. (2004) Tissue transglutaminase is a multifunctional BH3-only protein, *J. Biol. Chem. 279*, 54783–54792.

217. Thomázy, V. and Fésüs, L. (1989) Differential expression of tissue transglutaminase in human cells. An immunohistochemical study, *Cell Tissue Res. 255*, 215–224.

218. Upchurch, H. F., Conway, E., Patterson, M. K., and Maxwell, M. D. (1991) Localization of cellular transglutaminase on the extracellular matrix after wounding: characteristics of the matrix bound enzyme, *J. Cell Physiol. 149*, 375–382.

219. Akimov, S. S., Krylov, D., Fleischman, L. F., and Belkin, A. (2000) Tissue transglutaminase is an integrin-binding adhesion coreceptor for fibronectin, *J. Cell. Biol. 148*, 825–838.

220. Akimov, S. S. and Belkin, A. M. (2001) Cell surface tissue transglutaminase is involved in adhesion and migration of monocytic cells on fibronectin, *Blood 98*, 1567–1576.

221. Hang, J., Zemskov, E. A., Lorand, L., and Belkin, A. M. (2005) Identification of a novel recognition sequence for fibronectin within the NH2-terminal beta-sandwich domain of tissue transglutaminase, *J. Biol. Chem. 280*, 23675–23683.

222. Janiak, A., Zemskov, E. A., and Belkin, A. M. (2006) Cell surface transglutaminase promotes RhoA activation via integrin clustering and suppression of the Src-p190RhoGAP signaling pathway, *Mol. Biol. Cell 17*, 1606–1619.

223. Fésüs, L., Thomázy, V., and Falus, A. (1987) Induction and activation of tissue transglutaminase during programmed cell death, *FEBS Lett. 224*, 104–108.

224. Szende, B., Schally, A. V., and Lapis, K. (1991) Immunocytochemical demonstration of tissue transglutaminase indicative of programmed cell death (apoptosis) in hormone sensitive mammary tumours, *Acta. Morphol. Hung. 39*, 53–58.

225. Jiang, H. and Kochhar, D. M. (1992) Induction of tissue transglutaminase and apoptosis by retinoic acid in the limb bud, *Teratology 46*, 333–340.

226. Guenette, R. S., Corbeil, H. B., Léger, J., Wong, K., Mézl, V., Mooibroek, M., and Tenniswood, M. (1994) Induction of gene expression during involution of the lactating mammary gland of the rat, *J. Mol. Endocrinol. 12*, 47–60.

227. Piacentini, M. and Autuori, F. (1994) Immunohistochemical localization of tissue transglutaminase and Bcl-2 in rat uterine tissues during embryo implantation and post-partum involution, *Differentiation 57*, 51–61.

228. Amendola, A., Gougeon, M. L., Poccia, F., Bondurand, A., Fésüs. L., and Piacentini M. (1996) Induction of "tissue" transglutaminase in HIV pathogenesis: evidence for high rate of apoptosis of CD4+ T lymphocytes and accessory cells in lymphoid tissues, *Proc. Natl. Acad. Sci. U.S.A. 93*, 11057–11062.

229. Fésüs, L., Thomázy, V., Autuori, F., Ceru, M. P., Tarcsa, E., and Piacentini, M. (1989) Apoptotic hepatocytes become insoluble in detergents and chaotropic agents as a result of transglutaminase action, *FEBS Lett. 245*, 150–154.

230. Piacentini, M., Autuori, F., Dini, L., Farrace, M. G., Ghibelli, L., Piredda, L., and Fésüs, L. (1991) "Tissue" transglutaminase is specifically expressed in neonatal rat liver cells undergoing apoptosis upon epidermal growth factor-stimulation, *Cell Tissue Res. 263*, 227–235.

231. Fésüs, L., Mádi, A., Balajthy, Z., Nemes, Z., and Szondy Z. (1996) Transglutaminase induction by various cell death and apoptosis pathways, *Experientia 52*, 942–949.

232. Szondy, Z., Molnár, P., Nemes, Z., Boyiadzis, M., Kedei, N., Tóth, R., and Fésüs, L. (1997) Differential expression of tissue transglutaminase during in vivo apoptosis of thymocytes induced via distinct signalling pathways, *FEBS Lett. 404*, 307–313.

233. Szegezdi, E., Szondy, Z., Nagy, L., Nemes, Z., Friis, R. R., Davies, P. J., and Fésüs, L. (2000) Apoptosis-linked in vivo regulation of the tissue transglutaminase gene promoter, *Cell Death Differ. 7*, 1225–1233.

234. Kiss, I., Rühl, R., Szegezdi, E., Fritzsche, B., Tóth, B., Pongrácz, J., Perlmann, T., Fésüs, L., and Szondy, Z. (2008) Retinoid receptor-activating ligands are produced within the mouse thymus during postnatal development, *Eur. J. Immunol. 38*, 147–155.

235. Nagy, L., Saydak, M., Shipley, N., Lu, S., Basilion, J. P., Yan, Z. H., Syka, P., Chandraratna, R. A., Stein, J. P., Heyman, R. A., and Davies, P. J. (1996) Identification and characterization of a versatile retinoid response element (retinoic acid receptor response element-retinoid X receptor response element) in the mouse tissue transglutaminase gene promoter, *J. Biol. Chem. 271*, 4355–4365.

236. Ritter, S. J. and Davies, P. J. (1998) Identification of a transforming growth factor-beta1/bone morphogenetic protein 4 (TGF-beta1/BMP4) response element within the mouse tissue transglutaminase gene promoter, *J. Biol. Chem. 273*, 12798–12806.

237. Folk, J. E. (1980) Transglutaminases, *Annu. Rev. Biochem. 49*, 517–531.

238. Fésüs, L., Tarcsa, E., Kedei, N., Autuori, F., and Piacentini, M. (1991) Degradation of cells dying by apoptosis leads to accumulation of epsilon(gamma-glutamyl)lysine isodipeptide in culture fluid and blood, *FEBS Lett. 284*, 109–112.

239. Oliverio, S., Amendola, A., Di Sano, F., Farrace, M. G., Fésüs, L., Nemes, Z., Piredda, L., Spinedi, A., and Piacentini, M. (1997) Tissue transglutaminase-dependent posttranslational modification of the retinoblastoma gene product in promonocytic cells undergoing apoptosis, *Mol. Cell. Biol. 17*, 6040–6048.

240. Malorni, W., Farrace, M. G., Matarrese, P., Tinari, A., Ciarlo, L., Mousavi-Shafaei, P., D'Eletto, M., Di Giacomo, G., Melino, G., Palmieri, L., Rodolfo, C., and Piacentini, M. (2009) The adenine nucleotide translocator 1 acts as a type 2 transglutaminase substrate: implications for mitochondrial-dependent apoptosis, *Cell Death Differ. 16*, 1480–1492.

241. Piacentini, M. (1995) Tissue transglutaminase: a candidate effector element of physiological cell death, *Curr. Top. Microbiol. Immunol. 200*, 163–175.

242. Nemes, Z., Jr, Adány, R., Balázs, M., Boross, P., and Fésüs, L. (1997) Identification of cytoplasmic actin as an abundant glutaminyl substrate for tissue transglutaminase in HL-60 and U937 cells undergoing apoptosis, *J. Biol. Chem. 272*, 20577–20583.

243. Nishiura, H., Shibuya, Y., and Yamamoto, T. (1998) S19 ribosomal protein cross-linked dimer causes monocyte-predominant infiltration by means of molecular mimicry to complement C5a, *Lab. Invest. 78*, 1615–1623.

244. Sarang, Z., Mádi, A., Koy, C., Varga, S., Glocker, M. O., Ucker, D. S., Kuchay, S., Chishti, A. H., Melino, G., Fésüs, L., and Szondy, Z. (2007) Tissue transglutaminase (TG2) facilitates phosphatidylserine exposure and calpain activity in calcium-induced death of erythrocytes, *Cell Death Differ. 14*, 1842–1844.

245. Schroff, G., Neumann, C., and Sorg, C. (1981) Transglutaminase as a marker for subsets of murine macrophages, *Eur. J. Immunol. 11*, 637–642.

246. Murtaugh, M. P., Mehta, K., Johnson, J., Myers, M., Juliano, R. L., and Davies, P. J. (1983) Induction of tissue transglutaminase in mouse peritoneal macrophages, *J. Biol. Chem. 258*, 11074–11081.

247. Zemskov, E. A., Mikhailenko, I., Strickland, D. K., and Belkin, A. M. (2007) Cell-surface transglutaminase undergoes internalization and lysosomal degradation: an essential role for LRP1, *J. Cell Sci. 120*, 3188–3199.

248. Tóth, B., Garabuczi, E., Sarang, Z., Vereb, G., Vámosi, G., Aeschlimann, D., Blaskó, B., Bécsi, B., Erdφdi, F., Lacy-Hulbert, A., Zhang, A., Falasca, L., Balajthy, Z., Birge, R., Melino, G., Fésüs, L., and Szondy, Z. (2009) Transglutaminase 2 is needed for the formation of an efficient phagocyte portal in macrophages engulfing apoptotic cells, *J. Immunol. 182*, 2084–2092.

249. Tóth, B., Sarang, Z., Vereb, G., Zhang, A., Tanaka, S., Melino, G., Fésüs, L, and Szondy Z. (2009) Over-expression of integrin beta3 can partially overcome the defect of integrin beta3 signaling in transglutaminase 2 null macrophages, *Immunol. Lett. 126*, 22–28.

250. Rose, D. M., Sydlaske, A. D., Agha-Babakhani, A., Johnson, K., and Terkeltaub, R. (2006) Transglutaminase 2 limits murine peritoneal acute gout-like inflammation by regulating macrophage clearance of apoptotic neutrophils, *Arthritis Rheum. 54*, 3363–3371.

251. Feng, J. F., Readon, M., Yadav, S. P., and Im, M. J. (1999) Calreticulin down-regulates both GTP binding and transglutaminase activities of transglutaminase II, *Biochemistry 38*, 10743–1079.

252. Moore, K. G. and Sartorelli, A. C. (1992) Annexin I and involucrin are cross-linked by particulate transglutaminase into the cornified cell envelope of squamous cell carcinoma Y1, *Exp. Cell Res. 200*, 186–195.

253. Falasca, L., Iadevaia, V., Ciccosanti, F., Melino, G., Serafino, A., and Piacentini M. (2005) Transglutaminase type II is a key element in the regulation of the anti-inflammatory response elicited by apoptotic cell engulfment. *J. Immunol. 174*, 7330–7340.

254. Nunes, I., Shapiro, R. L., and Rifkin, D. B. (1995) Characterization of latent TGF-beta activation by murine peritoneal macrophages, *J. Immunol. 155*, 1450–1459.

255. Nishimura, S. L. (2009) Integrin-mediated transforming growth factor-beta activation, a potential therapeutic target in fibrogenic disorders, *Am. J. Pathol. 175*, 1362–1370.

256. Yehualaeshet, T., O'Connor, R., Green-Johnson, J., Mai, S., Silverstein, R., Murphy-Ullrich, J. E., and Khalil, N. (1999) Activation of rat alveolar macrophage-derived latent transforming growth factor beta-1 by plasmin requires interaction with thrombospondin-1 and its cell surface receptor, CD36, *Am. J. Pathol. 155*, 841–851.

257. Harpel, J. G., Schultz-Cherry, S., Murphy-Ullrich, J. E., and Rifkin, D. B. (2001) Tamoxifen and estrogen effects on TGF-beta formation: role of thrombospondin-1, alphavbeta3, and integrin-associated protein, *Biochem. Biophys. Res. Commun. 284*, 11–14.

258. Vila, J., Isaacs, J. D., and Anderson, A. E. (2009) Regulatory T cells and autoimmunity, *Curr. Opin. Hematol. 16*, 274–279.

259. Letterio, J. J. and Roberts, A. B. (1998) Regulation of immune responses by TGF-beta, *Annu. Rev. Immunol. 16*, 137–161.

260. Sarang, Z., Köröskényó, K., Pallaí, A., Duró, E., Melino, G., Griffin, M., Fésüs, L., and Szondy, Z: (2011) Transglutaminase 2 null macrophages respond to lipopolysaccharide stimulation by elevated proinflammatory cytokines production due to an enhanced α (v) β (3) integrin-induced sue tyrosine kinase signaling. Immunol Lett., *138*, 71-78.

261. Rébé, C., Raveneau. M., Chevriaux, A., Lakomy, D., Sberna, A. L., Costa, A., Bessède, G., Athias, A., Steinmetz, E., Lobaccaro, J. M., Alves, G., Menicacci, A., Vachenc, S., Solary, E., Gambert, P., and Masson, D. (2009) Induction of transglutaminase 2 by a liver X receptor/retinoic acid receptor alpha pathway increases the clearance of apoptotic cells by human macrophages, *Circ. Res. 105*, 393–401.

262. Matsuura, E., Kobayashi, K., Matsunami, Y., Shen, L., Quan, N., Makarova, M., Suchkov, S. V., Ayada, K., Oguma, K., and Lopez, L. R. (2009) Autoimmunity, infectious immunity, and atherosclerosis, *J. Clin. Immunol. 29*, 714–721.

263. Ait-Oufella, H., Pouresmail, V., Simon, T., Blanc-Brude, O., Kinugawa, K., Merval, R., Offenstadt, G., Lesèche, G., Cohen, P. L., Tedgui, A., and Mallat, Z. (2008) Defective mer receptor tyrosine kinase signaling in bone marrow cells promotes apoptotic cell accumulation and accelerates atherosclerosis, *Arterioscler. Thromb. Vasc. Biol. 28*, 1429–1431.

264. Kratzer, A., Buchebner, M., Pfeifer, T., Becker, T. M., Uray, G., Miyazaki, M., Miyazaki-Anzai, S., Ebner, B., Chandak, P. G., Kadam, R. S., Calayir, E., Rathke, N., Ahammer, H., Radovic, B., Trauner, M., Hoefler, G., Kompella, U. B., Fauler, G., Levi, M., Levak-Frank, S., Kostner, G. M., and Kratky, D. (2009) Synthetic LXR agonist attenuates plaque formation in apoE$^{-/-}$ mice without inducing liver steatosis and hypertriglyceridemia, *J. Lipid Res. 50*, 312–326.

EFFECTS AND ANALYSIS OF TRANSGLUTAMINATION ON PROTEIN AGGREGATION AND CLEARANCE IN NEURODEGENERATIVE DISEASES

ZOLTÁN NEMES

CONTENTS

Advances in Enzymology and Related Areas of Molecular Biology, Volume 78.
Edited by Eric J. Toone.

I. INTRODUCTION

Transglutaminases (TGs; Enzyme Commission 2.3.2.13.) are Ca^{2+}-dependent enzymes that catalyze the transfer of the γ-glutamyl residue in protein-bound glutamine to a primary amine through the formation of a high-energy thioester intermediate [1]. As a result of the catalytic mechanism, these enzymes are capable of transferring the protein-bound γ-glutamyl onto alcohols with longer aliphatic carbon chains or water in the absence of other utilizable nucleophylic functional group [2]. These reactions esterify [3] or deamidate glutamine moieties to glutamic acid, respectively [4, 5]. The scientific relevance of these side-reaction pathways is yet largely unexploited, although their physiological relevance might be as important as that of the best known amine transfer catalysis.

The amine onto which the γ-glutamyl moiety is transferred by TGs may theoretically be any primary amine with straight carbon chain [6]. However, in biological systems, most abundantly available amine substrates are the ε-amino groups of protein-bound lysines or eventually polyamines [7]. Even in specialized cells with high polyamine turnover, the availability of lysyl substrates and the formation of N^{ε}(γ-glutamyl)lysine cross-links between proteins is thought to be dominant over the formation of protein-(γ-glutamyl)polyamine modifications or N^1, N^8-*bis*(γ-glutamyl)spermidine bonds [8].

The specificity of various TGs for lysyl-donor substrate proteins is determined by the type and steric position of residues that surround the exposed lysine on the surface of the protein and whether the amine group is engaged in the formation of an ionic bond or is protonated or not [6, 9]. In case such favorable lysine is found, the enzyme will form a covalent N^{ε}(γ-glutamyl)lysine isopeptide bond between the two protein chains. Proteins harboring both reactive glutamine as well as reactive lysine residues can form large protein polymers by transglutamylation.

True glutamyl- and lysyl-substrate residues appear to be rare and occur only in a specialized set of proteins for most eukaryotic TGs, except for microbial TGs used in industrial food processing [10]. Nevertheless, even the most "choosy" enzymes are willing to take nearly any glutaminyl and lysyl moiety from unfolded random coil sequences of denatured proteins, provided those remain soluble [11, 12]. This mechanism has resulted in the generation of numerous in vitro cross-linking papers making speculative claims of in vivo transglutamylation of far more proteins than it could have been focused into a few plausible, coherent, and converging biological models of what these enzymes are doing in health and disease. In spite of 30 years of rampant TGase research, the true biological role and significance of specific TGs seems to be as obscure as it used to be.

The lucrative generation of TGase-dependent cross-linking artifacts in the 1990s made the verification of alleged TGase substrate properties inevitable from in vivo experimental material. Working with in vivo cross-linked proteins is a technically tedious task, which is hindered through numerous methodological challenges and pitfalls. Therefore, this field of research has rarely offered competitive yields for most newcomers lately.

TGase-product protein polymers can reach several million Daltons in size and are practically insoluble in buffers used for chromatography or electrophoresis. Despite a good deal of promising innovation and pioneering to deliver labeled amine or glutamine substrates to living cells and label native proteins in living cells in situ, currently available techniques yield very modest signals and a bulk of starting material is necessary to fish out manifest amounts of TGase-labeled protein [13, 14].

II. THE TGASE ENZYME FAMILY

Several TGs have been identified in prokaryotes, fungi, plants, and diverse *species* of the animal kingdoms [8, 15]. In humans, nine evolutionarily related transglutaminase genes have been described.

TGase family homologs in humans are named TG1 (keratinocyte type), TG2 (tissue type), TG3 (epidermal), TG4 (prostatic type), TG5 (TGX), TG6 (neuronal type), TG7 (TGZ), coagulation factor XIII a subunit, and erythrocyte membrane band 4.2 protein, the last lacks TGase activity [16, 17].

TGase activity was implicated in several phenomena of cell biology, including cell proliferation, differentiation, and death, extracellular matrix assembly, integrin signal transduction, and phagocytosis, and complex physiological processes, like blood coagulation, sperm immunosupression, and keratinocyte barrier function have been shown to depend on transglutaminase activity (see other chapters of this volume for review). Huge amount of data highlight the key role of TGs in celiac disease, where TG2 might act both as autoantigen and trigger immune response by deamidating wheat gliadin peptides in the gut (see [18] for review). Clearly, TGs are present and active in all these critical phenomena, but the flaws of previous research methods necessitates a careful reevaluation of previous data in the light of recent findings.

Up to the mid-1990s, it was widely agreed that TG2 is the ubiquitous enzyme present in most tissues and cell lines. At the turn of the millennium it was shown that while some members of the TGase family are expressed in specific tissues (e.g., FXIIIa, band 4.2 protein, and TG4), other TGs besides TG2 are expressed ubiquitously, and several other *tg* gene products coexist in most cells, including enzymes that have previously been thought to have very specific functions, like fibrin cross-linking or keratinization. Depending on species, brain region, and cell type, TG1, TG2, TG3 TG5, TG6, and FXIIIa were shown to be expressed in neurons [19–21]. The coexistence of several, by structure as well as by antigenicity, closely related TGs seems to be functionally redundant, no wonder the *tg* knockout mice showed only a mild phenotype, as compared to the number and importance of cellular processes, in which TG2 was implicated [22]. Previous experimental works attributed cross-linking to "tissue transglutaminase" activity, wherever TG2 enzyme was detected and did not care for other types. Moreover, in addition to its transamidation and deamidation activity, TG2 can also isomerize disulfide bonds, bind, and hydrolyze ATP and GTP, and may also have intrinsic kinase activity and can regulate the function of numerous proteins by binding to them. Similar activities of TGase homologs have occasionally been shown and are expected, though experimental evidence is far less abundant than for TG2. The knowledge of the entire human genome sequence has finalized the number of homologous *tgs* [17]. Thus, it is unlikely that further TGs could be discovered, though evolutionarily unrelated proteins may also catalyze transglutamination of proteins under special conditions. In a milieu expected in a living cell, TGs are thought to be strongly inhibited by the low calcium and high nucleotide triphosphate concentrations [23]. Therefore, numerous authors share the notion that transglutamination of proteins might only be possible in dying cells, and thus all TGase activities inside normal (healthy) cells must be unrelated to transglutamination.

In this review, divergent models and data on the role of TGs in the CNS under normal and pathological conditions are overviewed in an attempt to reconcile extant data in a timely model of intraneuronal inclusion formation.

A. TG2 STRUCTURE

TG2, the most investigated TGase homolog, is a protein varying in full length from 685 to 691 amino acids in vertebrates. The human *tg2* gene was mapped to chromosome 20 and includes 13 exons and 12 introns. The transcription of *tg2* is induced by retinoids, the biological relevance of which is unknown. Besides retinoids, the expression of TG2 is regulated by generic transcription activators, possibly binding to the AP-2- and SP-1-binding sites in its promoter sequence [24, 25]. The X-ray crystal structure of human FXIIIa and TG2 complexed with GDP at 2.8 Å resolution showed that TG2 (and obviously all TGs) consist of four distinct domains [26]. These include an *N*-terminal β-sandwich domain, a core domain (where the residues responsible for TGase activity are located), and two *C*-terminal β-barrel domains. These structures revealed the structural basis for the negative regulation of transamidation activity by the bound nucleotide and the positive regulation of transamidation by calcium ion binding (Figure 1). Bound guanidine nucleotides fit into the binding groove between the core and barrel 1 domains and compact the domain structure, which keeps the

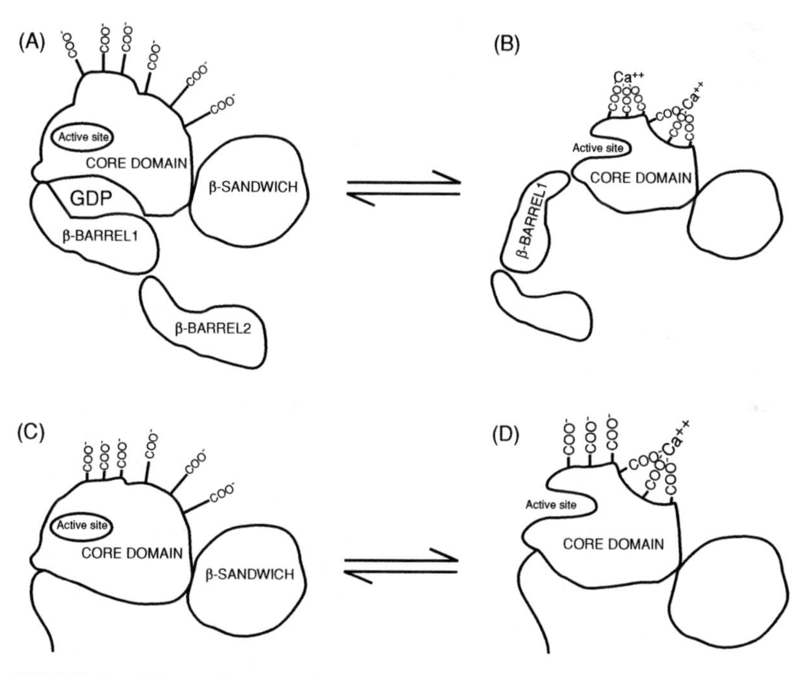

FIGURE 1. Scheme of antagonistic regulation of TG2 by purine-nucleotide di- or triphosphate and calcium binding. The catalytically inactive, closed conformation (A) of the enzyme is stabilized by the nucleotide di- or triphosphate binding, whereas the ligation of calcium ions opens up the catalytic site and eliminates the nucleotide phosphate binding pocket (B). The short transcript variant of TG2 (C) lacks the two barrel domains present in full-length TG2. Thus, its catalytic activity is supposed to appear at lower calcium levels (D).

catalytic (transglutaminase) binding site inaccessible to glutamine–substrate binding, whereas the release of nucleotide and the chelation of calcium ions opens up the four domains and makes the core domain accessible to both γ-glutamyl and ε-lysyl substrate proteins [27]. Nevertheless, the energetic mechanisms that permit the release of the product have not yet been explained on the basis of extant (static) structural models [28], because by all certainty, large domain movements and conformational changes are required for binding, cross-linking, and then releasing the two substrate proteins during the catalytic cycle [29].

Domain truncation experiment revealed that the presence of the β-sandwich domain is needed for TGase catalytic activity, whereas the deletion of the two *C*-terminal β-barrels disinhibits TGase activity by disrupting the nucleotide-binding site and eliminating the bulky and spatially inhibitory barrels [30]. A natural barrel-deleted variant of TG2 was found in neuronal cells that result from intronic read-through, leading to early termination of the mRNA translation [31]. mRNAs indicating the existence of alternatively spliced TG2 products were also reported in cells, though their expression has not been shown and characterized in the nervous system [24, 32].

B. STRUCTURES OF OTHER TGS

The main structural differences of other TGase isoenzymes that distinguish them from TG2 are outlined here.

Factor XIIIa and TG1 have diverged earlier from the phylogeny of *tg*s than the other isotypes; therefore, they share more common features with one another than with the rest of the family. FXIIIa and TG1 are larger (~100 kDa) proteins due to an exon that is not present in other TGs. The extra *N*-terminal sequence in FXIIIa contains the thrombin-activation peptide and in TG1 the ~10-kDa part that anchors the protein to the membrane through cystein fatty acylation. TG1 protein is proteolytically cleaved into three fragments, which increases transamidating activity and eventually results in the partition of TG1 from the membrane anchorage [33].

TG2 and TG3 reveal similar domain architectures and both bind inhibitory nucleotides. The domain structure of TG4-7 and band 4.2 protein is also predicted to be the same as that of TG2, though no X-ray coordinates have been measured thus far. TG3 and TG5 are activated by proteolytic cleavage of a flexible loop sequence between the core and barrel domains [34]. Neither protease-mediated activation nor nucleotide-mediated inhibition of TG4, TG6, or TG7 has been reported, though it might be expected on the basis of sequence similarities.

III. BRAIN TGS IN NOT DISEASED NERVE CELLS

TGase activity was first detected in the giant cholinergic neurons of Aplysia [35]. TGase activity was demonstrated in isolated synaptosomal preparations [36], cerebellar granule neurons [37], and in cultured astrocytes [38, 39]. Measurable levels of TGase activity were detected in all parts of CNS and peripheral nerves (see [37] for review). TG2 was demonstrated by immunocytochemistry in human brain, particularly in neurons [21, 40–42]. Significant TGase activity has been found in the frontal and temporal neocortex, hippocampus, cerebellum, and white matter in human and rat brain [43, 44].

Two TGs of 45 and 29 kDa, respectively, were purified from rat brain, which were immunologically slightly different from full-size TG2 (~77 kDa) [45]. One of the TGs from rat brain was found to be associated with the synaptic vesicles, whereas the other was cytosolic [46]. As the active form of TG3 consists of ~25- and ~50-kDa subunits [47], and TG1 is active in a membrane-bound form [48], previous findings in rats might have harbingered the discovery that TG1, TG2, and TG3 are expressed in different parts of human cortex and basal ganglia [21]. The presence of factor XIIIa in human nervous system has not yet been investigated in detail [49]; however, this transglutaminase was found to be present in fish nerves [50]. TG2 was found to be expressed in two isoforms in human brain, and the shorter form that results from an intronic read-through was associated with Alzheimer's disease (AD) [31].

An approximately 2.5-fold increase of TGase activity was found in mouse brain from birth to adulthood [51], whereas TGase activity of rat brain and spinal cord homogenates was found to be highest shortly before birth [52].

There are several reports pointing to a role of TG2 in neurite outgrowth [51, 53]. TG2 has been proposed to play a role in the stabilization of focal adhesions as well as cytoskeletal stabilization [37, 54, 55], processes that are crucial for dendrite and axon formation. Neurofilaments [56–58], actin [59], and vimentin [60] were shown to be modified by TG2. As actin and intermediate filaments are determinants of cytoskeletal motility and shape, modification of filament monomers by TGs is likely to have profound effects on cell functions and survival. In developing mouse brain, tubulin was found to be abundantly modified, and it was suggested that cross-linking of cytoskeletal components might be important for the formation and stabilization of neuritic processes [55].

In cultured neuronal and neuron-like cells, neurite outgrowth associated with induced expression of TG2 [51] and induced expression or transfection of TG2 (but not catalytically inactive TG2) increased neuritic process outgrowth in SH-SY5Y neuroblastoma cells [53, 61, 62], whereas antisense or competitive inhibition of TG2 inhibited differentiation into a neuron-like phenotype. TG2 was shown to appear on the cell surface of primary rat cerebellar granule cells, where it also promoted neuritic process formation [63, 64]. Some reports suggested that neural TG2 activity might be important for dimerizing midkine. Midkine is TGase substrate in vitro [65] and was shown to be cross-linked by cerebellar granule neurons [63]. Midkine is a growth factor that, in dimerized form, enhances neurite outgrowth and neuronal survival during development [66]. It has also been demonstrated that TGase-mediated cross-linking of midkine might be necessary for biological effect [65].

A role for TG2 was proposed in facilitating axonal regrowth by cross-linking and stabilization of interleukin-2 dimers and thus eliminating growth-inhibitory oligodendrocytes [56].

TGase-mediated protein cross-linking was shown to appear during the long-term potentiation induced by high-frequency electric stimulation of rat hippocampal slices, suggesting a role for TGase activity in early memory consolidation [67]. TGase activity was shown to potentiate the effects of tetanus toxin in inhibiting neurotransmitter release by inactivating synapsin I [46].

The expression of TG5 was also noted in neurons, and autoantibodies to TG6 in gluten ataxia revealed the broad expression of TG6 in the CNS [20, 68, 69].

To date, studies have only been published on the effects of TG2 on neurons or neuron-like cultured cells.

A. TG2 IN APOPTOSIS OF NERVE CELLS

Another intensively studied aspect of TG2 involvement was apoptotic cell death in neuron-like cells. In SK-NE-BE cells, overexpression of TG2 sensitized cells to apoptotic stressors, whereas antisense inhibition of TG2 promoted survival [70, 71]; nevertheless, TG2 overexpression alone did not cause increased cell death. In vivo, *tg2* knockout mice do not display any apparent deficits in the major apoptotic pathways, though the clearance of apoptotic cells is impaired, which leads to an apparent accumulation of apoptotic cells in dextamethasone-treated thymus [72]. All these results demonstrate that TG2, similar to nonneuronal cell types, facilitates or attenuates apoptosis depending on cell type and cytopathic stimulus. We find it noteworthy to mention that induced expression of nucleotide-disinhibited mutants of TG2 were associated with cell death per se, whereas the full-length (wild type) variant was found to decrease cell death in various apoptosis-induction paradigms [73, 74].

In SK-N-BE neuroblastoma cells, the apoptosis-bound cells were found to express high TG2 levels during retinoic acid-induced differentiation [70], whereas in SH-SY5Y cells treated with retinate, the induction of TGase appeared in successfully differentiating cells too [62]. It is obvious that TG2 is activated in cells executing their natural and active suicide program, which is morphologically recognized as apoptosis [75].

In the CNS, mature postmitotic neurons are tightly surrounded by glial elements and therefore cannot swell, form blebs, and fall into pieces like individual lymphocytes in other organs. Thus, the term apoptosis is applied rather functionally than morphologically in brain, where the sole (unspecific) morphological manifestation of programed cell death is the pycnosis (condensation) of nuclei. As the morphology of acute and chronic neuronal and glial lesions had been established long before the term "apoptosis" was introduced [76, 77], neuronal apoptosis usually refers either to the organized and controlled way of cell death involving the activation of enzymes that hallmark apoptosis in in vitro apoptotic systems or to the physiological nature of cell elimination during brain development. Several lines of evidence indicates that TGs are activated in neural cells coinciding with apoptosis meant by either way; however, gene expression and induction is not needed for several forms of neuronal death, like neural anoxia [78]. The activation of TG2 during apoptosis was regarded as a salvage mechanism that fixes cellular organelles and prevents the leakage and exposure of intracellular antigens [68]. Whether this teleology is applicable for the CNS, where large necrotic areas (emollitions) can be resorbed without obvious systemic immune reactions, is questionable.

IV. TGS IN DISEASE

A. TGS IN INJURY

Several reports have shown that acute stress to neurons altered TGase activity or TG2 expression in nerve cells. TGase activity is rapidly, but transiently increased after nerve injury in rat superior cervical ganglia [43, 79]. Similar transient surge was detected in the vagus nerve after crush injury [80]. A contusion to the rat spinal cord and forebrain elevated TG2 mRNA expression at the site of lesion. This increase was associated with the increased splicing and translation of both the full-length TG2 and the nucleotide-resistant short-transcript variant of TG2 [81, 82]. Ischemic stress to the spinal cord resulting from the ligation of the dorsal spinal artery branch resulted in a lasting elevation of TGase activity within the spinal cord [83]. Temporary occlusion of the common carotid arteries in the gerbil resulted in elevated TGase activity in the hippocampus, and occlusion of the rat middle carotid artery was shown to elevate TG2 protein expression in the hippocampus. Again, both short- and full-length TG2 forms were found induced [84].

TG2 can also be induced by cerebral inflammation [85] and brain stress induced by glutamate excitotoxicity, calcium influx, oxidative stress, UV exposure, and inflammatory cytokines [86].

While it seems that ischemic stress and traumatic injury elevates TG2 expression in the brain and spinal cord tissue, it is largely unclear which type of cells produces the enzyme and what for. Since common inflammatory mediators TNF-alpha and IL-1beta also induce the expression of the short TG2 variant in astrocytes [38], elevated TG2 expression might also be a marker of tissue repair.

B. NEURODEGENERATIVE DISEASES

AD is the most common cause of dementia in the elderly and is characterized by progressive neuron loss that results in impaired cognitive functions and memory decline. The onset of AD is predominantly sporadic although the mutations of presenilin and APP genes have been linked to rare genetic forms of AD [87]. AD is a very severe social, economic, and health problem for aging postindustrial societies; therefore, research on AD is of high priority worldwide. Nevertheless, in spite of tremendous efforts, research data thus far failed to provide a consensus model for the pathogenesis of this condition, neither found a cure for it. The obvious inefficacy of efforts is moving increasing mass of researchers to challenge the dogmas on AD and to find new interpretations to pathological phenomena. In order to understand the paradigmatic questions on AD, we have to outline the evolving of AD concept over the last hundred years.

Emil Kraepelin first coined "senile dementia of Alzheimer's type" in his renowned textbook [88] after Alzheimer described his classic case report of a demented patient (Frau Auguste D.) with detailed neuropathological examinations [89]. Alois Alzheimer classified amyloid plaques and neurofibrillary tangles (NFT) as pathognostic neuropathological hallmarks of this disease, and both lesions were easily recognizable by contemporary silver staining methods [90]. Since then AD has grown to one of the largest public health problems of developed societies [91]. Hundred years after Kraepelin, AD is still defined according to its clinical and neuropathological resemblance on the case of Frau Auguste D. Neither clinical nor imaging or laboratory data verify AD before the histological examination to the brain, thus in living patients, only the term "probable AD" is warranted, though some antemortem CSF diagnostic markers have been reported to show to predict findings of the neurohistopathologic evaluation with up to 85% predictive accuracies [92]. Diagnostic guidelines and consensus formulas were created by NINCDS-ADRDA workgroup [93] for clinicians and by CERAD [94] for pathologists when to claim possibility or likelihood of causative relationship between Alzheimer's neuropathological lesions and the cognitive decline in order to make scientific data comparable. These guidelines propose to diagnose AD if the patient was clinically demented, senile plaques and NFT reached a vaguely specified threshold in histological examination, and all other possible causes of dementia were possibly ruled out.

Again, similar diagnostic formulas are offered for another common cause of dementia caused by multiple brain infarcts: the vascular dementia [95] as well as dementia with Lewy bodies [96].

Still, no applicable diagnostic algorithm is available for rarer types of primary neurodegenerative dementias, like frontotemporal dementia (Pick's disease), and there are no guidelines to classify the most frequently encountered mixed cases [97]. The different pathological types of dementia are confluent and unabridged from one another. In a population-based study from the UK, the neuropathological diagnosis of vascular or Alzheimer-type pathology was shown to be of modest value in predicting whether dementia was present or not and most vascular cases revealed Alzheimer's pathology and vice versa [98], revealing that neither the neuropathologic evaluation can be consequently reproducible nor does the presence of lesions mean clinical dementia per se. Dementia with Lewy bodies was also shown to coincide mostly with Alzheimer-type and vascular pathology; some papers mention this disease as "Lewy body variant of AD" [99, 100].

Compelling evidence supports the notion that Alzheimer's neuropathological lesions might be caused by chronic hypoxic stress (see [101] for references). In this model, Alzheimer type neuropathological lesions would be the sequel of microangiopathy caused by amyloid deposition, atherosclerosis, and insufficient perfusion of smaller arteries, whereas neurologically assessable cerebrovascular disease, stroke, and vascular ("multi-infarct") dementia would arise from the same reasons affecting larger arteries 102.

Keeping in mind that all scientific reports on sporadic ("idiopathic") AD as well as parkinsonism/Lewy body-related dementias are flawed by the mentioned diagnostic ambiguities and confounding effects of comorbidity, here we try to review findings about TGase involvement in AD and other distinct type of dementias, while keeping in mind that "clean" AD cases probably do not exist.

Senile plaques and NFT are formed from abnormally interacting proteins in the brain, and both these lesions are highly insoluble structures. Purification and sequencing has shown that the main, birefringent component of amyloid plaques is the amyloid-β-protein (Aβ), a 40–43 amino acid peptide cleaved out from a large transmembrane glycoprotein of unknown function, the amyloid precursor protein (APP). NFT are argyrophylic bundles of fibrous material in dystrophic neurites. Ultrastructurally, NFT were shown to be built from bundles of paired helical filaments. Paired helical filaments contain several accidental proteins, but their core is probably composed of the microtubule-associated protein tau. Paired helical filament tau is abnormally hyperphosphorylated, glycated, ubiquitinated, and oxidized, although none of these posttranslational modifications is necessary to force tau to assemble into paired helical filament-like structures in vitro. Even so, hyperphosphorylation is thought to be prior to fibril formation as underphosphorylated tau associates strongly with axonal microtubules and therefore cannot interact with oneself (see [103] for references).

1. TG2 and Its Substrates in AD

The involvement of TGase-mediated cross-linking in the formation of insoluble protein deposits of AD was first predicted by Dennis Selkoe et al. [58] in 1982 by demonstating in vitro cross-linking of neurofilament proteins into larger aggregates by endogeneous TGase activity inherent in neurons. This experiment, and a later study [57], provided the first evidence that brain TGs are capable of cross-linking cytoskeletal proteins. Nevertheless, the in vitro aggregates formed by TG2 did not form structures similar to ex vivo NFT. Since these pivotal experiments, a reasonable amount of evidence accumulated to support the notion that TGase activity must be involved in the pathogenesis of neurodegenerative diseases characterized by the formation of large and extremely insoluble protein complexes [104, 105].

A significant elevation of TGase activity was measured in postmortem AD brain when compared to age-matched normal controls [44]. Elevated cross-linking activity was noted in the frontal cortex but not cerebellum, suggesting that TGase activation is associated with brain regions associated with AD pathology. Increased expression of TG2 mRNA was detected in AD cerebral cortex, which was accompanied by the appearance of the short transcript of TG2 that encodes for the nucleotide-uninhibited variant [31, 106]. It has been postulated that the short TG2 transcript variant may exhibit increased and dysregulated TGase activity in AD, since, unlike the full-length protein, it is unleashed from purine-nucleotide inhibition inside the cytosol. Nevertheless, the predicted higher cross-linking activity of the shorter transcript has not been demonstrated neither in vitro nor in living cells, and a structurally analogous shorter rat TG2 splice variant did not show higher catalytic rates [38].

Immunoreactivity for TG2 was also found elevated in the frontal cortex from AD cases by immunoblotting [44] and in the hippocampus by immunohistochemistry [40], suggesting that induction of TG2 expression might account for elevated cross-linking activity [21, 31, 44]. Elevated concentrations of TG2 protein were detected in cerebrospinal fluid (CSF) samples of AD patients [107]. TG1 protein expression was found to be increased in AD cerebella, though this part of the brain is spared form specific pathology in AD [21].

In vitro data provided support for the theory of TGase involvement in the protein aggregation by analyzing substrate properties of disease-related proteins. As TG2 can be externalized from cells [108], the possibility that TG2 may cross-link Aβ peptide and thus contribute to senile plaque formation was also investigated. Aβ 1–28 peptide was shown cross-linked by TG2 through Lys16 [109], whereas Aβ42 [110–114] and APP [115] were found to react through their Lys16, Lys28, and Gln15 residues [111, 115, 116].

Tau protein is also an excellent substrate for TG2 in vitro [31, 117–122] and possibly in vivo [123]. TG2 can cross-link tau to form filamentous polymers [112, 117]. The presence of transglutaminase-produced glutamyl-lysine cross-links was also shown by immunohistochemistry in situ, and it was suggested that detergent-insoluble tau might be directly cross-linked to itself through residues located to its microtubule-binding repeat motifs [120].

2. Huntington's and Other Polyglutamine Expansion Diseases

An expansion of CAG/CTG trinucleotide repeats coding for polyglutamine tracts is the underlying cause of nine neurodegenerative diseases, including five rare spinocerebellar ataxias, spino-bulbar muscular atrophy, dentatorubral-pallidoluysian atrophy (DRPLA), and the most common type, Huntington's disease (HD) (see [124] for references). All these diseases were thought to be inherited in an autosomal dominant fashion before the actual mode of non-Mendelian inheritance was elucidated. In each of these diseases, CAG/CTG triplet expansions occur in different genes that are specific to this disease. Even if the defected gene product is different, these diseases share common features in clinical appearance, namely the progressive impairment of motor coordination and/or motor neuron loss and can associate with variable degree of cognitive decline.

HD is a progressive, inherited, and fatal neurodegenerative disorder that leads to motor and psychic symptoms in the affected individuals. The neuronal degeneration in HD affects first the basal ganglia and at later stages of the disease the cerebral cortex as well. The gene that causes HD encodes for a protein of unclear function called huntingtin. The meiotic replication of the CAG-repeat tract of this gene often results in an expansion of the length of the CAG repeat tracts and thus to a longer polyglutamine sequence in the translated protein. In the nonaffected population, the polyglutamine sequence ranges from 6 to 36 repeats, whereas subjects with more than 39 glutamines in a row will develop the disease.

Obviously the polyglutamine tract above a certain length results in a gain of pathological function and leads to progressive neuronal degeneration in course of several decades [125].

One common neuropathological hallmark of all polyglutamine expansion diseases, including HD, is the appearance of insoluble intraneuronal protein inclusions that contain the disease-specific polyglutamine protein, often in the nucleus and in ubiquitylated form [126]. Despite the knowledge of the genetic cause of polyglutamine expansion diseases, the mechanisms by which the affected proteins produce neurodegeneration remain elusive.

The first connection between HD and TGase was noted when it was demonstrated that TGase activity is increased in HD lymphocytes [127]. TG2 expression and TGase activity was found to be elevated in neurons in affected areas of HD brains [42, 128]. TGase activity was also found upregulated in brains of the R6/2 huntingtin knock-in mice in the nuclear fraction of cells, where inclusions are preferentially formed [129].

It was first hypothesized by Green [130] that the gain of function might be accomplished by unusual TGase substrate properties of proteins with polyglutamine tract expansions. It was shown that TGase substrate properties of "naked" polyglutamine tracts improve more than linearly as the number of glutamines increases above 20 [131]. However, other studies found no significant improvement in substrate properties if the length was increased over ten glutamines [132, 133]. Though TG2 catalyzes the formation of intramolecular cross-links between proteins containing polyglutamine stretches, this modification actually results in the formation of high molecular weight, soluble oligomers, and the cross-linking was actually shown to decrease the propensity for the oligomers to form insoluble aggregates [134].

Max Perutz speculated that there is no need to implicate protein cross-linking into the genesis of these inclusions, as polyglutamine tracts spontaneously associate with one another forming a "polar zipper" held together by hydrogen bonds between the β-sheets [135]. This interaction might make the pathological proteins as insoluble as cellulose, or silk is insolubilized by hydrogen bonding of parallel β-sheets [136]. In addition, polyglutamine–huntingtin self-aggregation may be driven by the relative hydrophobicity of the polyglutamine stretch, regardless of the conformation of this sequence [137]. It was shown in a Tet-conditional transgenic expression model that the deposition of inclusion-forming proteinaceous material can be reverted if the continuous supply of aberrant protein is stopped [138]. Inclusions associated with HD were shown to form independently of TG2 in SH-SY5Y cells transfected with mutant huntingtin [139]. Chun et al. proposed that might rather be the huntingtin-associated proteins that react with TG2 to stabilize the primary aggregates that might be held together by secondary forces.

3. *TG2 Knockout and Inhibition in Mouse Models of HD*

Genetic deletion of TG2 in two mouse models of HD resulted in increased aggregate formation within the basal ganglia and in the neocortex [140, 141] and prolonged survival. This indicates that TG2 reduces the formation of (protective) aggregates and enhances the neurotoxic properties of mutant huntingtin.

Administration of TGase-inhibitor cystamine decreased the number of protein inclusions in transfected cells expressing the mutant DRPLA protein [142]. In another series of experiments, the intraperitoneal administration of TGase inhibitor cystamine has been shown to reduce pathological extrapyramidal symptoms and extend survival without altering the microscopic appearance of intraneuronal inclusions [129, 143].

Whether TGase-mediated cross-linking is involved in the series of pathobiochemical events finally causing the death of neurons or TGs are involved in some sort of salvage program of protein sequestration or contribute to the apoptotic process is still unclear (for review see [144, 145]).

4. TG2 and α-Synuclein in Parkinson's Disease (PD) and Lewy Body Dementias

PD is the second most common neurodegenerative condition after AD and affects more than 1% of the elderly population. Akin to AD, the majority of the PD cases is also sporadic and may be explained by the aging-dependent destruction of substantia nigra dopamine-producing cells that pursue highly oxidative metabolic processes during the synthesis of dopamine neurotransmitter [146]. Considering that no species in biology evolved to acquire properties that would explicitly favor the survival of its individuals after their reproductive period, the destruction of substantia nigra neurons can be regarded as a natural process of neuronal wear off. By the time clinically significant parkinsonian symptoms (rigidity, tremor, bradikynesia) appear, 90% of substantia nigra neurons are lost.

Neuropathology of PD involves the degeneration of dopaminergic neurons within the pars compacta of the substantia nigra, which is accompanied by the formation of cytoplasmic protein aggregates known as Lewy bodies. As with aggregates in AD and polyglutamine diseases, neither the mechanism of Lewy body formation nor their role in the PD disease process has been elucidated. Lewy bodies were initially thought to hallmark the degenerating dopaminergic neurons in PD; however, they are abundant in cortical neurons and basal ganglia in distinct neurodegenerative dementias (such as AD and Lewy body dementia) and are often encountered as accidental findings in neurons of elderly people without any clinical record of parkinsonism or dementia [147].

TG2 and its mRNA were found to be induced in PD substantia nigra [106, 148]. TGase activity has been implicated in the formation of Lewy bodies.

The nonamyloid component of senile plaques, α-synuclein, is the key component of Lewy bodies. α-Synuclein is a small protein of 140 amino acids and consists of an *N*-terminal lipid-binding amphipathic α-helix, a *C*-terminal acidic tail, and amyloid-forming repetitive sequences in between.

α-Synuclein is natively unfolded in solution, but may acquire α-helical conformation in the presence of phospholipid vesicles [149] or β-pleated sheet conformation in aggregates [150]. Owing to its natively unfolded conformation, the protein readily aggregates with itself and coaggregates with other proteins, a property that may be necessary for Lewy body formation and its cytopathic effects.

α-Synuclein is also a substrate for TG2 in vitro [148, 151–156], in cultured cells [152], and in human brain [148, 154].

The formation of intracytoplasmic α-synuclein aggregates was shown to depend on the cross-linking activity of TG2 in COS-7 cells [152] TG2 was shown to colocalize and coimmunopercipitate with α-synuclein in cotransfection experiments. In the substantia nigra, α-synuclein and TG2 were found to associate in the cytosol of PD neurons, but not in controls, though TG2 was not found to colocalize with Lewy bodies by immunohistochemistry [148]. Nevertheless, a recent study detected TG2 in the penumbra of Lewy bodies by mass spectrometry [154]. Jensen et al. reported that Gln80 and Lys79 in the highly aggregation-prone mid portion of α-synuclein is modified by TG2 in vitro [151]. A more detailed mass spectrometric analysis identified three glutamines and four lysine residues in α-synuclein cross-linked by TG2 in vitro [155]. In our hands, three glutamines and eight lysines were utilized for cross-linking by soluble TG1, TG2, TG3, and TG5 in solution; however, only one single cross-link was formed when α-synuclein was reacted with membrane-anchored TG2 or TG1. It was shown by thioflavine-dye fluorescence monitoring that promiscuous cross-linking of α-synuclein created dominantly oligomers and blocked amyloid formation, whereas the membrane-directed reaction led to the formation of intramolecularly cross-linked α-synuclein, which accelerated amyloid growth [154].

C. TGASE-MEDIATED CROSS-LINKING OF BRAIN PROTEINS IN NEURODEGENERATION IN VIVO

1. Detection of Cross-Links by Antibodies

Compelling evidence has been accumulated from the literature to verify that some TGs are expressed in the nervous system both in health and in neurodegenerative diseases, and that key disease-related proteins are substrates of TG2 under in vitro conditions. Nevertheless, all these indirect evidences are far from being convincing without a demonstration of activity and disease-modifying roles of TGs in disease models or ex vivo samples.

In vivo cross-linking and investigation of ex vivo cross-linked material has though rarely been reported. This is largely due to the restricted availability of patient samples, interpretation problems of animal disease models, and to limited sensitivity of analysis techniques.

The development of methods to detect isopeptide cross-links in situ has provided support for the validation of in vitro findings and speculations about their pathological relevance. Raising antibodies against the cross-link isodipeptides was the first attempt to verify the presence of previously in vitro demonstrated cross-linking in tissues.

Polyclonal antibodies developed against the glutamyl–lysine isodipeptide were first utilized by Aeschlimann et al. to demonstrate the presence of isopeptide bonds in osteonectin [157]. Later, monoclonal antibodies targeted against the cross-link were commercialized. Because of the similarity of the recognized epitope to abundant and chemically homologous structures, such as ε-acetyl-lysine of histones, specificity has always been of concern with the antibody-based detection methods. A strong background signal in immunoblotting applications may compromise the conclusions drawn solely from the labeling of proteins with these antibodies. Immunostaining with these antibodies was used to demonstrate the presence of GGEL cross-links in HD inclusions [158, 159], Lewy bodies [148, 152], and NFT [119, 160, 161]. Some studies used preadsorption control experiments with the isodipeptide to demonstrate the specificity of the anticross-link antibodies in their particular experimental systems. Nevertheless, the size and the chemical structure of the γ-glutamyl-ε-lysine antigen implicates that the specific affinity of the antibody, i.e., the bonding forces that link the antigen binding sites to the (methylene) groups of the compound that are not part of lysine, ε-acetyl-lysine, glutamine, or glutamic acid, cannot be expected to be higher than a few kJ/mol. This low binding power inevitably means that either the abundant natural amino acids or the acetyl-lysines will necessarily compete for antibody binding. If we assume a 10^5-fold preference for glutamyl-lysine bonds over lysine and glutamine residues (which is close to the physicochemical limits) and that a minimum 100-fold signal-to-noise ratio is needed for immunocytochemical detection, the threshold frequency of cross-links must be in the range of 0.5–5 cross-links in 100 proteins of average size and amino acid composition. An immunoblot study by Johnson and LeShoure [162] found that the commercial anticross-link monoclonal antibodies applied in the abovementioned studies were unable to detect the cross-links when tau protein was cross-linked. The lack of recognition of the cross-linked tau in this experiment may reflect a lower sensitivity of the antibody when it is used for immunoblotting applications, as compared to immunostaining of tissues, cultured cells, or immunoprecip-tiation of cross-linked proteins [152, 163] This noted difference in antibody affinity depending on whether native, unfolded, or variably denatured (mixed) proteins are used suggests that the epitope recognized by the monoclonal immunoglobulins may be larger than just the isopeptide bond and adjacent methylene chains and that it involves some of the functional groups attached to the α-carbon atoms. It is likely that the unrestricted rotation of the cross-link is required for attaining the proper epitope conformation, as the binding avidity of the antibody to soluble cross-linked oligopeptides and free isodipeptide was found to be satisfactory for quantitative ELISA applications and was able to detect one cross-link in ten atamol amino acids ($\sim 10^8$ Lys+Glx/cross-link) from digests of biological specimens [164, 165].

2. *Free Glutamyl-Lysine Isopeptide in Neurodegenerative Conditions*

Cross-linked proteins are broken down to amino acids by lysosomal enzymes in the same or in the phagocyting cells. The hydrolysis of glutamyl-lysine isopeptide bonds is thought to proceed in the extraneural tissues by γ-glutamyl-transpeptidase in the liver and γ-glutamyl cyclotransferase in the kidneys and in the liver [166, 167]. Thus, cross-linked proteins broken down in cells of the central nervous system result in the release of free glutamyl-lysine isodipeptide to the interstitial fluid, from where it diffuses to the CSF. From the CSF, the isopeptide may be resorbed to the circulation by generic peptide transport mechanisms of the arachnoid granulations.

We found that the concentration of glutamyl-lysine isodipeptide is significantly elevated in the CSF of patients with clinical diagnoses of both Alzheimer and vascular-type dementias; however, these values did not differ significantly between the two diagnostic groups [168]. Free CSF cross-link-levels were also found elevated in CSF of PD-associated (Lewy body-related) dementia cases [169].

The glutamyl-lysine isodipeptide was shown to accumulate in lumbar CSF of HD patients [170, 171].

3. *Cross-Links in Brain Tissue*

Total homogenized brain tissue contains 8–734 cross-links/10^9 amino acids depending on age and brain region. The mean frequency of cross-links was found to be 6–13-fold higher in the non-AD controls of a mean age of 75 years than in the control group with 34 years of mean age. Brain regions typically affected by specific neurodegenerative pathologies (such as the hippocampus by NFT and the basal ganglia by Lewy bodies in Lewy body dementia or diffuse Lewy body disease) revealed elevations by 2- to 4-fold in AD and Lewy body dementia patients as compared to the age-matched controls, both when an ELISA method with 81D4 cross-link antibody or stable isotope-labeling and mass spectrometric analysis was applied to analyze the isodipeptide contents from exhaustively digested brain homogenates [154, 164, 165].

4. Cross-Linked Proteins in Neurodegenerative Inclusions

Mass spectrometry was used to identify Alz-50 as a phospho-tau antigen in NFT in 1990 [172] Since then, large advances in proteome analysis were achieved at the turn of the millennium due to the development and availability of mass spectrometric instruments and availability of proteome-wide bioinformatics resources for analysis of mass data. Nevertheless, the proteomic analysis of cross-linked proteins is still a poorly developed field of methodology, where only positive findings have predictive value. Especially with complex protein samples from cells or tissues, the overlap between the proteins and peptides identified from the same complex protein samples by using different instruments and sample pretreatment methods is usually modest. Some proteins appear ubiquitously in mass chromatograms in samples, whereas other protein components produce less ionized fragments and remain suppressed and undetected, even if abundantly present in the source material.

A proteomic analysis of laser-microdissected NFT revealed 72 proteins [173] (Figure 2), 18 of which were identified from us among 66 proteins

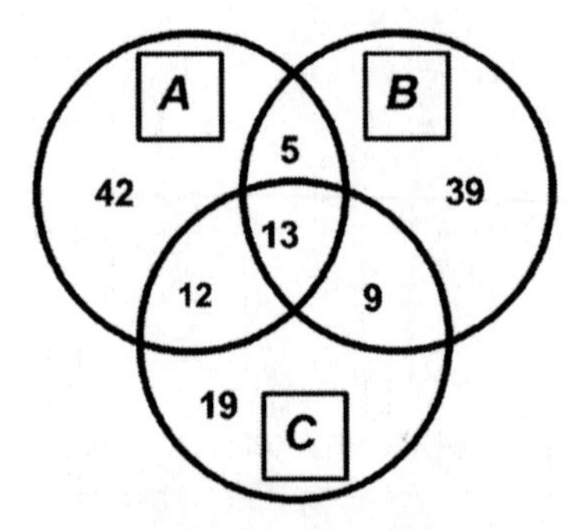

FIGURE 2. Venn diagram comparing the number of proteins identified by proteomics studies carried out on neurofibrillary tangles (A and B [173]) and on Lewy bodies [154]. The graph reflects the representativity of mass identification methods in the first decade of the millennium.

from chaotrop-insoluble affinity-isolated NFT proteins and 25 of which were also detected (among 53 proteins) in detergent and chaotrop-insoluble Lewy body proteins [154]. Indeed, most of these proteins are abundant cytoskeletal or regulatory proteins probably unrelated to specific aggregation-related pathologies and neither of the proteins related to specific pathogenetic pathways were found in these structures.

Detergent- and chaotrop-insolubility of such a complex mixture of cytosolic proteins may result from (chemical) cross-linking of these proteins (or fragments thereof) into large meshwork. Alternatively, it is sufficient to cross-link a few of the components, and the not cross-linked ones simply stick into the mesh. Given that the latter is more likely to happen, the majority of these identifications may just represent a random sampling of cytoplasmic proteins.

Chaotrop- and detergent-insoluble material from brain tissue contains overwhelmingly basement-membrane and extracellular matrix-derived proteins that are covalently cross-linked by lysine oxidation or (iso)desmosine cross-link formation. In order to isolate intracellular aggregates, a component of these must be chosen and insoluble particles be selected by flow sorting or affinity capture techniques. Ubiquitin affinity isolation of the insoluble material did not yield microscopically detectable amounts of insoluble particles from brains of patients who died of neurodegeneration-unrelated causes at a younger age, though in older patients, the appearance of such fraction was detectable, just like in AD patients of similar ages. Ubiquitin-immunoreactive particles were found to contain more than 50% of total brain cross-links, which might just as well had been considerably higher owing to loss of cross-linked material during purification [164]. Remarkably, similar results were obtained from chaotrop-insoluble material from Lewy body dementia cases that was affinity-isolated with antibodies to α-synuclein [154].

Analysis of isopeptide cross-linked peptides is a methodologically tedious task, even with one known protein that has been surely cross-linked in vitro. Shotgun sequencing of complex protein mixtures of unknown composition still offers little chance to find isopeptide cross-linked peptide pairs. In a set of analysis, we could identify cross-linked peptide pairs from tryptic fragments of NFT and Lewy body proteins by affinity-enrichment of cross-linked peptides using anti-isodipeptide antibodies. Peptides bound to the monoclonal antibodies were sequenced in front of the background of the antibody-derived tryptic peptides in order to reduce peak-suppression effects from more complex mixtures. Clearly, these identification methods are far from being representative or exclusive, partially due to the unclear

TABLE 1

Ex Vivo Identified TGase Cross-linked Proteins and Residue Positions in AD and Lewy Body Dementia

Alzheimer's Disease [164]	Lewy Body Dementia [154]
Reactive glutamine residues	
HSPB1	Gln190
HSPB1	Gln31
α-synuclein	Gln99
Reactive lysine residues	
Ubiquitin	Lys29
Ubiquitin	Lys48
Parkin[a]	Lys48
α-Synuclein	Lys58

Identifications were done from chaotrope-insoluble ubiquitin or α-synuclein immunoreactive proteins by trypsine digestion and affinity isolation to 81D4 monoclonal antibody to glutamyl-lysine isodipeptide cross-link followed by mass sequencing.

[a] Identified sequence is common in all Parkin isoforms.

specificity of the antibody. Another concern is imposed by the practical constraints of MS/MS analysis: the utilization of tryptic peptides and a mass window, where too short and too long sequences are missed and many other methodological factors that make some peptides more detected than others.

While bearing in mind all these limitations, the identification of cross-linked proteins from neurodegenerative inclusions identified five cross-links to date, four of which were found both in AD and Lewy body dementia [154, 164], Table 1. α-Synuclein was found in inclusions from AD brains, just as ubiquitin is present in all known neurodegenerative inclusions [164, 173]. Again, α-synuclein is present in extracellular amyloid plaques, where it is called nonamyloid component of plaques.

The cross-links identified from ex vivo AD and Lewy body inclusions were found between small heat shock protein HSPB1 and α-synuclein as glutamine and ubiquitin and parkin as lysine donor proteins. Small heat shock proteins are ubiquitous molecular chaperones that attach to cytoplasmic proteins exposing hydrophobic regions. Some of them is also involved in carrying misfolded proteins to sites of refolding or degradation (see [174] for references). Ubiquitin is a small and evolutionarily very much conserved protein that acts as a molecular label when attached to ε-amino groups of lysine moieties in various proteins. Attachment of a stretch of multiple

FIGURE 3. Pipe-cleaner model for the structure of neurodegnenerative inclusions. Pipe-cleaners have a steel wire core and a fuzzy cotton coat. Neurodegenerative inclusions consist of a core of amyloid fibrils surrounded by a fuzzy halo of glycyl-lysine isopeptide cross-linked polyubiquitine chains linked together via HSPB1 through TGase-produced glutamyl-lysine isopeptide bonds.

ubiquitin moieties to a protein through the formation of an isopeptide bond between the free *C*-terminal glycine of the polyubiquitin chain to a lysine amino group generally means that the protein is designated for transport to and breakdown through the proteasomes. Polyubiquitin chains are linked together by Gly76 to Lys48/Lys29/Lys11 or Lys6 isopeptide cross-links, though the exact meanings of differentially cross-linked polyubiquitin ribbons is not clear.

From the generic association of these proteins with protein misfolding and clearance mechanisms, it was proposed that (Figure 3) cross-linking of these proteins might be tertiary to protein misfolding, recruitment of small heat shock protein chaperones, and designation of the misfolded proteins for proteasomal breakdown by the attachment of polyubiquitin chains. In inclusions from AD 79% [164] and in Lewy bodies 71% of the total tissue glutamyl-lysine content was confined to HSPB1-ubiquitin cross-links (Nemes et al., unpublished data). This indicates that it is unlikely that the methodological bias of detection would miss that the main site of cross-linking would be between tau protein residues in AD and α-synuclein in Lewy bodies.

In order to reconcile mainstream hypotheses on the formation of protein inclusions from soluble proteins by misfolding and amyloid-type assembly into protofibrils and fibrils with the finding that obviously inclusions consist of many more generic proteins, some of which get cross-linked by TGs, it was proposed that eventually cross-linked HSPB1-ubiquitin is on the periphery of the inclusions, whereas misfolded proteins nucleate these structures. In the core of these lesions, misfolded proteins may reside in homophylic interactions stabilized by amyloid-type hydrogen bonding between their β-pleated sheet repeat motifs. This theory was underpinned by showing that Lewy bodies consist of two different compartments: (1) a protease-digestible less compacted outer zone ("halo"), which is rich in TGase cross-links and (2) an amyloid core, which apparently consists of pure α-synuclein. Low frequency of intrasynuclein cross-linking was detected in the core amyloid, and it was proposed that it promotes, or even might trigger, amyloid assembly, an effect contrasted to the marked anti-aggregation effects of other intra- and intersynuclein crossbridges, respectively.

To date, no such analyses have been published on inclusions from HD or other polyglutamine expansion diseases.

D. CROSS-LINKING AND CLEARANCE OF AGGREGATES

Abundant ubiquitination of neurodegenerative inclusions marks that waste proteins were marked for proteasomal breakdown as well as a failure of this clearance pathway to eliminate these structures.

Cross-linking of inclusion-associated proteins was suggested to contribute to disease by blocking proteasomal breakdown pathways either by thwarting the carriage of misfolded proteins to the proteasomes or by jamming feeding machineries into the proteasomal multiproteinase cavity. Some studies indicate that a failure of proteasomal breakdown mechanisms may be an early event in the formation of Lewy bodies [175, 176].

Nevertheless, the ubiquitin proteasome pathway is not the only means for neurons to degrade noxious protein waste. Macroautophagy is an alternative clearance mechanism to proteasomes. During the process, large membrane-leaflets pinch off from the endoplasmic reticulum and surround parts of the cytoplasm so that the autophaged compartment is isolated from the cytosol by a continuous membrane. Then this autophagic vesicle fuses with lysosomes to form an autophagolysosome, and lysosomal enzymes break down its contents.

There is evidence that intracytoplasmic aggregates elicit their macroautophagy in neurodegenerative diseases (see [177] for review). Ubiquitylation of proteins also is a signal for macroautophagy [178]. In TG2 knockout mice, the number of autophagic vacuoles is increased, while the maturation of those into autophagolysosomes is blocked [179], a finding in agreement with the observation that the number of intranuclear inclusions is elevated in TG2 knockout and R6/2 huntingtin knock-in mice [141]. Despite more intranuclear inclusions, these mice showed less brain damage and slower progression, indicating that the wrapping up of protein aggregates into vesicles shields from their cytotoxicity.

ACKNOWLEDGMENT

I am indebted to Mate A. Demeny and Laszlo Fesus for careful reading of the manuscript and helpful comments. This work was supported by the Hungarian Scientific Research Fund (OTKA TS 044798) and EU grants MRTN-CT-2006-035624 and LSHB-CT-2007-037730.

REFERENCES

1. Folk, J. E. and Chung, S. I. (1973) Molecular and catalytic properties of transglutaminases, *Adv. Enzymol. Relat. Areas Mol. Biol. 38*, 109–191.
2. Folk, J. E. and Finlayson, J. S. (1977) The epsilon-(gamma-glutamyl)lysine crosslink and the catalytic role of transglutaminases, *Adv. Protein Chem. 31*, 1–133.
3. Nemes, Z., Marekov, L. N., Fesus, L., and Steinert, P. M. (1999) A novel function for transglutaminase 1: attachment of long-chain omega-hydroxyceramides to involucrin by ester bond formation, *Proc. Natl. Acad. Sci. U.S.A. 96*, 8402–8407.
4. Boros, S., Ahrman, E., Wunderink, L., Kamps, B., de Jong, W. W., Boelens, W. C., and Emanuelsson, C. S. (2006) Site-specific transamidation and deamidation of the small heat-shock protein Hsp20 by tissue transglutaminase, *Proteins 62*, 1044–1052.
5. Stamnaes, J., Fleckenstein, B., and Sollid, L. M. (2008) The propensity for deamidation and transamidation of peptides by transglutaminase 2 is dependent on substrate affinity and reaction conditions, *Biochim. Biophys. Acta 1784*, 1804–1811.
6. Folk, J. E. (1983) Mechanism and basis for specificity of transglutaminase-catalyzed epsilon-(gamma-glutamyl) lysine bond formation, *Adv. Enzymol. Relat. Areas Mol. Biol. 54*, 1–56.
7. Folk, J. E. and Chung, S. I. (1985) Transglutaminases, *Methods Enzymol. 113*, 358–375.
8. Del Duca, S. and Serafini-Fracassini, D. (2005) Transglutaminases of higher, lower plants and fungi, *Prog. Exp. Tumor Res. 38*, 223–247.

9. Grootjans, J. J., Groenen, P. J., and de Jong, W. W. (1995) Substrate requirements for transglutaminases. Influence of the amino acid residue preceding the amine donor lysine in a native protein, *J. Biol. Chem. 270*, 22855–22858.
10. Mariniello, L. and Porta, R. (2005) Transglutaminases as biotechnological tools, *Prog. Exp. Tumor Res. 38*, 174–191.
11. Csosz, E., Bagossi, P., Nagy, Z., Dosztanyi, Z., Simon, I., and Fesus, L. (2008) Substrate preference of transglutaminase 2 revealed by logistic regression analysis and intrinsic disorder examination, *J. Mol. Biol. 383*, 390–402.
12. Groenen, P. J., Grootjans, J. J., Lubsen, N. H., Bloemendal, H., and de Jong, W. W. (1994) Lys-17 is the amine-donor substrate site for transglutaminase in beta A3-crystallin, *J. Biol. Chem. 269*, 831–833.
13. Nemes, Z., Madi, A., Marekov, L. N., Piacentini, M., Steinert, P. M., and Fesus, L. (2001) Analysis of protein transglutamylation in apoptosis, *Methods Cell Biol. 66*, 111–133.
14. Nemes, Z., Petrovski, G., and Fesus, L. (2005) Tools for the detection and quantitation of protein transglutamination, *Anal. Biochem. 342*, 1–10.
15. Piacentini, M., Rodolfo, C., Farrace, M. G., and Autuori, F. (2000) "Tissue" transglutaminase in animal development, *Int. J. Dev. Biol. 44*, 655–662.
16. Beninati, S. and Piacentini, M. (2004) The transglutaminase family: an overview: minireview article, *Amino Acids 26*, 367–372.
17. Grenard, P., Bates, M. K., and Aeschlimann, D. (2001) Evolution of transglutaminase genes: identification of a transglutaminase gene cluster on human chromosome 15q15. Structure of the gene encoding transglutaminase X and a novel gene family member, transglutaminase Z, *J. Biol. Chem. 276*, 33066–33078.
18. Schuppan, D., Junker, Y., and Barisani, D. (2009) Celiac disease: from pathogenesis to novel therapies, *Gastroenterology 137*, 1912–1933.
19. Candi, E., Paradisi, A., Terrinoni, A., Pietroni, V., Oddi, S., Cadot, B., Jogini, V., Meiyappan, M., Clardy, J., Finazzi-Agro, A., and Melino, G. (2004) Transglutaminase 5 is regulated by guanine-adenine nucleotides, *Biochem. J. 381*, 313–319.
20. Hadjivassiliou, M., Aeschlimann, P., Strigun, A., Sanders, D. S., Woodroofe, N., and Aeschlimann, D. (2008) Autoantibodies in gluten ataxia recognize a novel neuronal transglutaminase, *Ann. Neurol. 64*, 332–343.
21. Kim, S. Y., Grant, P., Lee, J. H., Pant, H. C., and Steinert, P. M. (1999) Differential expression of multiple transglutaminases in human brain. Increased expression and cross-linking by transglutaminases 1 and 2 in Alzheimer's disease, *J. Biol. Chem. 274*, 30715–30721.
22. Iismaa, S. E., Mearns, B. M., Lorand, L., and Graham, R. M. (2009) Transglutaminases and disease: lessons from genetically engineered mouse models and inherited disorders, *Physiol. Rev. 89*, 991–1023.
23. Smethurst, P. A. and Griffin, M. (1996) Measurement of tissue transglutaminase activity in a permeabilized cell system: its regulation by Ca2+ and nucleotides, *Biochem. J. 313(Pt 3)*, 803–808.
24. Fraij, B. M. and Gonzales, R. A. (1997) Organization and structure of the human tissue transglutaminase gene, *Biochim. Biophys. Acta 1354*, 65–71.

25. Szegezdi, E., Szondy, Z., Nagy, L., Nemes, Z., Friis, R. R., Davies, P. J., and Fesus, L. (2000) Apoptosis-linked in vivo regulation of the tissue transglutaminase gene promoter, *Cell Death Differ. 7*, 1225–1233.
26. Liu, S., Cerione, R. A., and Clardy, J. (2002) Structural basis for the guanine nucleotide-binding activity of tissue transglutaminase and its regulation of transamidation activity, *Proc. Natl. Acad. Sci. U.S.A. 99*, 2743–2747.
27. Pinkas, D. M., Strop, P., Brunger, A. T., and Khosla, C. (2007) Transglutaminase 2 undergoes a large conformational change upon activation, *PLoS Biol. 5*, e327.
28. Weiss, M. S., Metzner, H. J., and Hilgenfeld, R. (1998) Two non-proline cis peptide bonds may be important for factor XIII function, *FEBS Lett. 423*, 291–296.
29. Nemes, Z., Petrovski, G., Csosz, E., and Fesus, L. (2005) Structure-function relationships of transglutaminases–a contemporary view, *Prog. Exp. Tumor Res. 38*, 19–36.
30. Iismaa, S. E., Chung, L., Wu, M. J., Teller, D. C., Yee, V. C., and Graham, R. M. (1997) The core domain of the tissue transglutaminase Gh hydrolyzes GTP and ATP, *Biochemistry 36*, 11655–11664.
31. Citron, B. A., SantaCruz, K. S., Davies, P. J., and Festoff, B. W. (2001) Intron-exon swapping of transglutaminase mRNA and neuronal Tau aggregation in Alzheimer's disease, *J. Biol. Chem. 276*, 3295–3301.
32. Lai, T. S., Liu, Y., Li, W., and Greenberg, C. S. (2007) Identification of two GTP-independent alternatively spliced forms of tissue transglutaminase in human leukocytes, vascular smooth muscle, and endothelial cells, *Faseb J. 21*, 4131–4143.
33. Candi, E., Melino, G., Lahm, A., Ceci, R., Rossi, A., Kim, I. G., Ciani, B., and Steinert, P. M. (1998) Transglutaminase 1 mutations in lamellar ichthyosis. Loss of activity due to failure of activation by proteolytic processing, *J. Biol. Chem. 273*, 13693–13702.
34. Pietroni, V., Di Giorgi, S., Paradisi, A., Ahvazi, B., Candi, E., and Melino, G. (2008) Inactive and highly active, proteolytically processed transglutaminase-5 in epithelial cells, *J. Invest. Dermatol. 128*, 2760–2766.
35. Ambron, R. T. and Kremzner, L. T. (1982) Post-translational modification of neuronal proteins: evidence for transglutaminase activity in R2, the giant cholinergic neuron of Aplysia, *Proc. Natl. Acad. Sci. U.S.A. 79*, 3442–3446.
36. Pastuszko, A., Wilson, D. F., and Erecinska, M. (1986) A role for transglutaminase in neurotransmitter release by rat brain synaptosomes, *J. Neurochem. 46*, 499–508.
37. Perry, M. J., Mahoney, S. A., and Haynes, L. W. (1995) Transglutaminase C in cerebellar granule neurons: regulation and localization of substrate cross-linking, *Neuroscience 65*, 1063–1076.
38. Monsonego, A., Shani, Y., Friedmann, I., Paas, Y., Eizenberg, O., and Schwartz, M. (1997) Expression of GTP-dependent and GTP-independent tissue-type transglutaminase in cytokine-treated rat brain astrocytes, *J. Biol. Chem. 272*, 3724–3732.
39. Reichelt, K. L. and Poulsen, E. (1992) gamma-Glutamylaminotransferase and transglutaminase in subcellular fractions of rat cortex and in cultured astrocytes, *J. Neurochem. 59*, 500–504.

40. Appelt, D. M., Kopen, G. C., Boyne, L. J., and Balin, B. J. (1996) Localization of transglutaminase in hippocampal neurons: implications for Alzheimer's disease, *J. Histochem. Cytochem. 44*, 1421–1427.

41. Hand, D., Campoy, F. J., Clark, S., Fisher, A., and Haynes, L. W. (1993) Activity and distribution of tissue transglutaminase in association with nerve-muscle synapses, *J. Neurochem. 61*, 1064–1072.

42. Lesort, M., Chun, W., Johnson, G. V., and Ferrante, R. J. (1999) Tissue transglutaminase is increased in Huntington's disease brain, *J. Neurochem. 73*, 2018–2027.

43. Gilad, G. M. and Varon, L. E. (1985) Transglutaminase activity in rat brain: characterization, distribution, and changes with age, *J. Neurochem. 45*, 1522–1526.

44. Johnson, G. V., Cox, T. M., Lockhart, J. P., Zinnerman, M. D., Miller, M. L., and Powers, R. E. (1997) Transglutaminase activity is increased in Alzheimer's disease brain, *Brain Res. 751*, 323–329.

45. Ohashi, H., Itoh, Y., Birckbichler, P. J., and Takeuchi, Y. (1995) Purification and characterization of rat brain transglutaminase, *J. Biochem. (Tokyo) 118*, 1271–1278.

46. Facchiano, F., Benfenati, F., Valtorta, F., and Luini, A. (1993) Covalent modification of synapsin I by a tetanus toxin-activated transglutaminase, *J. Biol. Chem. 268*, 4588–4591.

47. Kim, H. C., Lewis, M. S., Gorman, J. J., Park, S. C., Girard, J. E., Folk, J. E., and Chung, S. I. (1990) Protransglutaminase E from guinea pig skin. Isolation and partial characterization, *J. Biol. Chem. 265*, 21971–21978.

48. Steinert, P. M., Chung, S. I., and Kim, S. Y. (1996) Inactive zymogen and highly active proteolytically processed membrane-bound forms of the transglutaminase 1 enzyme in human epidermal keratinocytes, *Biochem. Biophys. Res. Commun. 221*, 101–106.

49. Akiyama, H., Kondo, H., Ikeda, K., Arai, T., Kato, M., and McGleer, P. L. (1995) Immunohistochemical detection of coagulation factor XIIIa in postmortem human brain tissue, *Neurosci. Lett. 202*, 29–32.

50. Monsonego, A., Mizrahi, T., Eitan, S., Moalem, G., Bardos, H., Adany, R., and Schwartz, M. (1998) Factor XIIIa as a nerve-associated transglutaminase, *FASEB J. 12*, 1163–1171.

51. Maccioni, R. B. and Seeds, N. W. (1986) Transglutaminase and neuronal differentiation, *Mol. Cell Biochem. 69*, 161–168.

52. Perry, M. J. and Haynes, L. W. (1993) Localization and activity of transglutaminase, a retinoid-inducible protein, in developing rat spinal cord, *Int. J. Dev. Neurosci. 11*, 325–337.

53. Tucholski, J., Lesort, M., and Johnson, G. V. (2001) Tissue transglutaminase is essential for neurite outgrowth in human neuroblastoma SH-SY5Y cells, *Neuroscience 102*, 481–491.

54. Chowdhury, Z. A., Barsigian, C., Chalupowicz, G. D., Bach, T. L., Garcia-Manero, G., and Martinez, J. (1997) Colocalization of tissue transglutaminase and stress fibers in human vascular smooth muscle cells and human umbilical vein endothelial cells, *Exp. Cell Res. 231*, 38–49.

55. Maccioni, R. B. and Arechaga, J. (1986) Transglutaminase (TG) involvement in early embryogenesis, *Exp. Cell Res. 167*, 266–270.

56. Eitan, S. and Schwartz, M. (1993) A transglutaminase that converts interleukin-2 into a factor cytotoxic to oligodendrocytes, *Science 261*, 106–108.

57. Miller, C. C. and Anderton, B. H. (1986) Transglutaminase and the neuronal cytoskeleton in Alzheimer's disease, *J. Neurochem. 46*, 1912–1922.

58. Selkoe, D. J., Abraham, C., and Ihara, Y. (1982) Brain transglutaminase: in vitro crosslinking of human neurofilament proteins into insoluble polymers, *Proc. Natl. Acad. Sci. U.S.A. 79*, 6070–6074.

59. Nemes, Z., Jr., Adany, R., Balazs, M., Boross, P., and Fesus, L. (1997) Identification of cytoplasmic actin as an abundant glutaminyl substrate for tissue transglutaminase in HL-60 and U937 cells undergoing apoptosis, *J. Biol. Chem. 272*, 20577–20583.

60. Clement, S., Velasco, P. T., Murthy, S. N., Wilson, J. H., Lukas, T. J., Goldman, R. D., and Lorand, L. (1998) The intermediate filament protein, vimentin, in the lens is a target for cross-linking by transglutaminase, *J. Biol. Chem. 273*, 7604–7609.

61. Tucholski, J., Roth, K. A., and Johnson, G. V. (2006) Tissue transglutaminase overexpression in the brain potentiates calcium-induced hippocampal damage, *J. Neurochem. 97*, 582–594.

62. Zhang, J., Lesort, M., Guttmann, R. P., and Johnson, G. V. (1998) Modulation of the in situ activity of tissue transglutaminase by calcium and GTP, *J. Biol. Chem. 273*, 2288–2295.

63. Mahoney, S. A., Perry, M., Seddon, A., Bohlen, P., and Haynes, L. (1996) Transglutaminase forms midkine homodimers in cerebellar neurons and modulates the neurite-outgrowth response, *Biochem. Biophys. Res. Commun. 224*, 147–152.

64. Mahoney, S. A., Wilkinson, M., Smith, S., and Haynes, L. W. (2000) Stabilization of neurites in cerebellar granule cells by transglutaminase activity: identification of midkine and galectin-3 as substrates, *Neuroscience 101*, 141–155.

65. Kojima, S., Inui, T., Muramatsu, H., Suzuki, Y., Kadomatsu, K., Yoshizawa, M., Hirose, S., Kimura, T., Sakakibara, S., and Muramatsu, T. (1997) Dimerization of midkine by tissue transglutaminase and its functional implication, *J. Biol. Chem. 272*, 9410–9416.

66. Iwasaki, W., Nagata, K., Hatanaka, H., Inui, T., Kimura, T., Muramatsu, T., Yoshida, K., Tasumi, M., and Inagaki, F. (1997) Solution structure of midkine, a new heparin-binding growth factor, *Embo J. 16*, 6936–6946.

67. Friedrich, P., Fesus, L., Tarcsa, E., and Czeh, G. (1991) Protein cross-linking by transglutaminase induced in long-term potentiation in the Ca1 region of hippocampal slices, *Neuroscience 43*, 331–334.

68. Fesus, L., Davies, P. J., and Piacentini, M. (1991) Apoptosis: molecular mechanisms in programmed cell death, *Eur. J. Cell Biol. 56*, 170–177.

69. Hadjivassiliou, M., Maki, M., Sanders, D. S., Williamson, C. A., Grunewald, R. A., Woodroofe, N. M., and Korponay-Szabo, I. R. (2006) Autoantibody targeting of brain and intestinal transglutaminase in gluten ataxia, *Neurology 66*, 373–377.

70. Melino, G., Annicchiarico-Petruzzelli, M., Piredda, L., Candi, E., Gentile, V., Davies, P. J., and Piacentini, M. (1994) Tissue transglutaminase and apoptosis: sense and antisense transfection studies with human neuroblastoma cells, *Mol. Cell Biol. 14*, 6584–6596.

71. Tucholski, J. and Johnson, G. V. (2002) Tissue transglutaminase differentially modulates apoptosis in a stimuli-dependent manner, *J. Neurochem. 81*, 780–791.
72. Sarang, Z., Toth, B., Balajthy, Z., Koroskenyi, K., Garabuczi, E., Fesus, L., and Szondy, Z. (2009) Some lessons from the tissue transglutaminase knockout mouse, *Amino Acids 36*, 625–631.
73. Antonyak, M. A., Jansen, J. M., Miller, A. M., Ly, T. K., Endo, M., and Cerione, R. A. (2006) Two isoforms of tissue transglutaminase mediate opposing cellular fates, *Proc. Natl. Acad. Sci. U.S.A. 103*, 18609–18614.
74. Ruan, Q., Quintanilla, R. A., and Johnson, G. V. (2007) Type 2 transglutaminase differentially modulates striatal cell death in the presence of wild type or mutant huntingtin, *J. Neurochem. 102*, 25–36.
75. Melino, G. and Piacentini, M. (1998) 'Tissue' transglutaminase in cell death: a downstream or a multifunctional upstream effector? *FEBS Lett. 430*, 59–63.
76. Wyllie, A. H., Beattie, G. J., and Hargreaves, A. D. (1981) Chromatin changes in apoptosis, *Histochem. J. 13*, 681–692.
77. Young, C., Tenkova, T., Dikranian, K., and Olney, J. W. (2004) Excitotoxic versus apoptotic mechanisms of neuronal cell death in perinatal hypoxia/ischemia, *Curr. Mol. Med. 4*, 77–85.
78. Fesus, L., Madi, A., Balajthy, Z., Nemes, Z., and Szondy, Z. (1996) Transglutaminase induction by various cell death and apoptosis pathways, *Experientia 52*, 942–949.
79. Ando, M., Kunii, S., Tatematsu, T., and Nagata, Y. (1993) Rapid and transient alterations in transglutaminase activity in rat superior cervical ganglia following denervation or axotomy, *Neurosci. Res. 17*, 47–52.
80. Tetzlaff, W., Gilad, V. H., Leonard, C., Bisby, M. A., and Gilad, G. M. (1988) Retrograde changes in transglutaminase activity after peripheral nerve injuries, *Brain Res. 445*, 142–146.
81 Festoff, B. W., SantaCruz, K., Arnold, P. M., Sebastian, C. T., Davies, P. J., and Citron, B. A. (2002) Injury-induced "switch" from GTP-regulated to novel GTP-independent isoform of tissue transglutaminase in the rat spinal cord, *J. Neurochem. 81*, 708–718.
82. Tolentino, P. J., DeFord, S. M., Notterpek, L., Glenn, C. C., Pike, B. R., Wang, K. K., and Hayes, R. L. (2002) Up-regulation of tissue-type transglutaminase after traumatic brain injury, *J. Neurochem. 80*, 579–588.
83. Fujita, K., Ando, M., Yamauchi, M., Nagata, Y., and Honda, M. (1995) Alteration of transglutaminase activity in rat and human spinal cord after neuronal degeneration, *Neurochem. Res. 20*, 1195–1201.
84. Tolentino, P. J., Waghray, A., Wang, K. K., and Hayes, R. L. (2004) Increased expression of tissue-type transglutaminase following middle cerebral artery occlusion in rats, *J. Neurochem. 89*, 1301–1307.
85. Kim, S. Y. (2006) Transglutaminase 2 in inflammation, *Front. Biosci. 11*, 3026–3035.
86. Ientile, R., Caccamo, D., and Griffin, M. (2007) Tissue transglutaminase and the stress response, *Amino Acids 33*, 385–394.
87. Williamson, J., Goldman, J., and Marder, K. S. (2009) Genetic aspects of Alzheimer disease, *Neurologist 15*, 80–86.

88. Kraepelin, E. (1910) *Psychiatrie: Ein Lehrbuch für Studierende und Aertzte*, Johann Abrosius Barth Verlag, Lepzig.

89. Alzheimer, A. (1907) Über eine eigenartige Erkrankung der Hirnrinde, *Allgem. Zeitschr. Psychat. Psych. Gericht. Med. 64*, 146–158.

90. Alzheimer, A. (1911) Über eigenartige Krankheitsfaelle des spaeteren Alters, *Zeitschr. Gesamte Neurol. Psychiatr. 4*, 356–385.

91. Jonsson, L. and Wimo, A. (2009) The cost of dementia in Europe: a review of the evidence, and methodological considerations, *Pharmacoeconomics 27*, 391–403.

92. Mitchell, A. J. (2009) CSF phosphorylated tau in the diagnosis and prognosis of mild cognitive impairment and Alzheimer's disease: a meta-analysis of 51 studies, *J. Neurol. Neurosurg. Psychiatry 80*, 966–975.

93. McKhann, G., Drachman, D., Folstein, M., Katzman, R., Price, D., and Stadlan, E. M. (1984) Clinical diagnosis of Alzheimer's disease: report of the NINCDS-ADRDA Work Group under the auspices of Department of Health and Human Services Task Force on Alzheimer's Disease, *Neurology 34*, 939–944.

94. Mirra, S. S., Heyman, A., McKeel, D., Sumi, S. M., Crain, B. J., Brownlee, L. M., Vogel, F. S., Hughes, J. P., van Belle, G., and Berg, L. (1991) The Consortium to Establish a Registry for Alzheimer's Disease (CERAD). Part II. Standardization of the neuropathologic assessment of Alzheimer's disease, *Neurology 41*, 479–486.

95. Roman, G. C., et al. (1993) Vascular dementia: diagnostic criteria for research studies. Report of the NINDS-AIREN International Workshop, *Neurology 43*, 250–260.

96. McKeith, I. G. (2006) Consensus guidelines for the clinical and pathologic diagnosis of dementia with Lewy bodies (DLB): report of the Consortium on DLB International Workshop, *J. Alzheimers Dis. 9*, 417–423.

97. Jellinger, K. A. (2008) A critical reappraisal of current staging of Lewy-related pathology in human brain, *Acta Neuropathol. 116*, 1–16.

98. MRC-CFAS (2001) Pathological correlates of late-onset dementia in a multicentre, community-based population in England and Wales. Neuropathology Group of the Medical Research Council Cognitive Function and Ageing Study (MRC CFAS), *Lancet 357*, 169–175.

99. Benecke, R. (2003) Diffuse Lewy body disease – a clinical syndrome or a disease entity? *J. Neurol. 250(Suppl 1)*, I39–I42.

100. Trojanowski, J. Q., Goedert, M., Iwatsubo, T., and Lee, V. M. (1998) Fatal attractions: abnormal protein aggregation and neuron death in Parkinson's disease and Lewy body dementia, *Cell Death Differ. 5*, 832–837.

101. Rocchi, A., Orsucci, D., Tognoni, G., Ceravolo, R., and Siciliano, G. (2009) The role of vascular factors in late-onset sporadic Alzheimer's disease. Genetic and molecular aspects, *Curr. Alzheimer Res. 6*, 224–237.

102. Kril, J.J. and Halliday, G.M. (2001) Alzheimer's disease: its diagnosis and pathogenesis, *Int Rev Neurobiol 48*, 167–217.

103. Friedhoff, P., von Bergen, M., Mandelkow, E. M., and Mandelkow, E. (2000) Structure of tau protein and assembly into paired helical filaments, *Biochim. Biophys. Acta 1502*, 122–132.

104. Muma, N. A. (2007) Transglutaminase is linked to neurodegenerative diseases, *J. Neuropathol. Exp. Neurol. 66*, 258–263.

105. Ruan, Q. and Johnson, G. V. (2007) Transglutaminase 2 in neurodegenerative disorders, *Front. Biosci. 12*, 891–904.

106. Citron, B. A., Suo, Z., SantaCruz, K., Davies, P. J., Qin, F., and Festoff, B. W. (2002) Protein crosslinking, tissue transglutaminase, alternative splicing and neurodegeneration, *Neurochem. Int. 40*, 69–78.

107. Bonelli, R. M., Aschoff, A., Niederwieser, G., Heuberger, C., and Jirikowski, G. (2002) Cerebrospinal fluid tissue transglutaminase as a biochemical marker for Alzheimer's disease, *Neurobiol. Dis. 11*, 106–110.

108. Collighan, R. J. and Griffin, M. (2009) Transglutaminase 2 cross-linking of matrix proteins: biological significance and medical applications, *Amino Acids 36*, 659–670.

109. Ikura, K., Takahata, K., and Sasaki, R. (1993) Cross-linking of a synthetic partial-length (1–28) peptide of the Alzheimer beta/A4 amyloid protein by transglutaminase, *FEBS Lett. 326*, 109–111.

110. Boros, S., Kamps, B., Wunderink, L., de Bruijn, W., de Jong, W. W., and Boelens, W. C. (2004) Transglutaminase catalyzes differential crosslinking of small heat shock proteins and amyloid-beta, *FEBS Lett. 576*, 57–62.

111. Dudek, S. M. and Johnson, G. V. (1994) Transglutaminase facilitates the formation of polymers of the beta-amyloid peptide, *Brain Res. 651*, 129–133.

112. Dudek, S. M. and Johnson, G. V. (1993) Transglutaminase catalyzes the formation of sodium dodecyl sulfate-insoluble, Alz-50-reactive polymers of tau, *J. Neurochem. 61*, 1159–1162.

113. Hartley, D. M., Zhao, C., Speier, A. C., Woodard, G. A., Li, S., Li, Z., and Walz, T. (2008) Transglutaminase induces protofibril-like amyloid beta-protein assemblies that are protease-resistant and inhibit long-term potentiation, *J. Biol. Chem. 283*, 16790–16800.

114. Zhang, W., Johnson, B. R., Suri, D. E., Martinez, J., and Bjornsson, T. D. (1998) Immunohistochemical demonstration of tissue transglutaminase in amyloid plaques, *Acta Neuropathol. 96*, 395–400.

115. Ho, G. J., Gregory, E. J., Smirnova, I. V., Zoubine, M. N., and Festoff, B. W. (1994) Cross-linking of beta-amyloid protein precursor catalyzed by tissue transglutaminase, *FEBS Lett. 349*, 151–154.

116. Rasmussen, L. K., Sorensen, E. S., Petersen, T. E., Gliemann, J., and Jensen, P. H. (1994) Identification of glutamine and lysine residues in Alzheimer amyloid beta A4 peptide responsible for transglutaminase-catalysed homopolymerization and cross-linking to alpha 2M receptor, *FEBS Lett. 338*, 161–166.

117. Appelt, D. M. and Balin, B. J. (1997) The association of tissue transglutaminase with human recombinant tau results in the formation of insoluble filamentous structures, *Brain Res. 745*, 21–31.

118. Grierson, A. J., Johnson, G. V., and Miller, C. C. (2001) Three different human tau isoforms and rat neurofilament light, middle and heavy chain proteins are cellular substrates for transglutaminase, *Neurosci. Lett. 298*, 9–12.

119. Halverson, R. A., Lewis, J., Frausto, S., Hutton, M., and Muma, N. A. (2005) Tau protein is cross-linked by transglutaminase in P301L tau transgenic mice, *J. Neurosci.* *25*, 1226–1233.

120. Miller, M. L. and Johnson, G. V. (1995) Transglutaminase cross-linking of the tau protein, *J. Neurochem.* *65*, 1760–1770.

121. Murthy, S. N., Wilson, J. H., Lukas, T. J., Kuret, J., and Lorand, L. (1998) Cross-linking sites of the human tau protein, probed by reactions with human transglutaminase, *J. Neurochem.* *71*, 2607–2614.

122. Norlund, M. A., Lee, J. M., Zainelli, G. M., and Muma, N. A. (1999) Elevated transglutaminase-induced bonds in PHF tau in Alzheimer's disease, *Brain Res.* *851*, 154–163.

123. Tucholski, J., Kuret, J., and Johnson, G. V. (1999) Tau is modified by tissue transglutaminase in situ: possible functional and metabolic effects of polyamination, *J. Neurochem.* *73*, 1871–1880.

124. Orr, H. T. and Zoghbi, H. Y. (2007) Trinucleotide repeat disorders, *Annu. Rev. Neurosci.* *30*, 575–621.

125. Wanker, E. E. (2000) Protein aggregation and pathogenesis of Huntington's disease: mechanisms and correlations, *Biol. Chem.* *381*, 937–942.

126. Ross, C. A. (2002) Polyglutamine pathogenesis: emergence of unifying mechanisms for Huntington's disease and related disorders, *Neuron* *35*, 819–822.

127. Cariello, L., de Cristofaro, T., Zanetti, L., Cuomo, T., Di Maio, L., Campanella, G., Rinaldi, S., Zanetti, P., Di Lauro, R., and Varrone, S. (1996) Transglutaminase activity is related to CAG repeat length in patients with Huntington's disease, *Hum. Genet.* *98*, 633–635.

128. Karpuj, M. V., Garren, H., Slunt, H., Price, D. L., Gusella, J., Becher, M. W., and Steinman, L. (1999) Transglutaminase aggregates huntingtin into nonamyloidogenic polymers, and its enzymatic activity increases in Huntington's disease brain nuclei, *Proc. Natl. Acad. Sci. U.S.A.* *96*, 7388–7393.

129. Karpuj, M. V., Becher, M. W., Springer, J. E., Chabas, D., Youssef, S., Pedotti, R., Mitchell, D., and Steinman, L. (2002) Prolonged survival and decreased abnormal movements in transgenic model of Huntington disease, with administration of the transglutaminase inhibitor cystamine, *Nat. Med.* *8*, 143–149.

130. Green, H. (1993) Human genetic diseases due to codon reiteration: relationship to an evolutionary mechanism, *Cell* *74*, 955–956.

131. Gentile, V., Sepe, C., Calvani, M., Melone, M. A., Cotrufo, R., Cooper, A. J., Blass, J. P., and Peluso, G. (1998) Tissue transglutaminase-catalyzed formation of high-molecular-weight aggregates in vitro is favored with long polyglutamine domains: a possible mechanism contributing to CAG-triplet diseases, *Arch. Biochem. Biophys.* *352*, 314–321.

132. Cooper, A. J., Sheu, K. F., Burke, J. R., Onodera, O., Strittmatter, W. J., Roses, A. D., and Blass, J. P. (1997) Polyglutamine domains are substrates of tissue transglutaminase: does transglutaminase play a role in expanded CAG/poly-Q neurodegenerative diseases? *J. Neurochem.* *69*, 431–434.

133. Kahlem, P., Terre, C., Green, H., and Djian, P. (1996) Peptides containing glutamine repeats as substrates for transglutaminase-catalyzed cross-linking: relevance to diseases of the nervous system, *Proc. Natl. Acad. Sci. U.S.A. 93*, 14580–14585.

134. Lai, T. S., Tucker, T., Burke, J. R., Strittmatter, W. J., and Greenberg, C. S. (2004) Effect of tissue transglutaminase on the solubility of proteins containing expanded polyglutamine repeats, *J. Neurochem. 88*, 1253–1260.

135. Perutz, M. (1994) Polar zippers: their role in human disease, *Protein Sci. 3*, 1629–1637.

136. Perutz, M. F. (1999) Glutamine repeats and neurodegenerative diseases: molecular aspects, *Trends Biochem. Sci. 24*, 58–63.

137. Burke, M. G., Woscholski, R., and Yaliraki, S. N. (2003) Differential hydrophobicity drives self-assembly in Huntington's disease, *Proc. Natl. Acad. Sci. U.S.A. 100*, 13928–13933.

138. Martin-Aparicio, E., Yamamoto, A., Hernandez, F., Hen, R., Avila, J., and Lucas, J. J. (2001) Proteasomal-dependent aggregate reversal and absence of cell death in a conditional mouse model of Huntington's disease, *J. Neurosci. 21*, 8772–8781.

139. Chun, W., Lesort, M., Tucholski, J., Ross, C. A., and Johnson, G. V. (2001) Tissue transglutaminase does not contribute to the formation of mutant huntingtin aggregates, *J. Cell Biol. 153*, 25–34.

140. Bailey, C. D. and Johnson, G. V. (2005) Tissue transglutaminase contributes to disease progression in the R6/2 Huntington's disease mouse model via aggregate-independent mechanisms, *J. Neurochem. 92*, 83–92.

141. Mastroberardino, P. G., Iannicola, C., Nardacci, R., Bernassola, F., De Laurenzi, V., Melino, G., Moreno, S., Pavone, F., Oliverio, S., Fesus, L., and Piacentini, M. (2002) 'Tissue' transglutaminase ablation reduces neuronal death and prolongs survival in a mouse model of Huntington's disease, *Cell Death Differ. 9*, 873–880.

142. Igarashi, S., Koide, R., Shimohata, T., Yamada, M., Hayashi, Y., Takano, H., Date, H., Oyake, M., Sato, T., Sato, A., Egawa, S., Ikeuchi, T., Tanaka, H., Nakano, R., Tanaka, K., Hozumi, I., Inuzuka, T., Takahashi, H., and Tsuji, S. (1998) Suppression of aggregate formation and apoptosis by transglutaminase inhibitors in cells expressing truncated DRPLA protein with an expanded polyglutamine stretch, *Nat. Genet. 18*, 111–117.

143. Bailey, C. D. and Johnson, G. V. (2006) The protective effects of cystamine in the R6/2 Huntington's disease mouse involve mechanisms other than the inhibition of tissue transglutaminase, *Neurobiol. Aging 27*, 871–879.

144. Jeitner, T. M., Delikatny, E. J., Ahlqvist, J., Capper, H., and Cooper, A. J. (2005) Mechanism for the inhibition of transglutaminase 2 by cystamine, *Biochem. Pharmacol. 69*, 961–970.

145. Pepe, I., Occhino, E., Cella, G., Luongo, A., Guardascione, F., and Gentile, V. (2004) Biochemical mechanisms for a possible involvement of the transglutaminase activity in the pathogenesis of the polyglutamine diseases: minireview article, *Amino Acids 26*, 431–434.

146. Samii, A., Nutt, J. G., and Ransom, B. R. (2004) Parkinson's disease, *Lancet 363*, 1783–1793.

147. Colosimo, C., Hughes, A. J., Kilford, L., and Lees, A. J. (2003) Lewy body cortical involvement may not always predict dementia in Parkinson's disease, *J. Neurol. Neurosurg. Psychiatry 74*, 852–856.

148. Andringa, G., Lam, K. Y., Chegary, M., Wang, X., Chase, T. N., and Bennett, M. C. (2004) Tissue transglutaminase catalyzes the formation of alpha-synuclein crosslinks in Parkinson's disease, *FASEB J. 18*, 932–934.

149. Ulmer, T. S., Bax, A., Cole, N. B., and Nussbaum, R. L. (2005) Structure and dynamics of micelle-bound human alpha-synuclein, *J. Biol. Chem. 280*, 9595–9603.

150. Giasson, B. I., Murray, I. V., Trojanowski, J. Q., and Lee, V. M. (2001) A hydrophobic stretch of 12 amino acid residues in the middle of alpha-synuclein is essential for filament assembly, *J. Biol. Chem. 276*, 2380–2386.

151. Jensen, P. H., Sorensen, E. S., Petersen, T. E., Gliemann, J., and Rasmussen, L. K. (1995) Residues in the synuclein consensus motif of the alpha-synuclein fragment, NAC, participate in transglutaminase-catalysed cross-linking to Alzheimer-disease amyloid beta A4 peptide, *Biochem. J. 310(Pt 1)*, 91–94.

152. Junn, E., Ronchetti, R. D., Quezado, M. M., Kim, S. Y., and Mouradian, M. M. (2003) Tissue transglutaminase-induced aggregation of alpha-synuclein: Implications for Lewy body formation in Parkinson's disease and dementia with Lewy bodies, *Proc. Natl. Acad. Sci. U.S.A. 100*, 2047–2052.

153. Konno, T., Morii, T., Hirata, A., Sato, S., Oiki, S., and Ikura, K. (2005) Covalent blocking of fibril formation and aggregation of intracellular amyloidgenic proteins by transglutaminase-catalyzed intramolecular cross-linking, *Biochemistry 44*, 2072–2079.

154. Nemes, Z., Petrovski, G., Aerts, M., Sergeant, K., Devreese, B., and Fesus, L. (2009) Transglutaminase-mediated intramolecular cross-linking of membrane-bound alpha-synuclein promotes amyloid formation in Lewy bodies, *J. Biol. Chem. 284*, 27252–27264.

155. Schmid, A. W., Chiappe, D., Pignat, V., Grimminger, V., Hang, I., Moniatte, M., and Lashuel, H. A. (2009) Dissecting the mechanisms of tissue transglutaminase-induced cross-linking of alpha-synuclein: implications for the pathogenesis of Parkinson disease, *J. Biol. Chem. 284*, 13128–13142.

156. Segers-Nolten, I. M., Wilhelmus, M. M., Veldhuis, G., van Rooijen, B. D., Drukarch, B., and Subramaniam, V. (2008) Tissue transglutaminase modulates alpha-synuclein oligomerization, *Protein Sci. 17*, 1395–1402.

157. Aeschlimann, D., Kaupp, O., and Paulsson, M. (1995) Transglutaminase-catalyzed matrix cross-linking in differentiating cartilage: identification of osteonectin as a major glutaminyl substrate, *J. Cell Biol. 129*, 881–892.

158. Dedeoglu, A., Kubilus, J. K., Jeitner, T. M., Matson, S. A., Bogdanov, M., Kowall, N. W., Matson, W. R., Cooper, A. J., Ratan, R. R., Beal, M. F., Hersch, S. M., and Ferrante, R. J. (2002) Therapeutic effects of cystamine in a murine model of Huntington's disease, *J. Neurosci. 22*, 8942–8950.

159. Zainelli, G. M., Ross, C. A., Troncoso, J. C., and Muma, N. A. (2003) Transglutaminase cross-links in intranuclear inclusions in Huntington disease, *J. Neuropathol. Exp. Neurol. 62*, 14–24.

160. Singer, S. M., Zainelli, G. M., Norlund, M. A., Lee, J. M., and Muma, N. A. (2002) Transglutaminase bonds in neurofibrillary tangles and paired helical filament tau early in Alzheimer's disease, *Neurochem. Int. 40*, 17–30.

161. Zemaitaitis, M. O., Lee, J. M., Troncoso, J. C., and Muma, N. A. (2000) Transglutaminase-induced cross-linking of tau proteins in progressive supranuclear palsy, *J. Neuropathol. Exp. Neurol. 59*, 983–989.

162. Johnson, G. V. and LeShoure, R., Jr. (2004) Immunoblot analysis reveals that isopeptide antibodies do not specifically recognize the epsilon-(gamma-glutamyl)lysine bonds formed by transglutaminase activity, *J. Neurosci. Methods 134*, 151–158.

163. Zainelli, G. M., Ross, C. A., Troncoso, J. C., Fitzgerald, J. K., and Muma, N. A. (2004) Calmodulin regulates transglutaminase 2 cross-linking of huntingtin, *J. Neurosci. 24*, 1954–1961.

164. Nemes, Z., Devreese, B., Steinert, P. M., Van Beeumen, J., and Fesus, L. (2004) Cross-linking of ubiquitin, HSP27, parkin, and alpha-synuclein by gamma-glutamyl-epsilon-lysine bonds in Alzheimer's neurofibrillary tangles, *FASEB J. 18*, 1135–1137.

165. Sarvari, M., Karpati, L., Fesus, L., Deli, L., Muszbek, L., and Nemes, Z. (2002) Competitive enzyme-linked immonosorbent assay for N epsilon gamma-glutamyl lysine, *Anal. Biochem. 311*, 187–190.

166. Fink, M. L., Chung, S. I., and Folk, J. E. (1980) gamma-Glutamylamine cyclotransferase: specificity toward epsilon-(L-gamma-glutamyl)-L-lysine and related compounds, *Proc. Natl. Acad. Sci. U.S.A. 77*, 4564–4568.

167. Hultsch, C., Bergmann, R., Pawelke, B., Pietzsch, J., Wuest, F., Johannsen, B., and Henle, T. (2005) Biodistribution and catabolism of 18F-labelled isopeptide N(epsilon)-(gamma-glutamyl)-L-lysine, *Amino Acids 29*, 405–413.

168. Nemes, Z., Fesus, L., Egerhazi, A., Keszthelyi, A., and Degrell, I. M. (2001) N(epsilon)(gamma-glutamyl)lysine in cerebrospinal fluid marks Alzheimer type and vascular dementia, *Neurobiol. Aging 22*, 403–406.

169. Sarvari, M., Fesus, L., and Nemes, Z. (2002) Transglutaminase-mediated crosslinking of neural proteins in Alzheimer's disease and other primary dementias, *Drug Dev. Res. 56*, 458–472.

170. Jeitner, T. M., Bogdanov, M. B., Matson, W. R., Daikhin, Y., Yudkoff, M., Folk, J. E., Steinman, L., Browne, S. E., Beal, M. F., Blass, J. P., and Cooper, A. J. (2001) N(epsilon)-(gamma-L-glutamyl)-L-lysine (GGEL) is increased in cerebrospinal fluid of patients with Huntington's disease, *J. Neurochem. 79*, 1109–1112.

171. Jeitner, T. M., Matson, W. R., Folk, J. E., Blass, J. P., and Cooper, A. J. (2008) Increased levels of gamma-glutamylamines in Huntington disease CSF, *J. Neurochem. 106*, 37–44.

172. Ueda, K., Masliah, E., Saitoh, T., Bakalis, S. L., Scoble, H., and Kosik, K. S. (1990) Alz-50 recognizes a phosphorylated epitope of tau protein, *J. Neurosci. 10*, 3295–3304.

173. Wang, Q., Woltjer, R. L., Cimino, P. J., Pan, C., Montine, K. S., Zhang, J., and Montine, T. J. (2005) Proteomic analysis of neurofibrillary tangles in Alzheimer disease identifies GAPDH as a detergent-insoluble paired helical filament tau binding protein, *FASEB J. 19*, 869–871.

174. Arrigo, A. P. (2007) The cellular "networking" of mammalian Hsp27 and its functions in the control of protein folding, redox state and apoptosis, *Adv. Exp. Med. Biol. 594*, 14–26.

175. Ardley, H. C., Scott, G. B., Rose, S. A., Tan, N. G., Markham, A. F., and Robinson, P. A. (2003) Inhibition of proteasomal activity causes inclusion formation in neuronal and non-neuronal cells overexpressing Parkin, *Mol. Biol. Cell 14*, 4541–4556.

176. McNaught, K. S., Shashidharan, P., Perl, D. P., Jenner, P., and Olanow, C. W. (2002) Aggresome-related biogenesis of Lewy bodies, *Eur. J. Neurosci. 16*, 2136–2148.

177. Li, X., Li, H., and Li, X. J. (2008) Intracellular degradation of misfolded proteins in polyglutamine neurodegenerative diseases, *Brain Res. Rev. 59*, 245–252.

178. Olzmann, J. A. and Chin, L. S. (2008) Parkin-mediated K63-linked polyubiquitination: a signal for targeting misfolded proteins to the aggresome-autophagy pathway, *Autophagy 4*, 85–87.

179. D'Eletto, M., Farrace, M. G., Falasca, L., Reali, V., Oliverio, S., Melino, G., Griffin, M., Fimia, G. M., and Piacentini, M. (2009) Transglutaminase 2 is involved in autophagosome maturation, *Autophagy 5*, 1145–1154.

TRANSGLUTAMINASE-MEDIATED REMODELING OF THE HUMAN ERYTHROCYTE MEMBRANE SKELETON: RELEVANCE FOR ERYTHROCYTE DISEASES WITH SHORTENED CELL LIFESPAN

LASZLO LORAND
S. N. PRASANNA MURTHY
ANWAR A. KHAN
WEIHUA XUE
OKSANA LOCKRIDGE
ATHAR H. CHISHTI

CONTENTS

Advances in Enzymology and Related Areas of Molecular Biology, Volume 78.
Edited by Eric J. Toone.

I. INTRODUCTION

The human red blood cell transglutaminase (hRBC TG2) was the first in this family of enzymes for which an important role in cell–matrix interaction was found by demonstrating that the protein—when released from cells—could form an extremely tight complex with human fibronectin (FN). The binding, with a stoichiometry of 2TG2:FN (i.e., 1TG2 per constituent chain of FN), is independent of the catalytic activity of TG2 and occurs in the absence as well as in the presence of Ca^{2+} ions [1–3]. Residues 81–106 of TG2, located at the extended hairpin between antiparallel β strands 5 and 6 of the first domain of the protein, seem to be essential for binding to FN; mutations of Asp94 and Asp97 to Ala reduce the binding affinity of TG2 to FN significantly. A synthetic peptide, corresponding to the sequence 88WTATVVDQQDCTLSLQLTT106 in TG2, inhibits the TG2–FN interaction, and also TG2-dependent cell adhesion and spreading [4]. The complementary binding sites of FN are located in a 42-kDa collagen-binding domain of the protein, comprising motifs I_6-II_1-II_2-I_7-I_8-I_9. This fragment shows as high an affinity for TG2 as the individual parent FN chains themselves [5]; furthermore, the 42-kDa fragment of FN can neutralize the functions of TG2 on cell surfaces [6]. Binding to TG2 is so specific that an affinity column made by coupling the 42-kDa fragment of FN to a gel matrix can be used for isolating hRBC TG2 to the highest purity with a single passage of hemoglobin-depleted erythrocyte lysate [5] (Figure 1A). This procedure was employed for purifying the TG2 protein on which nucleotide-binding studies were carried out [7], and on which the large conformational change—attendant to binding GTP—could be demonstrated

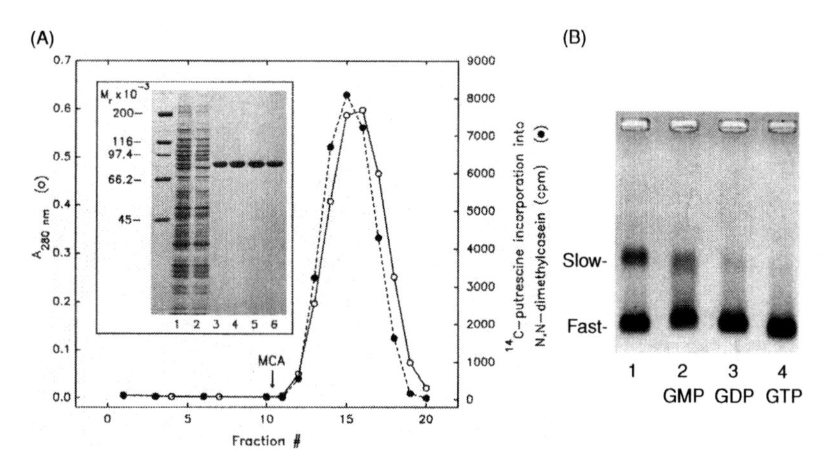

FIGURE 1. (A) Affinity purification of TG2 by single passage of the hemoglobin-depleted lysate of hRBCs through a column of the 42-kDa gelatin-binding fragment of human fibronectin. The Hb-depleted cell lysate was applied to the affinity column. After extensive washing, the tightly held enzyme was eluted with 0.25% monochloroacetic acid (MCA). Fractions were neutralized and analyzed for protein (left ordinate, open circle) and enzyme activity (right ordinate, closed circle). Inset: eluted samples were examined by SDS-PAGE, stained with Coomassie blue R. Lanes: 1, hemoglobin-depleted lysate; 2, nonretained material passing through the column; 3–6, fractions 14–17 eluted with MCA. Molecular masses in kDa are indicated on the left. For experimental details, see [5]. In later experiments, with better preservation of TG2 activity, MCA was replaced by 80% ethyleneglycol [8]. (B) Purified hRBC TG2 appears to be preponderantly in the GDP-bound form in the hydrodynamically compact, fast-moving electrophoretic conformation. Mobility shifts of the purified protein (lane 1) from slow to fast forms were examined upon mixing with GMP (lane 2), GDP (lane 3), and GTP (lane 4) by nondenaturing electrophoresis (in 3% agarose). For details, see [9].

by transition from a slow-moving, extended structure to a faster moving, compact configuration in nondenaturing electrophoresis [8] (Figure 1B).

TG2s of different species vary in sensitivities to inhibition by GTP, but hRBC TG2 binds tightly to the nucleotide (measured by a fluorescently labeled analog), with an association constant of 4×10^7 M^{-1} [7]. Even in the highly purified form, this TG2 seems to exist preponderantly in the closed compact, inactive configuration of the enzyme, corresponding to the electrophoretically fast-moving GDP-bound form (Figure 1B).

It is perhaps more relevant to the present discussion that human red cells provided the paradigm for showing that TG2—though inactive in the intracellular milieu—becomes rapidly converted by entry of Ca^{2+} to an active transamidase, producing profound alterations in the structural organization

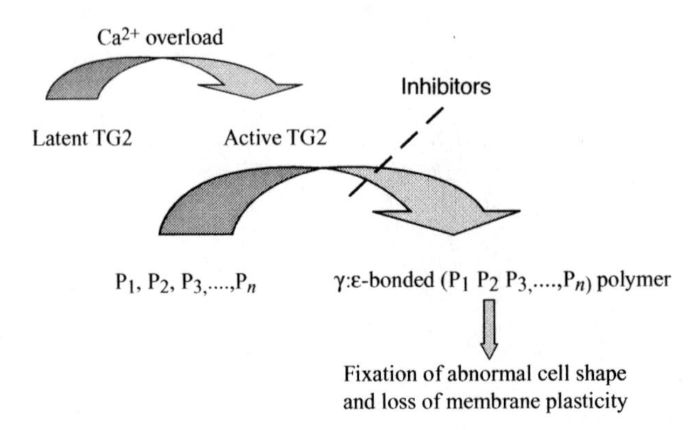

SCHEME 1. The Ca^{2+}-triggered, transglutaminase-mediated protein crosslinking cascade in cells.

and physical properties of the cell [10–14]. It is remarkable that the changes brought about by treating normal hRBCs with Ca^{2+} plus ionophore closely parallel those seen in some erythrocyte diseases in which the lifespans of the cells are appreciably shortened. Therefore, the sequence of events in the hRBC diseases, and also in the experimental model with Ca^{2+} overload, may be illustrated by Scheme 1.

In the resting cell, TG2 is kept in the inactive, latent form by virtue of its tight binding to GTP. However, the entry of Ca^{2+} ions removes the inhibition by GTP and allows expression of transamidating activity. The enzyme catalyzes the cross-linking of protein substrates ($P_1, P_2, P_3, \ldots, P_n$) by covalent $\gamma{:}\varepsilon$ isopeptide bridges and the concomitant formation of large-molecular-weight polymeric structures in the cell membrane. This activity causes a stiffening of the membrane and irreversible fixation of cell shape [15], which are thought to promote the premature removal of the affected hRBCs from the circulation. Moreover, competitive and noncompetitive inhibitors of TG2 prevent the protein cross-linking by the enzyme (i.e., the formation of abnormal protein polymers) and also block the physical consequences of membrane stiffening and fixation of cell shape.

It is important to bear in mind that the polymeric structures created by TG2 action are not conventional molecular aggregates; they cannot be separated into their original building blocks by protein-solubilizing agents (such as urea, weak acids and alkalis, ionic or nonionic detergents, including sodium dodecylsulfate, SDS) or their combinations. The portions of the polymers soluble in a mixture of SDS and a reducing agent, such as dithiothreitol (DTT), contain a fraction of the $N^{\varepsilon}(\gamma$-glutaminyl)lysine-bonded constituent chain framework—essentially the backbone of a branched structure with ill-defined geometry—onto which other polypeptides may be attached in the cell, for example, by disulfide interchain linkages.

II. Hb KOLN DISEASE, A PROTOTYPE OF ERYTHROCYTE DISORDERS CHARACTERIZED BY THE PRESENCE OF γ:ε-BONDED, ABNORMAL MEMBRANE SKELETAL POLYMERS

The primary defect in Hb Koln disease is a Val98Met mutation in the β subunits of hemoglobin near the heme pocket. This creates molecular instability, placing Hb Koln into the family of "unstable hemoglobin diseases," characterized by hemolytic anemia. The trait, unlike sickle cell disease, does not seem to be inherited; each case is thought to arise from fresh (parental gonadal?) mutation. Extensive metabolic, deformability, and survival studies have been carried out on Hb Koln erythrocytes [16, 17]. Among the many biochemical, morphological, and functional changes that have been reported, including a marked loss of cellular potassium, intracellular dehydration, low ATP concentration (around 66% of normal), decreased osmotic fragility, and decreased cell deformability are noteworthy. Because of the large reduction of intracellular ATP, it may be assumed that the TG2-inhibitory concentration of GTP would also be appreciably lower than normal; hence, expression of transamidating activity in the Hb Koln cells would be expected to occur with smaller augmentation of internal concentration of Ca^{2+} ions than in normal red cells. Membrane rigidity probably contributes to the enhanced splenic entrapment of Hb Koln erythrocytes and accounts for their drastically reduced lifespan in the circulation, which is only about one-fourth of normal. There is improvement of red cell lifespan after splenectomy, from about 31 days to 47 days.

Because membrane rigidity appears to be responsible for the premature death of erythrocytes in Hb Koln disease, our research focused on the biochemistry of the membrane compartment of this abnormal cell [18]. Erythrocyte ghosts, isolated from a patient with Hb Koln anemia, were uniformly of higher density than normal (1.18 g/mL versus 1.16 g/mL; Figure 2A). In another case, the density distribution of erythrocyte ghosts showed a dual profile, with a main peak at near the normal density of 1.15 g/mL and a smaller peak at 1.21 g/mL.

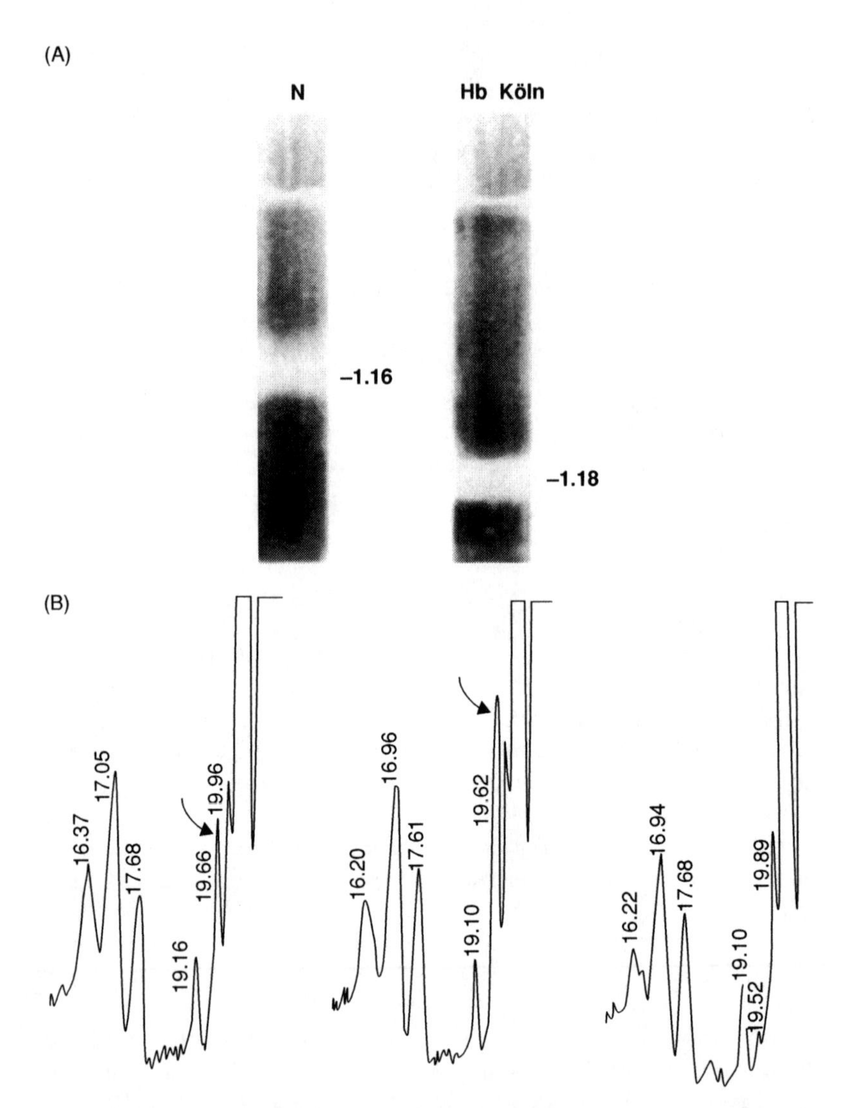

FIGURE 2. (A) Increased density of the red cell membranes from a patient with Hb Koln disease (right) compared with that of a normal individual (left), measured by sucrose density centrifugation. Density values, marked in g/mL, were computed from the refractive indices of the solutions at the points of highest turbidities of cell membranes. For details, see [18]. (B) Skeletal proteins of the Hb Koln erythrocyte membrane are modified by N^{ε}(γ-glutaminyl)lysine

Since the metabolically impaired Hb Koln cells probably allowed entry of Ca^{2+} ions, which activated the resident TG2, we undertook a search for high-molecular-weight polymeric membrane skeletal products covalently bonded by $N^{\varepsilon}(\gamma$-glutaminyl)lysine side chain bridges, the "footprints" of TG2 activity. Three polymeric fractions were obtained from the following: (1) the 0.1 N NaOH extract of erythrocyte ghosts, comprising peripherally associated membrane proteins; (2) the 1% SDS–10% DTT extractable portion of the membrane, which contained the intrinsic membrane proteins; and (3) the remaining insoluble pellet. The preparations were subjected to complete enzymatic digestion, and quantitative analysis for $N^{\varepsilon}(\gamma$-glutaminyl)lysine isopeptide was performed. All three fractions contained substantial amounts of the isopeptide, but the highest cross-link content was found in the SDS–DTT-insoluble pellet with a frequency of 1 mole per 120,000 g of protein (Figure 2B). However, because some protease-resistant core remained unprocessed from the material, this is probably a significant underestimation of the true frequency of cross-links in this fraction of Hb Koln membranes. Using similar techniques, no detectable amounts of polymers or $N^{\varepsilon}(\gamma$-glutaminyl)lysine peptides were obtained from a comparable pool of normal erythrocyte membranes. In regard to its biological significance, we should recall that similar frequencies of $N^{\varepsilon}(\gamma$-glutaminyl)lysine cross-links, introduced by factor XIIIa into fibrin, would result in close to a fivefold increase in the elastic storage modulus (i.e., stiffness) of the clot network [19], and their absence could cause potentially life-threatening hemorrhage [20]. Hence, it is justified to conclude that TG2 activity in Hb Koln erythrocytes contributes significantly to the stiffening of membrane skeletal structure by catalyzing the formation of $N^{\varepsilon}(\gamma$-glutaminyl)lysine protein-to-protein side chain bridges.

Observations very similar to those described for the Hb Koln case were made with the membrane preparation from a sickle cell patient [21]; it still remains to be investigated whether the γ:ε-bonded polymers were derived

FIGURE 2. (*Continued*) side chain bridges, footprints of the activity of intracellular TG2. High-pressure liquid chromatography of the total proteolytic digest of SDS–DTT-insoluble membrane proteins reveals the presence of the isopeptide (left panel). The same is shown in the middle panel but with synthetic $N^{\varepsilon}(\gamma$-glutaminyl)lysine peptide added. The right-hand panel corresponds to the first panel on the left, following treatment by γ-glutamylaminecyclotransferase, an enzyme specific for cleaving the isopeptide. Positions of $N^{\varepsilon}(\gamma$-glutaminyl)lysine are marked by curved arrows. For details, see [18].

exclusively from irreversibly sickled cell population. Sickle cell is yet another disease in which red cell survival is markedly reduced on account of increased membrane rigidity and change in cell shape. The finding supports the notion that TG2-mediated formation of abnormal membrane protein polymers may be a common finding in anemias of different etiologies and that it might actually account for the premature death of red cells in such diseases.

III. THE Ca^{2+}-ENRICHED HUMAN RED BLOOD CELL AS MODEL FOR THE PHYSICAL AND BIOCHEMICAL ABNORMALITIES OBSERVED IN ERYTHROCYTE DISEASES WITH SHORTENED CELL LIFESPAN

A. MORPHOLOGICAL AND STRUCTURAL CHANGES IN HUMAN ERYTHROCYTES, MEDIATED BY TRANSGLUTAMINASE 2 (TG2), UPON INFLUX OF Ca^{2+} IONS: PERMANENT FIXATION OF ABNORMAL CELL SHAPE AND LOSS OF MEMBRANE PLASTICITY

Treatment of hRBCs with Ca^{2+} plus ionophore (e.g., A23187 or ionomycin) causes a rapid change in shape from discocyte to spheroechinocyte [15], i.e., spheres with sea urchin-like surface spicules (Figure 3), seen also in some anemias [16].

The crenated cells become somewhat dehydrated in spite of the presence of potassium in the medium for lowering the efflux of water. Following short periods of Ca^{2+} loading (around 30 min), replacement of the incubation medium with one containing bovine serum albumin (BSA) and the Ca^{2+}

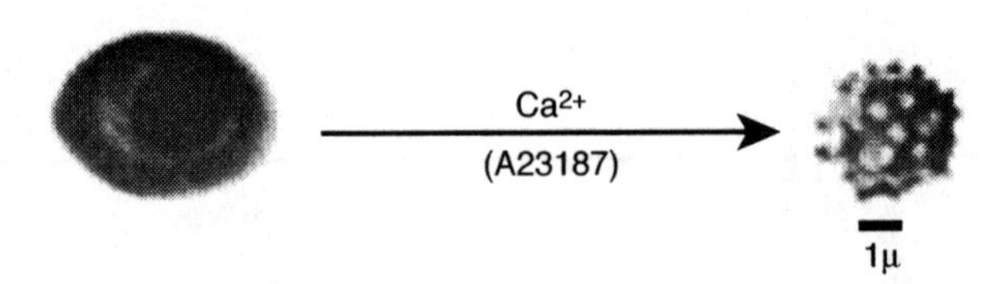

FIGURE 3. Change of shape in human red cells from discocyte to spheroechinocyte, brought about by Ca^{2+} overload, is illustrated by scanning electron micrography. The process of shape change is reversible only if the intracellular Ca^{2+} is removed within a short period of time; however, it becomes irreversible by longer exposure to the cation. Fixation of the abnormal spheroechinocytic shape—seen in some erythrocyte diseases—is due to the cross-linking of membrane skeletal proteins by $N^{\varepsilon}(\gamma$-glutaminyl)lysine bonds under the catalytic influence of TG2. Inhibitors of the enzymatic reaction prevent loss of membrane plasticity, which otherwise would irreversibly freeze the shape of the red cell. For details, see [15].

chelator ethylenediaminetetraacetate (EDTA) allows the majority of the cells to revert to the normal biconcave shapes. However, following longer periods of Ca^{2+} exposure, reversibility is lost and the cells retain their abnormal shapes. It is this irreversible fixation of shape that is caused by the covalent remodeling of the membrane skeleton by TG2, the slower kinetic step in the process of Ca^{2+}-induced shape change. The experimental evidence is based on findings with competitive and noncompetitive inhibitors of TG2, which specifically block protein cross-linking by the enzyme. For example, following a 2 h of exposure of hRBCs to Ca^{2+} plus ionophore, only 23% of the spheroechinocytes reverted back to discocytes upon removal of cellular Ca^{2+}; however, in the presence of 20 mM cystamine during Ca^{2+} treatment, reversibility to discocytes rose to 70% [15]. As a primary amine substrate of TG2, cystamine is a good inhibitor of protein cross-linking in erythrocytes [13]; however, it is known to act not only as a competitive inhibitor, but also by directly interfering with TG2 function [22] through formation of mixed disulfides with sulfhydryl groups, including that of the active center CysH residue when the enzyme is in the open configuration.

Primary amine substrates of TG2 serve as competitive inhibitors of protein cross-linking by virtue of the fact that they themselves become incorporated into the enzyme-reactive γ-glutamyl sites of acceptor proteins (Figure 4). Conversely, suitable Gln-containing short peptide substrates of the enzymes inhibit cross-linking reactions by blocking the ε-lysyl functionalities in donor proteins. Thus, with appropriate tags (isotope, fluorescent, and others), small substrates of transglutaminase can be employed for marking the potential cross-linking sites of proteins in biological systems in an enzyme-specific manner; this approach was first exploited to good advantage for probing the cross-linking sites of human fibrin by factor XIIIa [24]. However, in human red cells, only labeled primary amines were used with success, which allowed identification of some of the intracellular, potential acceptor proteins of cross-linking [13].

Concomitantly with the enzymatic remodeling of membrane skeletal structures by TG2, a loss of membrane deformability sets in [15]. The term is defined as the capacity for passive change of cellular configuration in response to shear forces. This depends on the viscoelastic properties of the membrane, the viscosity of the cytoplasm, and the shape of the cell. The biconcave disc shape of the erythrocyte has excess area compared with the minimum needed for enveloping the cell volume; thus, given its membrane elastic properties and the rather low viscosity of intracellular contents, passive change of shape occurs at relatively small forces. Factors that reduce

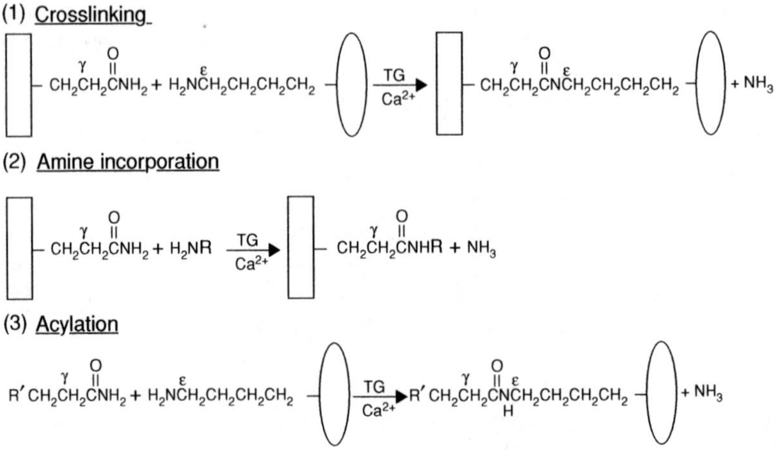

(1) Crosslinking

$$\text{—}CH_2CH_2\overset{\gamma}{}\overset{O}{\overset{\|}{C}}NH_2 + H_2N\overset{\varepsilon}{C}H_2CH_2CH_2CH_2\text{—} \xrightarrow[Ca^{2+}]{TG} \text{—}CH_2CH_2\overset{O}{\overset{\|}{C}}N\overset{\varepsilon}{C}H_2CH_2CH_2CH_2\text{—} + NH_3$$

(2) Amine incorporation

$$\text{—}CH_2CH_2\overset{O}{\overset{\|}{C}}NH_2 + H_2NR \xrightarrow[Ca^{2+}]{TG} \text{—}CH_2CH_2\overset{O}{\overset{\|}{C}}NHR + NH_3$$

(3) Acylation

$$R'CH_2CH_2\overset{O}{\overset{\|}{C}}NH_2 + H_2N\overset{\varepsilon}{C}H_2CH_2CH_2CH_2\text{—} \xrightarrow[Ca^{2+}]{TG} R'CH_2CH_2\overset{O}{\overset{\|}{C}}\underset{H}{N}\overset{\varepsilon}{C}H_2CH_2CH_2CH_2\text{—} + NH_3$$

FIGURE 4. Reactions catalyzed by hRBC TG2 and other transglutaminases. Cross-linking of two proteins by $N^{\varepsilon}(\gamma$-glutaminyl)lysine bonds (reaction 1) can be inhibited by the other two competing reactions shown. Incorporation of small primary amines (reaction 2) blocks the TG2-reactive γ-glutamyl cross-linking sites in the acceptor protein, whereas incorporation of small peptides with TG2-reactive glutaminyl residues blocks the ε-lysyl cross-linking sites of donor proteins (reaction 3). Labeled amines and glutamyl peptides are widely used for identifying and exploring potential protein substrates of transglutaminases; for a review, see [23].

the surface area, i.e., make the cell more spherical, have a pronounced effect on cellular deformability, particularly if deformation occurs at a rapid rate; increase in cytoplasmic viscosity or reduction of membrane elastic properties also reduces cellular deformability. The elastic shear modulus is estimated from the lengths of small extensions of membranes aspirated into micropipettes under conditions of low negative pressure. This procedure was used to evaluate the role played by TG2 in modifying loss of membrane deformability in hRBCs. After treatment with Ca^{2+} ions and ionophore, followed by removal of internal Ca^{2+} by BSA/EDTA, the red cells showed a significantly reduced membrane extension in comparison with both controls (incubated with Mg^{2+} ions and ionophore) and cells that were exposed to Ca^{2+} plus ionophore in the presence of histamine [15].

In view of these observations, the important issue was to identify the protein-remodeling reactions that appear to be the proximal cause of profound changes in the properties of the cell membrane in Ca^{2+}-enriched hRBCs.

B. POSTTRANSLATIONAL BIOCHEMICAL MODIFICATIONS OF PROTEINS TRIGGERED BY THE INFLUX OF Ca^{2+} IONS: PROTEOLYTIC CLEAVAGE OF TRANSMEMBRANE PROTEINS AND THE TG2-CATALYZED CROSS-LINKING OF MEMBRANE SKELETAL PROTEINS

1. Proteolytic Degradation

Though the focus of the present review is on protein cross-linking reactions by transamidation, as catalyzed by TG2 in erythrocyte diseases and in the Ca^{2+}-enriched cell model, it needs to be mentioned that elevation of intracellular Ca^{2+} also activates membrane and cytosolic proteases [25]. Whereas transamidation is the prime response of hRBCs to the influx of Ca^{2+}, proteolysis predominates in rat red cells, affecting mostly the band 4.1, the band 3 anion transporter, and the band 2.1 (ankyrin) protein substrates. Incidentally, this may explain the far greater ease of cell-to-cell fusion of rat erythrocyte membranes compared with human, which has been attributed to a much higher ratio of protease to endogenous protease inhibitor (i.e., calpain to calpastatin) in the rodent [26, 27]. As with the rat, mouse red cells also respond to the influx of Ca^{2+} by activating calpain, which causes protein degradation [28].

Two transmembrane proteins, anion transporter band 3 and glycophorin, are the major targets of proteolysis in Ca^{2+}-enriched human erythrocytes, and more than one protease may be involved in the process. Nevertheless, pepstatin alone can inhibit the degradation of both membrane proteins (see Figure 5). It is also noteworthy that the Ca^{2+}-dependent proteolytic phenomenon in hRBCs—though readily observable in freshly drawn cells—can no longer be elicited after a few days of blood bank storage [29] in the conventional preservative of CPDA-1 (a mixture of citric acid, sodium citrate, sodium biphosphate, dextrose, and adenine).

2. TG2-Catalyzed Cross-Linking of Skeletal Proteins by Transamidation

In addition to small primary amine substrates of TG2 (histamine, aminoacetonitrile, and cystamine) that can act as inhibitors of protein cross-linking [13] in hRBCs exposed to an overload of Ca^{2+} (see reaction 2 in Figure 4), other TG2 inhibitors can also effectively block membrane skeletal

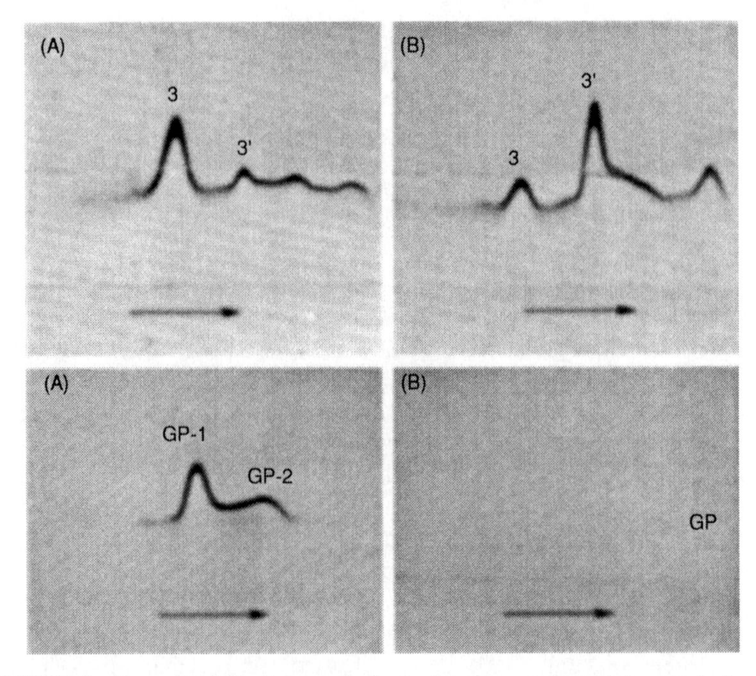

FIGURE 5. Pepstatin inhibits the proteolysis of anion transporter band 3 (top panels A and B) and glycophorin (bottom panels A and B), triggered in freshly isolated human erythrocytes by the influx of Ca^{2+}. Red cells were incubated (12 h, 37°C) in Ca^{2+} (1.5 mM) and ionophore A23187 (10 μM) in the presence (A) and the absence (B) of pepstatin (2 mM). Membranes were isolated; proteins were extracted with SDS–DTT and electrophoresed in polyacrylamide gels horizontally (see arrows). After removal of SDS, cross-immunoelectrophoresis was performed in agarose (in the perpendicular direction) into which monospecific antibodies against the band 3 protein (top panels) or glycophorins (bottom panels) were incorporated. Immunoprecipitates corresponding to the intact anion transporter are marked as 3 and to its main breakdown product as 3′, whereas the intact glycophorins are denoted GP-1 and GP-2. The barely recognizable, small glycophorin fragments remaining in the absence of pepstatin are marked GP. For further details, see [25].

alterations in these cells. Noteworthy among the findings is inhibition by 2-[3-(diallylamino)propionyl]benzothiophene or DAPBT [30]. The compound, originally synthesized as a "nontoxic" and noncompetitive inhibitor for coagulation factor XIIIa, was found to inhibit also the transamidase activity of the hRBC TG2. When tested on fresh red cells, it blocked both the protein-to-protein cross-linking reaction of TG2 and the proteolytic degradation of band 3 and glycophorin (Figure 6).

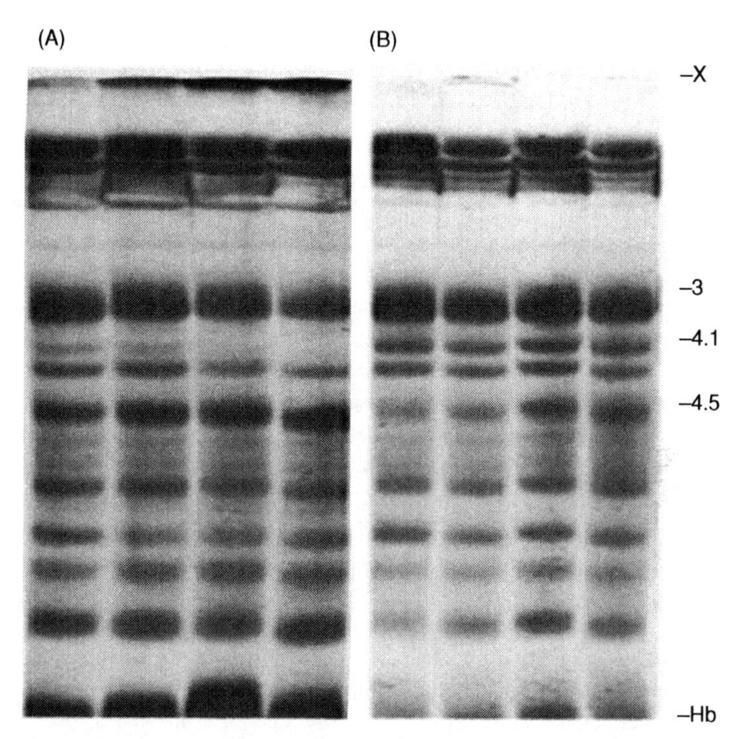

FIGURE 6. A direct, noncompetitive inhibitor of TG2, 2-[3(diallylamino)propionyl] benzothiophene, or DAPBT, effectively blocks the cross-linking of membrane skeletal proteins brought about by this enzyme when activated in the human red cell by an overload of Ca^{2+}. The inhibitor also hinders proteolytic degradation of the band 3 anion transporter. Cells were incubated at 37°C with 2-mM Ca^{2+} and 20-μM ionophore A23187 for 1, 2, 4 and 8 h (corresponding to four gel profiles in each set from left to right) in the presence of 0.1 mM (set A) or 0.6 mM DAPBT (set B). The higher concentration of the inhibitor is seen to have prevented the formation of high-molecular-weight membrane protein polymers, marked X on top of the SDS-PAGE profiles, whereas lower concentration of DAPBT still allowed polymer production to proceed (albeit at a much slower rate than it would have without the presence of the compound). In SDS-PAGE tests, a sensitive indicator of blocking the cross-linking reaction by TG2 is preservation of staining intensity of the 4.1 band, as in (B). This panel also shows, that—at a concentration of 0.6 mM—DAPBT hindered the proteolytic degradation of the anion transporter band 3 into its main fragment, appearing at the position of band 4.5 (which corresponds to that of 3' in cross-immunoelectrophoresis in Figure 5). For details, see [30].

IV. THE POLYDISPERSE NATURE OF HIGH-MOLECULAR-WEIGHT, γ:ε-BONDED MEMBRANE SKELETAL POLYMERS GENERATED IN HUMAN ERYTHROCYTES WITH Ca^{2+} OVERLOAD

The SDS–DTT-soluble high-molecular-weight (>10^6 Da) products of protein cross-linking, found in the membranes of the Hb Koln and sickle cell patient's erythrocytes, and also in the Ca^{2+}/ionophore-treated cell model, can be readily detected on one-dimensional polyacrylamide gels. Since they are too large to enter such gels, they are visualized on top of the separating and stacking gels (marked X1 and X2 in Figures 7 and 8) on the SDS-PAGE profiles of membrane proteins. A significant fraction remains with the cell membrane even after extraction with weak alkali (0.1 N NaOH), which removes peripherally associated proteins. The largest of these structures can be isolated by exclusion chromatography from the *a*lkali-*s*tripped *g*hosts (ASG, Figure 7A, X2) and also from the *a*lkali *e*xtract of membranes of Ca^{2+}-treated cells (AE, Figure 7B, X2). SDS electrophoresis in 2% agarose (a medium in which they can migrate; Figure 7C) shows them to be comprised of polypeptide chain assemblies estimated to range in size between 3 and 6×10^6 Da for those from the stripped ghosts (X2ASG) and 1 to 3×10^6 Da for those from the alkali extract (X2AE). Nevertheless, the diffuse staining pattern reflects the polydisperse nature of these products and contrasts with the distinct banding profile of nonreduced fibrin—cross-linked by coagulation factor XIIIa, used as a marker in Figure 7C—displaying a ladder of multiples of the approximately 330 kDa protein monomer.

V. DECIPHERING THE POLYPEPTIDE COMPOSITIONS OF THE SDS–DTT-SOLUBLE CORES OF CROSS-LINKED ERYTHROCYTE STRUCTURES: IMMUNOLOGIC AND PROTEOMIC ANALYSIS

A. IMMUNOBLOTTING

There are no methods available for separating the nondisulfide-bonded, covalent cores of the large protein structures into the polypeptide units from which they were assembled under the catalytic influence of TG2 in the human erythrocytes; hence, its constituents can be inferred only by employing cross-reactive antibodies [31] or by a proteomic approach, using mass spectroscopy. Since the most recent results have not yet been published, these data have to be presented and analyzed in considerable detail.

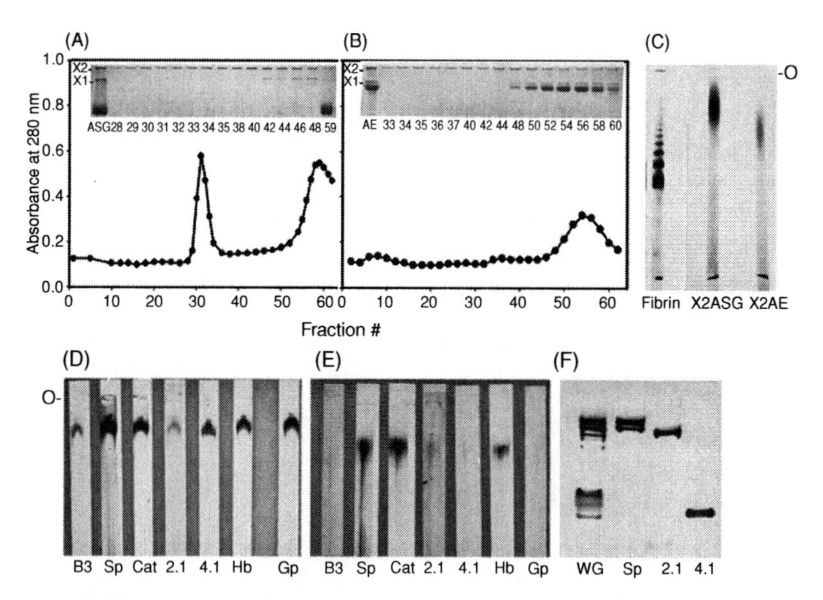

FIGURE 7. Properties of high-molecular-weight polymers isolated from Ca^{2+}-treated human erythrocytes. Red cells were exposed to Ca^{2+} plus ionophore for 3 h; after hypotonic lysis, membranes were harvested and treated with 0.1 N NaOH. Proteins of the alkali-stripped ghosts (ASG) and alkali extract (AE) were fractionated by chromatography on Sepharose CL-4B; see (A) and (B). Insets to the panels show the SDS-PAGE profiles for the ASG and AE preparations applied to the column and for the fractions that contained X1 and X2 polymers, visualized on top of the separating and stacking gels. The X2 fractions from the ASG (A; fractions #29–35) and from the AE materials (B; fractions #33–40) were pooled, concentrated, and examined by SDS-agarose electrophoresis (C; O indicates origin). A nonreduced fibrin preparation, cross-linked by factor XIIIa, was employed as a molecular weight marker [32]. (D) and (E) are immunoblots of the purified X2 polymers of the alkali-stripped ghosts and from the alkali extract, following SDS electrophoresis on agarose (O marks origin), as in (C). Nitrocellulose blots were probed with various dilutions of antibodies, i.e., antisera to band 3 or B3 (1:300), spectrins or Sp (1:80,000), ankyrin or 2.1 (1:800), protein 4.1 or 4.1 (1:700), hemoglobin or Hb (1:13,000), glycophorin or Gp (1:7000), and an IgG to catalase or Cat (1:800). (F) A rabbit IgG raised against the cross-linked polymers from Ca^{2+}-treated human erythrocytes [31], in 1:5000 dilution, reacted with several monomeric membrane skeletal protein on the SDS-PAGE profile (5% acrylamide) of whole ghosts of normal red cells (WG), which co-migrated with purified spectrins (Sp), ankyrin (2.1), and protein 4.1 (4.1), each of which was also recognized by the antibody. Bands on the immunoblot of lane WG above the position of 4.1 probably correspond to adducin and the anion transporter band 3 protein.

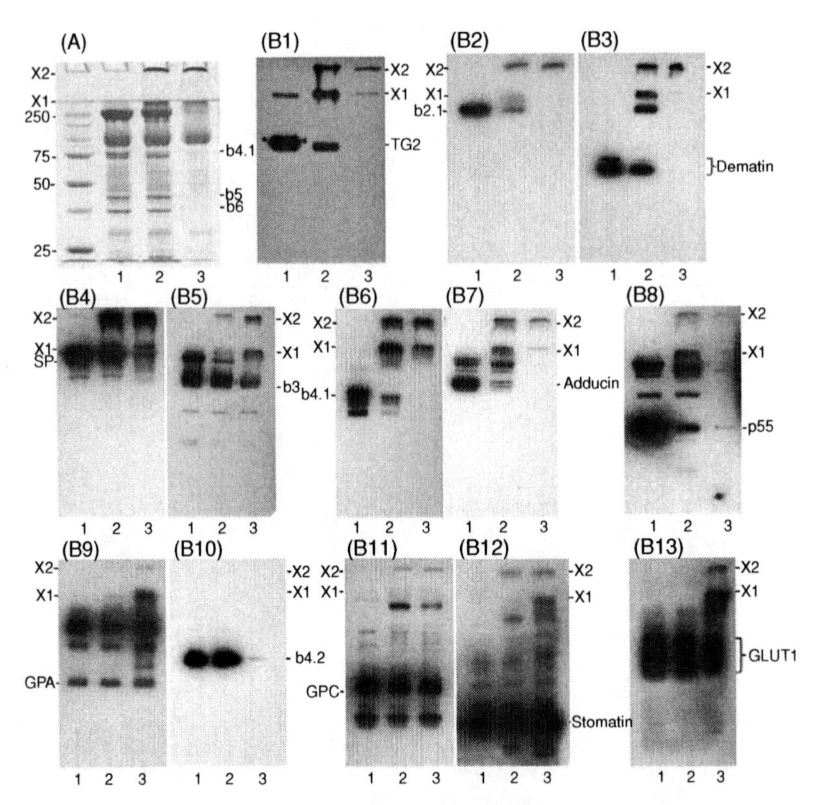

FIGURE 8. Cross-linked polymers, formed in human erythrocytes upon elevation of the concentration of internal Ca^{2+}, are recognized by several antibodies from our new antibody repertoire. The polymers on the SDS-PAGE protein profiles of membrane proteins at the top of separating and stacking gels are marked as X1 and X2. Part (A) shows the Coomassie blue-stained gels of whole ghosts of control erythrocytes (lane 1), those with Ca^{2+} overload (lane 2), and of the alkali-stripped membranes of the latter (lane 3); molecular weight marker values are given in kDa; the position of actin is marked as band 5 or b5 and that of glyceraldehyde-3-phosphate dehydrogenase as band 6 or b6. PVDF transblots of the gels were probed with various dilutions of antibodies targeting transglutaminase 2 (TG2, B1; 1:20000), ankyrin (b2.1, B2; 1:20,000), dematin (B3; 1:5000), spectrins (SP, B4; 1:10,000), band 3 (b3, B5; 1:20,000), band 4.1 (b4.1, B6; 1:4000), adducin (B7; 1:3000), p55 (B8; 1:4000), glycophorin A (GPA, B9; 1:1000), band 4.2 (b4.2, B10; 1:200,000), glycophorin C (GPC, B11; 1:1000), stomatin (B12; 1:100), and glucose transporter 1 (GLUT1, B13; 1:20,000). (See insert for color representation.)

Nitrocellulose blots of X2ASG and X2AE, after SDS electrophoresis on agarose (Figure 7), were probed with a variety of monospecific antibodies to individual human red cell proteins, including the anion transporter band 3 (B3), spectrins (Sp), catalase (Cat), ankyrin (2.1), band 4.1 (4.1), hemoglobin (Hb), and glycophorins (Gp). As shown in Figure 7D, the high-molecular-weight material from the stripped ghosts X2ASG was recognized by all of these antibodies, whereas positive cross-reactivity in the alkali extract X2AE was obtained only with antibodies to spectrins, catalase, and hemoglobin (Figure 7E). The absence of transmembrane proteins band 3 and glycophorins in X2B showed that the clusters recovered from the alkali extract were not anchored in the cell membrane.

A rabbit IgG raised against the polymeric material from Ca^{2+}-loaded human red cells [31] cross-reacted with several monomeric polypeptide constituents of the normal membrane skeleton (spectrins, ankyrin (2.1), and the band 4.1 [protein 4.1]) when these were used either as purified antigens or components found in the whole ghost preparation (WG). As revealed by the bands above that of protein 4.1, this antibody also reacted with two additional proteins on the electrophoretic profile of the ghosts at the normal positions of the anion transporter band 3 and adducin.

The Coomassie blue-stained protein profiles of ghosts from control and from Ca^{2+}-treated erythrocytes are shown in Figure 8A, lanes 1 and 2, respectively, and the alkali-stripped ghost profile of Ca^{2+}-treated cells in lane 3. A comparison of lanes 1 and 2 confirmed earlier findings [10, 13, 25, 29, 30] that intracellular TG2 activation by Ca^{2+} causes the essentially total disappearance of band 4.1, whereas the band 4.2 protein (seen below band 4.1 with a mass of about 75 kDa) and the approximately 42 kDa actin band staining intensities are unchanged from the control.

In the study of probing immunological cross-reactivities to the X1 and X2 high-molecular-weight polymers, a large panel of monospecific antibodies were used, and the results are presented in Figure 8. Figures 8B1–B13 are immunoblots for the three lanes of Figure 8A, with antibodies to transglutaminase (TG2; B1), ankyrin (b2.1, B2), dematin (B3), spectrin (alpha and beta, SP, B4), band 3 protein (b3, B5), band 4.1 protein (b4.1, B6), adducin (B7), p55 (palmitoylated membrane protein 1, B8), glycophorin A (GPA, B9), band 4.2 protein (b4.2, B10), glycophorin C (GPC, B11), stomatin (B12), and glucose transporter 1 (GLUT1, B13). Varying degrees of immunostaining of the polymers were obtained. While antibodies to spectrin (SP), band 4.1, and GPC recognized the X1 and X2 polymers essentially equally well in whole and alkali-stripped ghosts, those against band 2.1, dematin, adducin, and p55 seem to have reacted more strongly with the X1 polymers from whole ghosts. By contrast, immunostaining of polymers was considerably stronger in the alkali-stripped ghost preparations with antibodies to GPA and GLUT1. Variations of this kind may be due to differences in the concentrations of the building blocks of the polymers or due to differences in accessibilities of antigen epitopes. In the cross-linked polymeric structures, epitopes may become more buried or more exposed for antibody recognition than in the monomeric proteins. Limitations inherent in the immunological approach were further underscored by our experience with two monoclonal mouse anti-transglutaminase antibodies, MAb H23 and MAb G92. Binding of our MAb H23 is at the second Ca^{2+}-ion-binding site in TG2 [33], whereas that of G92 is at the *N*-terminal *N*-acetyl-AEDLILER peptide of mature guinea pig TG2 protein or its equivalent *N*-acetyl-AEELVLER sequence in human TG2. This *N*-terminal octapeptide segment of TG2 is a conformationally sensitive epitope, inaccessible to G92 in the native protein in solution, but exposed on nitrocellulose or PVDF blots and ELISA plates. Immunostaining of purified TG2 with the two MAbs is essentially identical; thus, both antibodies can be employed for identification of the monomeric form of TG2 on immunoblots of ghosts from Ca^{2+}-treated human erythrocytes, but—unlike H23—G92 barely recognizes the TG2 that is incorporated into the membrane-associated polymers of Ca^{2+}-treated cells.

B. PROTEOMICS

Proteomics were employed for analyzing the γ:ε-bonded cores of polymers generated in red cells upon exposure to Ca^{2+} overload. Gel slices at the locations of X1 and X2 polymer bands were cut from eight parallel SDS-PAGE runs of whole ghosts (Table 1) and of alkali-stripped ghosts (Table 2) preparations, and were processed for mass spectrometric analysis. Mascot software was used to identify proteins from the ms/ms spectra of tryptic digests, and each spectrum was checked separately. Criteria for acceptance of a protein were the presence of at least two peptides with Mascot scores in the homology/identity range [34]. Molecular weight search (MOWSE) scores, representing probabilities for correct assignment of proteins, ranged from 3181 to 147 in Table 1 and from 4103 to 118 in Table 2; a significant ($p < 0.05$) match is defined as a score of 60 or higher. Among the proteins identified by immunological testing (Figures 7D–F) as constituents of the red cell polymers, the proteomics data, summarized in Tables 1 and 2, confirmed the presence of spectrin, ankyrin, band 3, band 4.1, adducin (adducin

TABLE 1

Proteins in the X1 and X2 polymers of whole ghosts, identified by mass spectrometry analysis of tryptic peptides.

gi Number	Whole Ghost, X1 Polymer	Score	Peptide Count	GI Number	Whole Ghost, X2 Polymer	Score	Peptide Count
338438	Alpha spectrin	3479	189	119573202	Alpha spectrin	4188	135
134798	Beta spectrin	2678	152	134798	Beta spectrin	3181	115
226788	Ankyrin	2461	139	226788	Ankyrin	2234	93
68563369	Band 3	1406	109	68563369	Band 3	1390	57
119628069	Band 4.1	595	27	119628069	Band 4.1	580	19
115502394	Glucose transporter	392	14	115502394	Glucose transporter	469	10
29826323	Adducin 1 (alpha)	372	11	37588869	Ring finger protein	424	7
28175764	ADD1 protein	338	11	39777597	Transglutaminase 2	388	10
1200184	Stimulator of TAR RNA binding	312	9	95115683	Hemoglobin alpha 1-2 hybrid	312	19
7705925	Dicarbonyl/*L*-xylulose reductase	259	3	438069	Thiol specific antioxidant protein	294	6
19070472	p600	218	27	23268449	Hemoglobin beta	288	10
37588869	Ring finger protein 123	217	12	19070472	p600	278	27
119588998	Importin 7	209	9	4826878	Oxidative stress responsive 1	265	8
4503581	Dematin, band 4.9	202	10	4503581	Dematin, band 4.9	256	13
4502011	Adenylate kinase 1	197	12	4505237	Palmitoylated membrane protein 1	238	7
39777597	Transglutaminase 2	178	7	29826323	Adducin 1 (alpha)	219	8
438069	Thiol specific antioxidant protein	171	8	7705925	Dicarbonyl/*L*-xylulose reductase	217	6
68533125	ACLY variant protein	138	5	68533125	ACLY variant protein	212	14
4378804	Hemoglobin beta	139	6	4502011	Adenylate kinase 1	159	8
4557395	Carbonic anhydrase	127	4	6164624	F-box protein Fbx7	155	3
6225268	Protein diaphonous homolog 1	125	6	4827050	Ubiquitin specific protease 14	150	6
				4557014	Catalase	147	6

The gi number is the accession number in the NCBI protein database for the protein sequence.

A score greater than 60 indicates a significant match (<0.05). The peptide count is the number of times a peptide is found for a particular protein; a finding of two peptides is sufficient to identify a protein with high confidence, provided that the peptides have a combined score of 60, at least.

TABLE 2

Proteins in the X1 and X2 polymers of alkali-stripped ghosts, identified by mass spectrometry analysis of tryptic peptides.

gi Number	Alkali-Stripped Ghost, X1 Polymer	Score	Peptide Count	GI Number	Alkali-Stripped Ghost, X2 Polymer	Score	Peptide Count
338438	Alpha spectrin	4073	138	338438	Alpha spectrin	5145	152
134798	Beta spectrin	2815	113	134798	Beta spectrin	4103	115
226788	Ankyrin	2317	128	226788	Ankyrin	2854	92
68563369	*Band 3	1963	207	68563369	*Band 3	1727	103
119628069	Band 4.1	709	37	119628069	Band 4.1	570	27
115502394	*Glucose transporter	594	28	115502394	*Glucose transporter	459	17
95115683	Hemoglobin alpha 1-2 hybrid	325	13	114619118	Dematin, band 4.9	327	12
29826323	Adducin 1 (alpha)	313	8	39777597	Transglutaminase 2	307	6
114619118	Dematin, band 4.9	260	12	3114508	Hemoglobin alpha	280	14
25573100	*Rhesus blood group CcEe antigen ces	242	7	25573100	*Rhesus blood group CcEe antigen ces	255	5
4378804	Hemoglobin beta	223	9	4929993	Hemoglobin beta	251	6
37588869	Ring finger protein 123	204	5	29826323	Adducin 1 (alpha)	214	7
39777597	Transglutaminase 2	204	8	4502011	Adenylate kinase 1	193	10
4502011	Adenylate kinase 1	189	13	7705925	Dicarbonyl/*L*-xylulose reductase	193	4
4505237	Palmitoylated membrane protein 1	172	5	4826878	Oxidative-stress responsive 1	188	4
119587177	*hCG1980844, isoform CRA_i	146	12	14249955	F-box protein 7	185	7
				4505237	Palmitoylated membrane protein 1	167	5
				4758012	Clathrin heavy chain 1	138	8
				119572126	DDI1, DNA-damage inducible 1	119	5
				3211975	Putative glialblastoma cell differentiation-related protein	118	5

The gi number is the accession number in the NCBI protein database for the protein sequence.
A score greater than 60 indicates a significant match (<0.05). The peptide count is the number of times a peptide is found for a particular protein; a finding of two peptides is sufficient to identify a protein with high confidence, provided that the peptides have a combined score of 60 or higher. Transmembrane proteins are marked by an asterisk (*).

1 alpha and ADD1 are nearly identical proteins), and hemoglobin peptides. However, glycophorin, as detected by immunostaining (Figure 7D), was not found by proteomics, whereas catalase, recognized by antibodies, was found by the proteomics screen only in the largest polymer of whole ghosts preparation (X2, Table 1). On the other hand, proteomics furnished evidence that transglutaminase itself was incorporated into the polymers. Importantly, proteomics revealed some previously overlooked membrane skeletal components, notably dematin (also known as band 4.9 or protein 4.9), glucose transporter 1 (GLUT1), and the palmitoylated membrane protein 1 (also known as p55 or MPPI), as substrates for cross-linking by the enzyme.

VI. INFERENCES FROM THE ANALYSIS OF POLYPEPTIDE COMPONENTS OF γ:ε-BONDED CORES OF MEMBRANE POLYMERS FOUND IN HUMAN ERYTHROCYTES WITH Ca^{2+} OVERLOAD

The band 4.2 protein and TG2 both belong to the transglutaminase gene family and share considerable sequence similarities [35, 36], except that the Cys/His/Asp/Trp residues of the catalytic tetrad—essential for transamidating enzyme activity in TG2 [23]—is missing from band 4.2. Potentially, they compete for the same binding site at the *C*-terminal domain of the anion transporter band 3 protein [37]. Nevertheless, while TG2 was incorporated into the polymers of Ca^{2+}-enriched cells (Tables 1 and 2, and Figure 8B1), its close membrane skeletal protein relative, band 4.2, was not (Tables 1 and 2, and Figure 8B10).

Contrary to observations with immunoblotting (Figure 7D, Gp and Figures 8B9, B11, and B12), the proteomics data do not support the presence of glycophorins and stomatin in the polymers. We have no explanation for this discrepancy, except to note that earlier immunoblotting experiments also failed to identify glycophorin as a transglutaminase substrate [31] and to suggest that hydrophobic or glycosylated single-pass membrane proteins might be difficult to detect by proteomics approaches [38]. If significant proportions of these proteins had been incorporated into the polymers, a greater reduction of staining of their monomeric forms would have been expected than observed in Figures 8B9–B11. In spite of a report that actin was an amine-incorporating substrate for TG2 in mouse erythrocytes [28], there is no evidence that this protein is involved in the cross-linking reaction taking place in human red cells (Tables 1 and 2). Properties and reactions of the rodent erythrocyte membrane are known to differ significantly from those of human cells [39]. Moreover, it should be noted that amine incorporation merely identifies only potential rather than actual cross-linking substrates of transglutaminases in cells. Prior data also showed that the staining of actin band (at around 40 kDa) on the SDS-PAGE of membrane protein profiles of human erythrocytes did not change with Ca^{2+} treatment of the cells [10, 13, 25, 29, 30].

It is not unreasonable to assume that the proteomics peptide counts are proportional to the monomeric masses that make up the covalently linked membrane polymers. This allows estimating the relative abundance of building blocks in the high-molecular-weight materials in the Ca^{2+}-loaded erythrocytes. The normalized peptide count, adjusted for the molecular mass of the smallest constituent—i.e., alpha hemoglobin—among the twenty different protein constituents of the X2 polymer (Table 2), indicates that a structure containing at least one copy of all monomers would have a total mass of about 9×10^6 Da and would be made up of 93 units. A similar calculation for the X1 polymer in Table 2 would yield an approximate mass of 11×10^6 Da of 110 polypeptide monomers. Such estimates exceed the masses of actually detected erythrocyte polymers (Figure 7C). Hence, as illustrated in Figure 9,

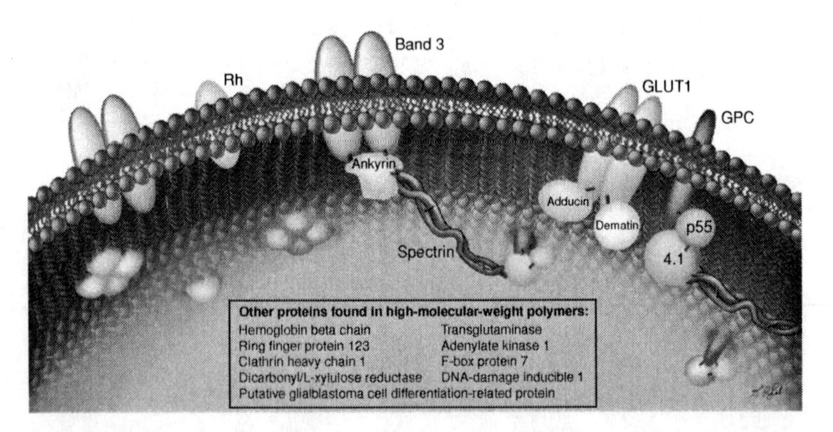

FIGURE 9. Illustration of a section of the inner surface of alkali-stripped ghosts from Ca^{2+}-loaded erythrocytes indicating scattered clusters of protein polymers, covalently linked to transmembrane proteins or to p55, which—through its palmitoyl moiety—is partially embedded in the lipid bi-layer. The $N^{\varepsilon}(\gamma$-glutamyl)lysine side chain bridges between constituent polypeptide chains of the polymer are shown by red lines. Without remodeling of the membrane skeleton (as in the left portion of the picture), only the cytoplasmic domains of proteins such as band 3, GLUT1, the Rhesus blood group CcEe antigen, and p55 are seen. (See insert for color representation.)

we suggest that a variety of smaller cross-linked clusters are produced, each of which—in case of the alkali-stripped materials—would be covalently anchored to an intrinsic membrane protein, mainly band 3, but also GLUT1 and the Rhesus blood group CcEe antigen or to p55 that, through its fatty acyl moiety, is partially embedded in the lipid bilayer. Band 3, previously identified as a major target for cross-linking by TG2 [13, 40], accounts for about 33% of monomers that make up the X1 polymers and about 16% of monomers in the X2 polymers, confirming that band 3 is mainly responsible for anchoring the polymers covalently into the membrane.

The proteomics data indicate that the reaction of TG2 in the human erythrocyte is rather specific because, from the myriad of proteins in the cells, only the relatively few listed in Tables 1 and 2 seem to serve as substrates for the Ca^{2+}-triggered remodeling of the membrane skeleton. Approximately half of the polypeptide components that make up the covalent polymeric products are well-known membrane constituents, such as the band 3 anion transporter, α and β spectrins, ankyrin, band 4.1, GLUT1, dematin, adducin, and p55. On the other hand, one should also note an essentially total absence of glycolytic enzymes. In spite of the fact that aldolase and glycerylaldehyde-3-phosphate dehydrogenase (GAPDH) are known to be functionally associated with band 3 [41–46] and GAPDH is an *in vitro* substrate for TG2 [47], neither was found to be incorporated into the cross-linked polymers.

Concerning the total number of polypeptide monomers built into the isopeptide-linked clusters of the cell, the best guideline seems to be the TG2-mediated disappearance of band 4.1, judged by the SDS-PAGE profiles of Ca^{2+}-treated human erythrocytes. Diminution of the staining intensity of band 4.1 on the electrophoretic protein profiles has long been considered to be a sensitive sign of TG2 activity in Ca^{2+}-loaded human cells, in comparison to controls [10,13] (see also Figure 6). Since there are 200,000 copies of band 4.1 per human erythrocyte [43] and if all of these became partners in the cross-linking reaction, approximately 20 million copies of polypeptides of various types per cell would be built into the X1 and X2 polymeric superstructures.

While the combination of immunologic and proteomic screens revealed the composition of the γ:ε-bonded cores of peripheral and membrane-anchored polymers in Ca^{2+}-loaded erythrocytes, there is no analytical methodology for identifying nearest neighbors for the arrangement of the original polypeptide building blocks within the polymer matrix. It is unlikely that the structure would arise as a collection of homopolymers, comprised of polyTG2, polyband 3, polyband 4.1, and other polypeptides. In this context, it may be mentioned that in the *in vitro* reaction of TG2 with purified spectrin as a substrate, mostly cross-linked dimers and virtually no higher order spectrin polymers were found. Thus, we suggest that TG2 functions by catalyzing the tight attachment of monomers to monomers according to the noncovalent binding partnerships in which the individual units were arranged within the membrane skeletal framework of the cell (Figure 9).

The enzymatic cross-linking reaction would merely freeze existing assemblies by introducing zero-distance $N^{\varepsilon}(\gamma$-glutamyl)lysine protein-to-protein cross-links without inserting extra mass between the participating Gln and Lys residues that are in close contact. Other cytoplasmic proteins, such as Hb, would then be recruited more slowly to the structure. Altogether, the process would be similar to the final step of blood coagulation whereby a prior assembly of fibrin molecules into half-staggered arrays of overlapping filaments—forming the urea or monochloroacetic acid soluble, noncovalent provisional clot structure—is necessary for efficient cross-linking by factor XIIIa [48]. However, the erythrocyte membrane polymer clusters would have a more complicated, branched geometry.

While the peripheral membrane proteins dematin and adducin were not previously detected in the proteomics profile of alkali-stripped vesicles from human erythrocytes [38], the present study shows that dematin and adducin, along with GLUT1, participate in forming the TG2-mediated high-molecular-weight material (Tables 1 and 2). This observation implies that these proteins must be in close contact to be amenable for cross-linking by TG2, a view that fits in with the suggestion that dematin and adducin provide an alternate mechanism for anchoring the spectrin–actin junctions to the plasma membrane via GLUT1 in human erythrocytes [38].

VII. CONCLUSION

A. THE ERYTHROCYTE CELL DEATH PROGRAM IN HB KOLN DISEASE AND IN OTHER ANEMIAS

This review examines the biochemical basis of the premature death of erythrocytes occurring in anemias such as Hb Koln, one of the unstable hemoglobin diseases. The condition is characterized by an irreversible change from the normal discocyte to echinocyte shape and also by a stiffening of the cell membrane (loss of plasticity or loss of membrane deformability). These altered physical properties are considered to be the cause of splenic entrapment and removal of the abnormal cells from the circulation

[16, 17]. As described in Section II, the membrane skeletal structure of Hb Koln cells is also modified by abundant, protein-to-protein $N^{\varepsilon}(\gamma$-glutaminyl)lysine crossbridges that are the footprints of transglutaminase action [18]. Hence, it is concluded that TG2—a latent enzyme present in red cells in an inactive state [10, 11, 13]—becomes activated in the Hb Koln erythrocytes and, based on the model studies presented in Section III, it appears to control a unique death program. Though change of cell shape and membrane rigidification [15] may be analogous to some events of apoptosis, inasmuch as human erythrocytes are devoid of nuclei and mitochondria, the biochemical pathway leading to early cell death in Hb Koln disease and in similar erythrocyte disorders must be different from apoptosis. The TG2-led program does not seem to fit in with any of the categories currently listed among the various cell death modalities [49]; it may be regarded perhaps as a truncated, short segment of the apoptotic process ("meroptosis"?).

Experimental evidence supports the notion that GTP/GDP is the prime, if not the only, physiological allosteric regulator of TG2 [50, 51]. Since the ATP concentration is significantly lower than normal in Hb Koln cells [16], the inhibitory GTP/GDP concentration must also be considerably reduced, that is, TG2 would be activated more easily than that in normal cells. TG2 activation is brought about by elevation of the intracellular concentration of Ca^{2+} ions (see Section III and also [52]). In erythrocytes, the entry and exit of Ca^{2+} are controlled by ion pumps [53], and the influx is also facilitated by the shear stress on the cells in the circulation [54]. It is not known whether Ca^{2+} accumulates in Hb Koln cells as they become energy depleted over their very brief lifespans, lasting only 31 days [16], or whether the cation is delivered in distinct pulses over time. Nor is it known what concentration of Ca^{2+} would trigger the TG2-mediated membrane skeletal remodeling reaction in the Hb Koln cells with the reduced GTP content. The Ca^{2+} requirement for eliciting half-maximal velocity of TG2 activity in normal hRBC lysates, as measured with the extraneous *N,N'*-dimethylcasein:putrescine substrate pair, is about 0.3 mM [13]. This figure is in the range of the total Ca^{2+} concentration found in sickle cells: 0.1–0.3 mmoles per liter of packed sickle cells versus 0.016–0.039 mM in normal [55]. Indeed, $N^{\varepsilon}(\gamma$-glutaminyl)lysine-bonded protein polymers could also be isolated from the erythrocyte membranes of a sickle cell patient (Section II; [21]).

Model experiments with normal erythrocytes—exposed to Ca^{2+} stress—offer a remarkably good reproduction of the events that lead to the premature death of the abnormal Hb Koln and sickle cells. Influx of Ca^{2+} triggers the TG2-mediated covalent cross-linking of the membrane skeleton [10, 11, 13, 29, 30]—and, as illustrated in Section III B1, also some degree of proteolysis [25, 29, 30]—that, in turn, cause an irreversible change in cell shape and loss of membrane plasticity [15]. In order to speed up the TG2-mediated remodeling of membrane skeletal structures and to secure maximal response for incorporating as many proteins as possible into the polymeric products, the outside concentration of Ca^{2+} was set to fully saturate the enzyme (1.5–3 mM, i.e., 5–10 times of half-saturation requirement).

The polypeptide compositions of two categories of N^{ε}(γ-glutaminyl)lysine-bonded, SDS–DTT-soluble cores of membrane skeletal polymers (peripheral and integrally anchored) are discussed in Section V and VI. Though the work with Hb Koln cells revealed that the SDS–DTT-insoluble material (see Figure 2) contains the highest frequency of N^{ε}(γ-glutaminyl)lysine cross-links [18], this compartment has not yet been studied by the methodologies employed for obtaining the data presented in Figures 7 and 8 and in Tables 1 and 2. In pathological situations, the intensity and duration of the Ca^{2+} stress is expected to be less severe; hence, it is to be expected that the covalent polymeric clusters—produced by TG2 action—would probably be smaller, containing fewer of the polypeptide building blocks that participate in the enzymatic reaction as cross-linking substrates in the model experiment.

Finally, note should be made of the observations that competitive and noncompetitive inhibitors of transglutaminase (histamine, cystamine, and 2-[3(diallylamino)propionyl]benzothiophene) can be employed to block the protein chemical remodeling of the membrane skeleton [13, 30] as well as fixation of abnormal cell shape and loss of membrane plasticity [15], induced by the Ca^{2+} overload in erythrocytes (see Section I and III). This finding provided the necessary "proof of principle" that transglutaminase activity can be interfered with inside erythrocytes and, also potentially, in other cell types; it carries far-reaching therapeutic implications. In relation to anemias, the effectiveness of a nontoxic inhibitor of TG2 in Hb Koln disease might conceivably be equivalent to what is currently accomplished by surgical splenectomy in lengthening erythrocyte survival by about 50% [16, 17].

ACKNOWLEDGMENTS

This article is dedicated to the memory of Dr. Joyce Bruner-Lorand who contributed significantly to the early phase of our human erythrocyte research project [10, 11, 29]. We also thank Ms. Deanna Rybak for illustrating

Figure 9. This work was partly supported by grants from thc National Institutes of Health (HL-051445 and HL-095050 to AHC) and from the US Army Medical Research and Materiel Command (W81XWH-07-2-0034 to OL).

REFERENCES

1. Lorand, L., Dailey, J. E., and Turner, P. M. (1988) Fibronectin as a carrier for the transglutaminase from human erythrocytes, *Proc. Natl. Acad. Sci. USA 85*, 1057–1059.
2. Turner, P. M. and Lorand, L. (1989) Complexation of fibronectin with tissue transglutaminase, *Biochemistry 28*, 628–635.
3. LeMosy, E. K., Erickson, H. P., Beyer, W. F. Jr., Radek, J. T., Jeong, J. M., Murthy, S. N., and Lorand, L. (1992) Visualization of purified fibronectin-transglutaminase complexes, *J. Biol. Chem. 267*, 7880–7885.
4. Hang, J., Zemskov, E. A., Lorand, L., and Belkin, A. M. (2005) Identification of a novel recognition sequence for fibronectin within the NH2-terminal beta-sandwich domain of tissue transglutaminase, *J. Biol. Chem. 280*, 23675–23683.
5. Radek, J. T., Jeong, J. M., Murthy, S. N., Ingham, K. C., and Lorand, L. (1993) Affinity of human erythrocyte transglutaminase for a 42-kDa gelatin-binding fragment of human plasma fibronectin, *Proc. Natl. Acad. Sci. U S A 90*, 3152–3156.
6. Akimov, S. S. and Belkin, A. M. (2001) Cell-surface transglutaminase promotes fibronectin assembly via interaction with the gelatin-binding domain of fibronectin: a role in TGFbeta-dependent matrix deposition, *J. Cell Sci. 114*, 2989–3000.
7. Murthy, S. N. and Lorand, L. (2000) Nucleotide binding by the erythrocyte transglutaminase/Gh protein, probed with fluorescent analogs of GTP and GDP, *Proc. Natl. Acad. Sci. USA 97*, 7744–7747.
8. Murthy, S. N., Velasco, P. T., and Lorand, L. (1998) Properties of purified lens transglutaminase and regulation of its transamidase/crosslinking activity by GTP, *Exp. Eye Res. 67*, 273–281.
9. Murthy, S. N., Lomasney, J. W., Mak, E. C., and Lorand, L. (1999) Interactions of G(h)/transglutaminase with phospholipase Cdelta1 and with GTP, *Proc. Natl. Acad. Sci. USA 96*, 11815–11819.
10. Lorand, L., Weissmann, L. B., Epel, D. L., and Bruner-Lorand, J. (1976) Role of the intrinsic transglutaminase in the Ca^{2+}-mediated crosslinking of erythrocyte proteins, *Proc. Natl. Acad. Sci. USA 73*, 4479–4481.
11. Lorand, L., Weissmann, L. B., Bruner-Lorand, J., and Epel, D. L. (1976) Role of Intracellular transglutaminase in Ca^{2+}-mediated cross-linking of erythrocyte-membrane proteins, *Biol. Bull. 151*, 419.
12. Anderson, D. R., Davis, J. L., and Carraway, K. L. (1977) Calcium-promoted changes of the human erythrocyte membrane. Involvement of spectrin, transglutaminase, and a membrane-bound protease, *J. Biol. Chem. 252*, 6617–6623.
13. Siefring, G. E. Jr., Apostol, A. B., Velasco, P. T., and Lorand, L. (1978) Enzymatic basis for the Ca^{2+}-induced cross-linking of membrane proteins in intact human erythrocytes, *Biochemistry 17*, 2598–2604.

14. Coetzer, T. L. and Zail, S. S. (1979) Cross-linking of membrane proteins of metabolically depleted and calcium-loaded erythrocytes, *Br. J. Haematol. 43*, 375–390.
15. Smith, B. D., La Celle, P. L., Siefring, G. E. Jr., Lowe-Krentz, L., and Lorand, L. (1981) Effects of the calcium-mediated enzymatic cross-linking of membrane proteins on cellular deformability, *J. Membr. Biol. 61*, 75–80.
16. Miller, D. R., Weed, R. I., Stamatoyannopoulos, G., and Yoshida, A. (1971) Hemoglobin Koln disease occurring as a fresh mutation: erythrocyte metabolism and survival, *Blood 38*, 715–729.
17. Pedersen, P. R., McCurdy, P. R., Wrightstone, R. N., Wilson, J. B., Smith, L. L., and Huisman, T. H. (1973) Hemoglobin Koln in a black: pre- and post-splenectomy red cell survival (DF32P and 51Cr) and the pathogenesis of hemoglobin instability, *Blood 42*, 771–781.
18. Lorand, L., Michalska, M., Murthy, S. N., Shohet, S. B., and Wilson, J. (1987) Cross-linked polymers in the red cell membranes of a patient with Hb-Koln disease, *Biochem. Biophys. Res. Commun. 147*, 602–607.
19. Shen, L. and Lorand, L. (1983) Contribution of fibrin stabilization to clot strength. Supplementation of factor XIII-deficient plasma with the purified zymogen, *J. Clin. Invest. 71*, 1336–1341.
20. Lorand, L. (2005) Factor XIII and the clotting of fibrinogen: from basic research to medicine, *J Thromb. Haemost. 3*, 1337–1348.
21. Lorand, L., Siefring, G. E. Jr., and Lowe-Krentz, L. (1980) Ca^{2+}-triggered and enzyme mediated cross-linking of membrane proteins in intact human erythrocytes. In: *Membrane Transport in Erythrocytes*, eds Lassen, U. V., Ussing, H. H., and Weith, J. O. Munksgaard, Copenhagen; Vol. *14*, pp. 285–299.
22. Lorand, L. and Conrad, S. M. (1984) Transglutaminases, *Mol. Cell. Biochem. 58*, 9–35.
23. Lorand, L. and Graham, R. M. (2003) Transglutaminases: crosslinking enzymes with pleiotropic functions, *Nat. Rev. Mol. Cell Biol. 4*, 140–156.
24. Lorand, L. (2001) Factor XIII: structure, activation, and interactions with fibrinogen and fibrin, *Ann. NY Acad. Sci. 936*, 291–311.
25. Lorand, L., Bjerrum, O. J., Hawkins, M., Lowe-Krentz, L., and Siefring, G. E. Jr. (1983) Degradation of transmembrane proteins in Ca^{2+}-enriched human erythrocytes. An immunochemical study, *J. Biol. Chem. 258*, 5300–5305.
26. Glaser, T. and Kosower, N. S. (1986) Calpain-calpastatin and fusion. Fusibility of erythrocytes is determined by a protease–protease inhibitor [calpain–calpastatin] balance, *FEBS Lett. 206*, 115–120.
27. Kosower, N. S., Glaser, T., and Kosower, E. M. (1983) Membrane-mobility agent-promoted fusion of erythrocytes: fusibility is correlated with attack by calcium-activated cytoplasmic proteases on membrane proteins, *Proc. Natl. Acad. Sci. USA 80*, 7542–7546.
28. Sarang, Z., Madi, A., Koy, C., Varga, S., Glocker, M. O., Ucker, D. S., Kuchay, S., Chishti, A. H., Melino, G., Fesus, L., and Szondy, Z. (2007) Tissue transglutaminase (TG2) facilitates phosphatidylserine exposure and calpain activity in calcium-induced death of erythrocytes, *Cell Death Differ. 14*, 1842–1844.

29. Lorand, L. and Michalska, M. (1985) Altered response of stored red cells to Ca^{2+} stress, *Blood 65*, 1025–1027.
30. Lorand, L., Barnes, N., Bruner-Lorand, J. A., Hawkins, M., and Michalska, M. (1987) Inhibition of protein cross-linking in Ca^{2+}-enriched human erythrocytes and activated platelets, *Biochemistry 26*, 308–313.
31. Bjerrum, O. J., Hawkins, M., Swanson, P., Griffin, M., and Lorand, L. (1981) An immunochemical approach for the analysis of membrane protein alterations in Ca^{2+}-loaded human erythrocytes, *J. Supramol. Struct. Cell. Biochem. 16*, 289–301.
32. Moroi, M., Inoue, N., and Yamasaki, M. (1975) Analysis of the fibrin-polymerizing reaction using sodium dodecylsulfate-agarose gel electrophoresis, *Biochim. Biophys. Acta 379*, 217–226.
33. Korponay-Szabo, I. R., Vecsei, Z., Kiraly, R., Dahlbom, I., Chirdo, F., Nemes, E., Fesus, L., and Maki, M. (2008) Deamidated gliadin peptides form epitopes that transglutaminase antibodies recognize, *J. Pediatr. Gastroenterol. Nutr. 46*, 253–261.
34. Perkins, D. N., Pappin, D. J., Creasy, D. M., and Cottrell, J. S. (1999) Probability-based protein identification by searching sequence databases using mass spectrometry data, *Electrophoresis 20*, 3551–3567.
35. Korsgren, C. and Cohen, C. M. (1991) Organization of the gene for human erythrocyte membrane protein 4.2: structural similarities with the gene for the a subunit of factor XIII, *Proc. Natl. Acad. Sci. USA 88*, 4840–4844.
36. Sung, L. A., Chien, S., Chang, L. S., Lambert, K., Bliss, S. A., Bouhassira, E. E., Nagel, R. L., Schwartz, R. S., and Rybicki, A. C. (1990) Molecular cloning of human protein 4.2: a major component of the erythrocyte membrane, *Proc. Natl. Acad. Sci. USA 87*, 955–959.
37. Gutierrez, E. and Sung, L. A. (2007) Interactions of recombinant mouse erythrocyte transglutaminase with membrane skeletal proteins, *J. Membr. Biol. 219*, 93–104.
38. Khan, A. A., Hanada, T., Mohseni, M., Jeong, J. J., Zeng, L., Gaetani, M., Li, D., Reed, B. C., Speicher, D. W., and Chishti, A. H. (2008) Dematin and adducin provide a novel link between the spectrin cytoskeleton and human erythrocyte membrane by directly interacting with glucose transporter-1, *J. Biol. Chem. 283*, 14600–14609.
39. Iismaa, S. E., Mearns, B. M., Lorand, L., and Graham, R. M. (2009) Transglutaminases and disease: lessons from genetically engineered mouse models and inherited disorders, *Physiol. Rev. 89*, 991–1023.
40. Murthy, S. N., Wilson, J., Zhang, Y., and Lorand, L. (1994) Residue Gln-30 of human erythrocyte anion transporter is a prime site for reaction with intrinsic transglutaminase, *J. Biol. Chem. 269*, 22907–22911.
41. Tsai, I. H., Murthy, S. N., and Steck, T. L. (1982) Effect of red cell membrane binding on the catalytic activity of glyceraldehyde-3-phosphate dehydrogenase, *J. Biol. Chem. 257*, 1438–1442.
42. Murthy, S. N., Liu, T., Kaul, R. K., Kohler, H., and Steck, T. L. (1981) The aldolase-binding site of the human erythrocyte membrane is at the NH2 terminus of band 3, *J. Biol. Chem. 256*, 11203–11208.

43. Kliman, H. J. and Steck, T. L. (1980) Association of glyceraldehyde-3-phosphate dehydrogenase with the human red cell membrane. A kinetic analysis, *J. Biol. Chem. 255*, 6314–6321.
44. Strapazon, E. and Steck, T. L. (1976) Binding of rabbit muscle aldolase to band 3, the predominant polypeptide of the human erythrocyte membrane, *Biochemistry 15*, 1421–1424.
45. Strapazon, E. and Steck, T. L. (1977) Interaction of the aldolase and the membrane of human erythrocytes, *Biochemistry 16*, 2966–2971.
46. Chu, H. and Low, P. S. (2006) Mapping of glycolytic enzyme-binding sites on human erythrocyte band 3, *Biochem. J. 400*, 143–151.
47. Orru, S., Ruoppolo, M., Francese, S., Vitagliano, L., Marino, G., and Esposito, C. (2002) Identification of tissue transglutaminase-reactive lysine residues in glyceraldehyde-3-phosphate dehydrogenase, *Protein Sci. 11*, 137–146.
48. Samokhin, G. P. and Lorand, L. (1995) Contact with the N termini in the central E domain enhances the reactivities of the distal D domains of fibrin to factor XIIIa, *J. Biol. Chem. 270*, 21827–21832.
49. Kroemer, G., Galluzzi, L., Vandenabeele, P., Abrams, J., Alnemri, E. S., Baehrecke, E. H., Blagosklonny, M. V., El-Deiry, W. S., Golstein, P., Green, D. R., Hengartner, M., Knight, R. A., Kumar, S., Lipton, S. A., Malorni, W., Nunez, G., Peter, M. E., Tschopp, J., Yuan, J., Piacentini, M., Zhivotovsky, B., and Melino, G. (2009) Classification of cell death: recommendations of the Nomenclature Committee on Cell Death 2009, *Cell Death Differ. 16*, 3–11.
50. Begg, G. E., Carrington, L., Stokes, P. H., Matthews, J. M., Wouters, M. A., Husain, A., Lorand, L., Iismaa, S. E., and Graham, R. M. (2006) Mechanism of allosteric regulation of transglutaminase 2 by GTP, *Proc. Natl. Acad. Sci. USA 103*, 19683–19688.
51. Pinkas, D. M., Strop, P., Brunger, A. T., and Khosla, C. (2007) Transglutaminase 2 undergoes a large conformational change upon activation, *PLoS Biol. 5*, e327, 2788–2796.
52. Bergamini, C. M., Signorini, M., and Poltronieri, L. (1987) Inhibition of erythrocyte transglutaminase by GTP, *Biochim. Biophys. Acta 916*, 149–151.
53. Andrews, D. A., Yang, L., and Low, P. S. (2002) Phorbol ester stimulates a protein kinase C-mediated agatoxin-TK-sensitive calcium permeability pathway in human red blood cells, *Blood 100*, 3392–3399.
54. Swislocki, N. I. and Tierney, J. M. (1985) Covalent modification of membrane components during erythrocyte aging, *Prog. Clin. Biol. Res. 195*, 195–211.
55. Eaton, J. W., Skelton, T. D., Swofford, H. S., Kolpin, C. E., and Jacob, H. S. (1973) Elevated erythrocyte calcium in sickle cell disease, *Nature 246*, 105–106.

IRREVERSIBLE INHIBITORS OF TISSUE TRANSGLUTAMINASE

JEFFREY W. KEILLOR
NICOLAS CHABOT
ISABELLE ROY
AMINA MULANI
OLIVIER LEOGANE
CHRISTOPHE PARDIN

CONTENTS

Advances in Enzymology and Related Areas of Molecular Biology, Volume 78.
Edited by Eric J. Toone.

I. INTRODUCTION

Tissue transglutaminase (TG2) is a Ca^{2+}-dependent enzyme that catalyses the cross-linking of proteins through a transpeptidation reaction involving acyl transfer from the γ-carboxamide group of a peptide-bound glutamine residue ("donor substrate") to the ε-amino group of a peptide-bound lysine residue ("acceptor substrate") [1, 2]. Studies based on the use of small molecule substrate analogs have demonstrated that the catalytic cycle proceeds through a modified ping–pong mechanism [1, 3, 4]. According to this mechanism, the active site thiol nucleophile of the free enzyme is initially acylated by the donor substrate. The resulting acyl-enzyme then binds and reacts with the nucleophilic acceptor substrate, leading to transamidation (Scheme 1). This mechanism has been confirmed for protein substrates as well [5]. Primary amines can also serve as acceptor substrates in TG2-mediated transamidation reactions, whereas in the absence of a suitable primary amine, hydrolysis of the acyl-enzyme thiolester occurs, leading to glutamine deamidation [2].

Many different activity assays have been developed for TG2-mediated processes, allowing detection of deamidation [6], esterolysis [4], transamidation [7], and transpeptidation [8]. On the basis of the comparison of rate constants measured by these different methods, it is apparent that acylation is rapid relative to both the hydrolysis and aminolysis modes of deacylation [4, 5]. TG2 is thus remarkably efficient at catalyzing the formation of its thiolester acyl-enzyme intermediate. This can perhaps be attributed to the activation of the γ-carboxamide group or stabilization of the tetrahedral

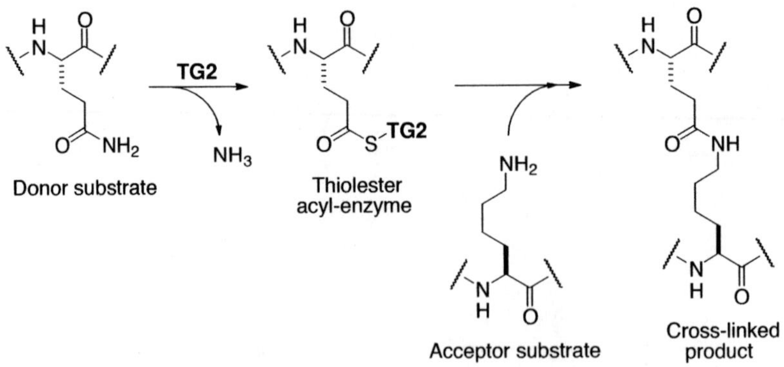

SCHEME 1. Transglutaminase-mediated protein cross-linking.

intermediate formed upon nucleophilic attack by the active site thiolate [9, 10]. Additionally, studies with donor substrate analogs of varied electrophilicity have confirmed the notable nucleophilicity of the active site thiolate [11]. This feature of TG2 has been exploited in the design of many of the irreversible inhibitors presented below, most of which react as affinity labels of the active site thiolate.

To provide a context for the application of irreversible inhibitors of TG2, it is instructive to mention some physiological roles, although detailed reviews can be found elsewhere [12–14]. TG2 has been shown to participate in endocytosis [15, 16], apoptosis [17], and cell growth regulation [18]. TG2 is mostly a cytosolic protein, but it can also be secreted from the cell. One of its most important biological roles is to catalyze the cross-linking of the extracellular matrix (ECM), thus making it less susceptible to proteolytic degradation [19]. TG2 displays some substrate selectivity, but many proteins present glutamine and lysine residues that can participate in TG2-mediated cross-linking. Adventitious cross-linking related to unregulated TG2 activity has been implicated in a number of physiological disorders, such as acne [20], the formation of cataracts [21], immune system diseases [22], psoriasis [23], Alzheimer's disease [24, 25], Huntington's disease [26, 27], Celiac disease [28], and cancer metastasis [29, 30].

Given the broad and growing interest in transglutaminases, many inhibitors have been prepared for this class of enzymes, recently reviewed by Siegel and Khosla [31]. Herein, we will focus on the irreversible inhibitors that have been developed and their application as mechanistic probes either for interrogation of the biological roles of TG2 or for their therapeutic potential.

In general terms, irreversible inhibitors can be characterized as affinity labels or as mechanism-based inactivators. Affinity labels are characterized by their ability to react directly with an active site residue, forming a stable bond with the enzyme, resulting in its inactivation (Scheme 2). By way

$$\mathrm{E} + \mathrm{I} \underset{k_{-1}}{\overset{k_1}{\rightleftharpoons}} \mathrm{E{\cdot}I} \xrightarrow{k_{inact}} \mathrm{E\text{-}I}$$

$$K_I = \frac{k_{-1}}{k_1}$$

SCHEME 2. Kinetic scheme for irreversible inhibition by enzyme affinity labels.

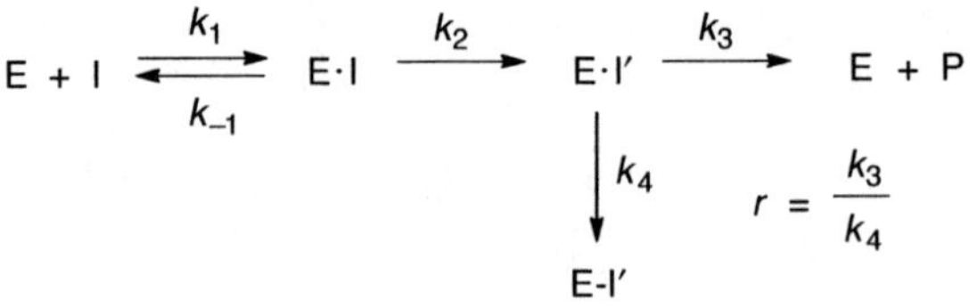

$$E + I \underset{k_{-1}}{\overset{k_1}{\rightleftharpoons}} E\cdot I \xrightarrow{k_2} E\cdot I' \xrightarrow{k_3} E + P$$

$$E\cdot I' \xrightarrow{k_4} E\text{-}I' \qquad r = \frac{k_3}{k_4}$$

SCHEME 3. Kinetic scheme for irreversible inhibition by suicide substrates.

of contrast, mechanism-based or “suicide” inactivation requires that the bound ligand must first be activated by the enzyme, in a step that resembles the mechanism of its native reaction, to generate an activated species E•I′ (Scheme 3). This activated intermediate can then react in either one of two ways. Further reaction as a substrate analog leads to regeneration of the active enzyme (k_3), whereas reaction in a different manner, often with an adjacent active site residue, leads to the formation of a stable covalent adduct resulting in enzyme inactivation (k_4). The ratio of the rate constants of these two reactions ($r = k_3/k_4$) provides a partitioning constant for the common intermediate E•I′, representing its intrinsic efficiency as an inhibitor [32, 33].

The vast majority of TG2 inhibitors presented herein can thus be classified as affinity labels. The efficiency of an affinity label is typically expressed as the ratio k_{inact}/K_I, derived from the hyperbolic Kitz and Wilson equation, analogs to the way that k_{cat}/K_M is used to express the efficiency of a substrate [34]. For example, a k_{inact}/K_I ratio greater than 10^6 M^{-1} min^{-1} may be considered to denote a highly efficient irreversible inhibitor. The separate kinetic parameters that characterize irreversible inhibitors, k_{inact} and K_I, can be measured in a number of different manners, either discontinuously, by measuring the residual activity of enzyme incubated with inhibitor, or in a continuous manner (Scheme 4), by incubating enzyme and inhibitor in the presence of substrate [34–36]. As kinetic constants, these parameters

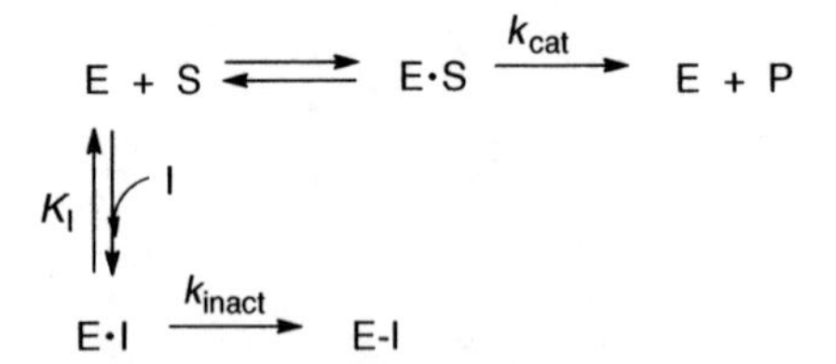

SCHEME 4. Kinetic scheme for continuous evaluation of irreversible inhibition by affinity labels.

can be used as a basis of comparison of the affinity and reactivity of different inhibitors, provided that important differences in experimental conditions (such as pH, temperature, and enzyme source) are duly taken into account. The irreversible inhibitors developed for TG2 have been compared on this basis.

II. IRREVERSIBLE INHIBITORS

A. INHIBITOR DESIGN

Affinity labels are often designed as donor substrate mimics, bearing functional groups that are sufficiently electrophilic to react with the highly nucleophilic active site thiolate, without being so reactive that they are subject to degradation in solution. Several functional groups have emerged in this regard. Given the phenomenological similarity between the mechanism of TG2-mediated transamidation and protease-mediated amide hydrolysis, it is instructive to consider the functional groups developed as "warheads" for the irreversible inhibition of proteases [37]. Of these, the cysteine proteases are the most pertinent to this review, since they possess the same catalytic Cys-His residues as the transglutaminases. Upon this comparison, it is unsurprising to note that epoxides, halomethyl ketones, diazomethyl ketones, and Michael acceptors, all of which have been used extensively as cysteine protease inhibitors [37], have also proven to be effective against TG2.

In addition to the "warhead" that characterizes each irreversible inhibitor, the scaffold bearing this functional group is another characteristic design feature. In general, irreversible inhibitors of TG2 can be categorized as having either (1) little to no framework, (2) a dipeptide scaffold, or (3) an extended peptide structure. For each class of irreversible inhibitor presented below, scaffold structure will also be analyzed.

B. HALOMETHYL CARBONYL DERIVATIVES

Iodoacetamide (Figure 1) is one of the simplest affinity labels used to irreversibly inhibit thiol-dependent enzymes such as cysteine proteases [38].

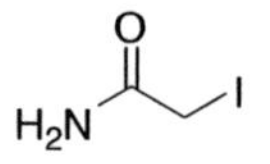

O
H_2N I

FIGURE 1. Structure of iodoacetamide.

SCHEME 5. Alkylation of TG2 by halomethyl ketones.

Although it is relatively stable in aqueous solution, its iodide leaving group can be displaced easily by a nucleophilic active site thiolate (Scheme 5). As such, it was one of the first irreversible inhibitors tested against TG2. Folk showed that iodoacetamide is an effective inhibitor of transglutaminase purified from guinea pig liver [39]. The reactivity of iodoacetamide with guinea pig liver TG2 was more recently studied in greater detail in the Keillor group [7]. Using a photometric transamidation assay, the irreversible inhibition was determined in a continuous fashion [35], providing an inhibition constant, $K_I = 75$ nM, an inactivation rate constant, $k_{inact} = 0.90$ min^{-1}, and an efficiency constant $k_{inact} = 12$ μM^{-1} min^{-1}. Normally, the simple structure of iodoacetamide precludes extensive interactions with an enzyme active site, allowing it to be used as a nonspecific inhibitor of cysteine-dependent enzymes. In this context, the nanomolar inhibition constant measured by de Macédo et al. is notable. Presumably, the simple structure of iodoacetamide resembles the side chain of the Gln donor substrate sufficiently to allow favorable interactions within the active site tunnel.

Subsequently, the same research group prepared a small series of chloromethyl amides [40]. As shown in Table 1, these were based on the simple dipeptide scaffold derived from Cbz-Phe and reported earlier by Khosla et al. [29] (see below). Kinetic parameters were measured for inhibition of recombinant guinea pig liver TGase, using a continuous chromogenic assay method, and are shown in Table 1. It is apparent from these

TABLE 1
Chloromethyl Amide Inhibitors of TG2 [40]

Structure	n	k_{inact} (min^{-1})	K_I (μM)	k_{inact}/K_I (M^{-1} min^{-1})
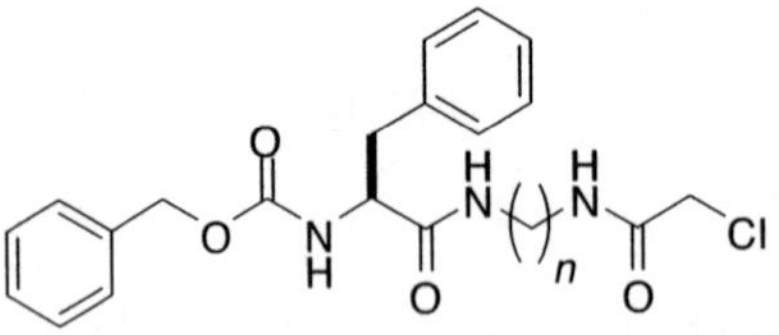	4	0.76	73	1.0×10^5
	6	0.32	6	5.3×10^5

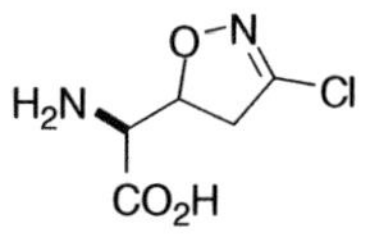

FIGURE 2. Structure of acivicin.

parameters that the chloromethyl amide group reacts nearly as quickly as iodoacetamide, having k_{inact} values of the same order of magnitude. The micromolar K_I values of the Cbz-Phe dipeptide scaffold also confirm its significant affinity for TG2, even though they are 2–3 orders of magnitude higher than that of iodoacetamide. Furthermore, the efficiency of the inhibition reaction ($k_{inact}/K_I > 10^4\ M^{-1}\ min^{-1}$) is ~6 orders of magnitude higher than the second order rate constant for the reaction with glutathione, attesting to the selectivity of the reaction of the inhibitors with TG2 rather than with adventitious thiol.

C. 3-HALO-4,5-DIHYDROISOXAZOLES

Following iodoacetamide, one of the next classes of compounds to be used as irreversible transglutaminase inhibitors were 3-halo-4,5-dihydroisoxazoles [41]. These inhibitors were designed as analogs of acivicin (Figure 2), a natural product that is considered to be a glutamine isostere and known to inhibit cysteine-dependent enzymes (Scheme 6) [42].

Initial derivatives were incorporated into a Cbz-Phe dipeptide framework and tested against epidermal TGase [41] as well as TG2 [43]. In the latter study, the authors used solid state NMR to characterize the imino thioether adduct formed upon irreversible inhibition by 3-halo-4,5-dihydroisoxazoles (Scheme 6). Later, Choi et al. extended the use of this class of inhibitors, preparing a large series of 3-bromo analogs of the original Cbz-Phe dipeptide inhibitors, and testing them against recombinant human TG2 [29], using a continuous coupled enzyme assay [6, 28]. Some of the most potent members of this extensive series are shown in Table 2.

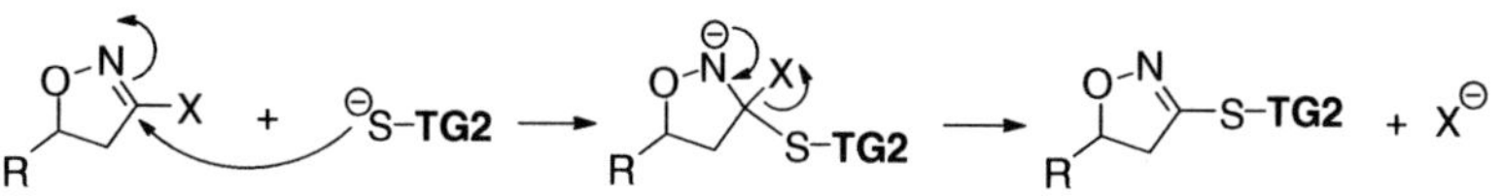

SCHEME 6. Inhibition of TG2 by 3-halo-4,5-dihydroisoxazole analogs.

TABLE 2
3-Bromo-4,5-Dihydroisoxazoles Inhibitors of TG2 [29]

Structure	R_1	R_2	k_{inact} (min^{-1})	K_I (mM)	k_{inact}/K_I (M^{-1} min^{-1})
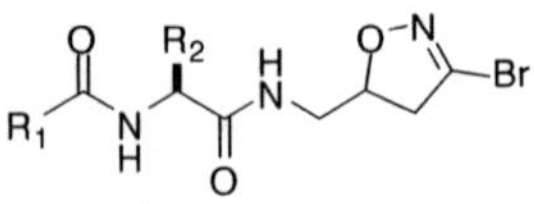	OBn	Bn	1.3	0.74	1.9×10^3
	OBn	$CH_2C_6H_4$-*p*-OH	0.86	0.42	2.0×10^3
	CH_3	Bn	0.60	2.7	2.2×10^2
	OBn	CH_3	0.41	0.91	4.5×10^2
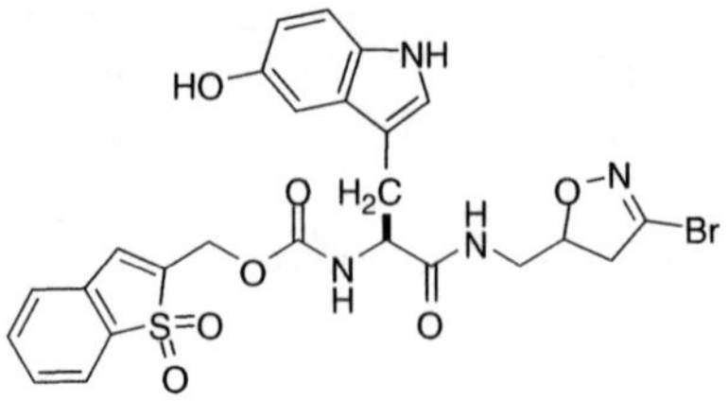			0.54	0.079	6.8×10^3

Analysis of the structure–activity profiles of these inhibitors reveals the importance of a hydrophobic aromatic group on the side chain of the central amino acid and on its *N*-terminal protecting group. The first entry in Table 2 demonstrates the efficacy of the inhibitor based on the parent Cbz-Phe scaffold or the analogous Cbz-Tyr scaffold. Replacing the bulky Cbz group with a small acetyl group caused a marked decrease in efficiency. Subsequent variation of the amino acid side chain revealed that alanine showed reduced activity, indicating a bulky side chain is also critical to affinity. Replacement of the benzyl group by dioxobenzothiophenyl and replacement of the central amino acid by a tryptophan derivative gave the most potent inhibitor in this category. Furthermore, upon testing these inhibitors for their reaction with glutathione, to assess their expected stability in physiological systems, it was found that all compounds were stable in the presence of glutathione, with the exception of the dioxobenzothiophene derivatives that decomposed to give multiple products.

Later, the same research group showed that only the 5-(*S*) stereoisomer of the dihydroisoxazole is active against TG2 and provided definitive mass spectrometry evidence that this activity takes place in the enzyme active site [44]. Further variation of both the amino acid side chain and the *N*-terminal protecting group revealed that quinolyl-protected 5-fluoroindole derivatives proved to be the most effective inhibitors (Table 3), presumably due to putative hydrogen bonding interactions formed within the substrate-binding

TABLE 3
3-Bromo-4,5-Dihydroisoxazole Tryptophan Inhibitors of TG2 [44]

Structure	Configuration	X	k_{inact} (min^{-1})	K_I (mM)	k_{inact}/K_I (M^{-1} min^{-1})
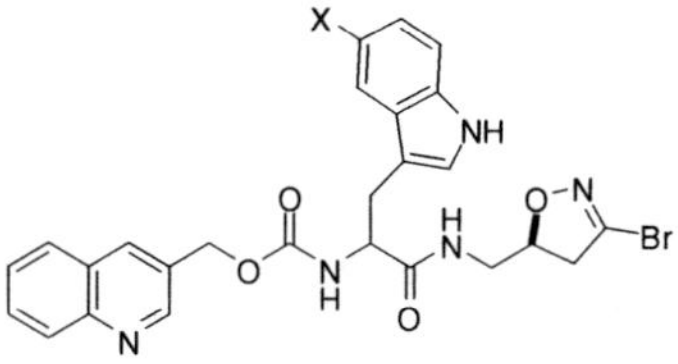 X, NH, O, N H, H N, O-N, Br, N	*R/S*	F	0.186	0.018	1.0×10^5
	S	F	0.189	0.004	4.5×10^5
	S	H	0.072	0.0013	5.7×10^5

site. The configuration of the central amino acid was also shown to be important. Over the course of the work presented in these two publications, the authors were able to increase the efficiency of their dihydroisoxazole inhibitors by a factor of ~30 over the original Cbz-Phe derivative.

The same group also prepared an extended peptide inhibitor (Figure 3), in which acivicin was substituted into a high-affinity peptide sequence [45]. The sequence PQPQLPY was previously identified as a peptide fragment of gluten protein that serves as a high-affinity substrate for TG2 [46]. Evaluation of this inhibitor as described above yielded kinetic parameters of k_{inact} = 0.014 min^{-1}, K_I = 0.780 mM, and k_{inact}/K_I = 18 M^{-1} min^{-1}. While this inhibitor may be 100-fold more efficient than unmodified acivicin, it remains relatively inefficient compared to inhibitors based on simpler dipeptide scaffolds (Tables 2 and 3). This suggests that upon binding of the extended heptapeptide backbone, the side chain dihydroisoxazole group of this specific inhibitor may not be optimally positioned to react with the active site thiol.

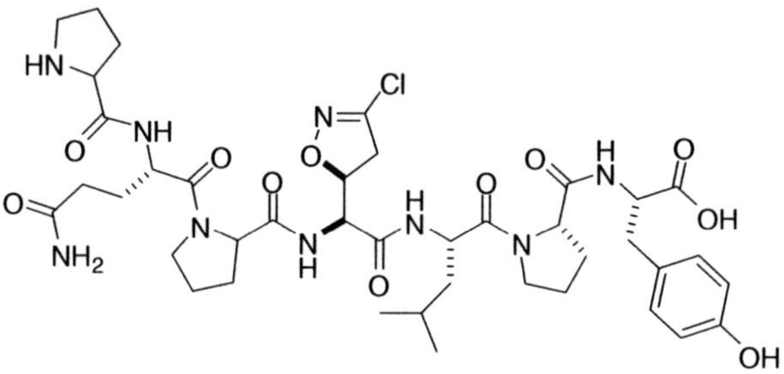

FIGURE 3. Extended peptide acivicin analog [45].

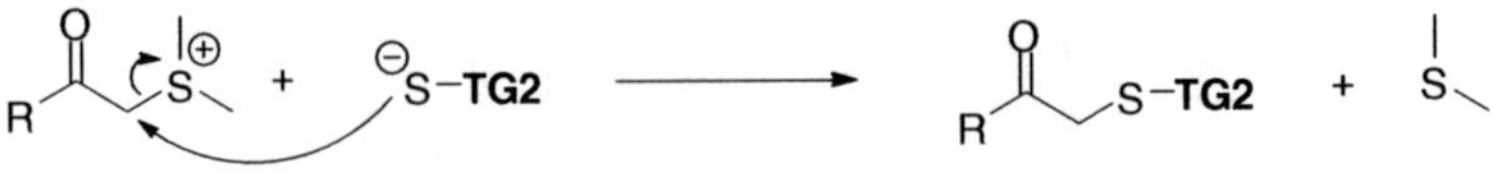

SCHEME 7. Dimethylsulfonium ketone inhibition of TG2.

D. SULFONIUM METHYL KETONES

Pliura et al. have prepared a series of inhibitors related to the halomethyl ketones, but bearing a dimethyl sulfonium group as a leaving group, rather than a halide (Scheme 7). Such warheads have been used successfully on cysteine proteases [37], so the authors explored the effect of the peptidic scaffold on the selectivity of inhibition of cathepsin B, papain, epidermal transglutaminase, and TG2 [47]. The design of their inhibitors was based on the familiar Cbz-Phe scaffold, to which the electrophilic warhead was attached via a spacer of varied length, as shown in Table 4.

The inhibitor from this work that demonstrates the greatest efficiency against TG2 comprises a γ-aminobutyric acid (GABA) spacer ($n = 2$, Table 4). It is noteworthy that this inhibitor also showed excellent selectivity, in its time-dependent inactivation of cathepsin B. Moreover, it is interesting to compare these results with those obtained for the chloromethyl ketone inhibitors shown in Table 1, which are based similarly on the Cbz-Phe scaffold and comprise spacers of variable lengths. In both cases, it was noted that inhibition efficiency increased with spacer length, although differences in the conformational nature of the spacers precludes a more quantitative comparison. This qualitative comparison of the most efficient inhibitors suggests that the hydrophobic, aromatic end may be bound by TG2 in a

TABLE 4
Dimethylsulfonium Ketone Inhibitors of TG2 [47]

Structure	n	k_{inact}/K_I (M^{-1} min^{-1})
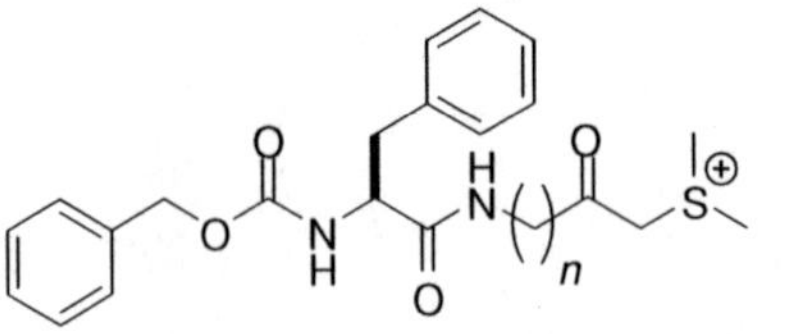	1	1.0×10^3
	2	1.54×10^4
	3	2.95×10^4
	4	1.90×10^4
	5	4.5×10^3

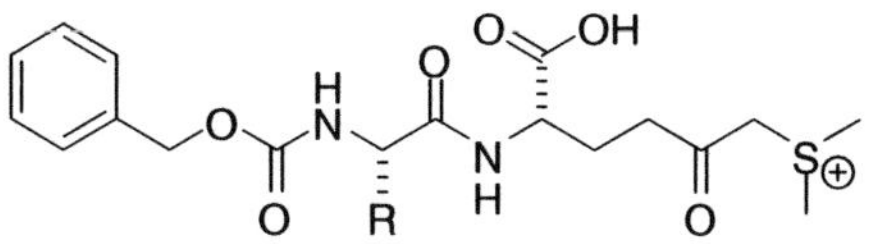

R = Bn, IC_{50} = (10 ± 2) μM
R = H, IC_{50} = (8 ± 1) μM

FIGURE 4. Dimethylsulfonium ketone TG2 inhibitor [48].

subsite distant from the active site, such that a spacer of sufficient length is required to allow reaction of the electrophilic warhead with the active site thiol.

Interestingly, this was also the conclusion drawn by the authors of a more recent article presenting similar sulfonium-based TG2 inhibitors. (As it will be shown below, this structure–activity relationship is common to other classes of inhibitors, as well, suggestive of a common binding mode.) Griffin et al. recently reported the dimethylsulfonium ketone inhibitors shown in Figure 4 [48]. They are similar to those shown in Table 4, but have higher aqueous solubility. Their potency was determined by incubation for 1 hour with either guinea pig or recombinant human TG2, followed by discontinuous evaluation of residual enzyme activity according to a plate-based transamidation assay [49] and reported as IC_{50} values (Figure 4). Irreversibility was confirmed by dilution experiments and by confirming the requirement of calcium for inhibition. Unfortunately, more precise kinetic parameters were not reported, making it difficult to compare the potency of these inhibitors to the others presented in this review.

E. THIOIMIDAZOLIUM DERIVATIVES

In 1991, the 2-[(2-oxopropyl)thio]imidazolium derivative known as L682777 (Table 5, R = CH_3) was reported as an inactivator of hepatocyte TG2, although no kinetic data were reported at the time. Later, the same group showed that L682777 also inhibits factor XIIIa and erythrocyte TGase, with efficiency constants of $k_{inact}/K_I = 3.8$ and 1.6 μM^{-1} min^{-1}, respectively [50]. Although limited structure–activity information was reported, the remarkable efficiency is noteworthy. L682777 is similar in structure and activity to iodoacetamide ($k_{inact}/K_I = 12$ μM^{-1} min^{-1} [7]), but it distinguishes itself in terms of its selectivity. The acetonyl group of L682777

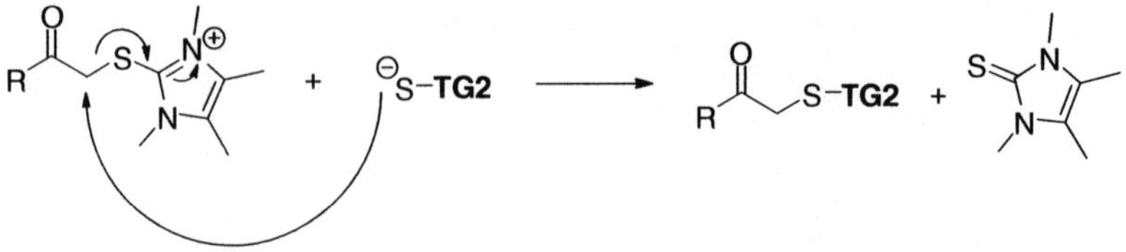

SCHEME 8. Inactivation of TG2 by thioimidazolium methyl ketones.

is less activated than the methylene group of iodoacetamide, such that the latter reacts 10^3-fold faster with glutathione [50]. While iodoacetamide still reacts 10^3- to 10^5-fold more efficiently with transglutaminases than with glutathione, L682777 reacts 10^7-fold more quickly with transglutaminases [50] (Scheme 8).

Later, Khosla et al. followed up on these promising preliminary results obtained with L682777, by incorporating the tetramethyl thioimidazolium warhead into dipeptide and extended peptide scaffolds [45]. These inhibitors, shown in Figure 5, were evaluated kinetically against human TG2, using a continuous coupled enzyme assay [6]. In this case, the peptide scaffolds did

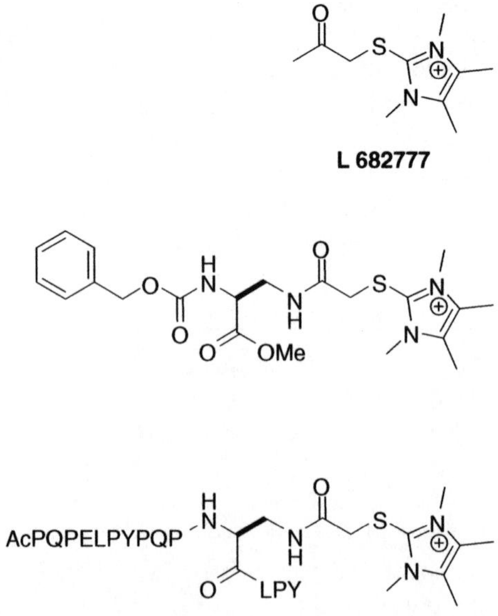

L 682777

FIGURE 5. Thioimidazolium ketone inhibitors of TG2 [45].

not confer greater affinity for TG2, but rather had the opposite effect. Given the reduced specificity of these inhibitors, relative to L682777, they were not characterized further.

F. 1,2,4-THIADIAZOLES

After 1991, the next class of inhibitors to be developed against TG2 were 1,2,4-thiadiazole derivatives [51, 52]. These heterocyclic warheads have been reported to react selectively with cysteine proteases, relative to other proteases, in either a reversible or irreversible fashion [37]. The mode of inhibition depends on the reactivity of the thiadiazole, which depends in turn on the nature of the substituent at the C-3 position (Scheme 9). X-ray crystallography of a papain-inhibitor complex revealed the formation of a disulfide bond, consistent with the proposed ring opening mechanism shown in Scheme 9 [52].

Given the selectivity of the thiadiazole warhead for cysteine nucleophile enzymes, leading to the formation of a stable disulfide bond, the Keillor group reasoned that thiadiazole inhibitors selective for TG2 could be developed through the design of an appropriate scaffold, to be substituted at C-5 (R, Scheme 9) [53]. The scaffold chosen for this work was based on the dipeptide sequence of Cbz-Gln-Gly, a commonly used substrate. Two synthetic strategies were used to attach the thiadiazole moiety, through either an amine or amide bond, to the amino acid serving as a glutamine mimic. The length of the side chain spacer and the sequence of the peptide backbone were also varied. A total of 14 compounds were prepared and tested against TG2 purified from guinea pig livers, using a continuous transamidation assay [7]. The five most interesting inhibitors of this series are shown in Table 5.

Cursory analysis of the kinetic parameters measured for these inhibitors shows the similarity of their efficiency. As many authors have observed, there appears to be a compensatory effect between affinity and reactivity of irreversible TG2 inhibitors. The manifestation of this putative relationship is such that overall efficiency varies little among these inhibitors. The length of the side chain appears to have little effect in the series of 3-methoxy-1,

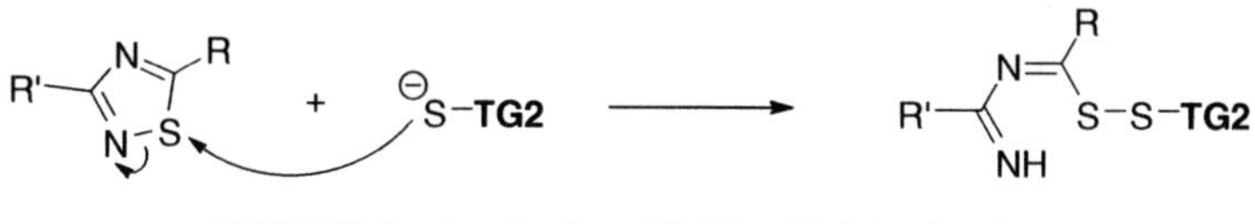

SCHEME 9. Inactivation of TG2 by 1,2,4-thiadiazoles.

TABLE 5
1,2,4-Thiadiazole Inhibitors of TG2 [53]

Structure	n	k_{inact} (min^{-1})	K_I (μM)	k_{inact}/K_I ($M^{-1}\ min^{-1}$)
Cbz-NH-CH(CO-NH-CH₂-CO₂H)-CH₂-(CH₂)$_n$-NH-(3-methoxy-1,2,4-thiadiazol-5-yl)	1	0.88	1.3	7.1×10^5
	2	1.2	2.3	5.5×10^5
	3	1.2	1.7	7.2×10^5
Cbz-NH-CH(CO-NH-CH₂-CO₂H)-CH₂-(CH₂)$_n$-CO-NH-(3-methyl-1,2,4-thiadiazol-5-yl)	1	4.7	14	3.3×10^5
Cbz-NH-CH(CO-OH)-CH₂-(CH₂)$_n$-CO-NH-(3-methyl-1,2,4-thiadiazol-5-yl)	1	0.69	0.77	8.9×10^5

2,4-thiadiazole inhibitors. Although a 100-fold drop in reactivity has been observed on passing from 3-methoxy to 3-methyl thiadiazole inhibitors of cathepsin B [37, 52], no such drop is observed here. No distinctive effect was observed on passing from an amine to an amide side chain linkage, and finally, the terminal glycine residue designed to confer greater affinity to the peptide scaffold appears to have a deleterious effect instead. Even so, the high efficiency of these inhibitors and the reputed selectivity of the thiadiazole group suggest that further work with this warhead may be justified [52].

G. MICHAEL ACCEPTORS

Michael acceptors have proven to be specific for cysteine proteases, as well as one of the most efficient classes of inhibitors, justifying their interest [37]. The efficiency of a Michael acceptor inhibitor typically increases with its reactivity, according to the addition reaction shown in Scheme 10. Activated derivatives such as vinyl sulfones, α,β-unsaturated carbonyl derivatives, and even more highly activated maleimide groups are among the most commonly applied. Inhibitors of the last two categories have been reported by Keillor et al., adopting the same design strategy as for the thiadiazole inhibitors shown above.

Their first Michael acceptor inhibitors of TG2 contained α,β-unsaturated (acryloyl) amides [53]. As can be seen in Table 6, this series of inhibitors was based on a dipeptidic scaffold intended to mimic the structure of Cbz-Gln-Gly, a commonly used substrate. The length of side chain of these inhibitors was varied, as was the sequence of the peptide backbone. The inhibitors were evaluated for their activity against purified guinea pig liver TG2, using a continuous transamidation assay [7]. Notably, all inhibitors were found to be excellent inhibitors of TG2, with second-order rate constants of inactivation (k_{inact}/K_I) around 10^6 M^{-1} min^{-1}. Within the series of inhibitors, few distinctive structure–activity relationships can be seen. All inhibitors showed roughly the same reactivity, as evidenced by their similar k_{inact} values. Furthermore, the K_I values varied very little, despite the significant variation of chain length, from one to four methylene units. In general, the inhibitors having the highest affinity also bore the longest side chains. The most efficient acryloyl inhibitor, Cbz-Lys(Acr)-Gly ($k_{inact}/K_I =$ 3.0 μM^{-1} min^{-1}) also had the highest apparent affinity, with a remarkably low K_I value of 150 nM.

The structure of the most efficient acryloyl inhibitor was also used as the basis of successive generations of irreversible TG2 inhibitors. For example,

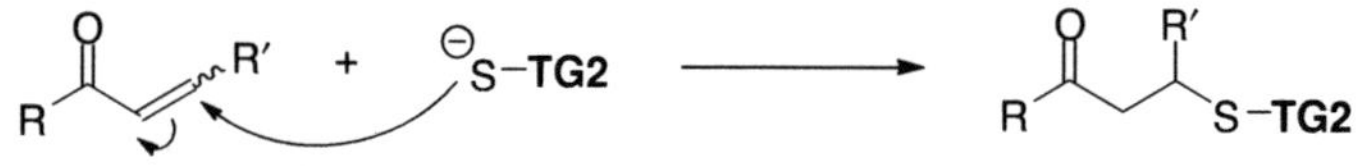

SCHEME 10. TG2 inactivation by Michael acceptors.

TABLE 6
Dipeptide Mimic Acryolyl Inhibitors of TG2 [53]

Structure	n	k_{inact} (min^{-1})	K_I (μM)	k_{inact}/K_I (M^{-1} min^{-1})
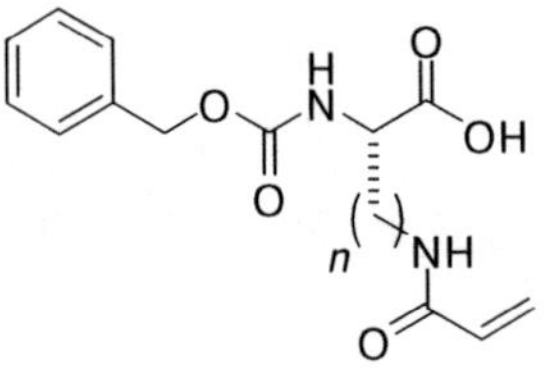 Cbz-NH-CH[(CH$_2$)$_n$-NH-acryloyl]-COOH	1	1.08	1.17	0.92×10^6
	2	0.73	0.81	0.90×10^6
	3	1.33	2.32	0.57×10^6
	4	0.67	0.52	1.28×10^6
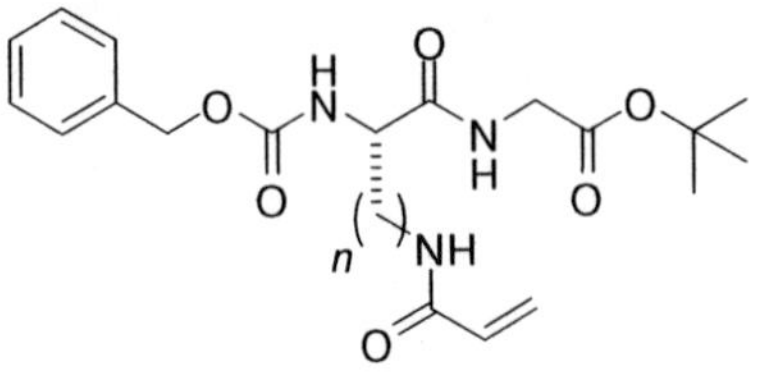 Cbz-NH-CH[(CH$_2$)$_n$-NH-acryloyl]-CO-NH-CH$_2$-CO-O-tBu	1	0.95	1.48	0.64×10^6
	2	0.60	0.51	1.18×10^6
	3	0.75	0.85	0.89×10^6
	4	0.49	0.23	2.20×10^6
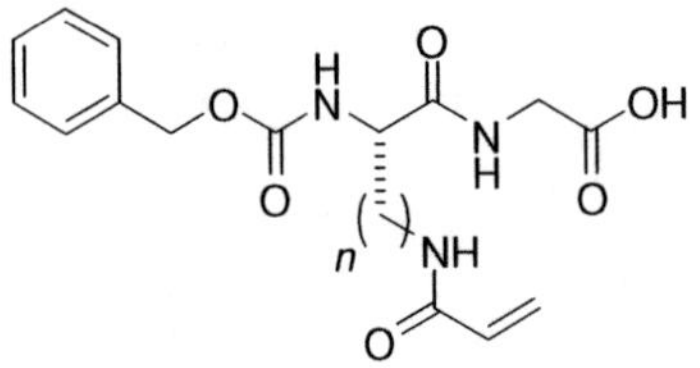 Cbz-NH-CH[(CH$_2$)$_n$-NH-acryloyl]-CO-NH-CH$_2$-COOH	1	1.34	2.75	0.49×10^6
	2	0.59	0.48	1.24×10^6
	3	0.54	0.28	1.95×10^6
	4	0.46	0.15	3.00×10^6

the TG2 inhibitors shown in Figure 6 were prepared by attaching a dansyl group to the Cbz-Lys(Acr) scaffold, via two different spacers [54]. The addition of the spacer and fluorophore resulted in a loss of roughly one order of magnitude of affinity and efficiency, as determined by the same method of kinetic evaluation. Nevertheless, these inhibitors remain reasonably efficient, and their fluorescence bodes well for their application as biological probes (see below).

The Keillor group also incorporated the acryloyl group into another commonly studied peptidic scaffold, namely Cbz-Phe [40]. These inhibitors were evaluated against recombinant guinea pig liver TG2 [55], using continuous chromogenic [4] or fluorogenic [56] activity assays. As shown in Table 7, these inhibitors are also reasonably efficient, displaying ~threefold more efficiency than analogous dimethyl sulfonium ketones that were also evaluated against liver TG2 (Table 4) [47]. Furthermore, comparison of

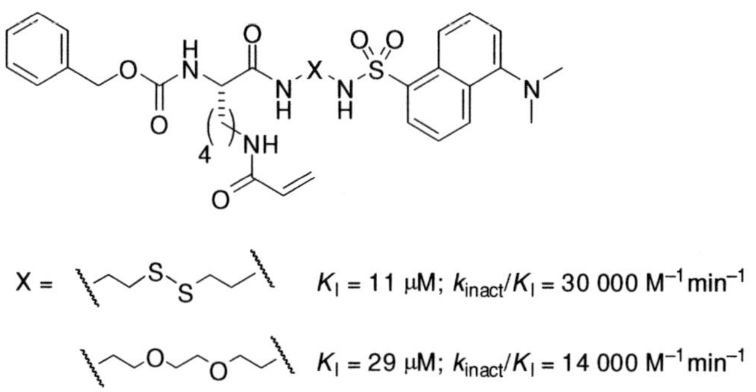

K_I = 11 μM; k_{inact}/K_I = 30 000 $M^{-1}min^{-1}$

K_I = 29 μM; k_{inact}/K_I = 14 000 $M^{-1}min^{-1}$

FIGURE 6. Fluorescent Cbz-Lys-based acryloyl inhibitors of TG2 [54].

the shortest chain derivative of Table 7 to an analogous 3-bromo-4,5-dihydroisoxazole derivative (Table 2) [29] shows that the acryloyl derivative is ~20-fold more active. However, it is important to note that different sources of TG2 were used for these two studies. Finally, the acryloyl derivatives of Table 7 are roughly eightfold more efficient than the analogous chloromethyl ketone derivatives described in the same study (Table 1) [40]. Conversely, the acryloyl inhibitors based on the Cbz-Phe scaffold (Table 7) are still roughly one order of magnitude less efficient than the acryloyl Cbz-Gln-Gly mimics shown in Table 6.

The second category of Michael acceptors that has been studied as TG2 inhibitors are the maleimides. Although initially presented in the same body of work as the acryloyl derivatives above, very inefficient inhibition was observed for the Cbz-Phe, $n = 4$ derivative [40]. Considering the apparent advantage of the Cbz-Gln-Gly scaffold over the Cbz-Phe scaffold, observed for the acryloyl derivatives shown in Tables 6 and 7, the Keillor group prepared a series of inhibitors bearing a maleimide group on the side chain

TABLE 7
Dipeptidic Acryloyl Inhibitors of TG2 [40]

Structure	n	k_{inact} (min^{-1})	K_I (μM)	k_{inact}/K_I (M^{-1} min^{-1})
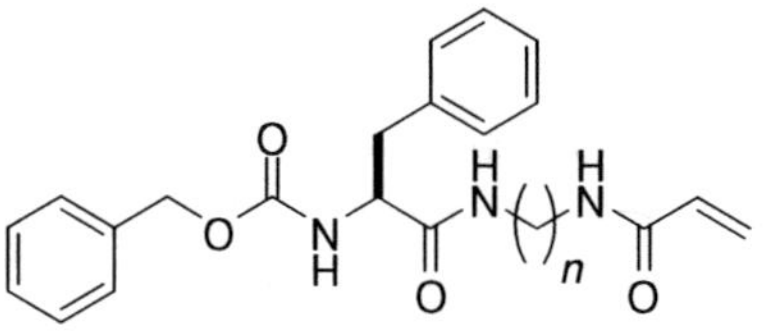	2	0.42	9.0	0.47×10^5
	4	0.36	4.5	0.80×10^5
	6	0.90	13	0.69×10^5
	8	0.38	3.5	1.09×10^5

TABLE 8
Dipeptide Mimic Maleimide Inhibitors of TG2 [57]

Structure	n	k_{inact}/K_I (M^{-1} min^{-1})
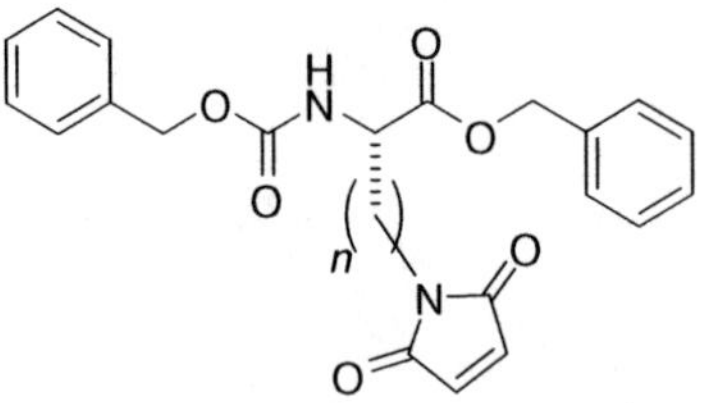	2	0.67×10^3
	3	6.26×10^3
	4	17.1×10^3
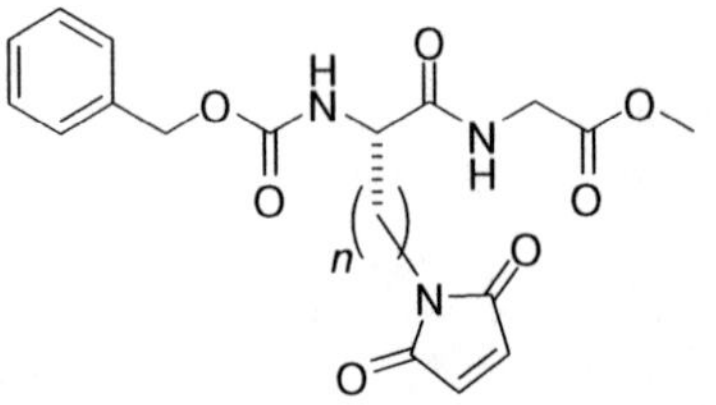	2	0.28×10^3
	3	0.43×10^3
	4	0.83×10^3

of the glutamine mimic [57]. As shown in Table 8, the benzyl esters tested were generally ~2- to 20-fold more efficient than the glycine dipeptides. This may suggest that the benzyl group can form specific interactions with the binding site that has been shown to interact favorably with hydrophobic, aromatic groups. It can also be noted that within each series, longer side chains result in higher efficiency, as has been observed for many other series of derivatives. Finally, it is obvious from the comparison of Tables 6 and 8 that the maleimide derivatives are ~10^3- to 10^4-fold less efficient than their acryloyl analogs. Given the inherently greater reactivity of the maleimide group relative to the acryloyl group, it is reasonable to assume that the lack of efficiency of these maleimide inhibitors is due to their decreased affinity for TG2. It is most likely that this is a result of the considerably larger volume of the maleimide group and the intrinsic barrier to its proper positioning in the sterically restricted environment of the active site tunnel of TG2 [10, 57].

H. EPOXIDES

Several α,β-epoxyketones have been used very successfully as cysteine protease inhibitors, with second-order rate constants of inactivation in the range of 10^5–10^6 M^{-1} min^{-1} [37]. Keillor et al. prepared a series

TABLE 9
Dipeptide Mimic Epoxide Inhibitors of TG2 [53]

Structure	n	k_{inact} (min^{-1})	K_I (μM)	k_{inact}/K_I (M^{-1} min^{-1})
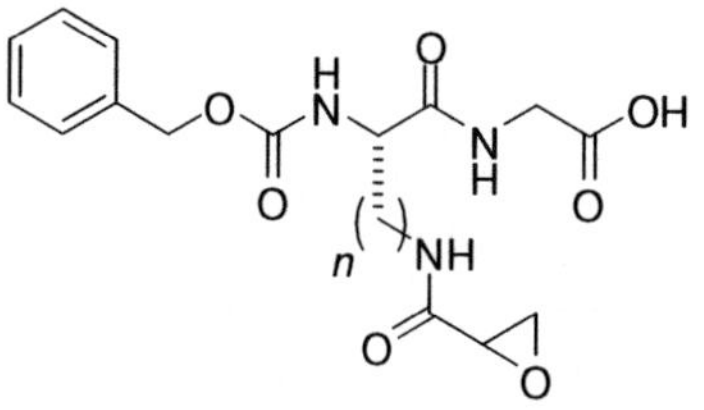	1	1.39	2.13	0.65×10^6
	2	0.61	0.50	1.23×10^6
	3	0.50	0.29	1.72×10^6
	4	0.56	0.28	2.03×10^6
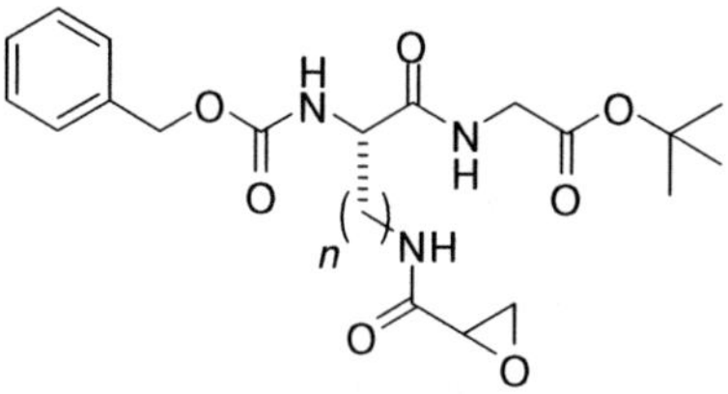	1	2.49	5.91	0.42×10^6
	2	2.02	4.20	0.48×10^6
	3	0.87	0.95	0.91×10^6
	4	1.45	2.42	0.60×10^6
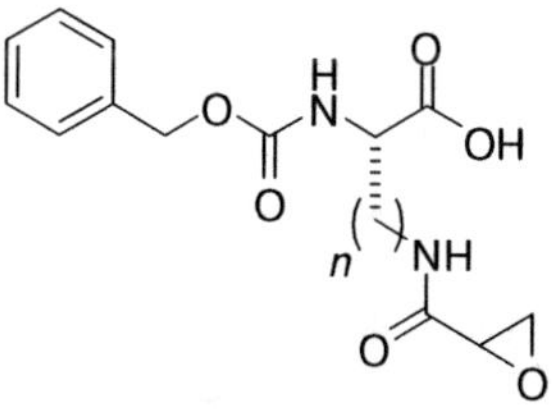	1	0.60	1.10	0.55×10^6
	2	0.81	1.23	0.66×10^6
	3	0.63	0.56	1.12×10^6
	4	0.59	0.59	1.00×10^6

of α,β-epoxyketone TG2 inhibitors by effecting the (racemic) epoxidation of the pendant double bond of the inhibitors shown in Table 6 [58]. The resulting epoxides are shown in Table 9.

Although the mechanism of inactivation of cysteine proteases by α,β-epoxyketones has not been studied in detail, the observed stereoselectivity of the inhibition of cruzain has been interpreted in terms of nucleophilic attack on the penultimate α carbon [37]. In contrast, the model reaction of an epoxide inhibitor with 2-aminoethanethiol led to addition on the terminal β carbon, leading to the proposed mechanism of inactivation shown in Scheme 11 [59].

The inhibitors shown in Table 9 were evaluated against purified guinea pig liver TG2 using a continuous transamidation assay, providing the kinetic

SCHEME 11. Proposed mechanism of TG2 inactivation by α,β-epoxyketones.

parameters shown [59]. All epoxides were excellent inhibitors of TG2, although not quite as efficient as the analogous Michael acceptors shown in Table 6, predominantly owing to higher K_I values. This may be due to a number of reasons, including steric clashes with the additional volume of the epoxide ring in the narrow active site tunnel, an unfavorable inhibitor conformation accompanying the change of hybridization of the pendant carbon atoms or reduced productive binding of the α-carbon epimers prior to what is potentially a stereoselective reaction.

I. DIAZOMETHYL KETONES

One of the most recent classes of irreversible TG2 inhibitors to be studied in detail are the diazomethyl ketones. This class of compounds has been known for some time to function as affinity labels of cysteine proteases. They exhibit reasonable specificity for cysteine-dependent enzymes and are highly efficient, with second-order inactivation rate constants in the range of 10^4–10^7 M^{-1} min^{-1} [37]. Although the mechanism of inactivation has not been studied thoroughly, it is thought to involve both nucleophilic attack by active site thiolate as well as protonation by active site imidazolium, analogous to the native mechanism of acylation of the cysteine proteases and transglutaminases (Scheme 12).

Among the diazomethyl ketones, the naturally occurring glutamine isostere 6-diazo-5-oxo-norleucine (DON) would appear to be well suited as a parent compound for inhibition of enzymes that operate on the side chain of glutamine, such as γ-glutamyl transpeptidase or TG2 [60]. The Khosla group tested DON as an inhibitor of recombinant human TG2, using

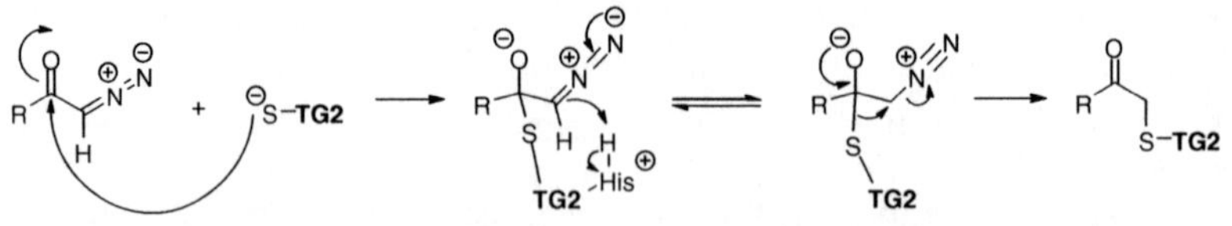

SCHEME 12. Proposed mechanism of inactivation of TG2 by diazomethyl ketones.

TABLE 10
Diazomethyl Ketone Inhibitors of TG2 [45, 61]

Structure	R_1	R_2	k_{inact} (min^{-1})	K_I (M)	k_{inact}/K_I (M^{-1} min^{-1})
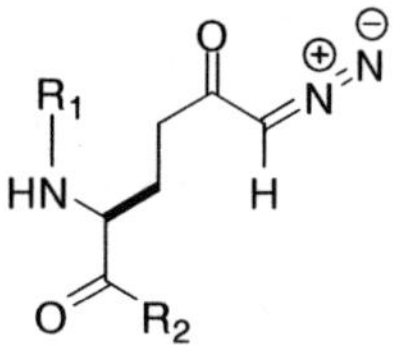	H	OH	0.025	0.13	0.2
	Cbz	OMe	0.12	1.35×10^{-4}	890
	Ac-PQP	LPF-NH_2	0.2	7×10^{-8}	2.9×10^6
	Ac-P	LPF-NH_2	0.5	6×10^{-8}	8.3×10^6

a coupled enzyme activity assay [6] in either continuous or discontinuous fashion [45]. As shown in Table 10 (R_1 = H, R_2 = OH), DON showed poor affinity for TG2 (K_I = 130 mM), but this is not surprising, given the low affinity displayed by TG2 for the unprotected amino acid glutamine. Therefore, the authors incorporated a DON residue as a glutamine mimic in a simple peptide scaffold, leading to an increase in affinity of nearly three orders of magnitude. Encouraged by this result, they then went on to incorporate a DON residue into the high-affinity peptide sequence derived from gluten peptide. The incorporation of a warhead-bearing residue into a peptidic scaffold has not always been a successful strategy and has often had even deleterious effects on affinity. However, in this case, this approach led to a further dramatic decrease in K_I of nearly four orders of magnitude. Interestingly, reactivity increased with increasing affinity, suggesting that DON is a particularly good glutamine mimic and can be properly positioned in active site, upon binding of the rest of the peptide backbone in the donor substrate binding site. The Khosla group would later verify this hypothesis by X-ray crystallography [61], using a slightly truncated version of this inhibitor (Ac-P-DON-LPF-NH_2) whose efficiency was even slightly improved relative to the longer peptide sequence (Table 10).

J. MISCELLANEOUS INHIBITORS

Most recently, Lai et al. have reported the discovery of several irreversible TG2 inhibitors [62]. They screened libraries containing >2000 pharmacologically pertinent compounds as inhibitors of recombinant human TG2, using a fluorescent transpeptidation assay [63]. Nine compounds having IC_{50} values of less than 20 μM were identified in this way, several of

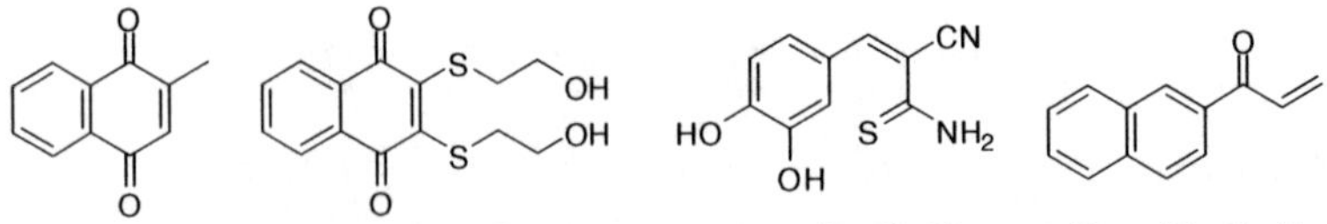

IC_{50} = (2.2 ± 0.8) μM IC_{50} = (1 ± 0.5) μM IC_{50} = (3 ± 3) μM IC_{50} = (5 ± 5) nM

FIGURE 7. Selected potential irreversible inhibitors of TG2 [62].

which are known to inhibit kinases or phosphatases and to target ATP- or GTP-binding sites. The authors reported that some of these inhibitors were irreversible, since activity was not recovered after dialysis, but no kinetic information was provided regarding potential time-dependent inactivation, so no comparison of potency is possible with the other inhibitors described in this review.

The nature of the observed inhibition also remains uncertain. For the structures shown in Figure 7, the authors reported observing inhibition in the absence of calcium, which rules out mechanism-based inhibition and makes active site-directed affinity labeling unlikely. As the authors noted, these compounds are known as kinase or phosphatase inhibitors and/or quinones. Inspection of the structures reveals the potential of these compounds to react as Michael acceptors or as active site thiol oxidizing agents. Indeed, when the authors tested these compounds in the presence of excess DTT, their IC_{50} values were found to increase by several orders of magnitude, effectively abolishing inhibition. The authors interpreted this as an indication of reactivity with active site thiol, but the published data do not rule out the possibility of simple degradation of the inhibitors by DTT itself. Nevertheless, as the authors point out, these simple molecules are potentially useful for further development as scaffolds, once their structure–activity relationships have been studied.

K. SUMMARY

Comparison of the tables of kinetic data presented above allows some generalizations to be made, regarding the efficiency, affinity, and reactivity of the most effective irreversible inhibitors of TG2. For example, of all the irreversible inhibitors that have been kinetically evaluated, very few display second-order inhibition constants over the arbitrarily chosen limit of 10^6 M^{-1} min^{-1}. These include the inhibitors reported by the Keillor group, where acryloyl and epoxide warheads are presented on the side chain of glutamine mimics incorporated in a simple peptidic scaffold, and those reported by the Khosla group, where a diazomethyl ketone warhead is presented on the side chain of the glutamine mimic DON, incorporated into an extended high-affinity peptide scaffold. Some of these inhibitors are shown again in Table 11. The similarity between them underlines the importance of placing the warhead on a side chain of sufficient length to reach down into the active site, within a minimal peptide sequence capable of properly positioning the inhibitor in the donor substrate-binding site.

Of course, the best inhibitors have high k_{inact} values as well as low K_I values, but for most inhibitors studied, there appears to be a compensatory, leveling effect between these two parameters. For this reason, caution must

TABLE 11
Highly Efficient Irreversible TG2 Inhibitors Selected from Tables 1–10

Structure	Ref.	k_{inact} (min^{-1})	K_I (μM)	k_{inact}/K_I (M^{-1} min^{-1})
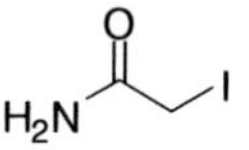 H_2N–C(=O)–CH_2–I	[7]	0.90	0.075	12×10^6
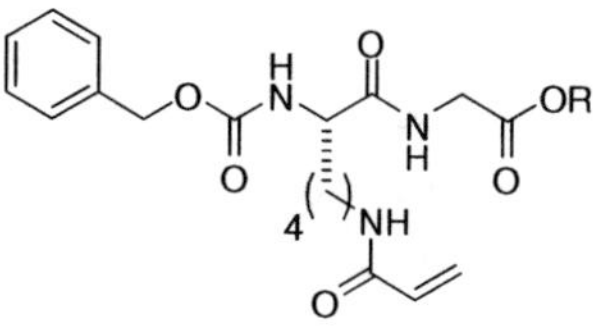 R = H, *t*Bu	[53]	0.46 0.49	0.15 0.23	3.00×10^6 2.20×10^6
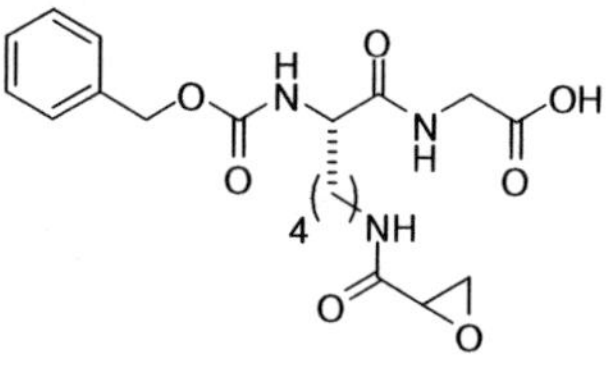	[45]	0.56	0.28	2.03×10^6
R = AcPQP, AcP; 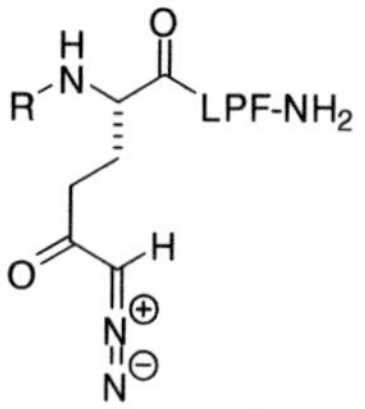 LPF-NH_2	[45, 61]	0.5 0.2	0.06 0.07	8.3×10^6 2.9×10^6

be shown in comparing individual kinetic parameters between classes of inhibitors. However, with this caveat in mind, it is interesting to note that most of the inhibitors reviewed above show similar reactivity, with $k_{inact} = 0.5\text{–}1.0\ min^{-1}$. Apart from the most efficient inhibitors shown in Table 11, others contain warheads that are particularly reactive and deserve mention. For example, certain 3-bromo-4,5-dihydroisoxazoles (Table 2) [29], 1,2,4-thiadiazoles (Table 5) [53], other Michael acceptors (Table 6) [53], and other epoxides (Table 9) [53] have all been shown to have k_{inact} values over 1 min^{-1}, and up to 4.7 min^{-1}. The similarity of the rate constants measured across several classes of electrophilic warheads probably reflects the intrinsic nucleophilicity of the active site thiolate, with the observed differences being due to proper positioning and additional activation by other active site residues.

The K_I values reported in Tables 1–10 vary considerably, but are typically in the low micromolar range for the best inhibitors. It is important to note that in addition to the inhibitors shown in Table 11, others have also shown nanomolar inhibition constants, namely Cbz-Gln-Gly mimics bearing either a 1,2,4-thiadiazole group (Table 5) [53] or an α,β-epoxyketone group (Table 9) [59]. This comparison reveals that irreversible inhibitors having the highest affinity for TG2 all bear warheads on the side chain of donor substrate mimic peptides, rather than on other scaffolds. For example, while the Cbz-Phe scaffold has afforded excellent success among cysteine protease inhibitors, it has not provided the highest affinity inhibitors of TG2.

Finally, it is instructive, if not humbling, to compare the lowly iodoacetamide to all other synthetic TG2 inhibitors. As shown in Table 11, it remains the most efficient irreversible inhibitor of TG2, with a nanomolar K_I value and a k_{inact} value approaching 1 min^{-1}. Of course, while the extended scaffolds of all other inhibitors, so carefully designed and so painstakingly constructed, may not confer additional affinity or reactivity compared to iodoacetamide, they are essential for providing selectivity for TG2, relative to other cysteine-dependent enzymes.

III. APPLICATIONS OF IRREVERSIBLE TG2 INHIBITORS

Some of the irreversible TG2 inhibitors shown above have already been applied to the study of its function. These studies can be loosely sorted into the following categories: in vitro mechanistic studies, studies of the biological roles of TG2 in cells, tissues or organisms, and exploration of the potential therapeutic role of these inhibitors for the treatment of TG2-related pathological conditions.

A. MECHANISTIC PROBES

One of the earliest mechanistic applications of an irreversible TG2 inhibitor was to identify the active site nucleophile. For example, Folk used a radiolabeled derivative of iodoacetamide to identify the active site nucleophile of transglutaminase purified from guinea pig liver [39]. However, in this study, it was noted that when iodoacetamide was used in great excess, it can also react with residues other than the active site cysteine, underlining the importance of designing more selective irreversible inhibitors.

Most recently an irreversible inhibitor designed to be highly selective for TG2 was used in a crystallographic investigation of TG2. Khosla et al. used their diazomethyl ketone inhibitor, based on the high-affinity gluten peptide sequence, to inhibit recombinant human TG2 [61]. They were then able to crystallize the inhibited enzyme, providing the first crystallographic evidence for an extended, "open" conformation. This structure is the only one available for TG2 with a ligand bound in its active site, so it represents a significant advance in our ability to correlate fine structure to catalytic function. For example, the crystallographic evidence for the extended conformation corroborates nicely with spectroscopic [64, 65] and electrophoretic [66, 67] evidence obtained by other investigators. At the very least, the crystallographic structure of TG2 in the open conformation represents the structure of the acyl-enzyme intermediate, if not the active conformation of the free enzyme.

In the context of the design of future inhibitors, it is instructive to study this X-ray structure and to compare it with the empirical structure–activity relationships noted above. For example, it has been noted above, for several classes of irreversible inhibitors, that those having the highest affinity for TG2 all possess an aromatic hydrophobic group remote from the electrophilic group that reacts with the active site thiol. Inspection of the structure of the ligand-bound TG2 structure reveals that the benzyl side chain of the Phe residue is bound in a complementary hydrophobic pocket that is well defined by residues A304, L312, I313, F316, I331, and L420 [61]. This binding pocket may also bind the aromatic hydrophobic groups of the highest affinity inhibitors while allowing their electrophilic groups to be properly positioned to react with the active site thiol. Furthermore, the coordinates of the irreversibly inhibited TG2 (PDB code 2Q3Z) could also be used in docking simulations, to visualize binding interactions or to refine this hypothetical binding model. This may provide a structural rationale, for example, for the preferred distance between the aromatic group and the pendant electrophilic group, observed for several of the classes of inhibitors presented above.

B. THERAPEUTIC POTENTIAL

Several commonly used drugs are irreversible inhibitors [32, 37]. For example, β-lactam antibiotics irreversibly inhibit the bacterial enzyme necessary for construction of its cell wall, and aspirin is an irreversible inhibitor of prostaglandin synthase. Many other enzymes have also been targeted for mechanism-based inactivation, for the treatment of disorders such as multiple sclerosis, cancer, and AIDS [68].

The therapeutic potential of certain irreversible TG2 inhibitors has also been considered. Early studies of thiolimidazolium derivative L682777, presented above, demonstrated its effectiveness as an inhibitor in a variety of different biological contexts [45, 69–71]. However, as mentioned above, L682777 also inhibits factor XIIIa. Since selectivity is a factor that is especially crucial in the application of *irreversible* inhibitors, L682777 and the other thioimidazolium derivatives presented to date are not considered to be appropriate for the treatment of diseases related to TG2 activity [31].

Khosla's DON-containing gluten peptides have also been evaluated in a biological environment. An activity assay based on the inhibition of differentiation of T84 human intestinal epithelial cells demonstrated that the inhibitor Ac-PQP-(DON)-LPF-NH_2 (Figure 8, Tables 10 and 11) is active in cell culture and shows limited cytotoxicity, which bodes well for its potential therapeutic treatment of Celiac sprue [45].

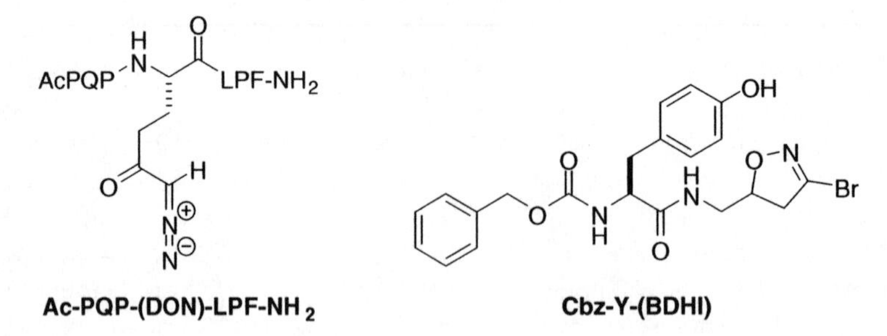

FIGURE 8. Irreversible inhibitors of TG2 tested for their therapeutic potential [29, 45].

Of all known irreversible TG2 inhibitors, those bearing a 3-bromo-4,5-dihydroisoxazole (BDHI) warhead have been evaluated the most extensively for their therapeutic potential [31]. For example, the inhibitor Cbz-Y-(BDHI) (Figure 8, Table 2) was shown to display good oral bioavailability and negligible long-term gross toxicity in mice [29]. Furthermore, this inhibitor proved to be effective in the chemosensitization of glioblastoma tumors. The weights of glioblastoma xenografts were found to decrease significantly and in a dose-dependent manner when the host mice were treated with this dihydroisoxazole inhibitor in combination with the clinically important chemotherapeutic agent *N*,*N*′-bis(2-chloroethyl)-*N*-nitrosourea [29, 72]. This combination therapy was later shown to extend the lifespan of mice with intracranial tumors [73], demonstrating its potential for the treatment of cancer [30].

C. BIOLOGICAL PROBES

An analogous derivative of this inhibitor has also been used as an imaging agent for probing the biological localization and function of TG2. The dansyl derivative shown in Figure 9 allowed the authors to visualize TG2 anchored in the ECM of HCT-116 cells, along with weaker labeling of granules in the cytosol [29].

As mentioned above, the Keillor group has prepared a different fluorescent probe (Figures 6 and 9) by attaching a dansyl group, via a short PEG spacer, to the acrylamide scaffold previously developed in our group [53]. This derivative, Cbz-K(Acr)-PEG-dansyl, was shown to be active in the ECM of osteoblast cell culture, effectively inhibiting TGase-mediated cross-linking of collagen, preventing its subsequent mineralization [54]. Currently, this dansyl inhibitor is being used in localization studies investigating the roles of factor XIIIa and TG2 in bone formation [74].

Most recently, a series of fluorescent inhibitors was prepared in the Keillor group, based on the high-affinity gluten peptide sequence and bearing the

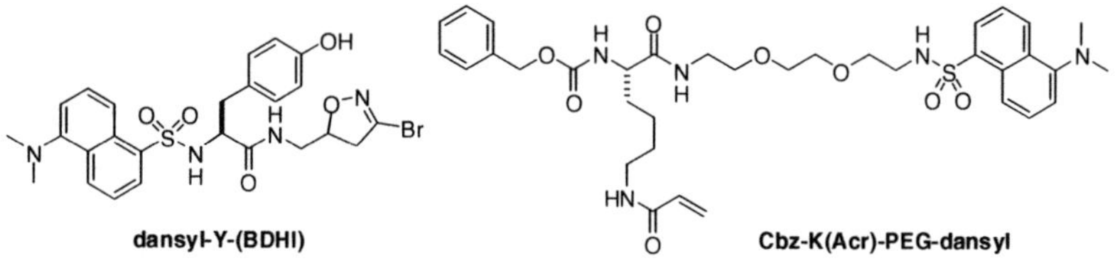

dansyl-Y-(BDHI)

Cbz-K(Acr)-PEG-dansyl

FIGURE 9. Fluorescent labeling agents of TG2 [29, 54].

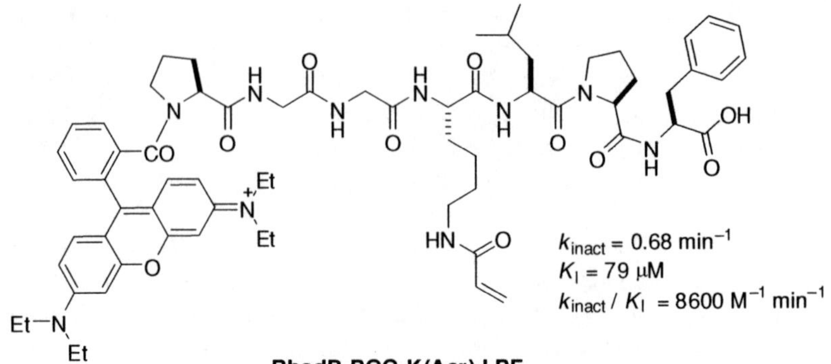

$k_{inact} = 0.68\ min^{-1}$
$K_I = 79\ \mu M$
$k_{inact}/K_I = 8600\ M^{-1}\ min^{-1}$

RhodB-PGG-K(Acr)-LPF

FIGURE 10. Fluorescent extended peptide TG2 inhibitor [75].

highly efficient acrylamide warhead (Figure 10). After optimization of the tether moiety linking the rhodamine B fluorophore to the peptide sequence, with respect to both affinity and fluorescence intensity, the derivative shown in Figure 10, RhodB-PGG-K(Acr)-LPF, was characterized in vitro against recombinant guinea pig liver enzyme, using a continuous chromogenic assay [4]. It was then used to stain cross sections of rat artery in a model for isolated systolic hypertension (ISH). Fluorescent staining was shown to be specific to TG2 activity, which increased over the four-week period of the ISH model, confirming the correlation between TG2-mediated cross-linking and arterial rigidification [75].

IV. CONCLUDING REMARKS

Several classes of effective irreversible TG2 inhibitors have been presented herein. In general, these inhibitors have been designed to bear an electrophilic warhead, already known to be reactive with cysteine proteases, on a peptidic scaffold intended to confer affinity, if not selectivity, for TG2. As our understanding of the donor substrate specificity of TG2 has increased, different scaffolds have been crafted, with varied levels of success.

The best irreversible TG2 inhibitors currently known have been shown to be remarkably effective, having efficiency constants over $10^6\ M^{-1}\ min^{-1}$ and inhibition constants in the nanomolar range. Some of these have also demonstrated in vivo efficacy, which bodes well for the development of inhibitors with therapeutic potential for the treatment of cancer, Celiac sprue, or neurodegenerative diseases. Furthermore, while new inhibitors may or may not prove useful for refining our understanding of the mechanisms of TG2-mediated transpeptidation, they will almost certainly be critically important in the near future as probes of the biological roles of this important enzyme.

REFERENCES

1. Folk, J. E. (1983) Mechanism and basis for specificity of transglutaminase-catalyzed epsilon-(gamma-glutamyl) lysine bond formation, *Adv. Enzymol. Relat. Areas Mol. Bio. 54*, 1–56.
2. Folk, J. E. and Chung, S. I. (1985) Transglutaminases, *Methods Enzymol. 113*, 358–375.
3. Folk, J. E. and Chung, S. I. (1973) Molecular and catalytic properties of transglutaminases, *Adv. Enzymol. Relat. Areas Mol. Biol. 38*, 109–191.
4. Leblanc, A., et al. (2001) Kinetic studies of guinea pig liver transglutaminase reveal a general-base-catalyzed deacylation mechanism, *Biochemistry 40(28)*, 8335–8342.
5. Case, A. and Stein, R. L. (2003) Kinetic analysis of the action of tissue transglutaminase on peptide and protein substrates, *Biochemistry 42(31)*, 9466–9481.
6. Day, N. and Keillor, J. W. (1999) A continuous spectrophotometric linked enzyme assay for transglutaminase activity, *Anal. Biochem. 274(1)*, 141–144.
7. de Macedo, P., et al. (2000) A direct continuous spectrophotometric assay for transglutaminase activity, *Anal. Biochem. 285(1)*, 16–20.
8. Case, A., et al. (2005) Development of a mechanism-based assay for tissue transglutaminase – results of a high-throughput screen and discovery of inhibitors, *Anal. Biochem. 338(2)*, 237–244.
9. Murthy, S. N. P., et al. (2002) Conserved tryptophan in the core domain of transglutaminase is essential for catalytic activity, *Proc. Natl. Acad. Sci. U.S.A. 99(5)*, 2738–2742.
10. Chica, R. A., et al. (2004) Tissue trans glutaminase acylation: proposed role of conserved active site Tyr and Trp residues revealed by molecular modeling of peptide substrate binding, *Protein Sci. 13(4)*, 979–991.
11. Gravel, C., et al. (2007) Acyl transfer from carboxylate, carbonate, and thiocarbonate esters to enzymatic and nonenzymatic thiolates, *Can. J. Chem.-Revue Canadienne De Chimie 85(3)*, 164–174.
12. Griffin, M., et al. (2002) Transglutaminases: nature's biological glues, *Biochem. J. 368*, 377–396.
13. Lorand, L. and Graham, R. M. (2003) Transglutaminases: crosslinking enzymes with pleiotropic functions, *Nat. Rev. Mol. Cell Biol. 4(2)*, 140–156.
14. Mehta, K. (2005) Mammalian transglutaminases: a family portrait, *Prog. Exp. Tumor Res. 38*, 1–18.

15. Davies, P. J., et al. (1980) Transglutaminase is essential in receptor-mediated endocytosis of alpha 2-macroglobulin and polypeptide hormones, *Nature 283(5743)*, 162–167.
16. Levitzki, A., et al. (1980) Evidence for participation of transglutaminase in receptor-mediated endocytosis, *Proc. Natl. Acad. Sci. USA 77(5)*, 2706–2710.
17. Fesus, L., et al. (1987) Induction and activation of tissue transglutaminase during programmed cell death, *FEBS Lett. 224*, 104.
18. Birckbichler, P. J., et al. (1983) Enhanced transglutaminase activity in transformed human lung fibroblast cells after exposure to sodium butyrate, *Biochim. Biophys. Acta 763*, 27.
19. Aeschlimann, D. and Thomazy, V. (2000) Protein crosslinking in assembly and remodelling of extracellular matrices: the role of transglutaminases, *Connect. Tissue Res. 41(1)*, 1–27.
20. De Young, L., et al. (1984) Transglutaminase activity in human and rabbit ear comedogenesis: a histochemical study, *J. Invest. Dermatol. 82*, 275.
21. Azari, P., et al. (1981) Transglutaminase activity in normal and hereditary cataractous rat lens and its partial purification, *Curr. Eye Res. 1(8)*, 463.
22. Fesus, L. (1982) Transglutaminase activation: significance with respect to immunologic phenomena, *Surv. Immunol. Res. 1*, 297.
23. Schroeder, W. T., et al. (1992) Type I keratinocyte transglutaminase: expression in human skin and psoriasis, *J. Invest. Dermatol. 99*, 27–34.
24. Selkoe, D. J., et al. (1982) Brain transglutaminase: in vitro crosslinking of human neurofilament proteins into insoluble polymers, *Proc. Natl. Acad. Sci. U.S.A. 79*, 6070.
25. Norlund, M. A., et al. (1999) Elevated transglutaminase-induced bonds in PHF tau in Alzheimer's disease, *Brain Res. 851*, 154.
26. Dedeoglu, A., et al. (2002) Therapeutic effects of cystamine in a murine model of Huntington's disease, *J. Neurosci. 22(20)*, 8942–8950.
27. Mastroberardino, P., et al. (2002) "Tissue" transglutaminase ablation reduces neuronal death and prolongs survival in a mouse model of Huntington's disease, *Cell Death Differ. 9*, 873–880.
28. Piper, J. L., et al. (2002) High selectivity of human tissue transglutaminase for immunoactive gliadin peptides: implications for Celiac sprue, *Biochemistry 41(1)*, 386–393.
29. Choi, K., et al. (2005) Chemistry and biology of dihydroisoxazole derivatives: selective inhibitors of human transglutaminase 2, *Chem. Biol. 12(4)*, 469–475.
30. Mehta, K. (2009) Biological and therapeutic significance of tissue transglutaminase in pancreatic cancer, *Amino Acids 36(4)*, 709–716.
31. Siegel, M. and Khosla, C. (2007) Transglutaminase 2 inhibitors and their therapeutic role in disease states, *Pharmacol. Ther. 115(2)*, 232–245.
32. Silverman, R. B. (1988) *Mechanism-Based Enzyme Inactivation: Chemistry and Enzymology*. CRC Press, Boca Raton, FL.
33. Silverman, R. B. (2000) *The Organic Chemistry of Enzyme-Catalyzed Reactions*. Academic Press, San Diego, CA.
34. Kitz, R. and Wilson, I. B. (1962) Esters of Methanesulfonic Acid as Irreversible Inhibitors of Acetylcholinesterase, *J. Biol. Chem. 237(10)*, 3245–3249.

35. Tian, W. X. and Tsou, C. L. (1982) Determination of the rate-constant of enzyme modification by measuring the substrate reaction in the presence of the modifier, *Biochemistry 21(5)*, 1028–1032.

36. Gray, P. J. and Duggleby, R. G. (1989) Analysis of kinetic data for irreversible enzyme inhibition, *Biochem. J. 257(2)*, 419–424.

37. Powers, J. C., et al. (2002) Irreversible inhibitors of serine, cysteine, and threonine proteases, *Chem. Rev. 102(12)*, 4639–4750.

38. Beynon, R. J. and Bond, J. S. (1989) *Proteolytic Enzymes: A Practical Approach*. IRL Press, Oxford, England.

39. Folk, J. E. and Cole, P. W. (1966) Identification of a functional cysteine essential for the activity of guinea pig liver transglutaminase, *J. Biol. Chem. 241(13)*, 3238–3240.

40. Pardin, C., et al. (2006) Synthesis and evaluation of peptidic irreversible inhibitors of tissue transglutaminase, *Bioorg. Med. Chem. 14(24)*, 8379–8385.

41. Castelhano, A. L., et al. (1988) Synthesis, chemistry, and absolute-configuration of novel transglutaminase inhibitors containing a 3-halo-4,5-dihydroisoxazole, *Bioorg. Chem. 16(3)*, 335–340.

42. Tso, J. Y., et al. (1980) Mechanism of inactivation of glutamine amidotransferases by the antitumor drug L-(alpha S, 5S)-alpha-amino-3-chloro-4,5-dihydro-5-isoxazoleacetic acid (AT-125), *J. Biol. Chem. 255(14)*, 6734–6738.

43. Auger, M., et al. (1993) Solid-state C-13 NMR-study of a transglutaminase inhibitor adduct, *Biochemistry 32(15)*, 3930–3934.

44. Watts, R. E., et al. (2006) Structure-activity relationship analysis of the selective inhibition of transglutaminase 2 by dihydroisoxazoles, *J. Med. Chem. 49(25)*, 7493–7501.

45. Hausch, F., et al. (2003) Design, synthesis, and evaluation of gluten peptide analogs as selective inhibitors of human tissue transglutaminase, *Chem. Biol. 10(3)*, 225–231.

46. Shan, L., et al. (2002) Structural basis for gluten intolerance in Celiac sprue, *Science 297(5590)*, 2275–2279.

47. Pliura, D. H., et al. (1992) Irreversible inhibition of transglutaminases by sulfonium methylketones—optimization of specificity and potency with omega-aminoacyl spacers, *J. Enzyme Inhib. 6(3)*, 181–194.

48. Griffin, M., et al. (2008) Synthesis of potent water-soluble tissue transglutaminase inhibitors, *Bioorg. Med. Chem. Lett. 18(20)*, 5559–5562.

49. Slaughter, T. F., et al. (1992) A microtiter plate transglutaminase assay utilizing 5-(biotinamido)pentylamine as substrate, *Anal. Biochem. 205(1)*, 166–171.

50. Freund, K. F., et al. (1994) Transglutaminase inhibition by 2-[(2-oxopropyl)thio]imidazolium derivatives: mechanism of factor XIIIa inactivation, *Biochemistry 33(33)*, 10109–10119.

51. Leung-Toung, R., et al. (2002) Thiol-dependent enzymes and their inhibitors: a review, *Curr. Med. Chem. 9(9)*, 979–1002.

52. Tam, T. F., et al. (2005) Medicinal chemistry and properties of 1,2,4-thiadiazoles, *Mini-Rev. Med. Chem. 5(4)*, 367–379.

53. Marrano, C., et al. (2001a) Synthesis and evaluation of novel dipeptide-bound 1,2,4-thiadiazoles as irreversible inhibitors of guinea pig liver transglutaminase, *Bioorg. Med. Chem. 9(12)*, 3231–3241.

54. Keillor, J. W., et al. (2008) The bioorganic chemistry of transglutaminase—from mechanism to inhibition and engineering, *Can. J. Chem.-Revue Canadienne De Chimie 86(4)*, 271–276.

55. Gillet, S. M. F. G., et al. (2004) Expression and rapid purification of highly active hexahistidine-tagged guinea pig liver transglutaminase, *Protein Expr. Purifi. 33(2)*, 256–264.

56. Gillet, S. M. F. G., et al. (2005) A direct fluorometric assay for tissue transglutaminase, *Anal. Biochem. 347(2)*, 221–226.

57. Halim, D., et al. (2007) Synthesis and evaluation of peptidic maleimides as transglutaminase inhibitors, *Bioorg. Med. Chem. Lett. 17(2)*, 305–308.

58. de Macedo, P., et al. (2002) Synthesis of dipeptide-bound epoxides and alpha,beta-unsaturated amides as potential irreversible transglutaminase inhibitors, *Bioorg. Med. Chem. 10(2)*, 355–360.

59. Marrano, C., et al. (2001b) Evaluation of novel dipeptide-bound a,b-unsaturated amides and epoxides as irreversible inhibitors of guinea pig liver transglutaminase, *Bioorg. Med. Chem. 9*, 1923–1928.

60. Tate, S. S. and Meister, A. (1985) Gamma-glutamyl-transferase transpeptidase from kidney, *Methods Enzymol. 113*, 400–419.

61. Pinkas, D. M., et al. (2007) Transglutaminase 2 undergoes a large conformational change upon activation, *Plos Biol. 5(12)*, 2788–2796.

62. Lai, T. S., et al. (2008) Identification of chemical inhibitors to human tissue transglutaminase by screening existing drug libraries, *Chem. Biol. 15(9)*, 969–978.

63. Case, A. and Stein, R. L. (2007) Kinetic analysis of the interaction of tissue transglutaminase with a nonpeptidic slow-binding inhibitor, *Biochemistry 46(4)*, 1106–1115.

64. Casadio, R., et al. (1999) The structural basis for the regulation of tissue transglutaminase by calcium ions, *Eur. J. Biochem. 262(3)*, 672–679.

65. Di Venere, A., et al. (2000) Opposite effects of Ca^{2+} and GTP binding on tissue transglutaminase tertiary structure, *J. Biol. Chem. 275(6)*, 3915–3921.

66. Begg, G. E., et al. (2006) Mechanism of allosteric regulation of transglutaminase 2 by GTP, *Proc. Natl. Acad. Sci. U.S.A. 103(52)*, 19683–19688.

67. Pardin, C., et al. (2009) Photolabeling of tissue transglutaminase reveals the binding mode of potent cinnamoyl inhibitors, *Biochemistry 48(15)*, 3346–3353.

68. Silverman, R. B. (1992) *The Organic Chemistry of Drug Design and Drug Action.* Academic Press, San Diego.

69. Barsigian, C., et al. (1991) Tissue (type II) transglutaminase covalently incorporates itself, fibrinogen, or fibronectin into high molecular weight complexes on the extracellular surface of isolated hepatocytes. Use of 2-[(2-oxopropyl)thio] imidazolium derivatives as cellular transglutaminase inactivators, *J. Biol. Chem. 266(33)*, 22501–22509.

70. Balklava, Z., et al. (2002) Analysis of tissue transglutaminase function in the migration of Swiss 3T3 fibroblasts, *J. Biol. Chem. 277(19)*, 16567–16575.

71. Maiuri, L., et al. (2005) Unexpected role of surface transglutaminase type II in celiac disease, *Gastroenterology 129(5)*, 1400–1413.

72. Yuan, L., et al. (2005) Tissue transglutaminase 2 inhibition promotes cell death and chemosensitivity in glioblastomas, *Molecular Cancer Therapeutics, 4(9)*, 1293–1302.

73. Yuan, L., et al. (2007) Transglutaminase 2 inhibitor, KCC009, disrupts fibronectin assembly in the extracellular matrix and sensitizes orthotopic glioblastomas to chemotherapy, *Oncogene 26(18)*, 2563–2573.

74. Al-Jallad, H. F., et al. (2009) Inhibition of osteoblast differentiation by a transglutaminase-specific inhibitor targets crosslinking activity of cell surface and extracellular factor XIIIA, but not transglutaminase 2, *Bone 44(2)*, S310–S310.

75. Chabot, N., et al. (2010) Fluorescent probes of tissue transglutaminase reveal its role in arterial stiffening, *Chem. Biol. 17(10)*, 1143–1150.

METHIONINE ADENOSYLTRANSFERASE (*S*-ADENOSYLMETHIONINE SYNTHETASE)

MARÍA A. PAJARES
GEORGE D. MARKHAM

CONTENTS

Advances in Enzymology and Related Areas of Molecular Biology, Volume 78.
Edited by Eric J. Toone.

I. INTRODUCTION

S-adenosylmethionine (abbreviated AdoMet or SAM) is the main methyl donor for the multitude of transmethylation reactions that take place in all organisms. However, the trivalent bonding of its cationic sulfur atom also allows donation of two other groups, hence broadening the catalog of reactions in which AdoMet participates; amongst these are decarboxylation followed by propylamine donation in polyamine biosynthesis and reductive cleavage of the bond between the sulfur and the ribosyl carbon to yield the 5′-deoxyadenosyl radical, which is catalyzed by radical SAM proteins (see Figure 1A) [1, 2]. It has been estimated that AdoMet participates in as many reactions as ATP. In contrast, AdoMet can be synthesized only through the process catalyzed by methionine adenosyltransferase (MAT also called SAM synthetase; E.C. 2.5.1.6) in the reaction shown in Figure 1B. The MAT family of enzymes has been shown to be present in all cell types through archaea, bacteria, and eucarya, except for a few obligatory parasites [3]. Moreover, a high level of sequence conservation exists between bacterial and eukaryotic MATs, whereas archaea express a distinct class of MAT. In all the cases, the preferred quaternary structure is an oligomer, at least a dimer. Amongst MATs, the *E. coli* (cMAT) and mammalian MATs (MAT I/III and MAT II) are the most studied, hence most of the reported structural and regulatory data have been obtained from these proteins.

MAT catalysis of AdoMet synthesis is the rate limiting step of the methionine cycle (Figure 2). The use of AdoMet in transmethylation reactions leads to production of *S*-adenosylhomocysteine (AdoHcy), which is a potent inhibitor of methyltransferases. AdoHcy is typically metabolized by *S*-adenosylhomocysteine hydrolase (SAHH), in a reversible reaction that favors resynthesis of AdoHcy. Thus, homocysteine (Hcy) and adenosine, the products of AdoHcy hydrolysis, have to be eliminated in a coordinated

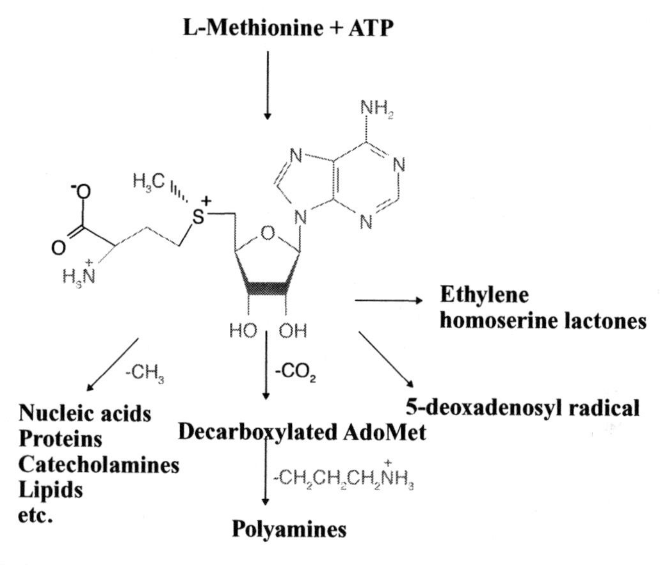

FIGURE 1. (A) MAT and AdoMet metabolism The many ways in which AdoMet is used contrast with the single way in which it is formed. (*Continued*)

manner in order to avoid accumulation of the transmethylation inhibitor, which would lead to a reduction in the methylation index (the AdoMet/AdoHcy ratio). Adenosine is removed by adenosine deaminase catalyzed conversion to inosine, whereas Hcy is consumed in several processes, which vary amongst tissues and organisms. In mammalian liver, there are three reactions involved in Hcy utilization, thus making a branch point in the cycle. These are (1) synthesis of cystathionine by cystathionine-β-synthase (CBS), which initiates the *trans*-sulfuration pathway; (2) remethylation of Hcy, catalyzed either by methionine synthase (MS) or by betaine homocysteine methyltransferase (BHMT); and (3) Hcy exportation to the plasma. Remethylation reactions allow recovery of methionine for AdoMet resynthesis, thus playing an important role when there are low levels of this amino acid. Moreover, MS links the methionine cycle to the folate cycle, and BHMT allows recovery of one of the methylation equivalents used for choline synthesis. On the other hand, export to the plasma and

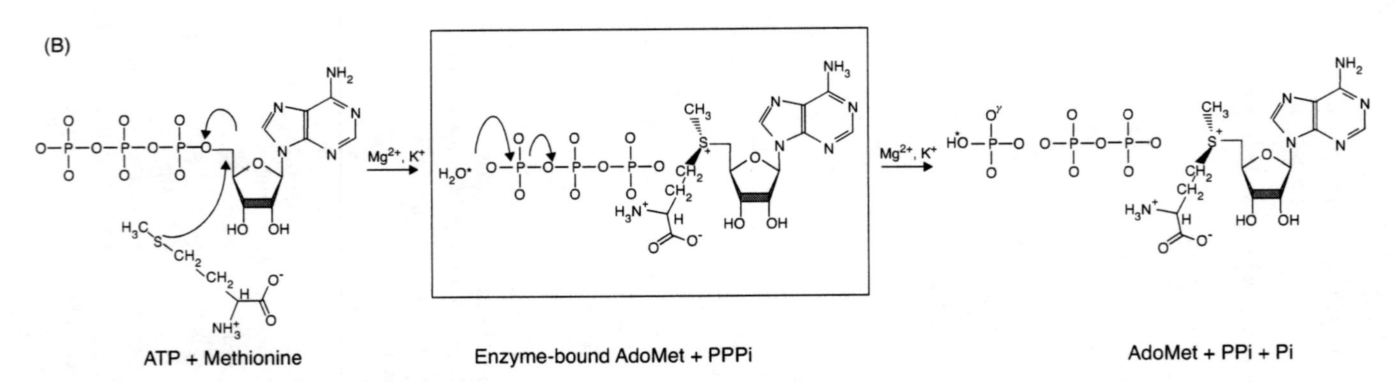

FIGURE 1. (*Continued*) (B) The MAT catalyzed reaction illustrating the intermediate complex with bound tripolyphosphate (PPPi).

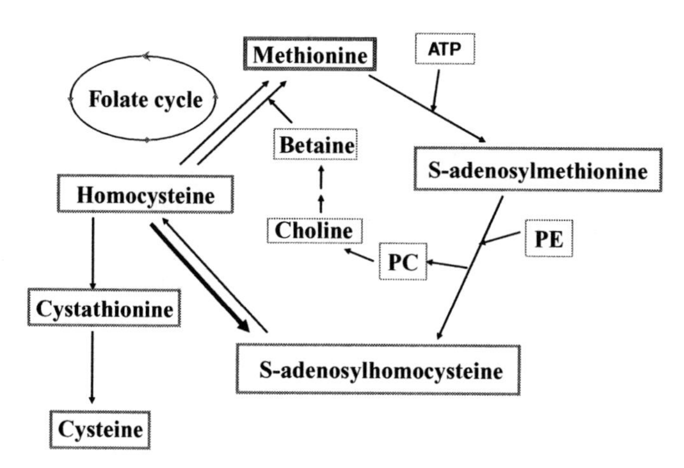

FIGURE 2. The methylation cycle in which is predominantly found in mammalian liver.

CBS-catalyzed entrance in the *trans*-sulfuration pathway allows elimination of excess Hcy, and in the last case, cysteine is formed, which can be used for protein and glutathione (GSH) production. The flux through the cycle is modulated at different levels by some of the metabolites mentioned above; see previous reviews such as [4].

The last decade has produced a large increase in the understanding of the structural, functional, and regulatory properties of MATs, including the elucidation of the 3-D structures of several of these enzymes and unexpected findings on evolutionary conservation and diversity. This review of the MAT family of enzymes attempts a complete overview of the achievements of the last decade. Lastly, important new results relating MAT function to the development of certain diseases are also discussed.

II. MAMMALIAN MATs

Early studies on mammalian MATs showed the existence of this activity in all the tissues studied, the highest being observed in liver. However, complex kinetics were obtained, which made the results difficult to interpret until the presence of several isoenzymes was realized. Separations carried out on phenyl Sepharose columns showed elution of three peaks with different hydrophobic character; these were named according to the order of elution as MAT I, II, and III. All these isoenzymes share some characteristics such as the Mg^{2+} dependence of their activity ($S_{0.5} \sim 1$ mM), stimulation by K^+ ions, and their AdoMet-inducible tripolyphosphatase activity [4, 5].

A. MAT I/III ISOENZYMES

Since their discovery, these isoenzymes were known as liver-specific forms of MAT, until Lu et al. in 2003 [6] showed their presence also in pancreas. Several laboratories have reported in depth studies of MAT I and III, which have been carried out using adult liver. The differences can be summarized as follows: (1) MAT I is a tetramer, whereas MAT III is a dimer, with respective Mr appearing as 200 kDa and 110 kDa upon gel filtration chromatography; (2) their K_m values for methionine are in the micromolar range for MAT I (60–120 μM) and in the millimolar range for MAT III (1 mM) [7, 8]; (3) MAT III can be activated by DMSO at physiological concentrations of methionine (60 μM) and is highly hydrophobic (eluting from phenyl Sepharose columns at 50% DMSO (v/v)); and (4) AdoMet inhibits MAT I and activates MAT III. Some of these differences are normally used to distinguish between isoenzymes in activity assays. Both purified proteins show a common single band (designated α1) of approximately 48 kDa on SDS-PAGE, and antibodies raised against the dimer form recognized both, thus suggesting that a single type of subunit arranged in two assemblies could explain the presence of both oligomers [7]. Further experiments that supported this hypothesis included peptide mapping of the purified isoenzymes [7], dissociation of the tetramer to the dimer using LiBr [9], or *N*-ethylmaleimide (NEM) modification [10]. The final confirmation came upon cloning of a single cDNA [11, 12] followed by overexpresion in *E. coli* cells [13], which showed in vivo production of dimers and tetramers. This single subunit type is a product of the *MAT1A* gene that contains nine exons and eight introns spanning ~20 kb [14] and has been localized by fluorescence in situ hybridization to the human chromosome 10q.22 [15].

Animal models, as well as the analysis of human samples, led to the identification of changes in the MAT I/III ratio in pathological conditions [16], which has stimulated interest in the mechanisms that regulate this interconversion. Several in vitro approaches have indicated the importance that cysteine residues have both in the regulation of the association state and in controlling their activity [8, 17–19]. NEM modification of two sulfhydryls/subunit leads to inactivation and dissociation of the tetramer to inactive dimers [10]. Moreover, only seven to eight out of the ten cysteines present in each subunit were modified by either NEM or nitric oxide (NO), even after denaturation [20, 21], but denaturation in the presence of

DTT allowed full labeling of the sulfhydryl groups [22]. These observations along with data on SDS-PAGE mobility changes after thiopropyl Sepharose chromatography, and the mechanisms of GSH/GSSG regulation involving the production of an intrasubunit disulfide [8], prompted deeper study of the oxidation state of MAT I/III cysteines. Thus, the presence of a disulfide bond linking C35 and C61 within a subunit was identified [22]. Once available, the MAT I crystal structure showed that these residues are the only cysteines properly oriented and at distance suitable for formation of an intrasubunit disulfide bridge [23]. Although it is not common for a cytosolic protein to contain a disulfide, examples exist, and their formation is highly dependent on a protein conformation that provides a high local concentration of sulfhydryls to allow formation of the disulfide even in the highly reducing environment of the cytosol [24]. The role of this disulfide in MAT I/III stability is discussed in Section X. In addition, site-directed mutagenesis studies demonstrated that changes of any of the cysteines located between residues 35 and 105 modify the MAT I/III ratio; most interesting was the C69S mutant, which appears mainly as dimers [19]. Within this region, C121 was found to be of great importance for control of activity, as deduced from the experiments using NO and hydrogen peroxide that are discussed in the Section II.C.

Early estimates of MAT I/III subunit size obtained by SDS-PAGE indicated molecular weights of 48 kDa for the $\alpha 1$ subunit [7]. This value was higher than the theoretical size calculated from the sequence and led several authors to suggest posttranslational modifications as responsible for these differences. Examination of the sequence revealed the presence of putative phosphorylation sites for several kinases, but only the case of PKC has been explored in vitro [25]. Incorporation of ^{32}P-phosphate onto threonine-342, the most exposed residue according to hydrophobicity profiles, was obtained mainly in the dimer with a concomitant delay in its elution on gel filtration chromatography. These results suggested that MAT III is a partially phosphorylated form and implied that complete phosphorylation was responsible for oligomer dissociation. These interesting data still await confirmation in vivo.

B. MAT II ISOENZYME

MAT II has been especially studied in lymphocytes and in cancerous cells in which it is located in the cytosol [26]. Immunohistochemistry revealed that MAT II is located in distal tubules of kidney [27] and in liver stellate cells [28]. MAT II was purified to apparent homogeneity from human chronic lymphocytic leukemia cells; the purified enzyme showed a single band on native PAGE gels, which was coincident with the MAT activity in gel slices [29]. Surprisingly, the preparation showed three bands in SDS-PAGE, with sizes of 53 ($\alpha2$), 51 ($\alpha'2$), and 38 (β) kDa. Peptide maps were identical for $\alpha2$ and $\alpha'2$, leading to the suggestion that they represent different post-translationally modified forms of the same subunit [29], the nature of this difference remains unknown to date. Subsequently, the *MAT2A* and *MAT2B* genes have been identified as encoding the $\alpha2$ and β subunits. The *MAT2A* gene consists of nine exons and eight introns covering ~6 kbp [30] and is located on human chromosome 2p.1.1 [31]. Further information about the β subunit and its gene is provided in Section II.C.1. The native molecular weight estimates obtained by both gel filtration chromatography and sedimentation velocity indicate an oligomer of approximately 185 kDa, a value compatible with several association arrangements ($\alpha2_2\beta_2$, $\alpha'2_2\beta_2$, or $\alpha2\alpha'2\beta_2$). To date, it has not been determined which of these defines the correct association state, but a tetramer is the accepted form. In addition, the presence of a 68 kDa precursor protein (λ) for $\alpha2$ and $\alpha'2$ has been proposed, based on recognition of such a band by monoclonal antibodies raised against $\alpha2$ subunits [32]. This protein, called λ, is not detectable in continuously proliferating cells and is lost upon induction of resting cells, wherein a parallel increase in $\alpha2$ subunits occurs. The λ subunit showed MAT activity in cells where the $\alpha2$, $\alpha'2$, and β proteins were undetectable. Such a precursor protein has not been observed in the nonmammalian organisms that express unusually large MAT α subunits [3], and its sequence has not been reported.

The kinetic properties of MAT II from human chronic lymphocytic leukemia cells have been characterized in detail, see Table 1 [29]. The enzyme has an optimal pH of 7–8.5 and is inhibited by AdoMet (with a K_i of 60 μM), PPP_i, PP_i, P_i, *S*-carbamyl-*L*-cysteine, polyamines, and 10% DMSO [29, 33–35]. The K_m values have been reported as 31 and 3.3 μM for ATP and methionine, respectively. The kinetic mechanism, deduced from product and dead-end inhibition patterns, is a steady-state ordered Bi-Ter with ATP adding first and AdoMet being the last product released [29].

1. The Regulatory β-subunit

The presence of a noncatalytic β-subunit in MAT II was found by Kotb et al. [29]. This subunit was cytosolically localized by immunofluorescence visualization in transiently transfected Huh7 cells [36]. Peptide maps [29]

TABLE 1
Enzymatic Properties of MAT from Different Sources

Source	Oligomeric State	K_m^{ATP} or $S_{0.5}$ (μM)	K_m^{Met} or $S_{0.5}$ (μM)	K_i^{AdoMet} (μM)	V_{max} (μmol/min/mg)	pH Optimum	Reference
E. coli	Tetramer	110	80	10	1.2	6–9	[121]
M. jannaschii	Dimer	290	240	120	3.0		[133]
Bakers yeast isozyme 1	Heterodimer	74	110		0.6		[115]
Bakers yeast isozyme 2	Heterodimer	47	140		0.36		[115]
Rat MAT I[a]	Tetramer	252	125			7.5–8.8	[8]
Rat MAT III[a]	Dimer	950	650			7.5–8.8	[8]
Human MAT II α/β	Heterotetramer	31	3.3	18	0.2	7–8.5	[29]
Human MAT II (no β)	Dimer		80				[42]
L. infantum	Dimer	370 PC	250 PC	1500	0.2		[145]
L. donovanii	Dimer	27 PC	250 PC	4000	0.2		[146]
P. falciparum	Dimer						[148]
T. brucei brucei mixed isoform	Partial purification	53–1750 NC	20–150 NC	240		7.5–10	[143]

Various assay methods have been used. Often radioactivity incorporation into AdoMet is measured, which requires separation of AdoMet from substrates by filtration either on ion exchange filters or on ion exchange columns. Phosphate production, either during AdoMet formation or in hydrolysis of added PPP_i, can also be monitored with purified enzymes. Much of the variation in literature values of kinetic constants can likely be traced to different assay conditions and methods of product determination.
C, cooperative; PC, positive cooperativity; NC, negative cooperativity.
[a]Data for rat MAT I and III refer to the reduced forms obtained after thiopropyl Sepharose purification.

and immunoreactivity [37] studies showed that the β protein is structurally unrelated to the α2 subunit. Expression of the β-subunit has been observed in all the extrahepatic tissues examined and at very low levels in normal adult liver [36]. Purified MAT II from bovine brain [38], Ehrlich ascites tumor cells [38] and human mature erythrocytes [39], showed two types of bands on SDS-PAGE gels, one of them being 38 kDa. The sequence of a full-length cDNA for the β-subunit of human MAT II isolated by LeGros et al. [40] showed that it encoded a protein of 334 amino acids (an ORF of 1002 bp), a calculated Mr of 37552 and a pI of 6.9. No sequence differences were found in the cDNA obtained from different cell types from several subjects; the ORF always started with the sequence Met-Val. Analysis of the deduced protein sequence revealed a 28% homology with bacterial enzymes that catalyze the reduction of TDP-linked sugars, several nucleoside-diphosphate sugar epimerases and other proteins involved in the synthesis of polysaccharides. Kyte-Doolittle hydrophobicity plots for β revealed two regions of prominent hydrophobicity with much higher scores than the three segments described for α2. Expression in *E. coli* of a His-tagged β showed a protein of the expected molecular weight, which was recognized by antibodies raised against *N*-terminal peptides of the β-subunit. These antibodies did not cross-react with α2 or any protein from either *E. coli* or yeast, confirming the absence of this type of subunit in these organisms.

The *MAT2B* gene consists of seven exons and six introns, spanning ~6.8 kbp [41]. Exon 1 contains 203 bp of the 5′-noncoding region in addition to 63 bp of the coding region, whereas exon 7 contains 117 bp of the coding sequence and 802 bp of the 3′-untranslated region. This gene was located in the long arm of the human chromosome 5, in an area that corresponds to the interphase between bands 5q34 and 5q35.1. A sequence of 1.1 kbp of the 5′-flanking region (along with several deletions) was tested for promoter activity in COS-1 and Jurkat T-cells, with little differences in behavior between the systems. Primer extension analysis allowed identification of the transcription start site 203 bp upstream the translation initiation site. The minimal promoter corresponds to the region +52/+93, a GC-rich sequence, the TATA box appears at −32/−8, and a Sp1 site was identified at +9/+15. Mutations of this Sp1 site and the TATA box reduced the activity 35–50% and 25%, respectively, whereas combination of both mutations rendered a 60% decrease. Supershift assays using anti-Sp1 and anti-Sp3 antibodies suggested Sp3 as one of the main factors that bind to the Sp1 site at +9/+15.

Moreover, chromatin immunoprecipitation showed the involvement of Sp1 and Sp3 in complexes at the *MAT2B* promoter [41].

The role of the β-subunit was not initially clear after its discovery. However, cloning of the α2 subunit and its expression in *E. coli* revealed important differences between the kinetic behavior of the homo-oligomer compared to the hetero-oligomer [42], leading to the suggestion of a regulatory role for the β-protein. This role was further supported by kinetic data obtained for MAT from peripheral blood mononuclear cells (PBMC) stimulated with superantigen (SEB), where the β-subunit expression was negligible and a threefold increase in the K_m for methionine (55–67 μM) was observed in addition to a decrease in the sensitivity to AdoMet feedback inhibition [43]. Moreover, transient transfections of COS-1 cells with either α2, β, or both subunits showed changes in the kinetics for methionine saturation [44]. Expression of a His-tagged α2 subunit showed two kinetic forms with K_m values of 15 and 75 μM; expression of the β-subunit did not have an effect on K_m, but increased the V_{max}. Moreover, coexpression of both subunits rendered an increase in specific activity along with a single K_m of 20 μM for methionine. Spontaneous association of the α2 and α'2 with β subunits was observed in both COS-1 [44] and Huh7 cells [36] and in cell-free systems [44], which was reflected in changes in sensitivity to AdoMet inhibition. The purified recombinant β-subunit shows no MAT activity, but is able to modulate α2 kinetics, as well as those of the cMAT and MAT I/III α1 subunits [40, 44].

C. MAMMALIAN MAT REGULATION

Studies on MAT regulation in mammals have been carried out at different levels using animal and cell models, as well as purified proteins. The main achievements in the last decade probably correspond to the understanding of transcriptional regulation as described below.

1. Transcriptional Regulation

Sequences for *MAT1A* [45], *MAT2A* [30, 46, 47], and *MAT2B* promoters [41] have been obtained, allowing their analysis in different cell types.

Promoter analysis. The genomic clone obtained for rat *MAT1A* contained 1557 bp of the promoter, showing 88% identity to that of the mouse [45]. Its main characteristics are depicted in Figure 3. Briefly, the transcription initiation site is located 251 nt upstream of the ATG, a possible TATA box

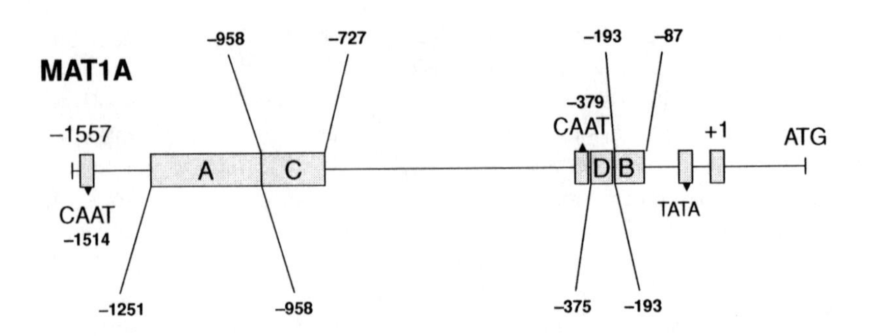

MAT2A

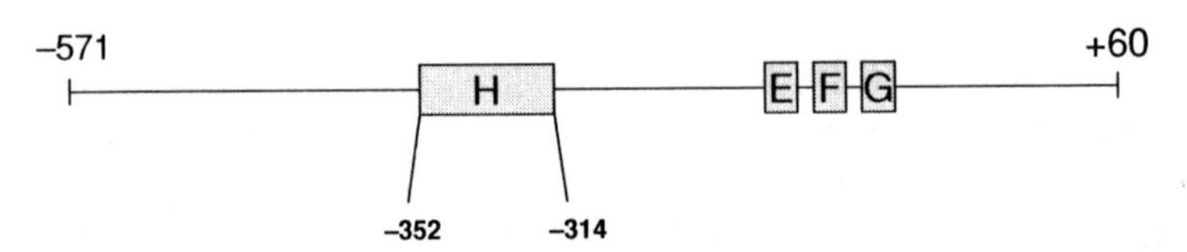

FIGURE 3. Schematic representation of MAT promoters. The upper scheme represents MAT1A promoter. The elements and DNase protection areas are identified as follows: +1 indicates the transcription initiation site, the TATA box is located at −29/−23, two CAAT box are also indicated. Boxes A–D indicate the mocations for positive acting elements, A and B identified in H35 cells and C and D in CHO cells. The lower scheme shows the MAT2A promoter, where E-G represent SpI sites and H the area of NFkB and AP-1 binding.

appears at −29/−23, and two canonical CAAT boxes are located at −379 and −1514. Several consensus sites for binding of specific *trans*-activating factors were also identified: (1) two putative AP-1 sites, (2) two PEA3 sites, (3) four glucocorticoid responsive elements, (4) two IL-6 binding sequences, (5) two NF-1 putative sites, and (6) one for each HNF-1, HNF-3, and HNF-4. Expression in hepatoma cells of promoter constructs linked to a luciferase reporter gene showed positive acting elements located at −1251/−958 and −193/−87, whereas regions −958/−727 and −375/−193 contain those elements in CHO cells. Footprinting analysis using rat liver nuclear proteins showed eight DNase protected areas; the fragment −1251/−996 contained binding elements for liver-enriched factors at −1132/−1229, as judged from comparison with the results obtained using kidney and lung proteins. Negative regulatory elements were also identified between −1154 and−1134 in both H35 and CHO cells, an area where gel mobility shift assays revealed HNF-3 and HNF-1 binding. Moreover, functional analysis of this promoter demonstrated its activity in hepatic and nonhepatic cells, suggesting that liver-restricted expression is probably mediated by DNA methylation or genomic organization [45]. Demonstration of *MAT1A* promoter hypermethylation in nonhepatic cells was obtained by Torres et al. [48], who also showed stronger histone acetylation in liver, as expected for methylation silencing. This same methylation pattern was observed in fetal hepatocytes, in livers from CCl_4-induced cirrhosis and transformed cells [48]. Treatment of hepatoma cells with demethylating agents or histone deacetylase inhibitors increased *MAT1A* expression [48].

Analysis of the human *MAT2A* promoter sequence revealed several putative sites for transcription factor binding. Among these factors, two sites for NF1 and AP-2, and one each for AP-1, c-Myb, NFκB, and SpI were found [49]. Footprinting of the −163/+5 fragment showed 3–5 DNase protected areas, which varied depending on the type of cell extract used; phytohemaglutinin (PHA) stimulation did not induce the same pattern. Four additional protected areas were identified using shorter promoter fragments, including three putative SpI sites in the −163/+5 fragment occupied as shown by supershift analysis [30, 46]. Positive elements responsive to TNFα have been also located in an area including NFκB and AP-1 consensus sites (−352/−314) [47]. Blocking NFκB with IκB superrepressor did not influence the capacity of TNFα to induce AP-1 binding to this promoter, but blockage of the AP-1 site with dominant negative cJun prevents induction of NFκB binding. Electrophoretic Mobility Shift Assays (EMSA) using normal liver and HCC proteins identified c-Myb binding (−354/−328) in both cases, whereas differences were observed in the −60/−40 area. One or two bands were shown in these cases for normal and malignant extracts, respectively, the differing band being ascribed to SpI binding [49]). Both c-Myb and SpI are known to increase their expression levels in HCC.

Effect of hormones and growth factors. Transcriptional regulation of MAT has been shown in several models. Thus, *MAT1A* and *MAT2A* expression is modified in the presence of growth factors, such as HGF and TNFα, as discussed in the Sections XII.C and XII.D, respectively [47, 50]. Partial hepatectomy (PH) also induced a decrease in *MAT1A* mRNA levels, whereas those of *MAT2A* increase due to a rise in transcriptional activity and in mRNA stability [51]. In addition, hormones have been shown to modulate expression; adrenalectomy induces a reduction in *MAT1A* mRNA, which correlates with decreases in protein and activity levels [52]. These effects were counteracted by administration of glucocorticoids, with synthetic forms being more effective [52]. Moreover, administration of triamcinolone to hepatoma cells increased *MAT1A* mRNA expression, with the protein levels following a similar trend. Increases in MAT activity were also observed upon alloxan treatment, whereas the combined action of insulin and triamcinolone on H35 cells reduced this parameter [52]. On the other hand, insulin alone had negligible effects on *MAT1A* mRNA levels. Confirmation of these effects was obtained using *MAT1A* promoter constructs and H35 cells treated with glucocorticoids; an increase in the decay rate of this mRNA was observed. Development provides another model where alterations in expression can be related to hormonal changes. In rat liver, *MAT2A* expression is reduced immediately before delivery, while a corresponding increase in *MAT1A* occurs [53].

Effects of methionine and related metabolites. The levels of methionine in the growth medium for hepatoma cells also influence expression. Thus, deficiency of this amino acid in the medium induces *MAT2A* mRNA expression and increases its half-life, while AdoMet levels decrease [54]. AdoMet and methylthioadenosine (MTA) downregulate *MAT2A* mRNA levels in the absence of transcription; the effects of MTA and methionine are carried out through their metabolites. The dependence of tumor cells on methionine for growth has been associated with defects in MTA phosphorylase expression, which suppress this salvage pathway [55–58]. In addition, the influence of *MAT1A* expression on other pathways has been studied using a knockout (KO) mouse [59]. Disappearance of *MAT1A* expression correlates with upregulation of genes involved in cell communication, control of cell growth/maintenance, cell death, and development, whereas downregulation of genes for metabolic functions occurs.

Effect of oxygen levels. Hypoxia can also modulate the glucocortocoid effect on *MAT1A* mRNA. It also produces decreases in *MAT1A* transcriptional activity, but not in mRNA stability. Concomitant GSH decreases and a reduction in MAT I/III protein levels, with no changes in isoenzyme ratio, were also observed [60, 61]. These effects are exerted through a heme-binding protein and not through respiratory chain production of ATP, as judged from the response obtained using appropriate inhibitors. H35 cells are resistant to *MAT1A* downregulation by hypoxia [62]. The role of oxygen tension is noteworthy among the factors that condition the functional heterogeneity of periportal and perivenous hepatocytes [63].

Other effects. SEB induced an increase in $\alpha2/\alpha'2$ mRNA and protein levels, whereas β-subunit expression was reduced and almost disappeared after 72 hours of treatment [43]. This downregulation in β-subunit expression occurs with a 6–10-fold increase in AdoMet intracellular levels [43]. The opposite effect on AdoMet levels upon stimulation of DNA synthesis was observed for *MAT2B* expression in Huh7 cells [54].

2. *Posttranscriptional Regulation*

Posttranscriptional regulation for MAT I/III was demonstrated in several models, with most studies confirming the important role of cysteines.

Regulation by GSH levels. In vivo studies using rats showed that changes in GSH levels produced either by carbon tetrachloride treatment or by inhibition of GSH synthesis by buthionine sulfoximine (BSO) were linked to decreases in MAT I/III activity [64, 65]. A direct effect of GSH on the enzyme was later demonstrated [8], showing inhibition by GSSG that could be modulated by GSH. The mechanism for this effect excluded production of a mixed disulfide, and detection of monomers after GSSG inhibition suggested formation of an intrasubunit disulfide. MAT III was shown to be more sensitive to GSH/GSSG action, and hence easier to oxidize. In vitro modulation occurs in the 0–50 range of the GSH/GSSG ratio, as expected for a strongly oxidizing environment, whereas physiological ratios between 10 and 300 are normal. Inclusion of thioltransferases in the assay revealed that a thiol-disulfide exchange catalyzed by a protein disulfide isomerase-type enzyme could permit MAT regulation by disulfide formation in conditions of mild to severe oxidative stress [66]. The question remains whether C35-C61 is the only disulfide formed in the protein or if other disulfides can partake in this type of regulation, and which thiol transferases are involved in such a regulatory mechanism in vivo.

Nitrosylation. Intraperitoneal injection of rats with bacterial lipopolysaccharide-induced serum accumulation of nitrites and nitrates, as well as expression of inducible nitric oxide synthase (iNOS) [18]. NO production induced by this treatment coincided with MAT I/III inhibition, but no reduction in protein levels or mRNA was observed. In hypoxia, *L*-NAME (an inhibitor of iNOS) and *N*-acetylcysteine prevented MAT inactivation, but not changes in GSH levels. Treatment of hepatocytes with BSO had no effect on iNOS protein or mRNA, but induced a time-dependent increase in MAT III nitrosylation and inhibition [67]. *L*-NNA (N^G-nitro-*L*-arginine), an inhibitor of NOS with higher affinity for the constitutive enzyme, did not exert any effect on MAT activity or nitrosylation, nor on the GSH depletion caused by BSO. However, recovery of GSH levels using GSH monoethyl ester (EGSH) caused the loss of most of the NO groups incorporated on MAT III and hence led to recovery of activity. Simultaneous administration of BSO and EGSH prevented the effect of BSO, thus suggesting that MAT inactivation induced by BSO takes place through *S*-nitrosylation of the protein.

Indirect data relate nitrosylation of MAT I/III with control of proliferation. Intracellular levels of AdoMet are related to the differentiation state of the hepatocyte, with low levels of this metabolite in growing cells. KO mice for iNOS show abnormal hepatic regeneration, and KO mice for MAT I/III have low AdoMet levels and induced expression of genes related to growth and differentiation [68]. The HGF and TGFα proliferative responses of isolated hepatocytes are inhibited by *L*-NAME, an effect which, for HGF, was overcome by concomitant administration of the NO donor SNAP. This effect of *L*-NAME was dependent on *L*-methionine concentration, and since inhibition of NO production activates MAT I/III, AdoMet levels also increased [69].

Using purified MAT I/III and NO donors such as SIN-1, *S*-nitrosylated glutathione (GSNO), and *S*-nitroso-*N*-acetylpenicillamine (which has no structural relationship to GSH), a direct inhibition of the enzyme was obtained, which did not affect the oligomerization state of the protein [18, 21]. This inhibition was reversed by either GSH or β-mercaptoethanol. GSNO reduced the V_{max} for both isoenzymes, and increased the $S_{0.5}$ of MAT III for both methionine and ATP [21]. In the search for NO-reactive residues, cysteine mutants of MAT were used, which revealed that C121S was resistant to SIN-1 action, whereas the C69S, C105S, and C150S mutants were more sensitive than the wild-type enzyme. This regulation by nitrosylation should be specific for MAT I/III, since the thiol of C121 has no equivalent in MAT II, which has glycine at that position.

The mechanism proposed for protein *S*-nitrosylation involves an acid–base catalyzed nitrosothiol exchange reaction, where the target cysteine is localized next to acidic and basic residues [70], requirements that were met by MAT I/III. Site-directed mutagenesis allowed identification of the residues involved, D355, R357, and R363, and confirmed their role [71]. Incubation of recombinant MAT III with peroxynitrite also inactivates MAT without nitrosylation, an effect that was prevented in the C121S mutant, whereas D355S, R357S, R363S, and R357S/R363S substitutions had no effect. Residue 121 is a glycine in MAT II; however, it was possible to create an equivalent nitrosylation site by the mutation G120C, as demonstrated using recombinant MAT II and GSNO [72]. Equivalent acidic and basic residues are preserved in MAT II (D354, R356, R362), thus allowing modification of position 120 in the G120C mutant. No effect of nitrosylation on the basal tripolyphosphatase activity of MAT I/III was detected, but its stimulation by AdoMet was reduced in the presence of NO donors [73]. These authors suggested the presence of two MAT III isoforms, one of low tripolyphosphatase activity, which is insensitive to NO, and another of high activity, which is inhibitable by this agent. Interconversion between the two conformations would be a slow process that results in a lag phase detected in kinetic studies.

Hydroxylation. Treatment of a stable CHO clone overexpressing MAT I/III (ST-C) with hydrogen peroxide showed enhanced cell death as compared to the wild type CHO cells [74]. Analysis of these cells showed that H_2O_2 further depletes the NAD and ATP levels that were already 50% reduced in the ST-C clone. This effect could be prevented using an inhibitor of poly (ADP-ribose) polymerase, 3-aminobenzamide, but not with the antioxidant *N,N′*-diphenylphenylenediamine (DPPD). In vitro, purified MAT I/III was inactivated by hydrogen peroxide in a dose-dependent fashion, the effect being exerted at the V_{max} level [75]. This inactivation was reversed in the presence of GSH and had no effect on the oligomeric state of the isoenzymes. However, the GSH concentration needed for this reversal of MAT I (25 mM) is far beyond the physiological levels of this metabolite ($\sim$5 mM), and hence oxidation was suggested as the cause for the low tetramer activity observed in cirrhosis. Site-directed mutagenesis allowed identification of C121 as the site of oxidation by hydrogen peroxide. However, inhibitory effects were also observed related to other cysteine residues (69, 105, 312, and 377). Hydrogen peroxide treatment of ST-C cells had no effect on the amount of MAT protein and reduced the activity in this clone, but not in wild-type CHO cells or CHO-T_{121}, a clone overexpressing the C121S MAT mutant. This effect was mediated by hydroxyl radicals as demonstrated by the use of desferrioxamine, an inhibitor of the Fenton reaction. Again this type of modification should be specific for the MAT I/III isoenzymes since C121 is not present in other MATs. Free-radical-mediated MAT inactivation was suggested as an adaptive response to a nitrosative or oxidative stress, aimed at preserving ATP cellular levels [61]. On the other hand, a long lasting interruption of MAT activity would stop many important cellular functions.

Other regulatory models. MAT1A changes upon PH are regulated at a posttranscriptional level through factors that do not need de novo protein synthesis. These alterations cause decreases in AdoMet levels, increases in those of AdoHcy, and hence reductions in the AdoMet/AdoHcy ratio. In addition, DNA methylation is decreased, whereas GSH levels rise; this last effect could help maintain MAT I/III activity [51]. This type of regulation is also observed as an early effect upon hepatocyte incubation under low oxygen concentrations [61]. The initial MAT inactivation with concomitant AdoMet and GSH reduction correlates with MAT I/III protein decrease after 24 hours. Inactivation under these conditions is mediated by NO, as judged from increases in iNOS transcripts and the blockage of MAT inhibition in the presence of *L*-NAME, an inhibitor of iNOS. The partial protection observed with *N*-acetylcysteine also suggests a role for GSH in this regulation. Ceramide stimulation of hepatocytes or H35 cells induces a reduction in MAT I/III expression [76]. Such induction is also observed upon IL-6 addition, either alone or in combination with ceramides. This decrease in protein correlates with the reduction in activity [76].

Regulation of MAT II isoenzyme. Changes in the relative amounts of the different MAT II subunits have been observed upon stimulation with T-cell mitogens [32]. PBMC cells incubated with PHA sequentially increase their IL-2 production, followed by their MAT activity and lastly DNA synthesis [32]. Analysis of the amount of MAT II subunits by Western blot revealed an increase in α2, and more dramatically in α'2, whereas β-subunit levels remained constant. Kinetic analysis by Hanes-Woolf plots indicated the presence of two enzymatic forms catalyzing AdoMet synthesis in activated cells. Increases in activity preceding protein synthesis could be due to post-translational modifications, leading to alterations in the relative amounts of α2/α'2, and hence in their oligomerization with β-subunits. These changes in the amount of MAT II subunits during activation were not observed in other cell lines, such as Jurkat or freshly isolated PBMC cells from acute lymphocytic leukemia ALL-2 patients. In those cases, activation induced IL-2 production, but no increase in MAT activity or change in the relative amount of MAT II subunits was observed [77]. Moreover, the use of SEB as stimulator also caused activation of PBMC cells and a twofold increase in MAT II activity, but the changes in the relative amounts of the subunits differed from those observed with PHA treatment [77].

III. MAT IN PLANTS

Plant cells need AdoMet to carry out a variety of specific AdoMet-dependent processes, such as ethylene formation [78] and lignin polymerization [79], as well as the functions essential for their viability that are common to other organisms [80]. To accomplish these needs, the presence of up to seven MAT genes has been described in some plants [81–91]. Some of these genes lack introns [87] and contain short leader sequences (9–13 nucleotides) [81, 82]. Most studies have been carried out using *Arabidopsis thaliana* as a model. Comparison of the *SAM1* and *SAM2* genes of this plant revealed a high divergence among the sequences flanking the ORFs [81]. However, the promoters possess three regions of conservation that are candidates for the presence of *cis*-acting elements responsible for organ-specific expression [82]. The ORFs encode proteins of approximately 394 residues and are highly homologous to each other (up to 96% identical at the amino acid level) and to the *E. coli* and yeast enzymes (50–60%) [3]. However, exceptions to this conservation were observed in wheat embryos and dwarf pea epicotyls, where subunits exhibited 84–87 kDa molecular weight on SDS-PAGE gels [92–94]. The native enzyme chromatographed normally as dimers [92, 94], except for the *Caranthus roseus* isoenzymes, which were recovered as monomers after gel filtration chromatography [87]. In addition, associated tripolyphosphatase activity has been observed in MATs purified from wheat embryos [92], and as occurs with other MATs, the importance of cysteine residues in three *Arabidopsis thaliana* isoenzymes has been shown using NEM and DTNB [95].

Expression varies among different parts of the plant, for example *SAM1* of *Arabidopsis* has higher mRNA levels in stem and root as compared to leaves, and this pattern is directly correlated with MAT activity levels [81]. Such data suggest that regulation takes place mainly at the transcriptional level. Moreover, expression is primarily located in vascular tissues, sclerenchyma, and root cortex. The strong cellular preference in the expression of the *SAM1* gene seems to be, at least partly, correlated with the extent of lignification the tissues are undergoing. In addition, several models showed upregulation of MAT gene expression, protein levels, and activity at different stages during development: (1) during germination [92, 96], (2) in early stages of development of leaf and root [85], (3) in young stems, sepals, and corollas [97], (4) during initiation of fruit development [86], (5) in steps preceding anthesis [98], (6) during adventitious root development [99], and (7) in the prestorage phase of endosperm development during early desiccation [100]. On the other hand, downregulation occurs in other steps, such as after anthesis [86, 101]. Hormonal, nutritional, and stress regulation of plant MATs is linked to the above mentioned changes. Thus, gibberellic acid induces expression of two isoenzymes in dwarf pea epicotyls, an effect blocked by cycloheximide [94] and reduces expression in ovaries after anthesis [86, 101]. Auxins induce MAT during fruit development [86]. Salt stress and drought induce differential expression of MAT isoenzymes in tomato seedlings, an effect that can be due to a combination of ion toxicity and osmotic stress, the latter inducing synthesis of abscisic acid (ABA) [84, 85, 102, 103]. Such differential expression is also obtained with mannitol and ABA treatment, as well as upon wounding. The reason for such induction may be the need for larger AdoMet amounts for either cell wall synthesis or modification. Moreover, these stimuli also increase ethylene biosynthesis, a common process for plant response to environmental stress. MAT genes are also among the genes whose expression increases in relation to pathogen defense upon exposure to fungal elicitor [104]. Several methylation reactions are known to be associated with pathogen defense, most notably the formation of diverse classes of phenylpropanoid derivatives. Ethylene biosynthesis is also stimulated in elicitor-treated leaves [104].

Regulation by nitrosylation has been studied in *Arabidopsis thaliana* isoenzymes [95]. Only recombinant SAM1 was inhibited by GSNO, the modified residue being cysteine 114. This residue is located close to the putative substrate binding site, at the loop of access to the active site according to a structural model, and flanked by amino acids that promote nitrosylation (TK*C*PEE). Cross-talk between ethylene and NO signaling has been proposed at the level of MAT regulation through nitrosylation of C114 of the SAM1 isoenzyme, whereas the other isoenzymes are not affected significantly [95, 105]. Production of NO regulates key plant defense networks, and thionitrosyl (SNO) formation and turnover are required for multiple modes of disease resistance. Finally, overexpression of Arabidopsis SAM1 in tobacco plants has phenotypic effects such as the appearance of dark green sectors in leaves [106]. The dark sectors have reduced MAT activity and silencing of the transgene. On the other hand, transgenic plants with suppressed MAT activity produce the methanethiol associated with cabbage odor, as do methionine-treated plants [106].

IV. MAT IN YEAST

The understanding of MAT has greatly benefited from the use of yeast as a model organism. The observation that some yeast strains accumulate large intracellular concentrations of AdoMet in UV-dense vacuoles [107–109], which also have high levels of polyphosphate as counter ion, has led to the use of yeast fermentation as a route to commercial production of AdoMet. The pioneering mechanistic work of Mudd in the 1960s utilized bakers yeast preparations [110–114]; it is unknown whether these contained mixtures of the isozymes that were only discovered a decade later by Chiang and Cantoni [115]. These authors obtained the first preparation of pure, catalytically active MATs, and found two isozymes in the genetically heterogeneous bakers yeast. SDS-PAGE gels showed that each isozyme had two distinct types of subunits with apparent molecular weights of 55 and 60 kDa, forming heterodimers with a native molecular weight of 110 kDa. Both MATs had tripolyphosphatase activity and displayed requirements for divalent and monovalent cations. The steady-state kinetic data showed downward curvature in double-reciprocal plots for both ATP and methionine; this was interpreted as negative cooperativity, although the same observations could result from the simultaneous action of two enzymes with different kinetic parameters. The kinetic parameters are listed in Table 1 along with those of other purified MATs.

Using the yeast *Saccharomyces cerevisiae*, Surdin-Kerjan and coworkers identified mutations in two genes that caused reductions in MAT activity; these genes are denoted *SAM1* and *SAM2* (originally called *ETH10* and *ETH2* because the mutations were selected by their conference of resistance to ethionine) [116, 117]. The *SAM1* and *SAM2* encoded MAT subunits are 92% identical in polypeptide sequence. The genes are differently regulated, with methionine inducing *SAM2* expression while repressing that of *SAM1* [118]. The same authors constructed a strain that combined mutations in *SAM1* and *SAM2*; this strain had no MAT activity and required AdoMet for growth, providing the first evidence that this metabolite is essential for any organism. The construction of deletion mutants that are totally defective in AdoMet synthesis was possible due to the ability of *S. cerevisiae* to import AdoMet from the media via the specific transport protein encoded by the *SAM3* gene [119]. More recently several obligate intracellular parasitic organisms have been found to have AdoMet transport systems that compensate for the absence of an encoded functional MAT (see Section VIII).

V. MAT IN BACTERIA

Most bacteria have a single copy of a structural gene for MAT, which is highly homologous to the eukaryotic form [3]. The protein appears as a homotetramer in which early studies demonstrated the importance of cysteine residues [120]. The modification of two cysteines per subunit of the *Escherichia coli* MAT (cMAT) led to inactivation and dissociation to inactive dimers, followed by aggregation. Identification of the modified cysteines revealed cysteines 90 and 240 of cMAT as the labeled residues. There are few biochemical data regarding the regulation of the activity of bacterial MATs, with the exception of the potent product inhibition of cMAT by AdoMet [121]. The transcriptional regulation of MAT in both Gram-positive and Gram-negative bacteria has been extensively studied. In both cases, the expression of the structural gene for MAT, generally denoted *metK*, is regulated in conjunction with the genes of the enzymes of methionine biosynthesis [122–125].

In Gram-positive bacteria, such as *Bacillus subtilis,* the "S-box" riboswitch metabolite-sensing RNA binds AdoMet and regulates the expression of at least 26 genes, including the *metK* gene [126–129]. The dissociation constant for AdoMet binding to this riboswitch RNA is 10 nM; the complex acts as a transcriptional terminator [129]. In contrast, Gram-negative bacteria regulate *metK* expression through use of the *metJ* repressor protein in a complex with AdoMet as corepressor [130]. Expression of the *metK* gene of *E. coli* is also repressed by *Lrp* (leucine regulatory protein) [131].

VI. ARCHAEAL MATs

The existence of an archaeal MAT activity was first demonstrated in *Sulfolobus solfataricus* extracts [132]. The subsequent purification of MAT from *Methanococcus jannaschii* enabled identification of the gene within the completed genome sequence [133]. The sequence of *M. jannaschii* MAT is widely diverged from that in eucarya or most bacteria, retaining ca. 18% sequence identity [133]. Close homologs of the *M. jannaschii* MAT are found in both *Crenarchaeota* and *Euryarchaeota.* Thus, a dramatic phylogenetic differentiation in MATs occurred early in evolution. It is unclear whether the difference in archaeal vs. bacterial/eucaryotic sequences reflects convergent or divergent evolution. The biochemical properties of archaeal MATs are generally similar to those from bacteria and eucarya providing no clue as to a functional meaning of the sequence divergence [132–134]. Thus, the *M. jannaschii* MAT is a dimer, has tripolyphosphatase activity, and requires both divalent and monovalent metal ions for activity [133, 134]. Recombinant, highly thermostable, *M. jannaschii* MAT is readily produced in *E. coli* and its ability to use a variety of nucleotides in addition to ATP may make it a useful synthetic tool [134]. How MAT activity is regulated in archaea remains unknown.

VII. BACTERIA WITH ARCHAEAL-TYPE MATs

The finding that the sequence of MATs in the archaea is substantially different from that in eubacteria or eucarya enabled computational identification of organisms with both types of MAT in their genomes [133]. The archaeal type of MAT has not been found in any eucarya, nor has the eucaryotic/bacterial class of MAT been discerned in any archaea (unpublished results). However, several bacteria have been identified as possessing both the expected type of MAT and the archaeal type. These bacteria are widely scattered throughout phlyogengy. These organisms include *Aquifex aeolicus*, a deeply branching bacterium, which encodes both types of MAT in its small 3 Mb genome; the Gram-positive human pathogen *Streptococcus pyogenes* (but not other *Streptococci*); several *Chlorobaicaes* such as *Chloribium tepidum* and *Pelodictyon lutelum*; the *Alphaproteobacterium Bradyrhizobium japonicum*; the *Betaproteobacteria Ralsotnia eutropham, Burkholderia vietnamiensis, and Pelobacter carbinolicus*; and several *Delatproteobacteria* (*Delta proteobacterium MLMS-1*, *Pelobacter* propionicus, *Syntrophus aciditrophicus,* and several members of the *Geobacter* group). Proteomic studies of *S. pyogenes* [135] and *G. sulfurreductans* [136] show that both types of MAT proteins are indeed expressed, implying functionality. The metabolic reason for the presence of two types of MATs in these organisms remains to be elucidated.

VIII. MAT IN PARASITES

Many parasitic organisms have MATs with atypical properties, and even a few cases have no MAT and are thus AdoMet auxotrophs (recently reviewed in [137]). The parasitic fungus *Pneumocystis carnii* is unable to synthesize AdoMet, but has two AdoMet transporters with K_m values of 4.5 and 333 μM [138, 139]. Trypanosomes of the *Trypanosoma brucei* group are amongst organisms that both have MAT and are able to transport AdoMet from the media via a specific transport system [140]. These organisms lack the capacity for de novo purine synthesis. It is relevant that the AdoMet level is reported to be 70 nM in human sera [141], whereas the K_m for the transporter is in the 4–10 millimolar range [142]. Thus, transport will be inefficient in vivo. Trypanosomal MATs are unusual in that they are not regulated by AdoMet, with K_i values in the millimolar range, in contrast to the micromolar values for AdoMet inhibition of MAT from most organisms [143]. Amongst the *Leishmania, Leishmania infantum,* a trypanosomatid protozoan parasite, has two MAT genes in a tandem array. Two AG dinucleotide *trans*-splicing sites are located in its 5′-UTR, and the +420/+905 region of the 3′-UTR contains *cis*-regulatory elements [144]. These genes are expressed as a single transcript [145]; the resultant protein forms a dimer. Surprisingly, selenomethionine and cycloleucine did not inhibit their MAT activity, and AdoMet had a remarkably high K_i of 1.5 mM. *Leishmania dovanii* also has two MAT genes [146]. The properties of the purified enzyme have been studied with particular attention to the importance of cysteine residues in folding and catalysis [137, 146]. These authors also showed that truncations at F382, D375, and F368 of MAT sequence resulted in loss of MAT activity and reduction of tripolyphosphatase activity [147]. An internal deletion, E376ΔF382, preserved a high tripolyphosphatase activity that is not stimulated by AdoMet [147].

The MAT from the protozoan parasite that causes malaria, *Plasmodium falciparum*, has been cloned and a molecular model was constructed based on the cMAT crystal structure [148]. The authors proposed this enzyme as a target for the development of new antimalarial drugs. The enzyme had a high K_i for cycloleucine as compared to a human hepatic MAT (17 mM vs. 10 mM); little additional characterization was reported.

The *Amoeba proteus* xD strain harbors symbiotic X-bacteria; these bacteria become obligatory endosymbionts in a process that is accompanied by a change in the MAT isozyme expressed by the amoeba [149, 150]. The lack of expression of the ameoba's primary MAT is due to lack of transcription rather than mutations to the gene itself. Apparently the bacterium becomes essential to the survival of the amoeba by modification of the expression of the host MAT gene. The mechanism of AdoMet exchange between the two organisms has not been reported.

Obligate intracellular bacteria have been found in which the *metK* gene encoding MAT is present but not functional [151, 152]. Among these are *Chlamydia trachomatis* [153], the α proteobacterium *Rickettsia prowazekii*, and other members of the spotted fever group of *Rickettsia*. The *metK* gene in these *Rickettsia* is regarded as a marker for the process of the degradation of genes that have become unnecessary as a result of changes either in the habitat of the organism or the gain of redundancy accompanying horizontal gene transfer [154]. In contrast, a functional MAT is present in *Rickettsia* of the typhus group [155]. *R. prowazekii* have a characterized transporter than can import AdoMet from the host [155], which is functional when expressed in *E. coli* [156]. Finally, *Nanoarchaeum equitans* is the only known archaeal parasite; it also lacks a gene for MAT in its small 0.5 Mb genome [157], but the nature of its acquisition of AdoMet remains unknown.

IX. STRUCTURE AND CATALYTIC MECHANISM

A. STRUCTURES OF MAT

Crystal structures have been reported for the rat MAT I (cf. pdb code 1QM4 (2.7 Å resolution) [23]), the human MAT IIa (pdb code 2PO2 (1.03 Å) [158]), and cMAT (cf. pdb codes 1XRA, 1XRB, and 1XRC, 1FUG) [159, 160]. As might be expected from their sequence similarity, the overall topology is the same for each protein (RMSD for alpha carbons = 1.3 Å between rat and human; 0.78 Å between *E. coli* and human). The structure of MAT is unique among all protein structures determined to date. The topology is described in the SCOP database (http://scop.berkeley.edu/data/scop.b.htm) as belonging to the Alpha + Beta protein class, and there are no other members of the MAT superfamily. Three repeats of *beta-alpha-beta*(2)-*alpha-beta* secondary structure are found; notably there is no significant sequence conservation within the structural repeats (illustrated for the human MAT IIa in Figure 4A). These repeats constitute the three domains of the

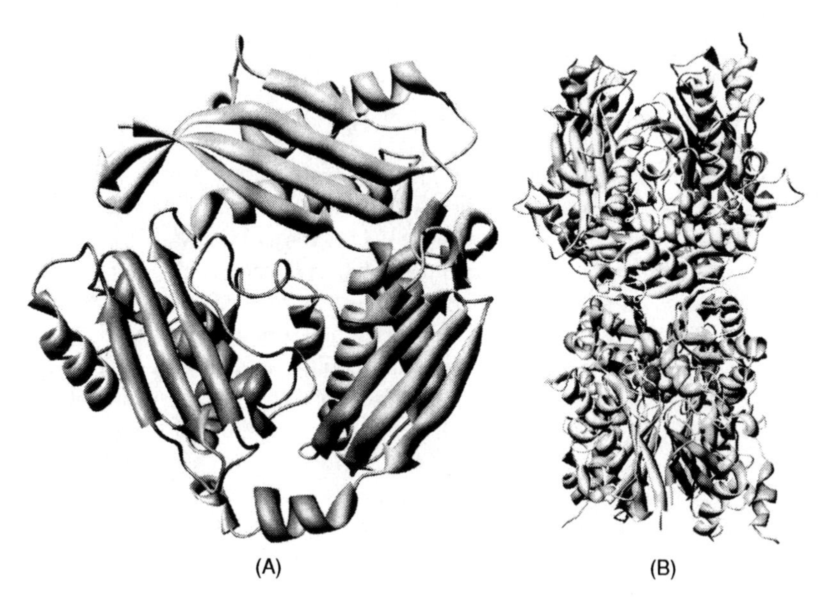

FIGURE 4. (A) Ribbon diagram of the monomer of the human MAT II illustrating the near three fold symmetry of the fold and the central loop (pdb code 2PO2); (B) the tetramer of cMAT (pdb code 1RL7) in complex formed with AMPPNP and methionine; half of the active sites had bound products AdoMet and PPNP. The active site is buried at a subunit interface. (See insert for color representation.)

subunit, which are formed by nonconsecutive stretches of the sequence. The dimer of the enzyme forms in an inverted arrangement so that the two active sites are located in a deep cavity between the subunits (Figure 4B). The tilt between the dimers of MAT I is distinct from that of cMAT; the human MAT II structure has not yet been analyzed in detail. A tetramer forms from the association of two dimers at right angles at their central tips (Figure 4B). The hydrophobic interface between monomers in the tight dimers is much more extensive than is the polar interface between pairs of dimers, consistent with interconversion of the dimeric and tetrameric forms of the mammalian MAT I and III and the stability of the dimer.

The majority of the reported MAT structures share a disordered loop segment (residues 117–128 in rat MAT I, 102–107 for cMAT) that forms a dynamic lid over the active site. The only structures in which this lid is ordered are of the human MAT II complex with AdoMet (pdb code 2PO2 [158]), the cMAT both in a ligand-free form (1FUG [160]) and in structures with the active site filled with methionine and the ATP analog AMPPNP or the products of the reaction in the crystal, AdoMet and PPNP (1RG9, 1P7L [161]). Other structures in which only some of the ligand binding sites are occupied have the lid segment disordered, suggesting that motion of the lid is involved in allowing access of substrate and products to the active site, an interpretation supported by kinetic and spectroscopic studies of cMAT [162, 163].

Structures for several MAT complexes have been published. These include (1) cMAT in complexes with ADP, BrADP, PP_i, and the ternary complex with AMPPNP and methionine [159, 161, 164]; (2) rat MAT I in binary complexes with *L-cis*-AMB or aminoethylphosphonate (AEP), and ternary complexes including ATP plus methionine, ATP plus AEP, and ADP plus *L-cis*-AMB [23, 165]; and (3) the AdoMet complex of human MAT II [158]. The ADP binding site observed in binary complexes with cMAT (1MXB, 1MXC) is essentially the same as that found in MAT I ternary complexes (1O93 and 1O9T). Moreover, both enzymes hydrolyze ATP to ADP and P_i, and hence in some of the complexes where ATP was added, ADP was observed. The methionine binding site determined for MAT I in complexes with the methionine analogs *L-cis*-AMB and AEP is also similar, with stacking against Phe-251 (Figure 5A). In addition, the three phosphates appear also in analogous positions for rat and *E. coli* enzymes [23, 159], the P_γ being oriented against a putative P-loop previously identified by photoaffinity labeling [166]. However, differences arise between ligand locations in ternary complexes. Thus, in the rat MAT I•ATP•methionine complex (1O9T,

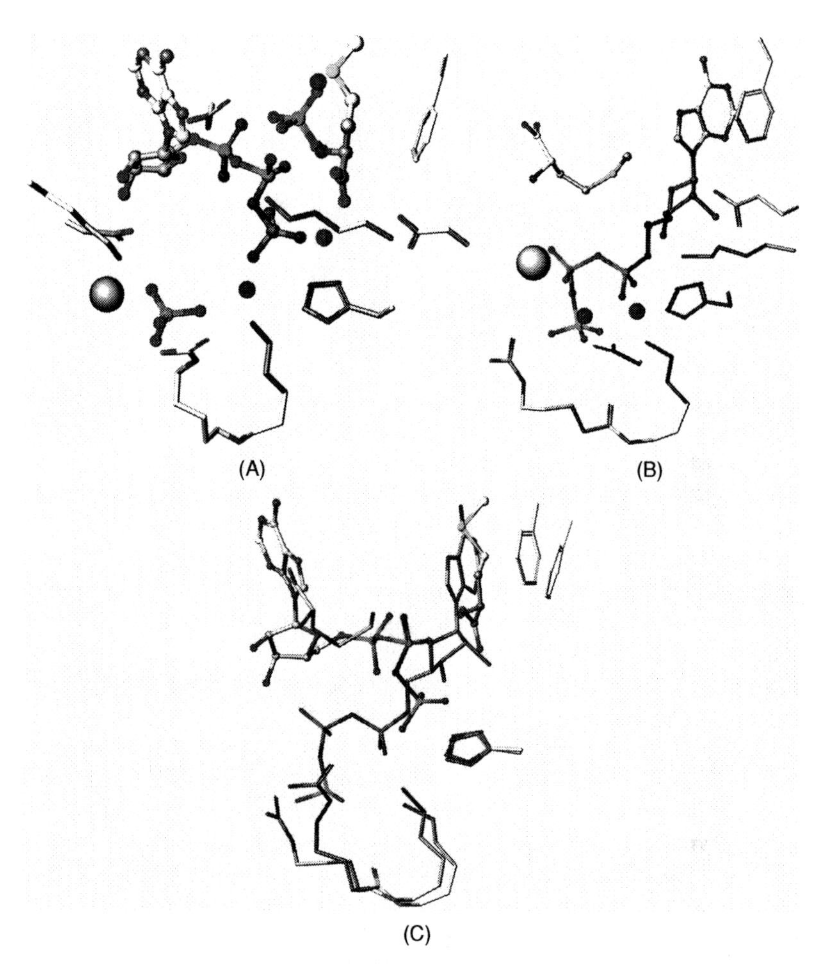

FIGURE 5. Active site structures of MATs: (A) MAT I B in complex with ATP and methionine; Mg^{2+}, K^{+}, and Pi are also in the active site (pdb code 1O9T); (B) *E. coli* MAT in complex with AMPPNP and methionine; the 2 $Mg^{2+}$and 1 K^{+} are shown (pdb code 1RL7); and (C) overlay of selected regions of the active sites of cMAT and rMAT. The exchanged orientation of the substrates is shown when the protein portion of the structures is aligned, illustrated by the position of the active site histidine. (See insert for color representation.)

2.70 Å), additional density compatible with AdoMet appears in a different orientation than ATP. The cMAT•AMPPNP•methionine complex (1P7L, 2.50 Å) shows interchanged positions for the methionine and adenine moieties, which is observed also in active sites containing AdoMet + PPNP (which were formed by reaction of AMPPNP and methionine in the crystal (Figure 5B; pdb codes 1PL7 and 1RG9). Figure 5C shows the ligand positions when the *E. coli* and rat MAT I protein structures are superimposed (see the active site histidine (residue 14 or 30, respectively) at lower right). This AdoMet location in the *E. coli* complex superimposes on that for AdoMet in the MAT II-AdoMet structure (2PO2) Figure 5B). In the structures of the MATII•AdoMet and cMAT•AdoMet•PPNP complexes, the adenine ring stacks against the phenylalanine in the position occupied by *L-cis*-AMB and AEP in the MAT I complex structures. The location of two phosphoryl groups (which might be expected to be relatively well located in the X-ray data due to the grouping of heavy atoms) is preserved throughout. At present, the origin of the differences between AdoMet locations in MAT I and *E. coli* or MAT II structures are unclear, but could either reflect the various other ligands bound (see the legend to Figure 5) or represent the positions for AdoMet as a product and as an inhibitor. Available mutagenesis data for the two proteins do not discern which is the functional configuration. Resolution of this difference will be of major importance in understanding the enzyme mechanism.

Among the questions that await a structural solution are the structure of the regulatory β subunit and its sites of interaction within MAT II isozyme. The high sequence similarity of the β subunit to nucleotide sugar reductases of known structure, such as dTDP-rhamnose reductase (26 % identity, 49 % similarity over 235 residues), suggests that the protein fold can be recognized by homology. The structure of dTDP-rhamnose reductase (pdb code 1vl0 with bound NADH, 2.05 Å resolution) shows a NAD-binding Rossman fold. No available data indicate the sites at which the MAT II α and β subunits interact or if NAD(H) is involved. It will also be intriguing to see the structure of an archaeal type MAT since the sequence similarity to the eukaryal/bacterial MATs is very low, and differences in circular dichroism spectra suggest that the proteins have different α-helical content, perhaps implying different topologies. Nevertheless, the few residues that are conserved in a sequence alignment are primarily located in the active site of the bacterial/eucaryotic MATs [134]. There are no reported structures of proteins with significant sequence homology to the archaeal MAT.

B. KINETICS AND MECHANISM

Studies of the kinetic mechanism of MAT catalysis date to the pioneering work of Mudd in the 1960s (reviewed in [167]. Many of the seminal reports are unfortunately complex because before the studies of Chiang and Cantoni [115], it was not realized that in many organisms, there are copurifying isoforms. Even in systems with pure isozymes, kinetic studies have been challenging, due to several unusual behaviors: (1) low concentrations of AdoMet activate its own synthesis by yeast MATs [112, 168], (2) the rat MAT III shows complex hysteretic behavior with a temporal lag phase before the steady state rate is attained [73], and (3) severe noncompetitive inhibition by the product AdoMet, particularly with cMAT [121]. Nevertheless, the overall picture that has emerged for MAT from various organisms is remarkably consistent. A common overall physical basis for the kinetic mechanism appears applicable; however, among MATs, there are physiologically significant variations in the detailed rate constants (and equally important from an overall perspective, the ratios of the rate constants and thus relative K_M values).

The two yeast MATs show negative cooperativity for both ATP and methionine [115]; the nonhepatic rat MAT II isozyme displays negative cooperativity for methionine [169], while the rat MAT III shows positive cooperativity with a Hill coefficient of 1.8 [169]. Hence in many cases, the concentration dependences of substrate saturation are referred to $S_{0.5}$ values rather than K_M values.

All studies point to a sequential kinetic mechanism, as illustrated in Figure 6. There is (1) formation of a E•ATP•Met complex, (2) reaction to form an E•AdoMet•PPP_i complex as an obligatory tightly enzyme-bound intermediate, and (3) the subsequent hydrolysis of PPP_i before regeneration of the free enzyme [112, 121, 170]; reviewed in [167, 171]. Variations in rate constants among MATs from different organisms, combined with a diversity of experimental conditions, can account for the observed differences in steady-state kinetic mechanisms. These differences are seen in the apparent order of substrate binding, for example, with ATP being first (most eucaryotic MATs and the archaeal *M. jannaschii* MAT) vs. random addition (e.g., *E. coli*, although in this case in the preferred order of binding ATP also binds first [170]). The order of product release is also variable with the most obvious path being taken by the *E. coli* enzyme, where the products are liberated in order of diminishing affinity, i.e., P_i before PP_i, before AdoMet.

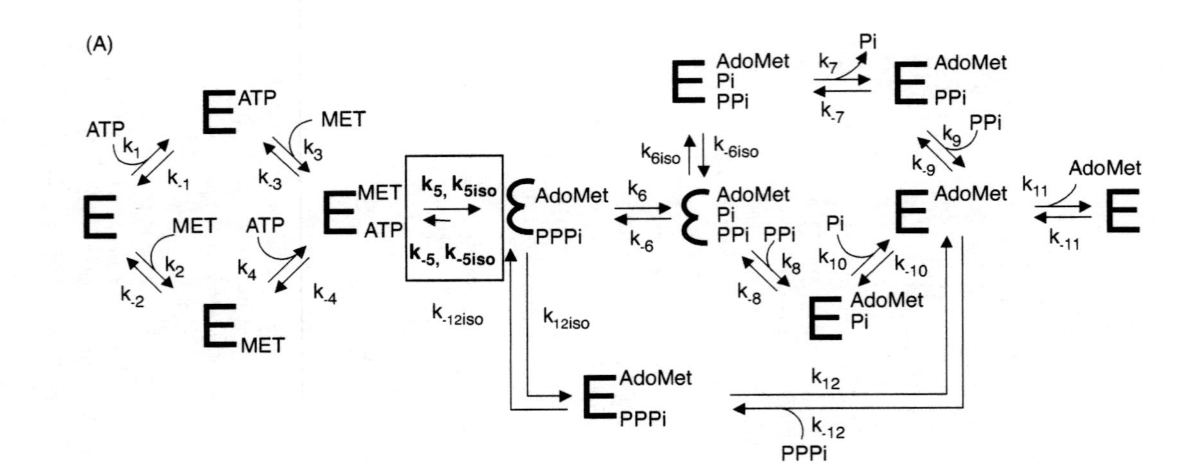

FIGURE 6. Kinetic mechanism. (A) The mechanism for the *E. coli* MAT is shown. Other MATs prefer to bind ATP first and have different orders of product release. The protein conformational change, probably due to loop movement, is shown by the change in font of **E** to "Ɛ." (*Continued*)

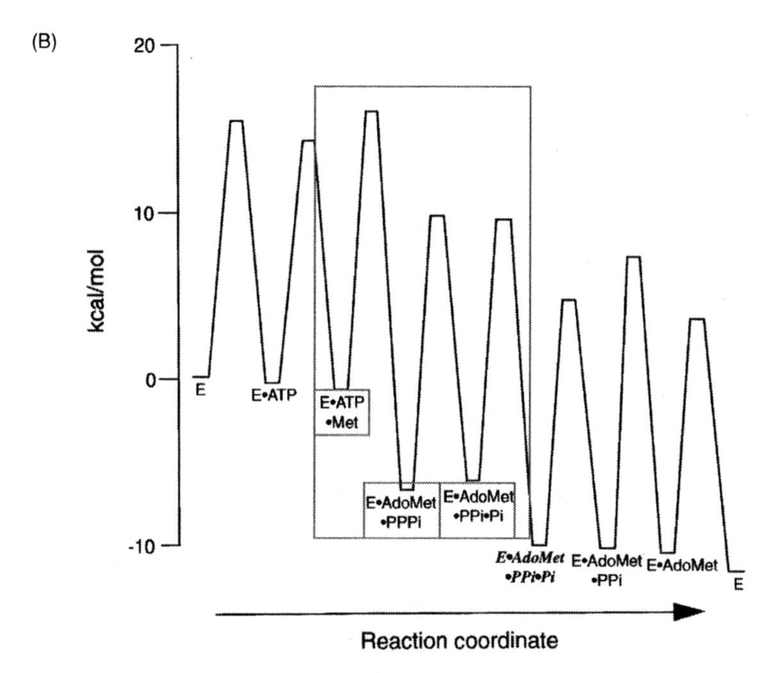

FIGURE 6. (*Continued*) (B) The free energy changes associated with each step are illustrated for physiological concentrations of substrates and products; the chemical reaction steps are included in the box.

In terms of interconversion of enzyme-bound substrates and products, extensive early work by Mudd and others demonstrated that the only detectable reaction intermediate was PPP_i [111]; no covalent enzyme-intermediates have been identified, either in AdoMet formation or in PPP_i hydrolysis. PPP_i is hydrolyzed in a predominantly oriented fashion; >95% of the P_i originates from the γ-phosphoryl group of ATP, and PP_i from the α, β groups [110]. This demonstrates that PPP_i is tightly bound during turnover so that the rate of escape from the enzyme, or even of reorientation of this symmetric compound within the active site, is slow compared to hydrolysis. The position of the γ-phosphate is preserved in the crystal structures to date, showing the active site structure at different steps of the reaction in which movement of the α- and β-phosphates, as well as on the ribose moiety, is observed [161, 164, 165]. The site of bond cleavage as $O_{\beta\gamma}$-$P\gamma$ was shown by the incorporation of ^{18}O from H_2O^{18} solely into P_i [110].

The AdoMet activation of its own synthesis in the reaction catalyzed by the yeast MATs [111, 168] and MAT III [169] deserves note: this phenomenon appears to reflect the escape of AdoMet from the E•AdoMet•PPP_i complex before PPP_i hydrolysis, which results in E•PPP_i hydrolysis becoming rate limiting in turnover. The rate of exogenous AdoMet binding to E•PPP_i becomes comparable to the rate of E•PPP_i hydrolysis and the hydrolysis rate accelerates after AdoMet rebinding to form E•AdoMet•PPP_i complex. In general, MATs catalyze the hydrolysis of added PPP_i, and the rate of this reaction is increased by added AdoMet. Most MATs show the anticipated product inhibition of the overall reaction by AdoMet.

C. CHEMISTRY OF CATALYSIS

A significant contribution to understanding the chemical mechanism of catalysis was the determination of the stereochemical course of the yeast MAT reaction using chiral [5′-^{2}H]-ATP [172]. The chiral [5′-^{2}H]-AdoMet was formed with inversion of stereochemical conversion at carbon-5′. This result demonstrated that an odd number of displacements take place at carbon-5′, which is most simply attributed to a direct displacement of the PPP_i chain from ATP by the sulfur of methionine. Subsequent kinetic isotope effect measurements with the *E. coli* enzyme showed a 13% reduction in catalytic rate when ^{14}C replaced ^{12}C at the C5′ of ATP, which demonstrates that the bonding change at C5′is involved in the rate-limiting step of the reaction [173]. The magnitude of the ^{14}C effect was near the theoretical maximum, and there was no detectable secondary isotope kinetic effect from 5′-^{3}H; the observed isotope effects from ^{35}S and [methyl-^{3}H]methionine showed the involvement of methionine in the displacement step. The combined data demonstrate that AdoMet is formed in a classical S_N2 displacement mechanism with simultaneous attack of the sulfur of methionine on the C5′ of ATP and displacement of the PPP_i chain [173].

The unique catalytic task of MAT is thus to facilitate the attack of methionyl sulfur on ATP to form AdoMet, a reaction in which charge separation occurs to yield the sulfonium cation and an additional negative charge on the polyphosphate chain. It is remarkable that when bound to the enzyme, the equilibrium for the formation of AdoMet and PPP_i lies far towards the products, i.e., at the active site, the PPP_i hydrolysis step is not needed for product accumulation [111, 121, 170]. Rather, PPP_i hydrolysis facilitates product release by the cleavage of the tightly bound PPP_i intermediate to the more weakly bound PP_i and P_i.

The mechanism of hydrolysis of the PPP_i intermediate appears to be similar to that of the myriad of metal-ion-dependent ATPases and other phosphatases [174]. However, the chemical means by which water is activated to react with the PPP_i and thus split the polyphosphate chain into PP_i and P_i is unknown. The hydrolysis reaction requires Mg^{2+}, and in many MATs, it is stimulated by K^+, although the effect of K^+ is typically not as large as is seen for AdoMet formation. It is not clear from the available crystal structures whether a metal-activated water (or hydroxide ion) is involved in this step of the reaction, or whether a protein group activates water. The active site arginine (R244 in cMAT, R265 in MAT I) is critically important for orienting the PPP_i for hydrolysis [175]. It is an intriguing possibility that one of the residues in the conserved Lys-Arg sequence may be unprotonated and act as the catalytic base; an unprotonated arginine has been recently proposed to be the catalytic base in other hydrolytic reactions [176]. However, the R244L mutant of cMAT had less than a tenfold decrease in tripolyphosphatase activity [175]; the K245M mutant had >100-fold decrease in PPPase activity, but CD spectra showed that it was incorrectly folded, precluding definitive interpretation [177].

Kinetic and spectroscopic studies of cMAT have revealed that protein conformational changes are coordinated with catalysis [163, 170, 178]. Presteady state and solution viscosity dependence kinetic studies of cMAT show that a conformational change occurs in conjunction with the formation of AdoMet and PPP_i, and a second change occurs after PPP_i hydrolysis [163, 170, 178]. These changes probably involve movement of the protein loop that covers the active site in some crystal structures but is disordered in others [23, 159–161, 165]. This loop is the least conserved region of the protein sequence among eucayal/bacterial MATs and varies in length from 6 to 19 residues [3]. The closure of the active site loop would prevent release of the PPP_i intermediate before hydrolysis. The movement of protein loops to sequester reaction intermediates while allowing access of substrates and release of products is a common feature in enzyme-catalyzed reactions [179–181]. Blockage of substrate access to the active site of MAT I/III by nitrosylation has been suggested as a physiological regulatory mechanism [18]. NO modification of C121 in this flexible loop may stabilize its closed conformation, thus impeding the entrance of the substrates for catalysis.

D. METAL ION ACTIVATION

MAT from all sources requires Mg^{2+} or another divalent metal ion for activity, and where it has been examined, a monovalent cation such as K^+ has been found to enhance the reaction rate on the order of 100-fold (studies that have specifically addressed the cation activation include those in [29, 113, 115, 121, 182]). The monovalent cation effect has not always been noted because other ions such as Na^+ and $NH_4{}^+$ can substitute for K^+, and these cations are commonly present as counter ions of ATP or buffer components. Crystal structures show that the active site of MAT is highly polar, with many charged residues. This is consistent with the course of the reaction in which additional charges accumulate as first AdoMet and PPP_i form, and then upon creation of PP_i and P_i, presumably with release of one proton from water. Thus, a role for cations in stabilizing the new negative charges is easily rationalized.

Binding studies in conjunction with EPR and NMR spectroscopic investigations of cMAT have shown the presence of two divalent cations and one monovalent cation at the active site [183–186]. EPR studies of the Mn^{2+}-activated enzyme showed that two Mn^{2+} ions bind very closely together, which crystallographic studies confirmed, showing two divalent cations 5.2 Å apart, with both binding to the phosphate groups; one ion coordinates all three α, β, and γ groups, while the second coordinates the α- and γ-phosphoryl groups. The coordination of two metal ions to the leaving α-phosphoryl group could play a substantial catalytic role by stabilizing the negative charge formed upon C–O bond rupture. A monovalent cation binds at distances of 3.8 and 8.1 Å from the two divalent ions. The sole protein ligand to this ion is the carboxylate of E42, and mutation of this residue to glutamine abolished monovalent cation activation [187]. Neither crystallographic, solution spectroscopic, nor mutagenesis studies have indicated that the monovalent cation directly binds to the substrates, and this ion appears to play an indirect role in organizing the active site structure [187]. The crystal structure of the *E. coli* enzyme revealed an additional monovalent cation bound between two subunits in the tight dimer; this ion appears to play a role in dimer stability and is conserved in the MAT I structures. The crystal structures of the rat MAT I enzyme have shown as many as three Mg^{2+} at the active site [23, 165]. One of these cations interacts with the carboxylate of the substrate methionine or an analog such as *L-cis*-AMB or AEP; such an interaction is not observed in any of the cMAT structures. The two additional PO_4 are each within 4 Å of the phosphates of ATP (1O9T) in the MAT I structures; the significance of these phosphate ions are unclear.

E. MUTAGENESIS PROBES OF THE MECHANISM

The crystal structures of MAT have directed mutagenesis studies of active site residues. In these, it has been advantageous to examine possible selective effects of the mutation on both AdoMet synthesis and on the hydrolysis of added PPP_i. The high evolutionary conservation of the active site residues implies little tolerance for variation; therefore, it is not surprising that most mutations impair the enzyme function. For the *E. coli* MAT, all mutations of polar active site residues have had larger effects on the AdoMet formation step than on the PPP_i hydrolysis reaction [163, 175, 177, 187, 188], which is consistent with the singular active site architecture known for the former activity and the variety of known ways that enzymes catalyze hydrolysis of polyphosphate chains. The only mutations that affected the tetrameric state of the cMAT were C89A and C90S from which separate and noninterconverting dimers and tetramers were isolated, with ca. tenfold reduction in k_{cat} for the tetramer and an additional 20- to 30-fold reduction in the dimer [189].

The most informative mutagenesis result for the *E. coli* enzyme may be that mutation of the active site arginine-244 to leucine (see Figure 5B) resulted in a reduction of 1000-fold in k_{cat} for AdoMet formation and a remarkable ability of the PPP_i intermediate to reorient within the active site before hydrolysis, showing the importance of the guanidinium in stabilizing the intermediate conformation [175]. In contrast to the R244L and R244H mutants of cMAT, which remained tetrameric and retained wild-type secondary structure as indicated by CD spectra, the R265H mutant of MAT I eluted in gel filtration chromatography with a calculated Mr of 41,000, corresponding to a monomer [190]. This mutant was reported to have no AdoMet synthesis capacity (2% of wild type), but preserved the tripolyphosphatase activity, its V_{max} for hydrolysis being similar to that shown by the wild-type MAT (128 vs. 156 nmol/minute/mg). A R265S mutant was also found to be a monomer, but in this case along with a 99% reduction in AdoMet synthesis, a fivefold decrease in tripolyphosphatase activity was also detected. These results together suggest the involvement of the positive charge at 265 in hydrolysis. No association of the mutant monomers was observed under assay conditions, as judged by gel filtration chromatography (41.6 kDa). However, coexpression with the wild-type protein produced hetero-oligomers (90 kDa on gel filtration chromatography) with tripolyphosphatase activity. This arginine 244 (or 265 in rat) residue is involved in a salt bridge with E58 of the opposite subunit, one of the few polar interactions formed at <3.5 Å between monomers in the dimer, and this interaction is preserved in MAT I crystal structure [23, 164].

Active site mutants (D180G, K182G, F251D, and F251G) prepared on rat MAT I and overexpressed in *E. coli* showed no AdoMet synthesis capacity, but preserved PPP_i hydrolysis activity. It is noteworthy that the PPPase activity of F251 mutants is double than that shown by the wild-type protein. A possible explanation for this fact may be that the reduction in size of the lateral chains in those mutants facilitates accessibility for PPP_i. Moreover, all these mutants were able to bind ATP with K_d values similar to that of the wild type (10.54 ± 0.85 mM) [23]. Thus, mutations at the methionine binding residue (F251) only prevent AdoMet synthesis.

F. SOLVENT EFFECTS ON MAT ACTIVITY—PRACTICAL CONSEQUENCES

A phenomenon that has aided discrimination among the mammalian MAT forms is the approximately tenfold DMSO activation of the rat MAT III isozyme [191, 192], the molecular basis for which has not been resolved. The ability of molar concentrations of mercaptoethanol, acetonitrile, and other compounds to alleviate the potent noncompetitive inhibition of the *E. coli* MAT by AdoMet has facilitated synthetic applications [193]. Significant changes in the far-UV CD spectra of cMAT are associated with the addition of these cosolvents, suggesting substantial alterations in protein folding. No analogous structural data have been reported for the mammalian MATs.

X. MAT FOLDING

Studies on MAT folding have been carried out mainly in vitro using rat MAT I/III with urea and thermal denaturation. The high conservation in sequence and structure found between α subunits of eucarya and bacteria make these results extrapolable to most of the known MATs, probably with the exception of the large subunits observed in some plants. In vivo data are restricted to the identification of cMAT as one of the proteins folded by the *E. coli* GroEL/GroES complex [194].

A. MAT III FOLDING UNDER EQUILIBRIUM CONDITIONS

Folding of MAT III, the rat liver MAT dimer, was studied using urea as the perturbing agent. The enzyme used for these studies was either the DTT-refolded protein from *E. coli* inclusion bodies at concentrations below 0.2 mg/mL [195] or the rat liver purified dimer [196]. In both cases, the equilibrium experiments showed the reversibility of the process that occurs through a three-state mechanism (Figure 7). Data were obtained by activity measurements, fluorescence spectroscopy, CD, sedimentation velocity, and gel filtration chromatography. These techniques allowed following of the process at different levels, i.e., the active site, the tertiary structure, the

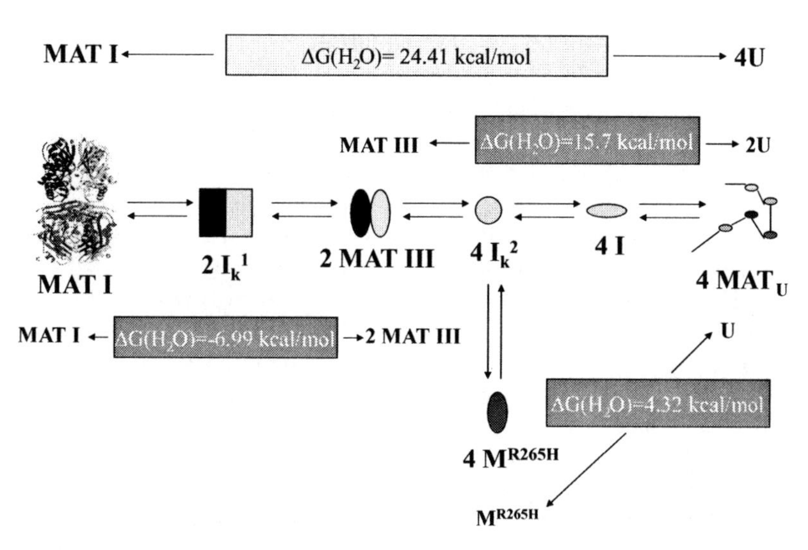

FIGURE 7. Folding pathways for MAT I/III. The figure shows a schematic representation of the data available for MAT I/III unfolding. Free energy changes calculated for each step are also included. The different intermediates are denoted I, I_k^1, and I_k^2, whereas M represents the monomeric mutant R265H and U the unfolded state. (See insert for color representation.)

secondary structure, and the subunit association level. The loss of activity preceded the changes detected by other techniques, and occurs earlier for AdoMet synthesis ($D_{50\%} = 0.3$ M) [195] than for tripolyphosphate hydrolysis ($D_{50\%} = 1$M) [196]. ANS binding showed a similar trend, exhibiting a single transition at these low urea concentrations ($D_{50\%}$ of 1 M) [196]. Both fluorescence spectroscopy and far-UV CD showed two transitions with a plateau at 1–2 M urea, an interval at which monomers were detected by both gel filtration chromatography and sedimentation velocity [195, 196]. Maximal fluorescence emission for MAT III appears at 334 nm, with intensity decreasing by denaturation as the maximum is displaced to 355 nm [195]. Thus, in the three-state mechanism for MAT III unfolding, the activity is lost before dissociation to a monomeric intermediate (I).

The structure of intermediate (I) allows limited proteolysis that produces a 3-kDa fragment, which as deduced from mass spectroscopy data, corresponds to an *N*-terminal cut at lysine 33 [196]. Other characteristics of intermediate (I) include a 2.5-fold decrease in tryptophan fluorescence intensity, preservation of a 70% of the secondary structure of native MAT III, and a 75% lower ANS binding as compared to that of the native dimer [195, 196].

All together, these data indicate that this intermediate (I) shows features of a molten globule. Accumulation of the intermediate was estimated to be maximal between 2 and 2.5 M urea as deduced from tryptophan fluorescence and CD data. Unfolding data obtained using the active site monomeric mutant R265H [190, 196] also show the presence of an intermediate, thus confirming its monomeric structure [196]. However, this intermediate seems less compact than monomeric R265H as judged from its larger hydrodynamic volume. Data obtained with this mutant, however, should be considered with care, since analogous mutants obtained for cMAT behave as tetramers [175].

The apparent stoichiometry of ANS binding was determined to be 1 mol ANS per MAT III subunit, suggesting its binding to a specific site [196]. This location appears to correspond to the start of the loop at entrance to the active site, where there is a similar environment to that found in the crystal structure of another ANS–protein complex [197]. Monomers in the crystal structure of MAT I [23] have a large flat hydrophobic contact surface, thus solvent exposure of this interface would lead to a rearrangement in the subunit for hydrophobic burial. Such a rearrangement is suggested by the different ANS-binding properties shown by the dimer and monomeric intermediate (I), as well as by the high value of the m_I constant calculated for the dissociation process in urea denaturation. Unfolding of the intermediate would further require minimal changes to render the unfolded protein.

Analysis of all these data using models for different unfolding mechanisms showed the best fit for a three-state process involving a monomeric intermediate [195, 196]. The global m coefficient for urea dependence was 7.25 kcal/mol/M, whereas the value obtained for the global free energy change was calculated to be 15.7 kcal/mol. This low dimer stability toward urea denaturation is mainly derived from the low $\Delta G_I(H_2O)$ value that represents 50% of the global stabilization energy for MAT III (subunit association). The importance of the monomeric intermediate (I) is highlighted by refolding experiments that include a two-step process in which the 2 M urea intermediate is favored before the total elimination of denaturant. Such a two-step process allows not only a 100% recovery of the activity, but also provides a better yield when using inclusion bodies as the enzyme source [198].

MAT III is one of the many proteins made up of two identical subunits. The proliferation of such an arrangement may involve some advantage for the cell. Recently, it has been proposed that the oligomerization process itself might tune the enzymatic function [199]. These authors indicate that the formation of intersubunit contacts influences the biological activity by allowing very subtle conformational changes in the active site in such a way that oligomerization can indeed activate the monomeric subunits. Moreover, the fact that a 50% of the free energy for unfolding represents dissociation processes is consistent with the idea that contacts at the surface hidden between the monomeric subunits plays a fundamental role in the stabilization of oligomeric proteins. For dimers of a given size, those proteins that have a folding intermediate displayed less empty volumes [199]. The role of dimerization in proteins of class B ($N_2 \leftrightarrow 2I \leftrightarrow 2U$), such as MAT, is mainly structural according to Mei et al. [199]. The dimerization process for this type of proteins might find a rationale in the protection and stabilization of those molten globule states that alone are not able to complete their self-assembly process.

B. KINETIC INTERMEDIATES

Two kinetic intermediates have been identified during MAT I unfolding by fluorescence spectroscopy (Figure 7). The first one (I_k^2) occurs during the MAT III-intermediate I transition [196]. The data fit to a single exponential, corresponding to a monomolecular process, thus indicating that the folding pathway includes at least one kinetic intermediate (Figure 7). More evidence for the presence of such intermediate was obtained by ANS binding in the presence of tripolyphosphate, where a transient binding of the dye was observed at 1–2.3 M denaturant. These results suggest that even though I_k^2 is not well populated, it must bind a large amount of dye to allow detection. No direct information about the association state of I_k^2 was obtained, and hence its monomeric state was inferred from other results [196]. Moreover, Sánchez del Pino et al. speculated that the shape of I_k^2 could be that of a folded or a well structured monomer and hence have a large solvent-exposed hydrophobic interface, which should be energetically unfavored, leading to a very fast association. In addition, such a conformation would also explain the high unspecific binding of ANS to I_k^2. This kinetic intermediate is also shared by the monomeric R265H mutant, prior to attaining its final folding, thus suggesting its monomeric character.

Another kinetic intermediate was identified in the tetramer–dimer association step using DTT-refolded MAT [17]. DTT-refolded tetramers and dimers appear in a concentration-dependent equilibrium and are not distinguishable by either tryptophan fluorescence or CD [198]. However, MAT I dissociation could be followed by ANS fluorescence [17], showing an exponential decay to ANS-dimer fluorescence in a process that yields the best fit as a single exponential. Calculations of the half-life of MAT I render a value of 14.69 ± 0.5 seconds [17], in contrast to the 858 seconds previously obtained for the cytosolically overexpressed oligomer [200], but the studies differed in the temperature used. Again, the association state of this new kinetic intermediate ($I_k{}^1$) cannot be determined directly due to its short lifetime. However, the fact that its ANS fluorescence intensity is the same as that shown by the tetramer suggests that it might be a dimer that has just separated from the tetramer, before undergoing the changes needed to achieve the MAT III conformation. These changes cause reduction of its affinity for methionine [7, 8], as well as acquisition of DMSO-activating capacity, and a higher hydrophobic character [191]. Moreover, in order to be detected by this technique, these structural changes should occur close to the ANS-binding site that, as postulated by Sánchez del Pino et al., is located close to the active site loop [196]. The free energy change for MAT I dissociation was calculated as −6.99 kcal/mol, which along with the data derived from MAT III unfolding, give a global free energy change for MAT I unfolding of 24.41 kcal/mol [17]. Further analysis looking for the presence of additional intermediates in this dissociation process will rely on the use of stopped-flow techniques.

C. THERMAL UNFOLDING

Two-dimensional infrared spectroscopy (IR), activity, and tryptophan fluorescence spectroscopy were used to follow thermal unfolding of MAT [201]. The high protein concentrations needed for IR precluded the study of dimers, since the equilibrium between DTT-refolded MAT forms was clearly displaced towards the tetramer. On the other hand, the low protein concentrations needed for activity and fluorescence measurements allowed analysis of the dimer behavior. IR spectra in H_2O and D_2O are dominated by two bands at 1652 cm^{-1} and 1636 cm^{-1} attributed to α-helix and β-sheet, respectively. A band at 1624 cm^{-1} and 1626 cm^{-1} in D_2O and water, respectively, agrees with MAT I existence as an oligomer. The secondary structure composition was calculated to be 44% α-helix, 37% β-sheet, 18% β-turns, and a very low percentage of unordered structure, in agreement with structural and CD data. Thermal unfolding was irreversible by all the techniques used in this study, showing a T_m of 47–51°C [201]. Activity and tryptophan fluorescence measurements revealed one transition for dimer unfolding that occurs in the same temperature range as the IR changes for the tetramer. Such a result indicates that MAT I stability is highly dependent on that of the dimer. Moreover, the IR data show no change in the β-sheet band

(1635 cm^{-1}) with temperature, whereas the corresponding α-helix band is split in two, with a reduction in the helical percentage [201].

Analysis of synchronous and asynchronous 2-D IR maps gives information about which structures are changing and the order of events taking place at each temperature, respectively. As deduced from this study, native MAT I starts its unfolding at 37°C, the dominant crosspeak being centered at 1622 cm^{-1}, which corresponds to changes in protein contacts and beginning of aggregation. The initial changes, according to asynchronous maps, correspond to peaks of α-helix, unordered structure and β-turns, structural elements located at the external surface of the subunits [23]. Moreover, changes in the 1624 cm^{-1} band indicate alterations in the oligomerization state that could be assigned either to T–D–M or T–M processes [201]. The unfolding continues, showing aggregation at the expense of the remaining secondary structures, in coincidence with changes in activity and tryptophan fluorescence that could be ascribed either to dimer dissociation and monomer aggregation or just aggregation of the unfolded dimer. No further changes are detectable after complete activity loss. Thus, MAT I thermal unfolding starts with changes in the structures of the hydrophilic surfaces, followed by loss of tertiary structure to render a dimer, possibly enzymatically active in a narrow range of temperature, and leading, finally, to unfolding and aggregation. A special characteristic shown by MAT is the ability of some structures to interconvert during the process, a new aspect of denaturation events [201].

D. ASSOCIATION OF MAT III TO RENDER MAT I: ROLE OF THE DISULFIDE C35–C61

As mentioned before, rat liver MAT appears in a protein concentration-dependent equilibrium upon overexpression in *E. coli* [17, 198, 200], a fact that has not been observed in the MAT I and III forms purified from liver. On the other hand, GSH/GSSG-refolding renders a mixture of tetramers and dimers that do not interconvert and that can be isolated as stable entities [198]. These results raised the question as to what feature modulates the association properties of MAT. The obvious difference among both refolding procedures is their different redox potential that might be crucial for a protein that contains ten cysteine residues per subunit. Thus, analysis of the number of free sulfhydryl groups for each species was performed by chemical modification, and mass spectrometry showed ten free sulfhydryls/subunit in DTT-refolded MAT, whereas only 8 were detected for GSH/GSSG-refolded MAT I and III [17]. The previous identification of the disulfide linking

C35–C61 in rat liver purified MAT [22] prompted the analysis of a possible role for such a disulfide in stability. Such a study was performed using both refolding systems and cysteine mutants (C35S, C57S, C61S, and C69S) located in the central domain of the subunit and at disulfide bonding distance according to the MAT I crystal structure [23]. Analysis of their behavior by analytical phenyl Sepharose chromatography showed persistence of the dimer–tetramer equilibrium for GSH/GSSG-refolded C35S and C61S [17], thus suggesting a role for the residues involved in the disulfide bond in the MAT III to MAT I association process. Mass spectrometry analysis of the oxidized residues in GSH/GSSG-refolded MAT confirmed the existence of the C35–C61 disulfide, and hence that this modification stabilizes both tetramers and dimers.

The question remained as which is the step of the folding process where this disulfide is established. Two structures could be suitable for this purpose, MAT III and $I_k{}^1$ (Figure 7). MAT III could undergo oxidation, in which case association would be precluded, or evolve to render $I_k{}^1$, which then evolves to the tetramer. This kinetic intermediate could also be the subject of oxidation, a process that would then preclude the conformational changes that lead to MAT III and hence produce the oxidized tetramer. All these events would take place at the central domain that establishes the contact between dimers in the tetramer structure. The tetrameric arrangement is maintained by only five polar interactions between dimers in rat liver MAT, with a small area in direct contact with the solvent. The dissociation constant for these proteins was calculated to be 10^5 M, a value comparable to those described for proteins showing association/dissociation behavior [200, 202], but much lower than that for cMAT (10^{10} M) [120]. Such a large difference in the constants is in agreement with all the data obtained to date with cMAT and its structure, which shows a much larger contact surface between dimers. Thus, the few interactions that maintain MAT I can be easily modified/avoided by just a small rearrangement in the central domain, or by a higher flexibility of its secondary structure. The presence of the C35–C61 disulfide would stabilize this domain and facilitate the correct pattern of interactions between dimers.

E. COMPLEMENTARY FOLDING DATA OBTAINED IN MAT PROTEINS OF DIFFERENT ORIGINS

The critical role of C69 in dimer association was shown by site-directed mutagenesis studies of the rat liver protein, where C69S appeared mostly as low activity dimers upon overexpression in *E. coli* cytosol [19]. However, DTT-refolding from inclusion bodies showed an increase in the specific activity of that mutant, reaching a 64% of that of the wild type, as well as production of tetramers [17]. These data indicate a blockage of the folding pathway for C69S in the bacterial cytosol that can be overcome in vitro. Cysteines at residues 61 and 69 are specifically found in liver MATs and hence can be related to the special behavior exhibited by these MAT forms. Both residues, as well as C57, are located in β-sheet B2, the position of C69 being close enough as to form a disulfide bond with C57 upon a slight torsion of the main protein chain [23]. However, it could be possible that under certain circumstances, such as during folding, this second disulfide is formed either alone or in concordance with the C35-C61 disulfide, a possibility that would be precluded for C69S. Thus, blockage of the normal folding pathway could lead to disulfide arrangements that would certainly render alterations in β-sheet B2, the contact area between dimers. Such changes would obviously alter the oligomerization pattern.

Folding studies carried out with *Leishmania donovani* MAT showed proper folding only under reducing conditions, with cysteines 22, 44, and 305 being crucial for the global process [146]. Conformational transitions were observed via fluorescence quenching during refolding from *E. coli* inclusion bodies in the presence of DTT. Such transitions generated fast and large events that took place during the first 2 hours of reaction and were found to be essential for activity recovery.

A mixture of tetramers and dimers was also described for the cMAT mutants C89A and C89S [189]; C89 is located at the border between dimers with no involvement in strong interactions [164]. However, C89 is a conserved cysteine residue for MAT proteins of different organisms, thus suggesting an important role for such amino acid in dimer production. If this is the case, the structure of the dimer interface might be changed substantially by cysteine to alanine replacement, leading to more labile liver-type arrangement.

XI. MAT INHIBITORS

The multitude of metabolic roles of AdoMet, and its involvement in modulation of cell growth, have resulted in an enduring search for MAT inhibitors with in vivo and in vitro utility. A selection of the most effective inhibitors produced to date, some of which have been used for structural studies, are

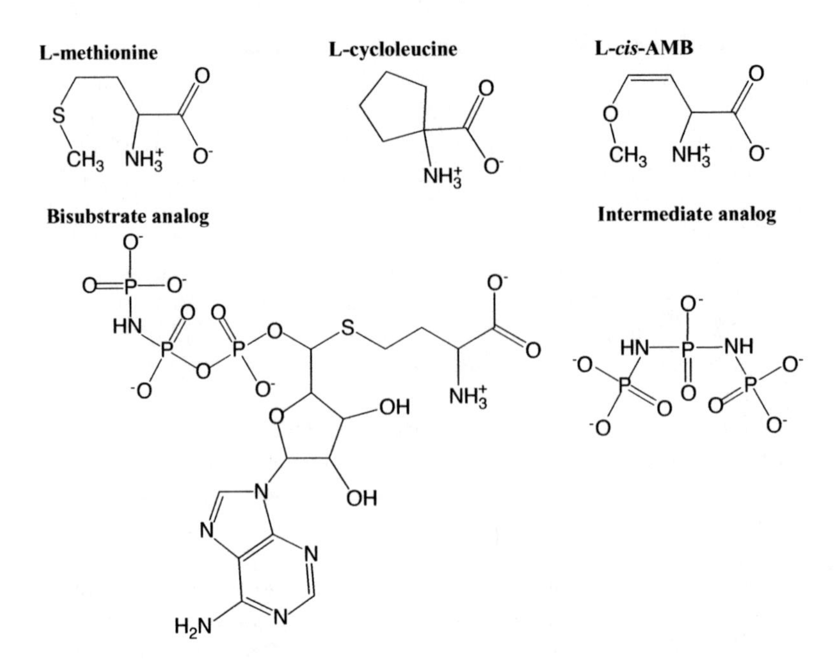

FIGURE 8. Structures of selected inhibitors of MATs. K_i values for various MATs are listed in Table 2.

shown in Figure 8, and their inhibitory properties are included in Table 2. The most widely used inhibitor has been the amino acid analog cycloleucine, since it is long established and commercially available [206]; with its K_i in the millimolar range and possible additional effects in vivo, it is clearly not an optimal inhibitor.

The inhibitor design process has included analogs of both substrates ATP and methionine, both which have been tested in cancerous cell lines. Much of this work has been carried out in the laboratories of Hampton and Sufrin (cf. [203, 204]); the two laboratories had significantly different approaches and have provided much information that we cannot analyze in depth in this review. Their results will be invaluable to the future development of potent, bioavailable, isozyme selective inhibitors. Hampton and coworkers strove for maximal inhibitory potency by linking the methionine methyl to the reactive 5′ position of the nucleotide substrate, forming bi-substrate analogs; isozyme selectivity was achieved by further modification of these inhibitors. These compounds had submicromolar affinity for

TABLE 2
Selected Inhibitors of MAT

Analog	Characteristics	K_i (μM)	References
Cycloleucine, 1-amino-cyclopentane carboxylic acid		209–1633	[206]
1-aminocyclobutane carboxylic acid	Less potent than cycloleucine	1.5–12.4 mM[a]	[206]
1-aminocyclo-hexane carboxylic acid		8.2–57.4 mM[a]	[206]
(±)-2-aminobicyclo [2.1.1]-hexane-2-carboxylic acid	K_i 2- to 7-fold lower than cycloleucine	80–680	[206, 287]
L-2-amino-4-hexynoic acid	As potent as cycloleucine against control enzyme or Novikoff hepatoma, and 2- to 3-fold more potent than cycloleucine against normal isoenzymes.	1500	[288]
Z-*L*-2-amino-5-chloro-*trans*-4-hexenoic acid		550	[288]
L-ethionine[a]	All of them have lower affinity than cycloleucine to the active site. [a]The most potent in this group, also substrate for normal and tumor isoenzymes, as well bacterial MAT	790–15.4 mM[a]	[206]
S-*n*-propyl-*DL*-homocysteine		39–113 mM[a]	[206]
S-*n*-butyl-*DL*-homocysteine		149–234 mM[a]	[206]
Se-*DL*-ethionine[a]		1.98–11.1 mM[a]	[206]
L-2-amino-4-methylthio-*cis*-but-3-enoic acid	Less potent than the ether analog due to nonplanrity of the sulfur with the olefin	5.7–21	[204]
L-2-amino-4-methyl-*cis*-but-3-enoic acid	The most potent methionine analog inhibitor	5.7–21[b]	[204, 289]
6-(n-Butylthio)-9-[5′(*R*)-*C*-(*L*-homocystein-*S*-yl-methyl]-β-*D*-ribofunanosyl]purine 5′-β,γ-imidotriphosphate	Bisubstrate analog, selective for tumor isozyme	0.1–1.7	[203]
Diimidotriphosphate	Slow tight binding intermediate analog; tested only with bacterial and archaeal enzymes	2 nM	[211]

Analogs of the substrates that have been tested as possible MAT inhibitors. Selected structures are shown in Figure 8. The table shows some of the most effective among them.
[a]The range indicates the lower and higher values for liver MAT isoenzymes.
[b]Assayed against L1210 murine leukemia cells.

the rat isozymes; however, the compounds are only available by complex synthetic routes, and this work has not been extended to yield bioavailable compounds.

Sufrin has focused on methionine analogs, which are permeable to many types of cells, and her *L-cis*-AMB (*L*-2-amino-4-methyl-*cis*-but-3-enoic acid) has proven to be a useful tool for in vivo inhibition of MAT, notably in trypanosomes [205]. A related analog, *L*-2-amino-4-methythio-*cis*-but-3-enoic acid (*L-cis*-AMTB), is a competitive inhibitor against the amino acid with K_i of 21 and 5.7 μM against MAT I and II, respectively. It is also a substrate with K_m values of 555 and 33 μM. This inhibitor is two- tenfold more potent against hepatoma MAT than against liver isoenzymes [204]. *L-cis*-AMTB has more structural fidelity to methionine than the oxygen analog *L-cis*-AMB. The highly planar conformation of the hydrophobic enol ether chain of *L-cis*-AMB was anticipated to be exhibited by *L-cis*-AMTB. However, X-ray crystallography demonstrated a slight torsion and a consequent deviation in planarity on the thioenol ether side chain [204], which may be reflected in the difference in affinity.

Among the methionine analogs, cyclic 5-membered ring amino acids bind with optimal affinity to all of the isozymes tested [206]. Recent studies with 4,5-epoxide and 4,5-epithio methionine analogs have provided inhibitors with affinity as high as 7 μM for recombinant rat liver MAT, although none were found to irreversibly inhibit the enzyme by covalent modification [207]. The 2′-hydroxyl of ATP and the dianionic form of the γ-phosphate of ATP may play a role in the reversible binding to these enzymes. The low V_{max} exhibited by 2′-deoxy-ATP in MAT II suggested a possible implication of the 2′-OH in catalytic events [208, 209]. All of the 2-, 3-, and 4-mono-*C*-methyl derivatives of methionine assayed are low affinity selective inhibitors of MAT II and tumor MAT, and the three exhibiting the highest inhibition are competitive inhibitors against methionine [210].

Diimidotriphosphate (O_3P-NH-PO_2-NH-PO_3), which is a nonhydrolyzable analog of the tripolyphosphate intermediate, is a slow binding inhibitor with nanomolar affinity for the *E. coli* and *M. jannaschii* enzymes; studies with eucaryotic MATs have not been reported [134, 211]. The slow binding and high affinity were associated with a protein conformational change after the initial binding, events that were attributed to ionization of one of the –NH- groups when the compound is bound to two metal ions at the active site [161]. The commercially available imididiphosphate (O_3P–NH–PO_3) is also a slow binding inhibitor of cMAT with a K_i of 0.8 μM [211].

XII. ROLE OF MAT IN DISEASES

Data from many studies link changes in MAT activity to the development of different types of disease. Such changes lead to the corresponding alterations in AdoMet levels and hence in the many reactions in which this compound is involved. Thus, in many cases, a direct relationship between the development of the disease and MAT is difficult to establish.

A. COGNITIVE AND NEURODEGENERATIVE DISEASES

Alterations in transmethylation mechanisms in neurodegenerative diseases such as Parkinson, Alzheimer, and subacute combined degeneration of the spinal cord have been demonstrated [212–215]. Similar defects have been suggested to occur in complex psychiatric disorders, such as schizophrenia and dementia [216]. In mice, monkeys, and pigs, an association has been found between a defect in methionine synthesis and/or MAT activity, and the development of myelopathies, ataxia, peripheral neuropathy, and subacute combined degeneration of the spinal cord [217–219]. Moreover, Charlton and Way showed in 1978 that AdoMet injection into the lateral ventricle of rats and mice produced tremors, rigidity, and abnormal posture [220]. Blood AdoMet and AdoHcy levels have been shown to decrease in Parkinson's patients, combining to yield a lower AdoMet/AdoHcy ratio, whereas erythrocyte MAT activity was found to be increased [215]. However, no association between AdoMet and/or AdoHcy levels, or MAT activity, and the age or the duration of Parkinson's disease has been described.

Parkinson's disease is a disorder marked by degeneration of nigrostriatal dopaminergic neurons and consequential depletion of both dopamine in the basal ganglia and melanin in the *susbtantia nigra pars compacta*. *L*-dopa administration is the major treatment for this disease, leading to an increase in hepatic and brain levels of AdoMet and AdoHcy in the short term [221]. However, AdoMet depletion occurs upon long-term therapy with this compound, which results in losses of efficacy [222–225]. Short-term treatments also produce increases in brain (48%) and liver (34%) MAT activities that parallel increments in the corresponding protein levels detected by Western blot [221]. Moreover, the levels of MAT and catechol *O*-methyltransferase (COMT) mRNA are increased by *L*-dopa, which also induces transcription factors such as NF-1 and cJun/AP-1, which are known to control MAT and COMT promoter activities. Clinical improvements (reduced tremor and rigidity) in these patients have been also reported after treatment with methionine [226]. In the presence of AdoMet, COMT methylates *L*-dopa to

3-O-methyldopa, and this treatment results in increased MAT activity in brain [227]. The frequency of administration and the duration of the treatment are related to the induction observed. A sudden rise in dopamine may be responsible for hyperkinesias that can occur following *L*-dopa administration, since motor activity has been associated with such increases [228]. On the other hand, the dimethylated product of dopamine is known to be hypokinetic [229]. Thus, it is proposed that patients from whom *L*-dopa control of the disease is long term may be resistant to the induction of MAT activity, whereas those in which the *L*-dopa treatment is ineffective may initially have enhanced MAT activity at the onset of therapy. There is also substantial evidence that indicates the neurotoxicity of *L*-dopa, for which the proposed mechanism is the generation of hydrogen peroxide and oxyradicals resulting from oxidation of dopamine in the nigral neurons and its nerve terminals. Thus, it is possible that *L*-dopa accelerates the rate of dopamine neuronal degeneration as a result of this direct neurotoxicity [230].

Schizophrenia has been also extensively studied, due to its high incidence (1% of the population worldwide). The rate of whole body methionine metabolism in these patients is reduced to one-third that of controls [231]. Significant increases in blood Hcy [216], low MAT activity in erythrocytes [232], regionally selective decreases in MAT's K_m for methionine [233], and low serine hydroxymethyltransferase [233] have also been reported. Some of these alterations were also found in individuals with dementia [212]. Exacerbation of schizophrenic symptoms was observed in response to large oral doses of methionine [234], whereas AdoMet is a reported antidepressant [235]. Profound defects in membrane phospholipids of lymphocytes have been also detected [236], phospholipid methylation being stimulated by dopamine, epinephrine, and norepinephrine, as well as serotonin in neuroblastoma SK-N-MC cells [237].

B. HUMAN MUTATIONS IN *MAT1A* GENE: HYPERMETHIONINEMIA AND DEMYELINATION

Mutations have been characterized by single-strand conformation polymorphism (SSCP) and DNA sequencing for patients showing isolated persistent hypermethioninemia (Table 3). However, MAT deficiency is not the only alteration related to the presence of high methionine levels in plasma. Such an increase could also take place transiently during the neonatal period, disappearing when the dietary intake of proteins is reduced [238, 239], or as a consequence of CBS deficiency, or due to deficiency in fumaryl

TABLE 3
Mutations of *MAT1A* Detected in Individuals with Hypermethioninemia

Exon	Gene	Protein	Character	Activity	Association State	Liver Activity	Reference
I	*C65T*	S22L	HO[a]				[255, 290]
	G113A	S38N	CHE[b]	None		68%	[244, 249]
II	*T125C*	L42P	HO				[255, 290]
	C164A	A55D	CHE				[14]
	G205A	G69S					[252]
	255ΔCA	92X	CHE	None			[244, 249]
III		G98S	HO	Normal			[249]
V	*C426T*		CHE				[14]
	539insTG	185X	HO, CHE			7%	[15, 250]
VI	*C595T*	R199C	CHE, HO	10%			[15]
	C745T	R249W					[252]
	G791A	R264H		<1%	Monomer	Low	[245, 247–249, 252]
	G791T	R264C	CHE	None			[249]
	G867A	R264C					[252]
		K289N					[252]
VII	*G870A*		CHE				[14]
	T914C	L305P	CHE				[14]
	827insG	351X	HO				[15]
VIII	*T966G*	I322M	HO, CHE	Low			[14, 249]
	G1006A	G336R	CHE	23%			[249]
	A1031C	E344A	CHE	Low			[249]
	1043, 1044del	350X	HO				[15, 240]
	G1067A	R356Q	CHE	53%		Low	[15, 291]
		R356P					[252]
	C1070T	P357L	CHE				[14, 252]
IX	*T1131C*		CHE				[14]
	G1132A	G378S	CHE	0.1%			[15, 251, 291]
	G1161A	W387X	CHE, HO	75%	Dimer		[251]
	G1188T	X396Yfs X464					[252]

[a]HO, homozygosis.
[b]CHE, compound heterozygosis.

acetoacetate hydrolase (tyrosinemia type I) or liver disease. Therefore, newborn children are routinely screened for hypermethioninemia [240]; a small percentage of those who show isolated persistent hypermethioninemia have deficiencies in hepatic MAT or glycine methyltransferase (GNMT) [241]. Definitive positive diagnosis, however, requires demonstration of reduced MAT activity in liver biopsies, a procedure that has been precluded in many cases due to the absence of pathological symptoms. Some patients present unusual breath odor, due to large increases in dimethylsulfide levels [242, 243], while others have neural demyelination [15, 214]. In cases where liver biopsies were available, a reduction in MAT's K_m for methionine was also observed [244, 245]. The plasma levels of methionine in this type of patients may be as high as 1.3 mM (35 μM for normal levels), but the presence of 50–60% MAT activity is enough for maintenance of a normal level of the amino acid [15, 240]. On the other hand, MAT activity in the erythrocytes, lymphocytes, or fibroblasts of the patients is normal [14, 243].

The first patient with MAT deficiency was an infant described by Gaull in 1974 [246], and subsequently since more than 60 individuals with mutations in the *MAT1A* gene have been identified (Table 3). Most of the mutations described were found to follow a recessive autosomal inheritance trait. However, gene tracking in families showing the G to A transition at nucleotide 791, leading to R264H, revealed a Mendelian-dominant inheritance [245, 247, 248]. Some of the mutations lead to stop codons in the ORF (Table 3) and hence to truncated forms of the protein. Subunits of 91 [249], 184 [15, 250], 349 [15, 240], 350 [15], and 386 residues [251] will result from these changes. In addition, a mutation at the wild-type stop codon has also been described, rendering a longer protein [252]. Only truncated forms of 349 and 350 amino acids are present in patients developing demyelination, a result that led to the hypothesis that the size of these anomalous subunits allows partial folding to an intermediate state that is able to associate with the wild-type subunits. Thus, these truncated proteins may sequester MAT2α subunits in inactive heterodimers, whereas shorter proteins (i.e., 184 residues) cannot. Heterodimerization was also suggested for the 386 residue truncated protein; however, no validation was provided [251]. The presence of truncated subunits will lead to a lack of methylated products, among them phosphatidylcholine and creatine are important for the myelin in neural sheath and neural structure [253]. These data, along with experiments carried out in rats that were given cycloleucine [254], suggested that AdoMet administration may overcome, at least in part, these problems. However, only one of the patients showing demyelination was treated with

AdoMet and responded to this therapy [214]. Kim et al. [251] also indicate that the physiological effects may be due to lack of AdoMet rather than to hypermethioninemia and suggested that elevation of methionine levels may promote a greater flux through the remaining MAT, thus alleviating AdoMet deprivation.

Polymorphisms such as that of T to C at nucleotide 426 have been also detected in *MAT1A* gene [15]. Other silent mutations in exons III (C225T rendering A75A) [248], VIII (C to T in intron 8 and C882T rendering A292A) [255], and IX (T1131C, leading to Y377Y)[248, 255] have also been observed. Another interesting mutation occurs in the last nucleotide of exon III. This change alters the splice–donor site and hence could either render normal (G98S) or anomalous splicing. Analysis of both possibilities was carried out, showing normal activity for the G98S mutant, whereas minigenes indicated the presence of abnormally large products [249].

Some of the mutant MATs have been expressed in COS-1 cells and/or *E. coli* in order to study their activity and association state to understand the behaviors shown by the patients. Thus, truncated proteins of 349 and 350 residues, as well as, the mutants R199C, R356Q, and G378S were prepared and shown to follow the same trend in both systems. Only R356Q (53%), R199C (10%), and G378S ($<0.1\%$) displayed MAT activity [15]. A more detailed analysis was carried out at the R264 position, the results suggesting the necessity of a positive charge at that position to attain maximal activity [248]. Moreover, some of the mutants at this position were able to homo- or heterodimerize [248, 249]. The hetero-oligomerization shown by R264H and wild-type subunits explained its dominant effect on MAT activity. Analogous mutants prepared on other MATs show differences in the oligomerization pattern [175, 190]. Thus, MAT I/III R265H mutant behaves as a monomer with tripolyphosphatase activity, whereas c-MAT R244H remained a tetramer. Moreover, R265H was shown to be able to dimerize with the wild-type enzyme, giving rise to an enzyme that is unable to synthesize AdoMet, but that hydrolyzes tripolyphosphate. On the basis of these results, Pérez-Mato et al. [190] suggested that R265H folds approximately to MAT's final structure before association takes place [190]; association then could occur in a late intermediate as deduced from the folding data [195, 196].

C. MAT AND CANCER

Cancer is one of the most common causes of death in the first world. Its basis depends on genetic and environmental factors, but in all cases, the common outcome is anomalous cell growth. For this to occur, cell cycle deregulation must take place. Moreover, another common property of human cancer cells is their dependence on exogenous methionine for growth, whereas normal cells can use Hcy instead [256]. Thus, there is interest in understanding the anomalies in the methionine cycle caused by carcinogenesis, as part of a search for chemotherapeutic drugs.

In this regard, the effects of carcinogens such as 2-acetyl-aminofluorene [257], thioacetamide [192, 258], *N*-2-fluorenylacetamide [259], and 3′-methyl-4-dimethylaminobenzene [260] on methionine metabolism have been studied in animal models. The collected data indicate reductions in *MAT1A* mRNA[257], protein [258, 261], and activity [192, 257, 258], whereas increases in these parameters were reported for *MAT2A* [258, 260, 262].

Other lines of research have taken advantage of samples from different types of tumors or tumor-derived cell lines [261, 263]. In all the cases, rapid growth has correlated with an increase in MAT II protein levels, while a corresponding decrease in MAT I/III occurs [5]. Moreover, an increase in MAT activity in malignant cells has been reported, using measurements at 50–80 μM methionine [5, 264, 265], thus indicating an alteration of the low K_m forms of the enzyme. Such an effect was confirmed in colorectal carcinoma samples where increases in MAT II protein and activity were observed [266]. Northern blots did not show signals for *MAT1A* mRNA in hepatoma cell lines, a fact that is not due to deletion or gene reorganization, but rather to lack of transcription [265]. Samples from cirrhotic patients and HCC show reduced *MAT1A* mRNA levels, while *MAT2A* is induced only in the carcinoma [267, 268]. These effects correlate with increases in transcription and low methylation of the *MAT2A* promoter only in cancer [268]. The effect of reduced methylation on the *MAT2A* promoter has been studied using SssI methylation of promoter constructs fused to a luciferase reporter gene, followed by transfection of HepG2 cells and AdoMet treatment of transfected cells. In both cases, a reduction in *MAT2A* promoter activity was observed, while no effect on *MAT1A* promoter was detected [268]. AdoMet treatment was shown to induce *MAT2A* promoter hypermethylation, a mechanism that has been shown to involve binding of histone deacetylase and methyl cytosine-binding proteins such as MeCP2 [269, 270]. Comparison of protein-binding patterns to the *MAT2A* promoter between normal and malignant tissues has been also carried out, and the observed differences were able to explain the transcriptional upregulation of *MAT2A* in HCC, as is discussed elsewhere in this review [47, 49].

The observed changes in MAT expression and activity are reflected in AdoMet levels. Thus, in hepatoma cells, an increase in AdoMet is observed upon overexpression of MAT I/III [271], whereas a decrease is detected upon thioacetamide treatment [272, 273]. Moreover, this rise in AdoMet is correlated with a boost of the AdoMet/AdoHcy ratio. Such effects, together with an increase in DNA methylation, can also be observed after AdoMet treatment. In fact, it has been suggested that the reduction in expression of several oncogenes during hepatic carcinogenesis may occur through changes in their DNA methylation [271]. Global DNA hypomethylation seems to be a common property of several diseases in which AdoMet levels and *MAT1A* expression are altered [268]. The role of AdoMet in cancer development was confirmed when a *MAT1A* KO mouse became available [59]; its chronic AdoMet deficiency led to the spontaneous appearance of HCC. Moreover, AdoMet therapy has been shown to be efficient in preventing HCC [274], since it inhibits HGF mitogenic activity and accelerates resynthesis of IκB, thus blocking NFκB activation of cytokine-stimulated hepatocytes [275]. On the other hand, antisense RNA against *MAT2A* produced cell death either in normal or *MAT1A* overexpressing hepatoma cells [271]. The presence of the regulatory β-subunit makes MAT II more susceptible to AdoMet inhibition [36]. *MAT2B* expression provides a growth advantage to human hepatoma cells, its expression being common in cancer and its induction is frequently associated with hepatic dysfunction [36]. AdoMet and 5-MTA induce apoptosis in cancer cells. This effect is mediated by induction of Bcl-x_s and increases in protein phosphatase 1 catalytic subunit, which would carry out the protein Ser-Arg dephosphorylation needed for alternative Bcl-x splicing [276]. Methionine deprivation imposes a metabolic stress that inhibits mitosis and induces cell cycle arrest and apoptosis. cDNA array analysis indicates that several families of transcription factors are affected by this type of stress, among them are AP-1 and NFκB. These factors are among those known to regulate *MAT2A* transcription, as discussed in Section II.C.1 [277].

The effect of several chemotherapies on methionine metabolism behavior has been also studied. Among them, fenretidine has been shown to counteract both the induction of NFκB mRNA and protein levels, as well as the reduction in levels of IκB upon 2-acetylaminofluorene administration. In parallel, the carcinogen induces iNOS while decreasing *MAT1A* expression [257]. NFκB activation is involved in the entry to the G1 phase of cells that are primed to proliferate, maintains cells in early G1, and ensures progression through S-phase by transcriptional activation of c-Myc and cyclin D1.

On the other hand, AdoMet increases IκB synthesis and hence raises the levels of the NFκB/IκB complexes and reduces MAT I/III activity. In addition, an increase in NO levels may also collaborate by inactivating MAT I/III and reducing AdoMet levels. However, NO can also inactivate NFκB [278] and hence lead to a more complicated picture. Resistance to chemotherapy has also been linked to methionine metabolism in some cases, probably through epigenetic effects on DNA methylation [279]. In others cases, i.e., doxorubicine resistance, overexpression of multidrug-resistance gene is the primary factor involved.

Fetal, regenerating, and transformed hepatocytes are able to proliferate. Their progression through the mid/late checkpoint at G0/G1 is regulated by growth factors such as HGF, insulin, EGF, and TGFα; among these HGF has been shown to induce *MAT2A* [50]. Moreover, inhibition of the MAPK and PI3K pathways prevents this upregulation with concomitant induction of *MAT1A*. Treatment with the growth arrest factor TGFβ attenuates mitogenesis and *MAT2A* upregulation, while inducing *MAT1A*. In addition, AdoMet has been shown to inhibit HGF induction of cyclin D1 and D2 expression, leading to the suggestion that this metabolite is a negative modulator of cell cycle progression [54]. Moreover, samples of human HCC always showed expression of the β-subunit at higher levels than those shown in cirrhotic samples [36]. The same is also true for HCC induced by diethylnitrosamine treatment. Analysis of expression in several hepatoma cell lines revealed the presence of the β-subunit in HepG2, PLC, and H3B, but its absence in Huh7. Expression of *MAT2A* in cultured hepatocytes led to an increase in *MAT2B* levels.

AdoMet inhibits the mitogenic activity of HGF and prevents *MAT2A* induction. The mitogenic activity of HGF depends on iNOS and on the *L*-methionine concentration in the medium, which exerts its effects through its conversion to AdoMet. A transitory decrease in AdoMet may liberate factors such as HGF from its inhibition, thus allowing the proliferative response, such as that observed during regeneration, whereas chronic AdoMet reduction would lead to malignant degeneration [276].

D. MAT AND ETHANOL

Intragastric ethanol feeding of rats in combination with a high-fat diet increased the steady-state levels of both *MAT1A* and *MAT2A* mRNAs [280]. In the rat model, only MAT II protein levels rose noticeably, and in concordance MAT activity (measured at 20 μM methionine) doubled [280].

Ethanol induction of *MAT2A* is mediated by TNFα stimulation, which exerts its action on the *MAT2A* promoter through NFκB and AP-1 binding [47]. On the other hand, micropigs showed a decrease in total MAT activity upon ethanol feeding (measured at 5 mM methionine) [281], and *MAT1A* transcripts decreased under these diets, either combined or not with folate deficiency [281]. Reductions in AdoMet levels were observed during ethanol feeding of rats [280], baboons [282], and micropigs [281], as well as in patients with alcohol hepatitis [283]. However, differences in the effects on AdoHcy levels were observed among the models. Thus, while rat intragastric ethanol feeding produced no change in AdoHcy [280], micropigs fed ethanol in combination with a folate-deficient diet showed an increase in the levels of this metabolite [281]. In both cases, the net result is a decrease in the AdoMet/AdoHcy ratio that in rats correlates with a reduction in the global DNA methylation. In addition, increases in c-Myc expression and genome-wide DNA strand break accumulation are observed.

Alcoholic liver cirrhosis is associated with increased risk of liver cancer [280]. A reduction in *MAT1A* mRNA levels was observed in samples from cirrhotic [267] and alcoholic hepatitis patients [283], while *MAT2A* is not induced. Hypermethylation of *MAT1A* promoter was also detected. A high percentage of samples of cirrhotic patients (84%) express the β-subunit, but no correlation between expression and etiology was established [36]. In addition, other enzymes of the methionine cycle, such as GNMT, MS, BHMT, and CBS have very low levels in these patients, the percentage of livers with low or very low levels is higher in HCV cirrhosis samples than in alcoholic cirrhosis specimens [267]. Moreover, Hcy can potentiate the fibrogenic effects of ethanol by inducing expression of tissue inhibitor metalloproteinase-1 and α1 (I) procollagen [4], and reducing GSH, plasma taurine, and urine sulfate levels. Thus, defects in Hcy remethylation may contribute to HCC development in cirrhosis [267]. Cellular levels of AdoMet seem to be related to the differentiation status of the hepatocyte, being lower in growing cells [271].

Liver damage during short- and long-term ethanol consumption [284–286] has been associated with a reduced oxygen supply to this organ. Sustained hypoxia such as that observed in liver cirrhosis and long-term ethanol consumption would induce downregulation of MAT expression; this situation may promote irreversible tissue damage through the impairment of methylation reactions and GSH-dependent detoxification capacity [61]. In this regard, the reversible inactivation of MAT regulated by the intracellular levels of GSH and NO [67] may exert a fundamental role in the effects derived from ethanol consumption. Both GSH reduction and increases in protein nitrosylation are known to occur in these cases. Other protein modifications, due to products of ethanol catabolism, such as the irreversible covalent binding of acetaldehyde to lysine residues, have not been explored yet and may also be involved in the described effects.

The *MAT1A* KO mouse displays alterations common to ethanol treatment, such as induction of CYP2E1; these effects make it more susceptible to injury (i.e., by CCl_4) and reduction in GSH levels [59].

XIII. CONCLUSIONS

This review has attempted to provide an overview of the methionine adenosyltransferase research, an area that has vastly expanded in the past decade. Numerous opportunities remain before a comprehensive understanding of this unique enzyme is obtained and can be exploited for therapeutic purposes. New crystal structures that clarify the mechanism of the reaction are needed, as well as structural data for dimeric and large-molecular weight isoenzymes. Moreover, discerning the role of MAT in diseases as complex as hepatoma or Parkinson's may relate not only to AdoMet synthesis, but also reveal interactions with other pathways that thus far remain elusive.

ACKNOWLEDGMENTS

The literature survey for this article was completed in the fall of 2008. Research in the Markham group was supported by National Institutes of Health Grants GM31186, CA06927 and also supported by an appropriation from the Commonwealth of Pennsylvania. The work carried out by the Pajares group was supported by grants of the Ministerio de Educación y Ciencia (PB94-0087, PM97-0064, BMC2002-00243, BFU2005-00050, BFU2009-08977) and Ministerio de Sanidad (FIS 01/1077, RCMN C05/08 and FIS PI05/0663) of Spain.

REFERENCES

1. Markham, G.D. (2002) S–Adenosylmethionine, *Nat. Encyclopedia Life Sci. 2002*: http://www.els.net/ [doi:10.1038/npg.els.0000662].
2. Fontecave, M., Atta, M., and Mulliez, E. (2004) S-adenosylmethionine: nothing goes to waste, *Trends Biochem. Sci. 29(5)*, 243–249.
3. Sanchez-Perez, G. F., Bautista, J. M., and Pajares, M. A. (2004) Methionine adenosyltransferase as a useful molecular systematics tool revealed by phylogenetic and structural analyses, *J. Mol. Biol. 335(3)*, 693–706.

4. Mato, J. M., et al. (1997) S-adenosylmethionine synthesis: molecular mechanisms and clinical implications, *Pharmacol. Ther. 73(3)*, 265–280.
5. Kotb, M. and Geller, A. M. (1993) Methionine adenosyltransferase: structure and function, *Pharmacol. Ther. 59(2)*, 125–143.
6. Lu, S. C., et al. (2003) Role of S-adenosylmethionine in two experimental models of pancreatitis, *FASEB J. 17(1)*, 56–58.
7. Cabrero, C., Puerta, J., and Alemany, S. (1987) Purification and comparison of two forms of S-adenosyl-L-methionine synthetase from rat liver, *Eur. J. Biochem. 170*(1–2), 299–304.
8. Pajares, M. A., et al. (1992) Modulation of rat liver S-adenosylmethionine synthetase activity by glutathione, *J. Biol. Chem. 267(25)*, 17598–17605.
9. Cabrero, C. and Alemany, S. (1988) Conversion of rat liver S-adenosyl-L-methionine synthetase from high-Mr form to low-Mr form by LiBr, *Biochim. Biophys. Acta. 952(3)*, 277–281.
10. Corrales, F., et al. (1990) Inactivation and dissociation of S-adenosylmethionine synthetase by modification of sulfhydryl groups and its possible occurrence in cirrhosis, *Hepatology 11*(2), 216–222.
11. Alvarez, L., et al. (1991) Analysis of the 5′ non-coding region of rat liver S-adenosylmethionine synthetase mRNA and comparison of the Mr deduced from the cDNA sequence and the purified enzyme, *FEBS Lett. 290(1–2)*, 142–146.
12. Horikawa, S., et al. (1989) Isolation of a cDNA encoding the rat liver S-adenosylmethionine synthetase, *Eur. J. Biochem. 184(3)*, 497–501.
13. Alvarez, L., et al. (1994) Expression of rat liver S-adenosylmethionine synthetase in *Escherichia coli* results in two active oligomeric forms, *Biochem. J. 301(Pt 2)*, 557–561.
14. Ubagai, T., et al. (1995) Molecular mechanisms of an inborn error of methionine pathway. Methionine adenosyltransferase deficiency, *J. Clin. Invest. 96(4)*, 1943–1947.
15. Chamberlin, M. E., et al. (1996) Demyelination of the brain is associated with methionine adenosyltransferase I/III deficiency, *J. Clin. Invest. 98(4)*, 1021–1027.
16. Cabrero, C., et al. (1988) Specific loss of the high-molecular-weight form of S-adenosyl-L-methionine synthetase in human liver cirrhosis, *Hepatology 8(6)*, 1530–1534.
17. Sanchez-Perez, G. F., et al. (2003) Role of an intrasubunit disulfide in the association state of the cytosolic homo-oligomer methionine adenosyltransferase, *J. Biol. Chem. 278(9)*, 7285–7293.
18. Avila, M. A., et al. (1997) Regulation of rat liver S-adenosylmethionine synthetase during septic shock: role of nitric oxide, *Hepatology 25(2)*, 391–396.
19. Mingorance, J., et al. (1996) Site-directed mutagenesis of rat liver S-adenosylmethionine synthetase. Identification of a cysteine residue critical for the oligomeric state, *Biochem. J. 315(Pt 3)*, 761–766.
20. Pajares, M. A., et al. (1991) The role of cysteine-150 in the structure and activity of rat liver S-adenosyl-L-methionine synthetase, *Biochem. J. 274(Pt 1)*, 225–229.
21. Ruiz, F., et al. (1998) Nitric oxide inactivates rat hepatic methionine adenosyltransferase In vivo by S-nitrosylation, *Hepatology 28(4)*, 1051–1057.
22. Martinez-Chantar, M. L. and Pajares, M. A. (2000) Assignment of a single disulfide bridge in rat liver methionine adenosyltransferase, *Eur. J. Biochem. 267(1)*, 132–137.

23. Gonzalez, B., et al. (2000) The crystal structure of tetrameric methionine adenosyltransferase from rat liver reveals the methionine-binding site, *J. Mol. Biol. 300*(2), 363–375.
24. Gilbert, H. F. (1994) *Mechanisms of Protein Folding*, ed Pain, R. H. IRL Press, Oxford, UK; pp. 104–136.
25. Pajares, M. A., et al. (1994) Protein kinase C phosphorylation of rat liver S-adenosylmethionine synthetase: dissociation and production of an active monomer, *Biochem. J. 303*(*Pt 3*), 949–955.
26. Sheid, B. and Bilik, E. (1968) S-adenosylmethionine synthetase activity in some normal rat tissues and transplantable hepatomas, *Cancer Res. 28*(*12*), 2512–2515.
27. Horikawa, S., et al. (1993) Immunohistochemical analysis of rat S-adenosylmethionine synthetase isozymes in developmental liver, *FEBS Lett. 330*(*3*), 307–311.
28. Shimizu-Saito, K., et al. (1997) Differential expression of S-adenosylmethionine synthetase isozymes in different cell types of rat liver, *Hepatology 26*(2), 424–431.
29. Kotb, M. and Kredich, N. M. (1985) S-Adenosylmethionine synthetase from human lymphocytes. Purification and characterization, *J. Biol. Chem. 260*(*7*), 3923–3930.
30. Halim, A. B., et al. (2001) Regulation of the human *MAT2A* gene encoding the catalytic alpha 2 subunit of methionine adenosyltransferase, MAT II: gene organization, promoter characterization, and identification of a site in the proximal promoter that is essential for its activity, *J. Biol. Chem. 276*(*13*), 9784–9791.
31. Kotb, M., et al. (1997) Consensus nomenclature for the mammalian methionine adenosyltransferase genes and gene products, *Trends Genet 13*(2), 51–52.
32. De La Rosa, J., et al. (1992) Changes in the relative amount of subunits of methionine adenosyltransferase in human lymphocytes upon stimulation with a polyclonal T cell mitogen, *J. Biol. Chem. 267*(*15*), 10699–10704.
33. Liau, M. C., G. W. Lin, and Hurlbert, R. B. (1977) Partial purification and characterization of tumor and liver S-adenosylmethionine synthetases, *Cancer Res. 37*(2), 427–435.
34. Finkelstein, J. D. (1990) Methionine metabolism in mammals, *J. Nutr. Biochem. 1*(*5*), 228–237.
35. Geller, A. M., et al. (1997) Inhibition of methionine adenosyltransferase by the polyamines, *Arch. Biochem. Biophys. 345*(*1*), 97–102.
36. Martinez-Chantar, M. L., et al. (2003) Methionine adenosyltransferase II beta subunit gene expression provides a proliferative advantage in human hepatoma, *Gastroenterology 124*(*4*), 940–948.
37. Kotb, M. and Kredich, N. M. (1990) Regulation of human lymphocyte S-adenosylmethionine synthetase by product inhibition, *Biochim. Biophys. Acta 1039*(2), 253–260.
38. Mitsui, K., H. Teraoka, and Tsukada, K. (1988) Complete purification and immunochemical analysis of S-adenosylmethionine synthetase from bovine brain, *J. Biol. Chem. 263*(*23*), 11211–11216.
39. Langkamp-Henken, B., et al. (1994) Characterization of distinct forms of methionine adenosyltransferase in nucleated, and mature human erythrocytes and erythroleukemic cells, *Biochim. Biophys. Acta 1201*(*3*), 397–404.

40. LeGros, H. L., Jr., et al. (2000) Cloning, expression, and functional characterization of the beta regulatory subunit of human methionine adenosyltransferase (MAT II), *J. Biol. Chem. 275(4)*, 2359–2366.

41. LeGros, L., et al. (2001) Regulation of the human *MAT2B* gene encoding the regulatory beta subunit of methionine adenosyltransferase, MAT II, *J. Biol. Chem. 276(27)*, 24918–24924.

42. De La Rosa, J., et al. (1995) Chromosomal localization and catalytic properties of the recombinant alpha subunit of human lymphocyte methionine adenosyltransferase, *J. Biol. Chem. 270(37)*, 21860–21868.

43. LeGros, H. L., Jr., Geller, A. M., and Kotb, M. (1997) Differential regulation of methionine adenosyltransferase in superantigen and mitogen stimulated human T lymphocytes, *J. Biol. Chem. 272(25)*, 16040–16047.

44. Halim, A. B., et al. (1999) Expression and functional interaction of the catalytic and regulatory subunits of human methionine adenosyltransferase in mammalian cells, *J. Biol. Chem. 274(42)*, 29720–29725.

45. Alvarez, L., et al. (1997) Characterization of rat liver-specific methionine adenosyltransferase gene promoter. Role of distal upstream cis-acting elements in the regulation of the transcriptional activity, *J. Biol. Chem. 272(36)*, 22875–22883.

46. Halim, A. B., et al. (2001) Distinct patterns of protein binding to the *MAT2A* promoter in normal and leukemic T cells, *Biochim. Biophys. Acta 1540(1)*, 32–42.

47. Yang, H., et al. (2003) Induction of human methionine adenosyltransferase 2A expression by tumor necrosis factor alpha. Role of NF-kappa B and AP-1, *J. Biol. Chem. 278(51)*, 50887–50896.

48. Torres, L., et al. (2000) Liver-specific methionine adenosyltransferase *MAT1A* gene expression is associated with a specific pattern of promoter methylation and histone acetylation: implications for *MAT1A* silencing during transformation, *Faseb J. 14(1)*, 95–102.

49. Yang, H., et al. (2001) The role of c-Myb and Sp1 in the up-regulation of methionine adenosyltransferase 2A gene expression in human hepatocellular carcinoma, *Faseb J. 15(9)*, 1507–1516.

50. Paneda, C., et al. (2002) Liver cell proliferation requires methionine adenosyltransferase 2A mRNA up-regulation, *Hepatology 35(6)*, 1381–1391.

51. Huang, Z. Z., et al. (1998) Changes in methionine adenosyltransferase during liver regeneration in the rat, *Am. J. Physiol. 275(1 Pt 1)*, G14–G21.

52. Gil, B., et al. (1997) Glucocorticoid regulation of hepatic S-adenosylmethionine synthetase gene expression, *Endocrinology 138(3)*, 1251–1258.

53. Gil, B., et al. (1996) Differential expression pattern of S-adenosylmethionine synthetase isoenzymes during rat liver development, *Hepatology 24(4)*, 876–881.

54. Martinez-Chantar, M. L., et al. (2003) L-methionine availability regulates expression of the methionine adenosyltransferase 2A gene in human hepatocarcinoma cells: role of S-adenosylmethionine, *J. Biol. Chem. 278(22)*, 19885–19890.

55. Hoffman, R. M. (1985) Altered methionine metabolism and transmethylation in cancer, *Anticancer Res. 5(1)*, 1–30.

56. Tang, B., Li, Y. N., and Kruger, W. D. (2000) Defects in methylthioadenosine phosphorylase are associated with but not responsible for methionine-dependent tumor cell growth, *Cancer Res. 60(19)*, 5543–5547.
57. Dumontet, C., Roch, A. M., and Quash, G. (1996) Methionine dependence of tumor cells: programmed cell survival? *Oncol. Res. 8(12)*, 469–471.
58. Leung-Pineda, V. and Kilberg, M. S. (2002) Role of Sp1 and Sp3 in the nutrient-regulated expression of the human asparagine synthetase gene, *J. Biol. Chem. 277(19)*, 16585–16591.
59. Martinez-Chantar, M. L., et al. (2002) Spontaneous oxidative stress and liver tumors in mice lacking methionine adenosyltransferase 1A, *FASEB J. 16(10)*, 1292–1294.
60. Chawla, R. K. and Jones, D. P. (1994) Abnormal metabolism of S-adenosyl-L-methionine in hypoxic rat liver. Similarities to its abnormal metabolism in alcoholic cirrhosis. *Biochim. Biophys. Acta. 1199(1)*, 45–51.
61. Avila, M. A., et al. (1998) Regulation by hypoxia of methionine adenosyltransferase activity and gene expression in rat hepatocytes, *Gastroenterology 114*(2), 364–371.
62. Graeber, T. G., et al. (1996) Hypoxia-mediated selection of cells with diminished apoptotic potential in solid tumours, *Nature 379(6560)*, 88–91.
63. Jungermann, K. and Kietzmann, T. (1996) Zonation of parenchymal and nonparenchymal metabolism in liver, *Annu. Rev. Nutr. 16*, 179–203.
64. Corrales, F., et al. (1991) Inhibition of glutathione synthesis in the liver leads to S-adenosyl-L-methionine synthetase reduction, *Hepatology 14(3)*, 528–533.
65. Corrales, F., et al. (1992) S-adenosylmethionine treatment prevents carbon tetrachloride-induced S-adenosylmethionine synthetase inactivation and attenuates liver injury, *Hepatology 16(4)*, 1022–1027.
66. Martinez-Chantar, M. L. and Pajares, M. A. (1996) Role of thioltransferases on the modulation of rat liver S-adenosylmethionine synthetase activity by glutathione, *FEBS Lett. 397*(2–3), 293–297.
67. Corrales, F. J., Ruiz, F., and Mato, J. M. (1999) In vivo regulation by glutathione of methionine adenosyltransferase S-nitrosylation in rat liver, *J. Hepatol. 31(5)*, 887–894.
68. Lu, S. C., et al. (2001) Methionine adenosyltransferase 1A knockout mice are predisposed to liver injury and exhibit increased expression of genes involved in proliferation, *Proc. Natl. Acad. Sci. U.S.A. 98(10)*, 5560–5565.
69. Garcia-Trevijano, E. R., et al. (2002) NO sensitizes rat hepatocytes to proliferation by modifying S-adenosylmethionine levels, *Gastroenterology 122(5)*, 1355–1363.
70. Stamler, J. S., et al. (1997) (S)NO signals: translocation, regulation, and a consensus motif, *Neuron 18(5)*, 691–696.
71. Perez-Mato, I., et al. (1999) Methionine adenosyltransferase S-nitrosylation is regulated by the basic and acidic amino acids surrounding the target thiol, *J. Biol. Chem. 274(24)*, 17075–17079.
72. Castro, C., et al. (1999) Creation of a functional S-nitrosylation site in vitro by single point mutation, *FEBS Lett. 459(3)*, 319–322.

73. Sanchez Del Pino, M. M., Corrales, F. J., and Mato, J. M. (2000) Hysteretic behavior of methionine adenosyltransferase III: methionine switches between two conformations of the enzyme with different specific activity, *J. Biol. Chem. 275(31)*, 23476–23482.
74. Sanchez-Gongora, E., et al. (1996) Increased sensitivity to oxidative injury in chinese hamster ovary cells stably transfected with rat liver S-adenosylmethionine synthetase cDNA, *Biochem. J. 319(Pt 3)*, 767–773.
75. Sanchez-Gongora, E., et al. (1997) Interaction of liver methionine adenosyltransferase with hydroxyl radical, *FASEB J. 11(12)*, 1013–1019.
76. Frago, L. M., et al. (2001) Short-chain ceramide regulates hepatic methionine adenosyltransferase expression, *J. Hepatol. 34(2)*, 192–201.
77. De La Rosa, J., et al. (1992) Induction of interleukin 2 production but not methionine adenosyltransferase activity or S-adenosylmethionine turnover in Jurkat T-cells, *Cancer Res. 52(12)*, 3361–3366.
78. Konze, J. R. and Kende, H. (1979) Interactions of methionine and selenomethionine with methionine adenosyltransferase and ethylene-generating systems, *Plant Physiol. 63(3)*, 507–510.
79. Higuchi, T. (1981) Biosynthesis of lignin. In: *Plant Carbohydrates II. Encyclopedia of Plant Physiology*, eds Tanner, W. and Loewus, F. A. Springer Verlag, Berlin; Vol. 13B, pp. 194–224.
80. Roje, S. (2006) S-Adenosyl-L-methionine: beyond the universal methyl group donor, *Phytochemistry 67(15)*, 1686–1698.
81. Peleman, J., et al. (1989) Strong cellular preference in the expression of a housekeeping gene of *Arabidopsis thaliana* encoding S-adenosylmethionine synthetase, *Plant Cell 1(1)*, 81–93.
82. Peleman, J., et al. (1989) Structure and expression analyses of the S-adenosylmethionine synthetase gene family in *Arabidopsis thaliana*, *Gene 84(2)*, 359–369.
83. Larsen, P. B. and Woodson, W. R. (1991) Cloning and nucleotide sequence of a S-adenosylmethionine synthetase cDNA from carnation, *Plant Physiol. 96(3)*, 997–999.
84. Espartero, J., Pintor-Toro, J. A., and Pardo, J. M. (1994) Differential accumulation of S-adenosylmethionine synthetase transcripts in response to salt stress, *Plant Mol. Biol. 25(2)*, 217–227.
85. Van Breusegem, F., et al. (1994) Characterization of a S-adenosylmethionine synthetase gene in rice, *Plant Physiol. 105(4)*, 1463–1464.
86. Gomez-Gomez, L. and Carrasco, P. (1996) Hormonal regulation of S-adenosylmethionine synthase transcripts in pea ovaries, *Plant Mol. Biol. 30(4)*, 821–832.
87. Schroder, G., et al. (1997) Three differentially expressed S-adenosylmethionine synthetases from *Catharanthus roseus*: molecular and functional characterization, *Plant Mol. Biol. 33(2)*, 211–222.
88. Van Doorsselaere, J., et al. (1993) A cDNA encoding S-adenosyl-L-methionine synthetase from poplar, *Plant Physiol. 102(4)*, 1365–1366.
89. Wen, C. M., et al. (1995) Cloning and nucleotide sequence of a cDNA encoding S-adenosyl-L-methionine synthetase from mustard (*Brassica juncea* [L.] Czern & Coss), *Plant Physiol. 107(3)*, 1021–1022.

90. Whittaker, D. J., Smith, G. S., and Gardner, R. C. (1995) Three cDNAs encoding S-adenosyl-L-methionine synthetase from *Actinidia chinensis*, *Plant Physiol. 108(3)*, 1307–1308.
91. Pavy, N., et al. (2005) Generation, annotation, analysis and database integration of 16,500 white spruce EST clusters, *BMC Genomics 6*, 144.
92. Mathur, M., Saluja, D., and Sachar, R. C. (1991) Post-transcriptional regulation of S-adenosylmethionine synthetase from its stored mRNA in germinated wheat embryos, *Biochim. Biophys. Acta 1078*(2), 161–170.
93. Mathur, M., Satpathy, M., and Sachar, R. C. (1992) Phytohormonal regulation of S-adenosylmethionine synthetase by gibberellic acid in wheat aleurones, *Biochim. Biophys. Acta 1137(3)*, 338–348.
94. Mathur, M., Sharma, N., and Sachar, R. C. (1993) Differential regulation of S-adenosylmethionine synthetase isozymes by gibberellic acid in dwarf pea epicotyls, *Biochim. Biophys. Acta 1162(3)*, 283–290.
95. Lindermayr, C., et al. (2006) Differential inhibition of Arabidopsis methionine adenosyltransferases by protein S-nitrosylation, *J. Biol. Chem. 281(7)*, 4285–4291.
96. Gallardo, K., et al. (2002) Importance of methionine biosynthesis for Arabidopsis seed germination and seedling growth, *Physiol. Plant 116*(2), 238–247.
97. Izhaki, A., Shoseyov, O., and Weiss, D. (1995) A petunia cDNA encoding S-adenosylmethionine synthetase, *Plant Physiol., 108*(2), 841–842.
98. Izhaki, A., Shoseyov, O., and Weiss, D. (1996) Temporal, spatial and hormonal regulation of the S-adenosylmethionine synthetase gene in petunia, *Physiol. Plant 97*, 90–94.
99. Lindroth, A. M., et al. (2001) Two S-adenosylmethionine synthetase-encoding genes differentially expressed during adventitious root development in *Pinus contorta*, *Plant Mol. Biol. 46(3)*, 335–346.
100. Radchuk, V. V., et al. (2005) The methylation cycle and its possible functions in barley endosperm development, *Plant Mol. Biol. 59*(2), 289–307.
101. Gomez-Gomez, L. and Carrasco, P. (1998) Differential expression of the S-adenosyl-L-methionine synthase genes during pea development, *Plant Physiol. 117*(2), 397–405.
102. Chang, S., et al. (1996) Gene expresión under water deficit in loblolly pine (*Pinus taeda*): isolation and characterization of cDNA clones, *Physiol. Plant 97(3)*, 139–148.
103. Sanchez-Aguayo, I., et al. (2004) Salt stress enhances xylem development and expression of S-adenosyl-L-methionine synthase in lignifying tissues of tomato plants, *Planta 220*(2), 278–285.
104. Kawalleck, P., et al. (1992) Induction by fungal elicitor of S-adenosyl-L-methionine synthetase and S-adenosyl-L-homocysteine hydrolase mRNAs in cultured cells and leaves of *Petroselinum crispum*, *Proc. Natl. Acad. Sci. U.S.A. 89(10)*, 4713–4717.
105. Wang, Y., et al. (2006) S-nitrosylation: an emerging redox-based post-translational modification in plants, *J. Exp. Bot. 57*, 1777–1784.
106. Boerjan, W., et al. (1994) Distinct phenotypes generated by overexpression and suppression of S-adenosyl-L-methionine synthetase reveal developmental patterns of gene silencing in tobacco, *Plant Cell 6(10)*, 1401–1414.

107. Jacquemin-Faure, I., et al. (1994) The vacuolar compartment is required for sulfur amino acid homeostasis in *Saccharomyces cerevisiae*, *Mol. Gen. Genet.* *244(5)*, 519–529.

108. Chan, S. Y. and Appling, D. R. (2003) Regulation of S-adenosylmethionine levels in *Saccharomyces cerevisiae*, *J. Biol. Chem.* *278(44)*, 43051–43059.

109. Nakamura, K. D. and Schlenk, F. (1974) Examination of isolated yeast cell vacuoles for active transport, *J. Bacteriol.* *118(1)*, 314–316.

110. Mudd, S. H. (1962) Activation of methionine for transmethylation. V. The mechanism of action of the methionine-activating enzyme, *J. Biol. Chem.* *237*, 1372–1375.

111. Mudd, S. H. (1963) Activation of methionine for transmethylation. VI. Enzyme-bound tripolyphosphate as an intermediate in the reaction catalyzed by the methionine-activating enzyme of Baker's yeast, *J. Biol. Chem.* *238*, 2156–2163.

112. Mudd, S. H. and Mann, J. D. (1963) Activation of methionine for transmethylation. VII. Some energetic and kinetic aspects of the reaction catalyzed by the methionine-activating enzyme of bakers' yeast, *J. Biol. Chem.* *238*, 2164–2170.

113. Mudd, S. H. and Cantoni, G. L. (1958) Activation of methionine for transmethylation. III. The methionine-activating enzyme of Bakers' yeast. *J. Biol. Chem.* *231(1)*, 481–492.

114. Mudd, S. H., Jamieson, G. A., and Cantoni, G. L. (1960) Activation of methionine for transmethylation. IV. The failure of 3,5′-cycloadenosine to replace adenosine triphosphate, *Biochim. Biophys. Acta.* *38*, 164–167.

115. Chiang, P. K. and Cantoni, G. L. (1977) Activation of methionine for transmethylation. Purification of the S-adenosylmethionine synthetase of bakers' yeast and its separation into two forms, *J. Biol. Chem.* *252(13)*, 4506–4513.

116. Cherest, H. and Surdin-Kerjan, Y. (1978) S-adenosyl methionine requiring mutants in *Saccharomyces cerevisiae*: evidences for the existence of two methionine adenosyl transferases, *Mol. Gen. Genet.* *163*(2), 153–167.

117. Thomas, D., et al. (1988) *SAM2* encodes the second methionine S-adenosyl transferase in *Saccharomyces cerevisiae*: physiology and regulation of both enzymes, *Mol. Cell Biol.* *8(12)*, 5132–5139.

118. Thomas, D., Cherest, H., and Surdin-Kerjan, Y. (1991) Identification of the structural gene for glucose-6-phosphate dehydrogenase in yeast. Inactivation leads to a nutritional requirement for organic sulfur, *Embo. J.* *10(3)*, 547–553.

119. Rouillon, A., Surdin-Kerjan, Y., and Thomas, D. (1999) Transport of sulfonium compounds. Characterization of the s-adenosylmethionine and s-methylmethionine permeases from the yeast *Saccharomyces cerevisiae*, *J. Biol. Chem.* *274(40)*, 28096–28105.

120. Markham, G. D. and Satishchandran, C. (1988) Identification of the reactive sulfhydryl groups of S-adenosylmethionine synthetase, *J. Biol. Chem.* *263(18)*, 8666–8670.

121. Markham, G. D., et al. (1980) S-Adenosylmethionine synthetase from *Escherichia coli*, *J. Biol. Chem.* *255(19)*, 9082–9092.

122. Sekowska, A., Kung, H. F., and Danchin, A. (2000) Sulfur metabolism in *Escherichia coli* and related bacteria: facts and fiction, *J. Mol. Microbiol. Biotechnol.* *2*(2), 145–177.

123. Old, I. G., et al. (1991) Regulation of methionine biosynthesis in the Enterobacteriaceae, *Prog. Biophys. Mol. Biol. 56(3)*, 145–185.

124. Sekowska, A. and Danchin, A. (2002) The methionine salvage pathway in *Bacillus subtilis, BMC Microbiol.* 2, 8.

125. Hondorp, E. R. and Matthews, R. G. (2006) Methionine. In: *EcoSal - Escherichia Coli and Salmonella: Cellular and Molecular Biology*, eds Böck, A., et al. ASM Press, Washington, DC; p. Module 3.6.1.7.

126. Grundy, F. J. and Henkin, T. M. (1998) The S box regulon: a new global transcription termination control system for methionine and cysteine biosynthesis genes in gram-positive bacteria, *Mol. Microbiol. 30(4)*, 737–749.

127. Corbino, K. A., et al. (2005) Evidence for a second class of S-adenosylmethionine riboswitches and other regulatory RNA motifs in alpha-proteobacteria, *Genome Biol. 6(8)*, R70.

128. Lim, J., et al. (2006) Molecular-recognition characteristics of SAM-binding riboswitches, *Angew. Chem. Int. Ed Engl. 45(6)*, 964–968.

129. Winkler, W. C., et al. (2003) An mRNA structure that controls gene expression by binding S-adenosylmethionine, *Nat. Struct. Biol. 10(9)*, 701–707.

130. Greene, R. C. (1996) Biosynthesis of methionine. In: *Escherichia coli and Salmonella: Cellular and Molecular Biology*, ed Neidhardt F. C. American Society for Microbiology Press, Washington, DC; pp. 542–560.

131. Chen, C. and Newman, E. B. (1998) Comparison of the sensitivities of two *Escherichia coli* genes to in vivo variation of Lrp concentration, *J. Bacteriol. 180(3)*, 655–659.

132. Porcelli, M., et al. (1988) S-adenosylmethionine synthetase in the thermophilic archaebacterium *Sulfolobus solfataricus*. Purification and characterization of two isoforms, *Eur. J. Biochem. 177*(2), 273–280.

133. Graham, D. E., et al. (2000) Identification of a highly diverged class of S-adenosylmethionine synthetases in the archaea, *J. Biol. Chem. 275(6)*, 4055–4059.

134. Lu, Z. J. and Markham, G. D. (2002) Enzymatic properties of S-adenosylmethionine synthetase from the archaeon *Methanococcus jannaschii, J. Biol. Chem. 277(19)*, 16624–16631.

135. Thompson, A. H. (2005) Plasma exposure alters the proteome of *S. pyogenes, J. Proteome Res. 4(6)*, 1901.

136. Ding, Y. H., et al. (2006) The proteome of dissimilatory metal-reducing microorganism *Geobacter sulfurreducens* under various growth conditions, *Biochim. Biophys. Acta 1764*(7), 1198–1206.

137. Perez-Pertejo, Y., et al. (2006) Characterization of a methionine adenosyltransferase over-expressing strain in the trypanosomatid *Leishmania donovani, Biochim. Biophys. Acta 1760(1)*, 10–19.

138. Merali, S., et al. (2000) S-adenosylmethionine and *Pneumocystis carinii, J. Biol. Chem. 275(20)*, 14958–14963.

139. Merali, S. and Clarkson, A. B., Jr. (2004) S-adenosylmethionine and pneumocystis, *FEMS Microbiol. Lett. 237*(2), 179–186.

140. Goldberg, B., et al. (1997) A unique transporter of S-adenosylmethionine in African trypanosomes, *FASEB J. 11(4)*, 256–260.

141. Stramentinoli, G. (1987) Pharmacologic aspects of S-adenosylmethionine. Pharmacokinetics and pharmacodynamics, *Am. J. Med. 83(5A)*, 35–42.

142. Goldberg, B., et al. (1999) Kinetics of S-adenosylmethionine cellular transport and protein methylation in *Trypanosoma brucei brucei* and *Trypanosoma brucei rhodesiense*, *Arch. Biochem. Biophys. 364(1)*, 13–18.

143. Yarlett, N., et al. (1993) S-adenosylmethionine synthetase in bloodstream *Trypanosoma brucei*, *Biochim. Biophys. Acta. 1181(1)*, 68–76.

144. García-Estrada, C., et al. (2006) Analysis of genetic elements regulating the methionine adenosyltransferase gene in *Leishmania infantum*, *Gene*, p. PMID: 17196769.

145. Reguera, R. M., et al. (2002) Cloning expression and characterization of methionine adenosyltransferase in *Leishmania infantum promastigotes*, *J. Biol. Chem. 277(5)*, 3158–3167.

146. Perez-Pertejo, Y., et al. (2003) *Leishmania donovani* methionine adenosyltransferase. Role of cysteine residues in the recombinant enzyme, *Eur. J. Biochem. 270(1)*, 28–35.

147. Perez-Pertejo, Y., et al. (2004) Mutational analysis of methionine adenosyltransferase from *Leishmania donovani*, *Eur. J. Biochem. 271(13)*, 2791–2798.

148. Chiang, P. K., et al. (1999) Molecular characterization of *Plasmodium falciparum* S-adenosylmethionine synthetase, *Biochem. J. 344(Pt 2)*, 571–576.

149. Jeon, T. J. and Jeon, K. W. (2003) Characterization of *SAMs* genes of *Amoeba proteus* and the endosymbiotic X-bacteria, *J. Eukaryot. Microbiol. 50(1)*, 61–69.

150. Jeon, T. J. and Jeon, K. W. (2004) Gene switching in *Amoeba proteus* caused by endosymbiotic bacteria, *J. Cell Sci. 117(Pt 4)*, 535–543.

151. Andersson, S. G., et al. (1998) The genome sequence of *Rickettsia prowazekii* and the origin of mitochondria, *Nature 396(6707)*, 133–140.

152. Andersson, S. G., et al. (2002) Comparative genomics of microbial pathogens and symbionts, *Bioinformatics 18 Suppl 2*, S17.

153. Stephens, R. S., et al. (1998) Genome sequence of an obligate intracellular pathogen of humans: *Chlamydia trachomatis*, *Science 282(5389)*, 754–759.

154. Andersson, J. O. and Andersson, S. G. (1999) Genome degradation is an ongoing process in Rickettsia, *Mol. Biol. Evol. 16(9)*, 1178–1191.

155. Tucker, A. M., et al. (2003) S-adenosylmethionine transport in *Rickettsia prowazekii*, *J. Bacteriol. 185(10)*, 3031–3035.

156. Driskell, L. O., et al. (2005) Rickettsial metK-encoded methionine adenosyltransferase expression in an *Escherichia coli* metK deletion strain, *J. Bacteriol. 187(16)*, 5719–5722.

157. Waters, E., et al. (2003) The genome of *Nanoarchaeum equitans*: insights into early archaeal evolution and derived parasitism, *Proc. Natl. Acad. Sci. U.S.A. 100(22)*, 12984–12988.

158. Papagrigoriou, E., et al., 2006. Crystal structure of the alpha subunit of human S-adenosylmethionine synthetase 2. http://www.pdb.org/pdb/results/results.do?outformat=&qrid=1EBA44CF&tabtoshow=Current

159. Takusagawa, F., Kamitori, S., and Markham, G. D. (1996) Structure and function of S-adenosylmethionine synthetase: crystal structures of S-adenosylmethionine synthetase with ADP, BrADP, and PPi at 2.8 angstroms resolution, *Biochemistry 35(8)*, 2586–2596.

160. Fu, Z., et al. (1996) Flexible loop in the structure of S-adenosylmethionine synthetase crystallized in the tetragonal modification, *J. Biomol. Struct. Dyn. 13(5)*, 727–739.
161. Komoto, J., et al. (2004) Crystal structure of the S-adenosylmethionine synthetase ternary complex: a novel catalytic mechanism of S-adenosylmethionine synthesis from ATP and Met, *Biochemistry 43(7)*, 1821–1831.
162. Taylor, J. C., Takusagawa, F., and Markham, G. D. (1996) A chimeric active site lid variant of S-adenosylmethionine synthetase, *FASEB J. 10*, A970.
163. Taylor, J. C., Takusagawa, F., and Markham, G. D. (2002) The active site loop of S-adenosylmethionine synthetase modulates catalytic efficiency, *Biochemistry 41(30)*, 9358–9369.
164. Takusagawa, F., et al. (1996) Crystal structure of S-adenosylmethionine synthetase, *J. Biol. Chem. 271(1)*, 136–147.
165. Gonzalez, B., et al. (2003) Crystal structures of methionine adenosyltransferase complexed with substrates and products reveal the methionine-ATP recognition and give insights into the catalytic mechanism, *J. Mol. Biol. 331*(2), 407–416.
166. Deigner, H. P., Mato, J. M., and Pajares, M. A. (1995) Study of the rat liver S-adenosylmethionine synthetase active site with 8-azido ATP, *Biochem. J. 308(Pt 2)*, 565–571.
167. Mudd, S. H. (1973) S-adenosylmethionine synthetase. In: *The Enzymes*, 3rd Edition. Academic Press, New York; pp. 21–154.
168. Chou, T. C. and Talalay, P. (1972) The mechanism of S-adenosyl-L-methionine synthesis by purified preparations of bakers' yeast, *Biochemistry 11(6)*, 1065–1073.
169. Sullivan, D. M. and Hoffman, J. L. (1983) Fractionation and kinetic properties of rat liver and kidney methionine adenosyltransferase isozymes, *Biochemistry 22(7)*, 1636–1641.
170. McQueney, M. S., Anderson, K. S., and Markham, G. D. (2000) Energetics of S-adenosylmethionine synthetase catalysis, *Biochemistry 39(15)*, 4443–4454.
171. Tabor, C. W. and Tabor, H. (1984) Methionine adenosyltransferase (S-adenosylmethionine synthetase) and S-adenosylmethionine decarboxylase, *Adv. Enzymol. Relat. Areas Mol. Biol. 56*, 251–282.
172. Parry, R. J. and Minta, A. (1982) Studies of enzyme stereochemistry. Elucidation of the stereochemistry of S-adenosylmethionine formation by yeast methionine adenosyltransferase, *J. Am. Chem. Soc. 104*, 871–872.
173. Markham, G. D., et al. (1987) A kinetic isotope effect study and transition state analysis of the S-adenosylmethionine synthetase reaction, *J. Biol. Chem. 262(12)*, 5609–5615.
174. Cleland, W. W. and Hengge, A. C. (2006) Enzymatic mechanisms of phosphate and sulfate transfer, *Chem. Rev. 106(8)*, 3252–3278.
175. Reczkowski, R. S., Taylor, J. C., and Markham, G. D. (1998) The active-site arginine of S-adenosylmethionine synthetase orients the reaction intermediate, *Biochemistry 37(39)*, 13499–13506.
176. Hedstrom, L. and Gan, L. (2006) IMP dehydrogenase: structural schizophrenia and an unusual base, *Curr. Opin. Chem. Biol. 10(5)*, 520–525.

177. Taylor, J. C. and Markham G. D. (2000) The bifunctional active site of S-adenosylmethionine synthetase. Roles of the basic residues, *J. Biol. Chem. 275(6)*, 4060–4065.

178. Taylor, J. C. and Markham G. D. (2003) Conformational dynamics of the active site loop of S-adenosylmethionine synthetase illuminated by site-directed spin labeling, *Arch. Biochem. Biophys. 415*(2), 164–171.

179. Hammes, G. G. (2002) Multiple conformational changes in enzyme catalysis, *Biochemistry 41*(*26*), 8221–8228.

180. Benkovic, S. J. and Hammes-Schiffer, S. (2003) A perspective on enzyme catalysis, *Science 301*(*5637*), 1196–1202.

181. Hammes-Schiffer, S. and Benkovic, S. J. (2006) Relating protein motion to catalysis, *Annu. Rev. Biochem. 75*, 519–541.

182. Lu, Z. J. and Markham, G. D. (2004) Catalytic properties of the archaeal S-adenosylmethionine decarboxylase from Methanococcus jannaschii, *J. Biol. Chem. 279*(*1*), 265–273.

183. Markham, G. D. (1981) Spatial proximity of two divalent metal ions at the active site of S- adenosylmethionine synthetase, *J. Biol. Chem. 256(4)*, 1903–1909.

184. Markham, G. D. (1984) Structure of the divalent metal ion activator binding site of S-adenosylmethionine synthetase studied by vanadyl(IV) electron paramagnetic resonance, *Biochemistry 23(3)*, 470–478.

185. Markham, G. D. (1986) Characterization of the monovalent cation activator binding site of S-adenosylmethionine synthetase by 205Tl NMR of enzyme-bound Tl+, *J. Biol. Chem. 261(4)*, 1507–1509.

186. Markham, G. D. and Leyh, T. S. (1987) Superhyperfine coupling between metal ions at the active site of S-adenosylmethionine synthetase, *J. Am. Chem. Soc. 109*, 599–560.

187. McQueney, M. S. and Markham, G. D. (1995) Investigation of monovalent cation activation of S-adenosylmethionine synthetase using mutagenesis and uranyl inhibition, *J. Biol. Chem. 270(31)*, 18277–18284.

188. Taylor, J. C. and Markham, G. D. (1999) The bifunctional active site of s-adenosylmethionine synthetase. Roles of the active site aspartates, *J. Biol. Chem. 274(46)*, 32909–32914.

189. Reczkowski, R. S. and Markham, G. D. (1995) Structural and functional roles of cysteine 90 and cysteine 240 in S-adenosylmethionine synthetase, *J. Biol. Chem. 270(31)*, 18484–18490.

190. Perez Mato, I., et al. (2001) Biochemical basis for the dominant inheritance of hypermethioninemia associated with the R264H mutation of the *MAT1A* gene. A monomeric methionine adenosyltransferase with tripolyphosphatase activity, *J. Biol. Chem. 276(17)*, 13803–13809.

191. Hoffman, J. L. and Kunz, G. L. (1977) Differential activation of rat liver methionine adenosyltransferase isozymes by dimethylsulfoxide, *Biochem. Biophys. Res. Commun. 77(4)*, 1231–1236.

192. Okada, G., et al. (1979) Differential effects of dimethylsulfoxide on S-adenosylmethionine synthetase from rat liver and hepatoma, *FEBS Lett. 106*(*1*), 25–28.

193. Park, J., et al. (1996) Enzymatic synthesis of S-adenosyl-L-methionine on the preparative scale, *Bioorg. Med. Chem. 4(12)*, 2179–2185.

194. Houry, W. A., et al. (1999) Identification of in vivo substrates of the chaperonin GroEL, *Nature 402(6758)*, 147–154.

195. Gasset, M., et al. (2002) Equilibrium unfolding studies of the rat liver methionine adenosyltransferase III, a dimeric enzyme with intersubunit active sites, *Biochem. J. 361(Pt 2)*, 307–315.

196. Sanchez del Pino, M. M., et al. (2002) Folding of dimeric methionine adenosyltransferase III: identification of two folding intermediates, *J. Biol. Chem. 277(14)*, 12061–12066.

197. Schonbrunn, E., et al. (2000) Structural basis for the interaction of the fluorescence probe 8-anilino-1-naphthalene sulfonate (ANS) with the antibiotic target MurA, *Proc. Natl. Acad. Sci. U.S.A. 97(12)*, 6345–6349.

198. Lopez-Vara, M. C., Gasset, M., and Pajares, M. A. (2000) Refolding and characterization of rat liver methionine adenosyltransferase from *Escherichia coli* inclusion bodies, *Protein Expr. Purif. 19(2)*, 219–226.

199. Mei, G., et al. (2005) The importance of being dimeric, *FEBS J. 272(1)*, 16–27.

200. Mingorance, J., et al. (1997) Recombinant rat liver S-adenosyl-L-methionine synthetase tetramers and dimers are in equilibrium, *Int. J. Biochem. Cell Biol. 29(3)*, 485–491.

201. Iloro, I., et al. (2004) Methionine adenosyltransferase alpha-helix structure unfolds at lower temperatures than beta-sheet: a 2D-IR study, *Biophys. J. 86(6)*, 3951–3958.

202. Friedman, F. K. and Beychok, S. (1979) Probes of subunit assembly and reconstitution pathways in multisubunit proteins, *Annu. Rev. Biochem. 48*, 217–250.

203. Kappler, F. and Hampton, A. (1990) Approaches to isozyme-specific inhibitors. 17. Attachment of a selectivity-inducing substituent to a multisubstrate adduct. Implications for facilitated design of potent, isozyme-selective inhibitors, *J. Med. Chem. 33(9)*, 2545–2551.

204. Sufrin, J. R., Lombardini, J. B., and Alks, V. (1993) Differential kinetic properties of L-2-amino-4-methylthio-cis-but-3- enoic acid, a methionine analog inhibitor of S-adenosylmethionine synthetase, *Biochim. Biophys. Acta. 1202(1)*, 87–91.

205. Goldberg, B., et al. (1998) Effects of intermediates of methionine metabolism and nucleoside analogs on S-adenosylmethionine transport by *Trypanosoma brucei brucei* and a drug-resistant *Trypanosoma brucei rhodesiense*, *Biochem. Pharmacol. 56(1)*, 95–103.

206. Lombardini, J. B. and Sufrin, J. R. (1983) Chemotherapeutic potential of methionine analogue inhibitors of tumor-derived methionine adenosyltransferases, *Biochem. Pharmacol. 32(3)*, 489–495.

207. Lavrador, K., et al. (1998) A new series of S-adenosyl-L-methionine synthetase inhibitors, *J. Enzyme Inhib. 13(5)*, 361–367.

208. Kappler, F., et al. (1986) Isozyme-specific enzyme inhibitors. 11. L-homocysteine-ATP S-C5′ covalent adducts as inhibitors of rat methionine adenosyltransferases, *J. Med. Chem. 29(6)*, 1030–1038.

209. Kappler, F., Hai, T. T., and Hampton, A. (1986) Isozyme-specific enzyme inhibitors. 10. Adenosine 5′-triphosphate derivatives as substrates or inhibitors of methionine adenosyltransferases of rat normal and hepatoma tissues, *J. Med. Chem. 29(3)*, 318–322.

210. Lim, H., et al. (1986) Isozyme-specific enzyme inhibitors. 12. C- and N-methylmethionines as substrates and inhibitors of methionine adenosyltransferases of normal and hepatoma rat tissues, *J. Med. Chem. 29(9)*, 1743–1748.

211. Reczkowski, R. S. and Markham, G. D. (1999) Slow binding inhibition of S-adenosylmethionine synthetase by imidophosphate analogues of an intermediate and product, *Biochemistry 38(28)*, 9063–9068.

212. Gomes Trolin, C., Regland, B., and Oreland, L. (1995) Decreased methionine adenosyltransferase activity in erythrocytes of patients with dementia disorders, *Eur. Neuropsychopharmacol. 5*(2), 107–114.

213. Gomes-Trolin, C., et al. (1998) Erythrocyte and brain methionine adenosyltransferase activities in patients with schizophrenia, *J. Neural. Transm. 105(10–12)*, 1293–1305.

214. Surtees, R., Leonard, J., and Austin, S. (1991) Association of demyelination with deficiency of cerebrospinal-fluid S-adenosylmethionine in inborn errors of methyl-transfer pathway, *Lancet 338(8782–8783)*, 1550–1554.

215. Cheng, H., et al. (1997) Levels of L-methionine S-adenosyltransferase activity in erythrocytes and concentrations of S-adenosylmethionine and S-adenosylhomocysteine in whole blood of patients with Parkinson's disease, *Exp. Neurol. 145*(*2 Pt 1*), 580–585.

216. Regland, B., et al. (1995) Homocysteinemia is a common feature of schizophrenia, *J. Neural. Transm. Gen. Sect. 100*(2), 165–169.

217. Lee, C. C., Surtees, R., and Duchen, L. W. (1992) Distal motor axonopathy and central nervous system myelin vacuolation caused by cycloleucine, an inhibitor of methionine adenosyltransferase, *Brain 115(Pt 3)*, 935–955.

218. Scott, J. M. (1992) Folate-vitamin B12 interrelationships in the central nervous system, *Proc. Nutr. Soc. 51*(2), 219–224.

219. Scott, J. M., et al. (1994) Effects of the disruption of transmethylation in the central nervous system: an animal model, *Acta. Neurol. Scand. Suppl. 154*, 27–31.

220. Charlton, C. G. and Way, E. L. (1978) Tremor induced by S-adenosyl-L-methionine: possible relation to L-dopa effects, *J. Pharm. Pharmacol. 30*(*12*), 819–820.

221. Zhao, W. Q., et al. (2001) L-dopa upregulates the expression and activities of methionine adenosyl transferase and catechol-O-methyltransferase, *Exp. Neurol. 171*(*1*), 127–138.

222. Melamed, E., et al. (1983) Chronic L-dopa administration decreases striatal accumulation of dopamine from exogenous L-dopa in rats with intact nigrostriatal projections, *Neurology 33*(*7*), 950–953.

223. Hunter, K. R., et al. (1973) Sustained levodopa therapy in parkinsonism, *Lancet 2*(*7835*), 929–931.

224. Marsden, C. D. and Parkes, J. D. (1977) Success and problems of long-term levodopa therapy in Parkinson's disease, *Lancet 1*(*8007*), 345–349.

225. Fahn, S. and Calne, D. B. (1978) Considerations in the management of parkinsonism, *Neurology 28*(*1*), 5–7.

226. Meininger, V., et al. (1982) L-Methionine treatment of Parkinson's disease: preliminary results, *Rev. Neurol. (Paris) 138(4)*, 297–303.

227. Benson, R., et al. (1993) The effects of L-dopa on the activity of methionine adenosyltransferase: relevance to L-dopa therapy and tolerance, *Neurochem. Res. 18(3)*, 325–330.

228. Gancher, S. T., Nutt, J. G., and Woodward, W. (1988) Response to brief levodopa infusions in parkinsonian patients with and without motor fluctuations, *Neurology 38(5)*, 712–716.

229. Barbeau, A., Trudeau, J. G., and Coiteux, C. (1965) Fingerprint patterns in Huntington's Chorea and Parkinson's disease, *Can. Med. Assoc. J. 92*, 514–516.

230. Olanow, C. W. (1997) Attempts to obtain neuroprotection in Parkinson's disease, *Neurology 49(1 Suppl 1)*, S26–S33.

231. Sargent, T., 3rd, et al. (1992) Tracer kinetic evidence for abnormal methyl metabolism in schizophrenia, *Biol. Psychiatry 32(12)*, 1078–1090.

232. Smythies, J. R., et al. (1986) Abnormalities of one-carbon metabolism in psychiatric disorders: study of methionine adenosyltransferase kinetics and lipid composition of erythrocyte membranes, *Biol. Psychiatry 21(14)*, 1391–1398.

233. Gomes-Trolin, C., et al. (1996) Influence of vitamin B12 on brain methionine adenosyltransferase activity in senile dementia of the Alzheimer's type, *J. Neural. Transm. 103*(7), 861–872.

234. Antun, F. T., et al. (1971) The effects of L-methionine (without MAOI) in schizophrenia, *J. Psychiatr. Res. 8*(2), 63–71.

235. Bressa, G. M. (1994) S-adenosyl-l-methionine (SAMe) as antidepressant: meta-analysis of clinical studies, *Acta Neurol. Scand. Suppl. 154*, 7–14.

236. Deth, R. C., et al. (1996) Lymphocyte phospholipid methylation is altered In schizophrenia, *Biol. Psychiatry 39*, 504–505.

237. Sharma, A., et al. (1999) D4 dopamine receptor-mediated phospholipid methylation and its implications for mental illnesses such as schizophrenia, *Mol. Psychiatry 4(3)*, 235–246.

238. Levy, H. L., et al. (1969) Hypermethioninemia with other hyperaminoacidemias. Studies in infants on high-protein diets, *Am. J. Dis. Child. 117(1)*, 96–103.

239. Komrower, G. M. and Robins, A. J. (1969) Plasma amino acid disturbance in infancy. I: hypermethioninaemia and transient tyrosinaemia, *Arch. Dis. Child. 44(235)*, 418–421.

240. Mudd, S. H., et al. (1995) Isolated persistent hypermethioninemia, *Am. J. Hum. Genet. 57(4)*, 882–892.

241. Mudd, S. H., et al. (2001) Glycine N-methyltransferase deficiency: a novel inborn error causing persistent isolated hypermethioninaemia, *J. Inherit. Metab. Dis. 24(4)*, 448–464.

242. Gahl, W. A., et al. (1987) Hepatic methionine adenosyltransferase deficiency in a 31-year-old man, *Am. J. Hum. Genet. 40(1)*, 39–49.

243. Gahl, W. A., et al. (1988) Transsulfuration in an adult with hepatic methionine adenosyltransferase deficiency, *J. Clin. Invest. 81*(2), 390–397.

244. Finkelstein, J. D., Kyle, W. E., and Martin, J. J. (1975) Abnormal methionine adenosyltransferase in hypermethioninemia, *Biochem. Biophys. Res. Commun. 66(4)*, 1491–1497.

245. Nagao, M. and Oyanagi, K. (1997) Genetic analysis of isolated persistent hypermethioninemia with dominant inheritance, *Acta Paediatr. Jpn. 39(5)*, 601–606.

246. Gaull, G. E. and Tallan, H. H. (1974) Methionine adenosyltransferase deficiency: new enzymatic defect associated with hypermethioninemia, *Science 186(4158)*, 59–60.

247. Blom, H. J., et al. (1992) Persistent hypermethioninaemia with dominant inheritance, *J. Inherit. Metab. Dis. 15(2)*, 188–197.

248. Chamberlin, M. E., et al. (1997) Dominant inheritance of isolated hypermethioninemia is associated with a mutation in the human methionine adenosyltransferase 1A gene, *Am. J. Hum. Genet. 60(3)*, 540–546.

249. Chamberlin, M. E., et al. (2000) Methionine adenosyltransferase I/III deficiency: novel mutations and clinical variations, *Am. J. Hum. Genet. 66(2)*, 347–355.

250. Hazelwood, S., et al. (1998) Normal brain myelination in a patient homozygous for a mutation that encodes a severely truncated methionine adenosyltransferase I/III, *Am. J. Med. Genet. 75(4)*, 395–400.

251. Kim, S. Z., et al. (2002) Methionine adenosyltransferase I/III deficiency: two Korean compound heterozygous siblings with a novel mutation, *J. Inherit. Metab. Dis. 25(8)*, 661–671.

252. Chien, Y. H., et al. (2005) Spectrum of hypermethioninemia in neonatal screening, *Early Hum. Dev. 81(6)*, 529–533.

253. Stockler, S., et al. (1996) Guanidinoacetate methyltransferase deficiency: the first inborn error of creatine metabolism in man, *Am. J. Hum. Genet. 58(5)*, 914–922.

254. Bianchi, R., et al. (1997) Role of methyl groups in myelination, *J. Peripher. Nerv. Syst. 2*, 84.

255. Linnebank, M., et al. (2005) Methionine adenosyltransferase (MAT) I/III deficiency with concurrent hyperhomocysteinaemia: two novel cases, *J. Inherit. Metab. Dis. 28(6)*, 1167–1168.

256. Halpern, B. C., et al. (1974) The effect of replacement of methionine by homocystine on survival of malignant and normal adult mammalian cells in culture, *Proc. Natl. Acad. Sci. U.S.A. 71(4)*, 1133–1136.

257. Simile, M. M., et al. (2005) Chemopreventive N-(4-hydroxyphenyl)retinamide (fenretinide) targets deregulated NF-{kappa}B and *MAT1A* genes in the early stages of rat liver carcinogenesis, *Carcinogenesis 26(2)*, 417–427.

258. Huang, Z. Z., et al. (1999) Differential effect of thioacetamide on hepatic methionine adenosyltransferase expression in the rat, *Hepatology 29(5)*, 1471–1478.

259. Tsukada, K. and Okada, G. (1980) S-Adenosylmethionine synthetase isozyme patterns from rat hepatoma induced by N-2-fluorenylacetamide, *Biochem. Biophys. Res. Commun. 94(4)*, 1078–1082.

260. Horikawa, S., et al. (1993) Expression of non-hepatic-type S-adenosylmethionine synthetase isozyme in rat hepatomas induced by 3′-methyl-4-dimethylaminoazobenzene, *FEBS Lett. 334(1)*, 69–71.

261. Liang, C. R., et al. (2005) Proteome analysis of human hepatocellular carcinoma tissues by two-dimensional difference gel electrophoresis and mass spectrometry, *Proteomics 5(8)*, 2258–2271.

262. Liau, M. C., Chang, C. F., and Becker, F. F. (1979) Alteration of S-adenosylmethionine synthetases during chemical hepatocarcinogenesis and in resulting carcinomas, *Cancer Res. 39(6 Pt 1)*, 2113–2119.

263. Abe, T. and Tsukada, K. (1981) S-adenosylmethionine synthetase isozymes in the liver of tumor-bearing mice, *J. Biochem. 90*(2), 571–574.

264. Akerman, K., Karkola, K., and Kajander, O. (1991) Methionine adenosyltransferase activity in cultured cells and in human tissues, *Biochim. Biophys. Acta 1097*(2), 140–144.

265. Cai, J., et al. (1996) Changes in S-adenosylmethionine synthetase in human liver cancer: molecular characterization and significance, *Hepatology 24*(5), 1090–1097.

266. Ito, K., et al. (2000) Correlation between the expression of methionine adenosyltransferase and the stages of human colorectal carcinoma, *Surg. Today 30*(8), 706–710.

267. Avila, M. A., et al. (2000) Reduced mRNA abundance of the main enzymes involved in methionine metabolism in human liver cirrhosis and hepatocellular carcinoma, *J. Hepatol. 33*(6), 907–914.

268. Yang, H., et al. (2001) Role of promoter methylation in increased methionine adenosyltransferase 2A expression in human liver cancer, *Am. J. Physiol. Gastrointest. Liver Physiol. 280*(2), G184–G190.

269. Boyes, J. and Bird, A. (1992) Repression of genes by DNA methylation depends on CpG density and promoter strength: evidence for involvement of a methyl-CpG binding protein, *Embo J. 11*(1), 327–333.

270. Jones, P. L., et al. (1998) Methylated DNA and MeCP2 recruit histone deacetylase to repress transcription, *Nat. Genet. 19*(2), 187–191.

271. Cai, J., et al. (1998) Differential expression of methionine adenosyltransferase genes influences the rate of growth of human hepatocellular carcinoma cells, *Cancer Res. 58*(7), 1444–1450.

272. Mangipudy, R. S., Chanda, S., and Mehendale, H. M. (1995) Hepatocellular regeneration: key to thioacetamide autoprotection, *Pharmacol. Toxicol. 77*(3), 182–188.

273. Dyroff, M. C. and Neal, R. A. (1981) Identification of the major protein adduct formed in rat liver after thioacetamide administration, *Cancer Res. 41*(9 Pt 1), 3430–3435.

274. Garcea, R., et al. (1989) Inhibition of promotion and persistent nodule growth by S-adenosyl-L-methionine in rat liver carcinogenesis: role of remodeling and apoptosis, *Cancer Res. 49*(7), 1850–1856.

275. Majano, P. L., et al. (2001) S-Adenosylmethionine modulates inducible nitric oxide synthase gene expression in rat liver and isolated hepatocytes, *J. Hepatol. 35*(6), 692–699.

276. Lu, S. C. and Mato, J. M. (2005) Role of methionine adenosyltransferase and S-adenosylmethionine in alcohol-associated liver cancer, *Alcohol 35*(3), 227–234.

277. Kokkinakis, D. M., et al. (2004) Modulation of gene expression in human central nervous system tumors under methionine deprivation-induced stress, *Cancer Res. 64*(20), 7513–7525.

278. Peng, H. B., Libby, P., and Liao, J. K. (1995) Induction and stabilization of I kappa B alpha by nitric oxide mediates inhibition of NF-kappa B, *J. Biol. Chem. 270*(23), 14214–14219.

279. Dwivedi, R. S., Wang, L. J., and Mirkin, B. L. (1999) S-adenosylmethionine synthetase is overexpressed in murine neuroblastoma cells resistant to nucleoside analogue inhibitors of S-adenosylhomocysteine hydrolase: a novel mechanism of drug resistance, *Cancer Res. 59(8)*, 1852–1856.

280. Lu, S. C., et al. (2000) Changes in methionine adenosyltransferase and S-adenosylmethionine homeostasis in alcoholic rat liver, Am. J. Physiol. Gastrointest. Liver Physiol. *279(1)*, G178–G185.

281. Villanueva, J. A. and Halsted, C. H. (2004) Hepatic transmethylation reactions in micropigs with alcoholic liver disease, *Hepatology 39(5)*, 1303–1310.

282. Lieber, C. S., et al. (1990) S-adenosyl-L-methionine attenuates alcohol-induced liver injury in the baboon, *Hepatology 11*(2), 165–172.

283. Lee, T. D., et al. (2004) Abnormal hepatic methionine and glutathione metabolism in patients with alcoholic hepatitis, *Alcohol. Clin. Exp. Res. 28(1)*, 173–181.

284. Arteel, G. E., et al. (1997) Chronic enteral ethanol treatment causes hypoxia in rat liver tissue in vivo, *Hepatology 25(4)*, 920–926.

285. Ji, S., et al. (1982) Periportal and pericentral pyridine nucleotide fluorescence from the surface of the perfused liver: evaluation of the hypothesis that chronic treatment with ethanol produces pericentral hypoxia, *Proc. Natl. Acad. Sci. U.S.A. 79(17)*, 5415–5419.

286. Tsukamoto, H. and Xi, X. P. (1989) Incomplete compensation of enhanced hepatic oxygen consumption in rats with alcoholic centrilobular liver necrosis, *Hepatology 9*(2), 302–306.

287. Sufrin, J. R., Coulter, A. W., and Talalay, P. (1979) Structural and conformational analogues of L-methionine as inhibitors of the enzymatic synthesis of S-adenosyl-L-methionine. IV. Further mono-, bi- and tricyclic amino acids, *Mol. Pharmacol. 15(3)*, 661–677.

288. Coulter, A. W., et al. (1974) Structural and conformational analogues of L-methionine as inhibitors of the enzymatic synthesis of S-adenosyl-l-methionine. 3. Carbocyclic and heterocyclic amino acids, *Mol. Pharmacol. 10*(2), 319–334.

289. Sufrin, J. R., Lombardini, J. B., and Keith, D. D. (1982) L-2-Amino-4-methoxy-cis-but-3-enoic acid, a potent inhibitor of the enzymatic synthesis of S-adenosylmethionine, *Biochem. Biophys. Res. Commun. 106*(2), 251–255.

290. Lagler, F., et al. (2000) Hypermethioninemia and hyperhomocysteinemia in methionine adenosyltransferase I/III deficiency, *J. Inherit. Metab. Dis. Suppl. 23*, 68.

291. Gaull, G. E., et al. (1981) Hypermethioninemia associated with methionine adenosyltransferase deficiency: clinical, morphologic, and biochemical observations on four patients, *J. Pediatr. 98(5)*, 734–741.

INDEX

Advances in Enzymology and Related Areas of Molecular Biology, Volume 78. Edited by Eric J. Toone.

B

H

O